Advanced Technologies in Cardiovascular Bioengineering

Jianyi Zhang • Vahid Serpooshan

Editors

Advanced Technologies in Cardiovascular Bioengineering

 Springer

Editors
Jianyi Zhang
Biomedical Engineering
University of Alabama at Birmingham
Birmingham, AL, USA

Vahid Serpooshan (iD)
Biomedical Engineering
Georgia Institute of Technology
Atlanta, GA, USA

ISBN 978-3-030-86139-1   ISBN 978-3-030-86140-7   (eBook)
https://doi.org/10.1007/978-3-030-86140-7

This Springer imprint is published by the registered company Springer Nature Switzerland AG
The registered company address is: Gewerbestrasse 11, 6330 Cham, Switzerland

Preface

In recent decades, the convergence of discoveries in biological sciences and engineering have resulted in the development of new industries that offer the promise of revolutionary changes in society, as part of the *Convergence Revolution (the Fourth Industrial Revolution)*. These advancements have offered the potential new management options for some of humanity's most intractable and deadly diseases. Within the cardiovascular sciences, many of the most provocative discoveries have emerged from studies of pluripotent stem cells, whose roles in the medical sciences and physiological and injury response are becoming increasingly acknowledged. In 2016, the National Institutes of Health (NIH) established the Progenitor Cell Translational Consortium (PCTC) to support research into the use of stem and progenitor cells for both biology and therapeutic applications. This book was inspired by the thought-provoking ideas and observations presented at the 2019 PCTC Cardiovascular Bioengineering (CVBE) Symposium and was written by leading scientists and physicians whose work in the CVBE field spans decades and was conducted on four different continents.

Cardiomyocytes in the hearts of humans and other mammals are largely incapable of self-replicating; thus, although advancements in the clinical management of cardiovascular conditions have led to substantial improvements in patient longevity and quality of life, the scarring caused by cardiac disease or injury is essentially permanent. Myocardial integrity can be fully restored via whole-heart transplantation surgery, but the supply of donated hearts is far smaller than the number of patients who require treatment, so alternative strategies for replacing the myocardial scar with functional contractile tissue are urgently needed. The lack of cell-cycle activity in adult mammalian cardiomyocytes also severely restricts their availability for investigational work, so early studies of myocardial cell therapy were frequently conducted with stem cells which, though obtained from a variety of sources (e.g., the bone marrow, adipose tissue), were expected to differentiate into cardiomyocytes after transplantation. However, the benefits observed in subsequent clinical trials were only marginal and likely evolved from the cells' paracrine activity, rather than through the production of new cardiomyocytes.

The scarcity of cardiomyocytes for therapeutic investigations was alleviated by the isolation of human embryonic stem cells (ESCs) and, especially, by the development of techniques for reprogramming somatic cells into induced-pluripotent stem cells (iPSCs). Both cell types can proliferate indefinitely and are capable of differentiating into diverse cellular lineage; however, direct stem cell transplantation can lead to tumor formation; so, ESCs and iPSCs must be differentiated into more specialized cell types before administration to patients, and only in recent years the differentiation protocols achieved the adequate efficiency to meet such demands. In general, the most effective protocols are modeled after the mechanisms that regulate cell specification during embryogenesis, when the four major lineages of cardiac cells evolve from progenitor cells of the first and second heart fields, the proepicardial organ, and the cardiac neural crest. These protocols may become even more efficient as researchers continue to refine and develop novel methods for determining the identity, ancestry, and progeny of progenitor cells during development and as the heart recovers from injury.

Only a small fraction of transplanted cells are engrafted within the native tissue and survives for more than a few days after administration, which is perhaps not surprising, since the cytotoxic conditions responsible for the loss of endogenous cells are likely to endure longer than the initial injury. One of the chief requirements of a more salubrious environment for transplanted cells is adequate perfusion. Both the size and thickness of engineered tissues are typically limited by the access of nutrients and signaling molecules to the cells within the tissue. Thus, the success of cell-based regenerative therapies for treatment of cardiac disease, as well as peripheral artery disease, critical limb ischemia, and other predominantly vascular conditions, will depend on understanding the mechanisms by which the vascular cell differentiation and proliferation can be manipulated to promote vessel growth.

Tissues constructed from human ESC- or iPSC-derived cells can also provide researchers with an entirely human-specific platform for studying the pathogenesis of disease and for testing new pharmaceutical products. Notably, iPSC-derived cell and tissue models are powerful tools for personalized therapies, because the iPSCs can be reprogrammed from the patient's own somatic cells and, consequently, recapitulate all of the genetic factors that regulate disease pathology and progression, as well as the patient's response to treatment. Autologous iPSC-derived cells are also expected to be minimally immunogenic when re-administered to the same patient for treatment of chronic conditions such as heart failure; however, the reprogramming and differentiation procedures take several weeks, so cell-based treatments for emergency situations, such as acute myocardial infarction, will require the use of allogeneic cells, which have rarely been studied. Furthermore, one of the primary concerns associated with cardiac cell therapy is the potential for arrhythmogenic complications caused by inadequate electromechanical coupling between the endogenous and transplanted cells. Thus, researchers continue to develop increasingly sophisticated tools for assessing the integration and electrophysiological function of engrafted cells and tissues, such as epicardial electrode arrays, genetically encoded fluorescent reporters, and catheter-based electroanatomic mapping.

Although the regenerative capacity of adult mammalian hearts is extremely limited, the hearts of at least some neonatal mammals (e.g., mice and pigs) can fully repair the damage caused by myocardial injury, provided that the injury occurs within the first few days after birth. Existing evidence suggests that this recovery is driven primarily by the proliferation of pre-existing cardiomyocytes, rather than the activity of stem or progenitor cells, which suggests that the cardiomyocytes of adult hearts may retain some latent proliferative capacity that could be therapeutically re-activated to improve cardiac performance in patients with heart disease. The mechanisms responsible for inducing proliferation in cardiomyocytes are just beginning to be explored. These works will be facilitated by advancements in single-cell genomics, which can characterize the gene expression profiles of thousands of individual cells; however, the resulting datasets are typically so enormous that they require the use of modern data science techniques, such as dimensionality reduction and clustering analysis, to identify the genes and pathways that are differentially activated in proliferating and non-proliferating cardiomyocytes. Machine-learning algorithms can even be applied to the text mined from the Medline database and other unstructured sources to identify relationships among specific genes, diseases, and disease symptoms, including those that may explain why outcomes of COVID-19 treatment are worse for patients with cardiovascular comorbidities.

In summary, many of the greatest advancements in science, and in civilization as a whole, have occurred when previously disparate lines of inquiry come together in unanticipated ways. The fields of personal and public health will soon reap the benefits of the unprecedented degree of synergy that has recently developed among the life and physical sciences, computing, and engineering. The authors of this book hope to foster these advancements by sharing their knowledge and expertise with the broader community of scientists, engineers, and clinicians.

Birmingham, AL, USA   Jianyi Zhang
Atlanta, GA, USA   Vahid Serpooshan

Contents

Part I
Cardiac Development and Morphogenesis

From Simple Cylinder to Four-Chambered Organ: A Brief Overview of Cardiac Morphogenesis

Carissa Lee, Sharon L. Paige, Francisco X. Galdos, Nicholas Wei, and Sean M. Wu

1 Introduction

CHDs are the most common type of birth defects, accounting for approximately 13% of deaths in the US in 2017, or 365,914 deaths [1]. In considering the origins of CHDs and related malformations, a fundamental understanding of cardiac growth and morphogenesis is requisite. This article reviews the salient morphogenic phases of the developing heart, beginning with the incipience of the two embryonic axes and concluding with the completion of complex septation and trabeculation processes that characterize the mature embryonic heart. We also discuss the origins of four major cardiac lineages, namely derivatives of the FHF, SHF, PEO, and cNCC progenitors. In this article, we have chosen to focus on the murine model of cardiac

C. Lee (✉) · F. X. Galdos · N. Wei
Stanford Cardiovascular Institute, Stanford University School of Medicine, Stanford, CA, USA
e-mail: smwu@stanford.edu

S. L. Paige
Stanford Cardiovascular Institute, Stanford University School of Medicine, Stanford, CA, USA

Department of Pediatrics, Division of Pediatric Cardiology, Stanford University School of Medicine, Stanford, CA, USA

S. M. Wu
Stanford Cardiovascular Institute, Stanford University School of Medicine, Stanford, CA, USA

Department of Medicine, Division of Cardiovascular Medicine, and Stanford University School of Medicine, Stanford, CA, USA

Institute of Stem Cell Biology and Regenerative Medicine, Stanford University School of Medicine, Stanford, CA, USA

© The Author(s), under exclusive license to Springer Nature Switzerland AG 2022
J. Zhang, V. Serpooshan (eds.), *Advanced Technologies in Cardiovascular Bioengineering*, https://doi.org/10.1007/978-3-030-86140-7_1

development due to its similarity to human models and popularity in recent and ongoing embryology research.

1.1 Early Gastrulation and Formation of the Cardiac Crescent (E5.0 – E7.5)

Prior to gastrulation at approximately embryonic day E5.0, the mouse embryo resembles an elongating cylinder consisting of a single proximal-distal axis. The TGF-beta signaling protein nodal growth differentiation factor (NODAL), expressed along a concentration gradient with its antagonists (Cer1 and Lefty1), in both embryonic and extraembryonic tissues induces patterning along a second anterior-posterior axis by E5.5 as shown in Fig. 1 [2].

Gastrulation is an essential process for embryogenesis, beginning at approximately E6.0 in mice [3]. During this period, a pluripotent group of embryonic cells called the epiblast ingresses through a strip-like structure known as the primitive streak (PS) to generate the three germ layers of the early embryo: endoderm, mesoderm, and ectoderm. While cells from these germ layers collectively give rise to the body and all its organs, the heart is specifically established by a group of myocardial

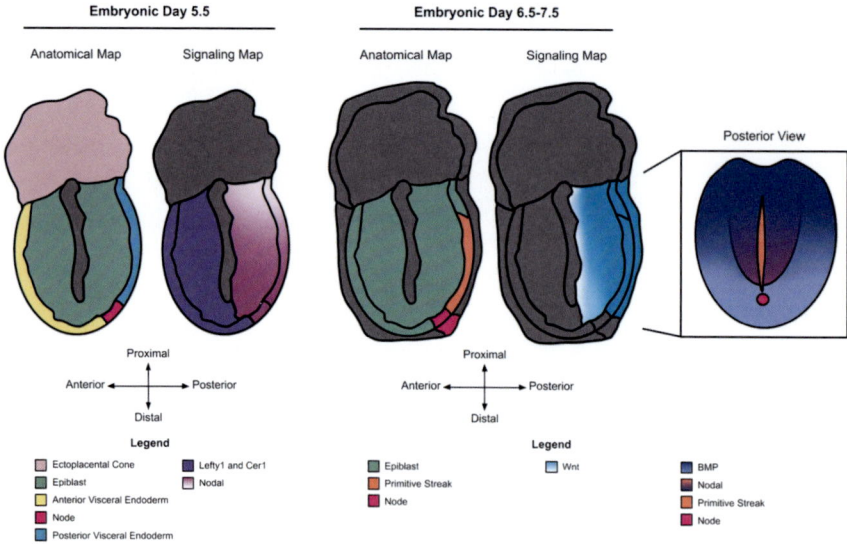

Fig. 1 At E5.5, NODAL expression originates from the Node, a structure located at the posterior side of the mouse embryo. NODAL antagonists Lefty1 and Cer1 are expressed anteriorly, allowing for formation of the primitive streak on the posterior side. Upon induction of the primitive streak around E6.5–7.5, NODAL, Bmp, and canonical Wnt signaling gradients direct commitment of migrating epiblast cells to various endoderm and mesodermal lineages, including cardiac mesoderm that gives rise to the heart. (Adapted from [2])

progenitor cells which derive from the mesoderm. Notably, the heart is the first functioning organ of the embryo, as it pumps blood carrying oxygen and nutrients necessary for embryonic development [4].

By E6.5, these precardiac mesodermal progenitors on the posterior end of the embryonic "cylinder" will migrate anteriorly and laterally to a region named the anterior lateral plate mesoderm (ALPM). Along the way, these cells acquire cardiac fates due to the patterned expression of bone morphogenic protein (BMP), wingless-related integration site (Wnt), fibroblast growth factor (FGF), and other signaling molecules that guide their organization at the ALPM into the two main cardiac progenitor cell populations: the first and second heart fields. These two cell populations contribute to distinct structures of the developing heart. Fate-mapping studies have shown that the FHF lineage primarily establishes the myocardium of the left ventricle (LV), while the SHF lineage gives rise to the right ventricle (RV) and outflow tract (OFT) [5]. Initially, FHF precursors differentiate rapidly, becoming beating cardiomyocytes (CMs) which form the early cardiac crescent [6]. As SHF precursors settle medially to the FHF, they together form the completed cardiac crescent, which is typically visible by E7.5 [2]. The temporal delay between the formation of FHF and SHF progenitors from the precardiac mesoderm, FHF preceding SHF, is the primary driver of organization into each heart field.

Recent findings have shown that coordinated flow of calcium ions between CMs triggers the first heartbeat [7]. This explains evidence of primitive pacemaker activity originating near the inflow tract and sinus venosus of the linear heart tube at this stage of development [8]. Furthermore, this portion of the heart tube includes the primordium of the sinus node, which later becomes the chief pacemaker of the mature cardiac conduction system [9].

1.2   The Linear Heart Tube (E8.0)

As development proceeds, the next few processes—heart tube formation, heart tube elongation, and early chamber establishment —all overlap in time across the embryonic heart. Additionally, these morphogenic stages from E8.0–11.0 contribute significantly to growth, facilitating a 100-fold increase in CM number and cardiac volume [10].

Following formation of the cardiac crescent, a process characterized by rapid differentiation of FHF progenitors into CMs, the embryonic heart enters a period of more extensive morphogenesis. Splanchnic mesoderm slides over endoderm, temporarily pausing CM differentiation as the heart tube (HT) is assembled. When differentiation resumes upon completion of the primitive HT, newly formed CMs instead contribute to SHF-derived regions of the HT and initiate closure of the dorsal side. The influx of SHF cells also strengthen the heartbeat such that by E8.25, the embryonic "cylinder" is transformed into a beating linear HT [6].

The linear HT is made up of three distinct layers, with the inner endocardial layer and outer myocardial layer separated by a thick band of extracellular matrix (ECM) referred to as the cardiac jelly [8]. The ECM plays a role in the maturation of cardiac cells into highly vascularized, densely compacted myocardium. Immunohistochemistry techniques have designated four essential ECM proteins—collagen types I and IV (COLI, COLIV), elastin (ELN), and fibronectin (FN)—that are found within the LV of the mouse heart [11]. The first beating CMs are found in the outer heart tube, while the cells of the inner heart tube retain an endothelial cell identity [2].

Subsequent migration of SHF progenitors to the arterial and venous poles facilitates gradual elongation of the heart tube as these cells undergo proliferation. The proepicardial organ (PEO), a transitory mesenchymal structure responsible for generating the embryonic epicardial cell lineage, also appears near the venous pole by E8.5 [12]. Importantly, this cluster of coelomic cells is highly conserved among vertebrates [13].

1.3   Cardiac Looping (E8.5)

As elongation slows, the linear heart tube undergoes a characteristic process of rightwards looping in which the posterior regions begin to move anteriorly as shown in Fig. 2. This establishes the structural basis of the heart's four distinct chambers: right atrium (RA), right ventricle (RV), left atrium (LA), and left ventricle (LV) [3].

Inversion of the "S" loop during this phase is a common malformation observed in patients with heterotaxy syndrome (HS), a disorder characterized by developmental abnormalities in the left-right axis [14]. Early defects in the looping process underscore its importance to proper heart formation as patients with HS may present with dextrocardia, a condition in which the apex of the heart points towards the right instead of the left.

Cardiac valve formation also initiates around E8.5 with the formation of the dorsal and ventral endocardial cushions following heart looping [15]. These "swollen" protrusions of the ECM are lined with endocardial cells, a portion of which will migrate into the cushion ECM and adopt mesenchymal identities through a process known as the endothelial-to-mesenchymal transition (EndMT). This thickening of the endocardial cushions provides the foundation for later remodeling into valve leaflets.

Finally, epicardium development proceeds simultaneously as vesicles from the PEO either directly adhere to the surface of the beating heart or gradually drift towards the myocardium following their release into the pericardial cavity [13]. Upon making contact, the attached cells will collapse and proliferate, generating a primitive layer of epicardium that covers the heart. These epicardially derived cells (EPDCs), a subset of which will undergo an epithelial-to-mesenchymal transition (EMT) and migrate into the myocardium, have the potential to differentiate into coronary smooth muscle cells and interstitial fibroblasts while reports of their differentiation into coronary endothelium and cardiomyocytes require further investigation [16].

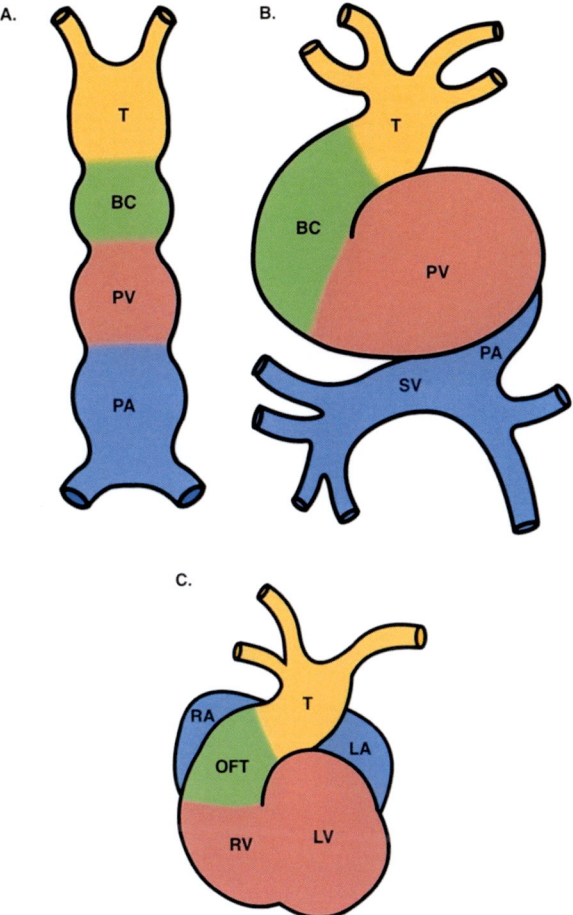

Fig. 2 Diagram illustrating the process of cardiac looping beginning at E8.5, whereby the linear heart tube is first transformed into an S-shaped curve before organization into primitive chambers. (**a**) The linear heart tube. (**b**) Looping. (**c**) The primitive 4-chambered heart. (Key: *T* truncus arteriosus, *BC* bulbus cordis, *SV* sinus venosus, *PV* primitive ventricle, *PA* primitive atrium, *pRV* primitive right ventricle, *pLV* primitive left ventricle, *pRA* primitive right atrium, *pLA* primitive left atrium, *OFT* outflow tract). (Adapted from [17])

1.4   The Four-Chambered Heart (E9.5)

The looped heart tube consists of four main structural elements: the atrium, atrioventricular canal (AVC), ventricle, and outflow tract [5]. Partitioning the left from right and the atrial from ventricular regions of the heart constitutes the most complex stage of heart morphogenesis, beginning around E9.5 and lasting approximately until E14.5 as depicted in Fig. 3. Proper chamber formation is crucial to the

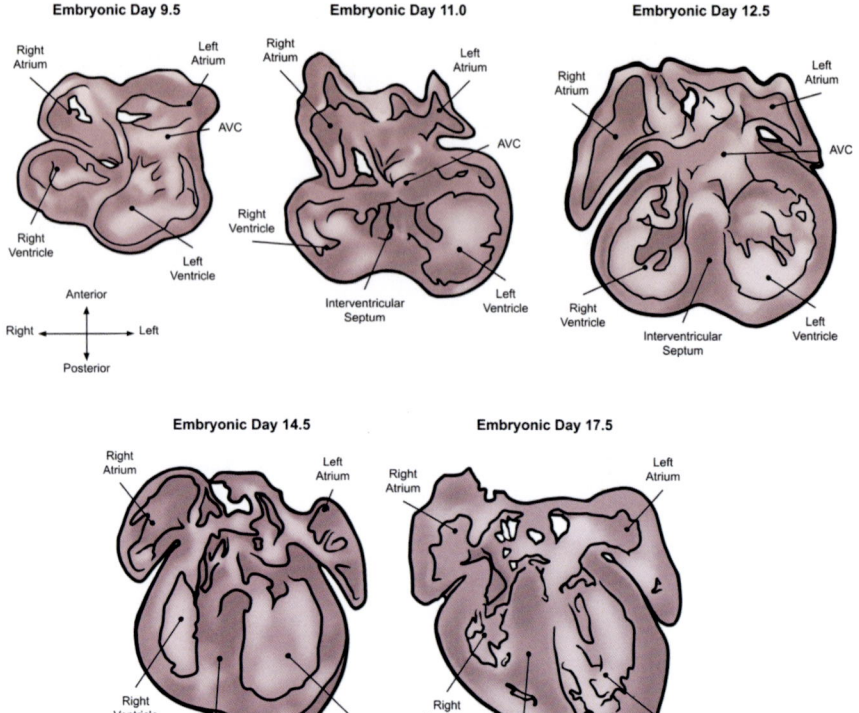

Fig. 3 Illustrations depicting ventral surface cuts of embryonic mouse hearts between E9.5 and E17.5. (Key: *AVC* atrioventricular canal). (Adapted from [10])

developmental pathway, as failure to establish structures capable of sustaining systemic circulation results in defects such as atrial septal defects (ASDs), ventricular septal defects (VSDs), and atrioventricular septal defects (AVSDs) that can progress to end-stage heart failure.

At E9.5, active proliferation and migration of the SHF leads to the formation and elongation of the outflow tract (OFT), a transient structure which, in its most primitive state, connects the developing right ventricle to the aortic sac [18].

Subsequent infiltration of cNCCs into the looped heart initiates septation of the elongated OFT into the aorta (Ao) and pulmonary trunk (PT) [19]. From their starting point in the cardiac neural crest, these cells migrate through the aortic arches and cluster near the distal OFT, forming truncal cushions. Interactions between these truncal cushions and the proximally-located conal cushions of the OFT creates a spiral septum that divides the OFT into the Ao and PT, allowing for separate systemic and pulmonary circulations [5]. Congenital malformations of the OFT may result in conotruncal defects, which include conditions such as Tetralogy of Fallot (TOF) and Transposition of the Great Arteries (TGA).

By E10.5, "well-defined chambers" are visible in the heart despite persistence of the primitive tubular structure [3]. Histological samples indicate the presence of a functioning sinoatrial node, which is responsible for initiating heart beats [8]. The epicardium is fully formed, creating a protective envelope around the heart. At this point, the atrial and ventricular chambers septate, a process that begins with the expansion of the mesenchymal cushions, leading to the formation of the right and left atria and ventricles.

The development of the two major atrioventricular (AV) cushions, the inferior and superior cushions, in the central AVC is facilitated by EndMT [20]. Endocardium derived cells (ENDCs) populate the cushions, displacing the existing ECM within the AV canal. As such, lineage tracing studies have shown that the majority of mesenchymal cells infiltrating the cushions are derived from endocardium [21].

As shown in Fig. 4, atrial septation occurs from E10.5–E13.5, beginning when the major AV cushions fuse at the AVC along with two mesenchymal structures: the vestibular spine and mesenchymal cap. Muscularization of the mesenchymal tissue results in two muscular tissue structures, the septum primum and septum secundum, which together septate the atrial chamber into right and left [22]. At E11.5, within the ventricular chamber, an outgrowth called the interventricular muscular septum fuses with the AV cushions to create distinct right and left ventricles [5]. The "minor" left and right lateral AV cushions, which form after the inferior and superior AV cushions, are also formed from ENDCs. These minor cushions become the

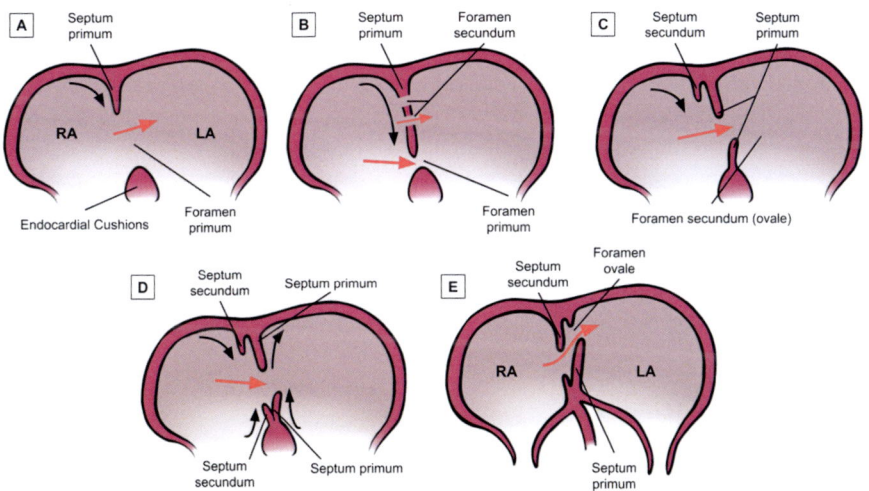

Fig. 4 Depiction of the atrial septation process, occurring approximately from E10.5–E13.5. (**a**) Formation of the first of two muscular septa, the septum primum, begins at the roof of the primitive atrial chamber. (**b**) As the septum primum elongates, the foramen primum and foramen secundum allow for continued communication between the right and left sides of the atrium. (**c**) The septum secundum grows to the right of the septum primum, forming an oval-shaped hole called the foramen ovale. (**d**) Both septa begin to fuse. (**e**) The foramen ovale remains open, allowing for blood flow from the right to left atrium. (Key: *RA* right atrium, *LA* left atrium). (Adapted from [25])

septal leaflets of the mitral and tricuspid valves, which are necessary to prevent retrograde flow of blood from the ventricles to the atria [23]. Failure of the tricuspid valve tissue to delaminate from the ventricular myocardium at this stage results in right ventricular myopathy and a apically-displaced tricuspid valve, which are characteristic of a rare congenital defect called Ebstein's anomaly [24].

Myocardial trabeculation, which also involves the endocardium, is another essential developmental step that begins at this timepoint, extending through the end of the embryonic stage. Within the heart wall, the myocardial layer projects into the cardiac jelly, the ECM layer between the endocardium and myocardium, as endocardial cells invaginate, the resulting finger-like structures are called trabeculae. This process aids in the gradual dissipation of the cardiac jelly as the trabeculae mature and eventually collapse to join with the compact myocardium, completing the inner wall of the heart.

1.5   The Mature Embryonic Heart (E15.0)

The conclusion of chamber and OFT septation around E15.0 prepares the heart for postnatal separation of the pulmonary and systemic circulatory pathways of the blood [3]. In the fetal heart, oxygenated blood flows from the placenta to the umbilical vein, entering the ductus venosus and passing through the inferior vena cava (IVC) before entering the RA. It then travels across the foramen ovale to the LA, down to the LV, and out the Ao to the brain and upper body. Deoxygenated blood from the superior vena cava (SVC) drains to the RA, down to the RV, through the pulmonary artery, and across the ductus arteriosus to the rest of the developing embryo. The fetal circulation pathway ensures that the most oxygenated blood in the fetus goes to the brain, with limited blood entering the lungs as oxygenation of this area occurs postnatally. In the final fetal morphogenic phase, the heart tissue undergoes "fine tuning" modifications that improve cardiac conduction, coronary circulation, and control of blood flow.

Cardiac Conduction System   The cellular origins of the cardiac conduction system have yet to be detailed in full. Currently, it is known that signaling from arterial endothelial cells induces the differentiation of Purkinje conduction cells from myocardium [26]. Fast-conducting chamber myocardium makes up the contractile fibers of the bundle of His, while slow-conducting myocardium from the inflow tract and AV canal creates the SA and AV nodes [4]. The sinus venosus region's important role as a primitive pacemaker provides evidence of SAN progenitors, but the mechanisms behind the formation of the AV node remain poorly understood.

Coronary Vessels   The appearance of coronary endothelial cells has been recorded as early as E12.5, with endocardium giving rise to coronary arterial endothelium and the sinus venosus generating coronary venous endothelium [27]. SV-derived "sprouts" of venous cells cover the fetal heart before proliferating to form the immature coronary plexus [28]. Recent single-cell RNA sequencing experiments found that a subset of these venous endothelial cells dedifferentiate and undergo pre-arterial specification via transcriptional changes that take place prior to the establishment of blood flow [29]. The connecting of the plexus to the Ao initiates blood flow and is crucial to arterial morphogenesis. Epicardial cells from the PEO also play a role in coronary vessel formation, assembling the smooth muscle wall of the coronary vasculature through EMT [30].

Valves The mitral and tricuspid valves are derived from the AV endocardial cushions. As the superior and inferior AV cushions fuse and divide the AVC, they give rise to the anterior mitral and septal tricuspid leaflets. Further remodeling of these primitive leaflets results in formation of the mature mitral and tricuspid valves that ensure unidirectional atrial to ventricular blood flow.

Atrioventricular EPDCs (AV-EPDCs) give rise to the AV sulcus, a transient mesenchymal structure that separates the atrial and ventricular myocardial walls. A portion of the AV-EPDCs within the sulcus infiltrate the AV myocardial junction where they begin to form the annulus fibrosus, a divider made up of fibrous tissue responsible for physically and conductively isolating the working atrial myocardium from its ventricular counterpart. Yet another group of AV-EPDCs will continue on from the annulus fibrosus to merge with the parietal AV valve leaflets and eventually become valve interstitial cells [23].

Trabeculation By E19.0, trabeculation concludes with the complete degradation of the cardiac jelly layer and resultant compaction of the ventricular wall [8]. This compaction is associated with greater strength of contraction, allowing the blood to penetrate deeper layers of the myocardium before the coronary vasculature fully develops in the post-embryonic stage. With the completion of trabeculation, the prenatal mouse heart is ready for postnatal modification following gestation, which typically occurs 20 days post-fertilization.

2   Challenges and Opportunities

While much of early heart development has been documented through fate-mapping and histological examination of mice, chick, and zebrafish embryos, a number of challenges still remain in documenting human heart morphogenesis, including lack of data, inconclusive literature, and poor imaging capabilities.

Due to a combination of technical, legal, and ethical complications, human fetal heart cells are exceedingly difficult to obtain for data collection. Use of human induced-pluripotent stem cells (hiPSCs), while generally considered more ethically sound, suffer from difficulties with chamber and cell type identification. Ongoing research focuses on single-cell RNA sequencing or lineage tracing-based solutions that enable determination of distinct genetic markers within the embryonic heart.

Recent studies have used sequencing data to identify candidate genes that may contribute to CHDs such as HS [14].

An emerging topic of research is the transcriptomic correlation between murine and human embryonic heart cells. A recent single-cell RNA sequencing (scRNA-seq) study of cardiac cells from 18 human embryos showed several differences between developing mouse and human cardiac cells in terms of certain gene expression levels, targets, and corresponding developmental timepoints, suggesting that correlating mouse transcriptomics to human may not be practical. However, preliminary research into using mouse scRNA-seq data in human cardiac cell classification models have yielded promising results. Ongoing research is looking into whether machine learning algorithms can be trained on mouse scRNA-seq data and fine-tuned on preliminary human data to identify human embryonic heart cells, as well as heart cells derived from induced pluripotent stem cells. The impact of this research would be two-fold. First, a mouse-to-human classification model can leverage the vast amount of mouse transcriptomic data that exists (as well as future epigenomic data) to create a highly useful tool providing insights into the mechanisms that control the signaling pathways of human cardiac development and regeneration. Second, a successful classification model could be used to identify cardiac cells derived from human induced pluripotent stem cells (hiPSCs), enabling scientists to use these findings in order to better study cardiac cells *in vitro*.

Another area of interest is improving existing functional and molecular definitions of certain cell types and processes. For instance, EndMT remains poorly understood in comparison to EMT [31] as cell culture conditions severely impact the process. Furthermore, endocardial precursors originate from a variety of regions, making it difficult to establish precise molecular criteria. Standardizing markers for both the presence of and definitive stages of ongoing EndMTs may improve understanding of endothelial dysfunctions that cause both CHDs and adult cardiovascular diseases.

The "fine tuning" steps that occur during the septation phase, most notably the fusion of the OFT and AVC, are complex and difficult to record. Additionally, while heart formation is thus far understood to unfold continuously, viewing morphogenetic events in real time may reveal new insight into the kinetics of development. Although limited by embryo survivability, whole-embryo live-imaging methods, such as the two-photon microscopy approach used by Ivanovitch and colleagues [6], may harbor potential to capture these nuances at cellular resolution.

In summary, cardiac morphogenesis involves an extensive process of looping, septation, and remodeling that transforms the early heart fields into a matured, four-chambered organ capable of systemic and pulmonary circulation. As the intricacies of the embryonic heart render it susceptible to disruption, the understanding of human fetal heart development will remain a popular topic that continues to generate novel research directions in hopes of finding solutions to prevalent the development of CHDs.

References

1. Virani, S., Alonso, A., Benjamin, E., Bittencourt, M., Callaway, C., Carson, A., et al.: Heart disease and stroke statistics—2020 update: a report from the American Heart Association. Circulation (New York, NY). **141**(9), e139–ee51 (2020). https://doi.org/10.1161/CIR.0000000000000757

2. Galdos, F.X., Wu, S.M.: Development of cardiac muscle. Elsevier Inc (2015)

3. Zaffran, S., Meilhac, S., Buckingham, M.: Building the mammalian heart from two sources of myocardial cells. Nat. Rev. Genet. **6**(11), 826–837 (2005). https://doi.org/10.1038/nrg1710

4. Hill, M.A., Aug. Cardiac Embryology (2020)

5. Lin, C.-J., Lin, C.-Y., Chen, C.-H., Zhou, B., Chang, C.-P.: Partitioning the heart: mechanisms of cardiac septation and valve development. Development. **139**(18), 3277–3299 (2012). https://doi.org/10.1242/dev.063495

6. Ivanovitch, K., Temiño, S., Torres, M.: Live imaging of heart tube development in mouse reveals alternating phases of cardiac differentiation and morphogenesis. Elife. **6** (2017). https://doi.org/10.7554/elife.30668

7. Tyser, R.C.V., Miranda, A.M.A., Chen, C.-M., Davidson, S.M., Srinivas, S., Riley, P.R.: Calcium handling precedes cardiac differentiation to initiate the first heartbeat. Elife. **5** (2016). https://doi.org/10.7554/elife.17113

8. Goodyer, W.R., Wu, S.M.: Fates aligned: origins and mechanisms of ventricular conduction system and ventricular wall development. Pediatr. Cardiol. **39**(6), 1090–1098 (2018). https://doi.org/10.1007/s00246-018-1869-9

9. Christoffels, V., Moorman, A.: Development of the cardiac conduction system: why are some regions of the heart more arrhythmogenic than others? Circ. Arrhythm. Electrophysiol. **2**(2), 195–207 (2009). https://doi.org/10.1161/CIRCEP.108.829341

10. de Boer, B.A., van den Berg, G., de Boer, P.A.J., Moorman, A.F.M., Ruijter, J.M.: Growth of the developing mouse heart: an interactive qualitative and quantitative 3D atlas. Dev. Biol. **368**(2), 203–213 (2012). https://doi.org/10.1016/j.ydbio.2012.05.001

11. Hanson, K.P., Jung, J.P., Tran, Q.A., Hsu, S.-P.P., Iida, R., Ajeti, V., et al.: Spatial and temporal analysis of extracellular matrix proteins in the developing murine heart: a blueprint for regeneration. Tissue Eng. Part A. **19**(9–10), 1132–1143 (2013). https://doi.org/10.1089/ten.tea.2012.0316

12. Vincent, S.D., Buckingham, M.E.: How to make a heart: the origin and regulation of cardiac progenitor cells. Curr. Top. Dev. Biol. **90**, 1–41 (2010). https://doi.org/10.1016/S0070-2153(10)90001-X

13. Simões, F.C., Riley, P.R.: The ontogeny, activation and function of the epicardium during heart development and regeneration. Development (Cambridge). **145**(7), dev155994 (2018). https://doi.org/10.1242/dev.155994

14. Liang, S., Shi, X., Yu, C., Shao, X., Zhou, H., Li, X., et al.: Identification of novel candidate genes in heterotaxy syndrome patients with congenital heart diseases by whole exome sequencing. Biochim. Biophys. Acta Mol. basis Dis. **2020**(12), 165906 (1866). https://doi.org/10.1016/j.bbadis.2020.165906

15. Kim, D.H., Xing, T., Yang, Z., Dudek, R., Lu, Q., Chen, Y.-H.: Epithelial mesenchymal transition in embryonic development, tissue repair and cancer: a comprehensive overview. J. Clin. Med. **7**(1), 1 (2017). https://doi.org/10.3390/jcm7010001

16. Zhou, B., Pu, W.T.: More than a cover: epicardium as a novel source of cardiac progenitor cells. Regen. Med. **3**(5), 633–635 (2008). https://doi.org/10.2217/17460751.3.5.633

17. Young, K.A., Wise, J.A., DeSaix, P., Kruse, D.H., Poe, B., Johnson, E., et al.: *Anatomy and physiology*, 1st edn, p. 1335. OpenStax (2013)

18. Anderson, R.H., Mori, S., Spicer, D.E., Brown, N.A., Mohun, T.J.: Development and morphology of the ventricular outflow tracts. World J. Pedia. & Congenit. Heart Surg. **7**(5), 561–577 (2016). https://doi.org/10.1177/2150135116651114

19. Mirzoyev, S., McLeod, C.J., Asirvatham, S.J.: Embryology of the conduction system for the electrophysiologist. Ind. Pacing & Electrophysiol J. **10**(8), 329–338 (2010)
20. Markwald, R.R., Fitzharris, T.P., Manasek, F.J.: Structural development of endocardial cushions. Am. J. Anat. **148**(1), 85–119 (1977). https://doi.org/10.1002/aja.1001480108
21. Snarr, B.S., Kern, C.B., Wessels, A.: Origin and fate of cardiac mesenchyme. Dev. Dyn. **237**(10), 2804–2819 (2008). https://doi.org/10.1002/dvdy.21725
22. Krishnan, A., Samtani, R., Dhanantwari, P., Lee, E., Yamada, S., Shiota, K., et al.: A detailed comparison of mouse and human cardiac development. Pediatr. Res. **76**(6), 500–507 (2014). https://doi.org/10.1038/pr.2014.128
23. Lockhart, M.M., van den Hoff, M., Wessels, A., Nakanishi, T., Markwald, R.R., Baldwin, H.S., et al.: The role of the epicardium in the formation of the cardiac valves in the mouse, pp. 161–170. Etiology and Morphogenesis of Congenital Heart Disease: From Gene Function and Cellular Interaction to Morphology. Springer Open (2016)
24. Possner, M., Gensini, F.J., Mauchley, D.C., Krieger, E.V., Steinberg, Z.L.: Ebstein's anomaly of the tricuspid valve: an overview of pathology and management. Curr. Cardiol. Rep. **22**(12), 157 (2020). https://doi.org/10.1007/s11886-020-01412-z
25. Obgynkey.com; Cardiac Development, Cardac Septation. Chapter 414.3.
26. Hatcher, C.J., Basson, C.T.: Specification of the cardiac conduction system by transcription factors. Circ. Res. **105**(7), 620–630 (2009). https://doi.org/10.1161/CIRCRESAHA.109.204123
27. Wu, B., Zhang, Z., Lui, W., Chen, X., Wang, Y., Chamberlain, A.A., et al.: Endocardial cells form the coronary arteries by angiogenesis through myocardial-endocardial VEGF signaling. Cell (Cambridge). **151**(5), 1083–1096 (2012). https://doi.org/10.1016/j.cell.2012.10.023
28. Red-Horse, K., Ueno, H., Weissman, I.L., Krasnow, M.A.: Coronary arteries form by developmental reprogramming of venous cells. Nature (London). **464**(7288), 549–553 (2010). https://doi.org/10.1038/nature08873
29. Su, T., Stanley, G., Sinha, R., D'Amato, G., Das, S., Rhee, S., et al.: Single-cell analysis of early progenitor cells that build coronary arteries. Nature (London). **559**(7714), 356–362 (2018). https://doi.org/10.1038/s41586-018-0288-7
30. Diman, N., Brooks, G., Kruithof, B., Elemento, O., Seidman, J.G., Seidman, C., et al.: Tbx5 is required for avian and mammalian epicardial formation and coronary Vasculogenesis. Circ. Res. **115**(10), 834–844 (2014). https://doi.org/10.1161/CIRCRESAHA.115.304379
31. Kovacic, J.C., Dimmeler, S., Harvey, R.P., Finkel, T., Aikawa, E., Krenning, G., et al.: Endothelial to mesenchymal transition in cardiovascular disease: JACC state-of-the-art review. J. Am. Coll. Cardiol. **73**(2), 190–209 (2019). https://doi.org/10.1016/j.jacc.2018.09.089

Lineage Tracing Models to Study Cardiomyocyte Generation During Cardiac Development and Injury

Kamal Kolluri, Bin Zhou, and Reza Ardehali

1 Introduction

Lineage tracing is a powerful method to mark a finite number of progenitors at a specific developmental stage and interrogate the progeny of the founder cell at later time points [1]. Lineage tracing has been particularly important in studying cardiovascular development by identifying cardiac progenitors that contribute to specific myocardial lineages and their clonal activities. Fundamental to understanding cardiac development is the ability to determine the identity of stem/progenitor cells, their ancestry and when and how their progeny move to reside in their final location. Lineage tracing, especially the clonal analysis of a single progenitor, is the main approach used to address these issues. Lineage tracing of cardiomyocytes is achieved by using cardiomyocyte-type specific marker genes that permanently label

K. Kolluri
Division of Cardiology, Department of Internal Medicine, David Geffen School of Medicine, University of California, Los Angeles, California, USA

B. Zhou
State Key Laboratory of Cell Biology, Shanghai Institute of Biochemistry and Cell Biology, Center for Excellence in Molecular Cell Science, Chinese Academy of Sciences, Shanghai, China

R. Ardehali (✉)
Division of Cardiology, Department of Internal Medicine, David Geffen School of Medicine, University of California, Los Angeles, California, USA

Eli and Edythe Broad Stem Cell Research Center, University of California, Los Angeles, California, USA

Molecular, Cellular and Integrative Physiology Graduate Program, University of California, Los Angeles, California, USA

Molecular Biology Institute, University of California, Los Angeles, CA, USA
e-mail: RArdehali@mednet.ucla.edu

© The Author(s), under exclusive license to Springer Nature
Switzerland AG 2022
J. Zhang, V. Serpooshan (eds.), *Advanced Technologies in Cardiovascular Bioengineering*, https://doi.org/10.1007/978-3-030-86140-7_2

any cell expressing those genes as well as all subsequent progeny. By using markers expressed by cardiomyocytes or cardiac progenitors, researchers can gain insight into how these cells progress over the course of development and growth, or in response to injury by analyzing their proliferative capacity as well as the expression profile of the generated clones. In this chapter we will review two powerful lineage tracing tools (Mosaic Analysis with Double Markers [MADM] and Rainbow) that have been successfully used for clonal analysis of cardiomyocytes during normal development and after injury. Both models allow for precise fate tracking and lineage tracing, making them applicable to many different types of studies.

2   Mosaic Analysis with Double Markers (MADM)

The extent of cardiomyocyte proliferation during development and in the postnatal heart remains an area of controversy. Recent studies suggest that there is a low turnover of cardiomyocytes that declines with age in mice and humans [2–6]. It has come to consensus in the field that newly generated cardiomyocytes originate from pre-existing cardiomyocytes through proliferation rather than differentiation from stem cells [5]. Studies of cardiomyocyte proliferation have been limited by reliance on indirect assays of markers for cell proliferation and on surrogates for cell division [7, 8]. These studies are particularly challenging to interpret, due to the confounding issues of cardiomyocyte polyploidy, multi-nucleation, and DNA repair after injury in adult hearts [9, 10]. Therefore, it is important to develop a system that differentiates between karyokinesis and cytokinesis in adult cardiomyocytes. Mosaic Analysis with Double Markers (MADM) uses Cre-Recombinase technology that induces genetic recombination upon cell division, which indelibly labels clones of proliferating cells with a single fluorescent marker. Analysis of clonal expansion using MADM has numerous applications in studying cellular proliferation in development, stem cell biology, and regenerative medicine.

2.1   Methodology

The MADM strategy uses two reciprocally chimeric genes that are knocked into the same location on homologous chromosomes, with each containing the N-terminus of one reporter and the C-terminus of the other reporter, interrupted by a lox-P site. After DNA replication, genetic recombination induced by Cre-LoxP creates functional reporter genes (green fluorescent protein [GFP] or red fluorescent protein [RFP]). Upon G2-X-segregation during cell division, the divided cells express either GFP or RFP, and this feature allows MADM to be used for genetically recording cytokinesis events upon G2-X segregation. Considering that inter-chromosomal Cre-LoxP recombination after S phase is a rare event, labeling a fraction of cells of interest allows for clonal analysis of marked cells and their progeny. Since G2-X

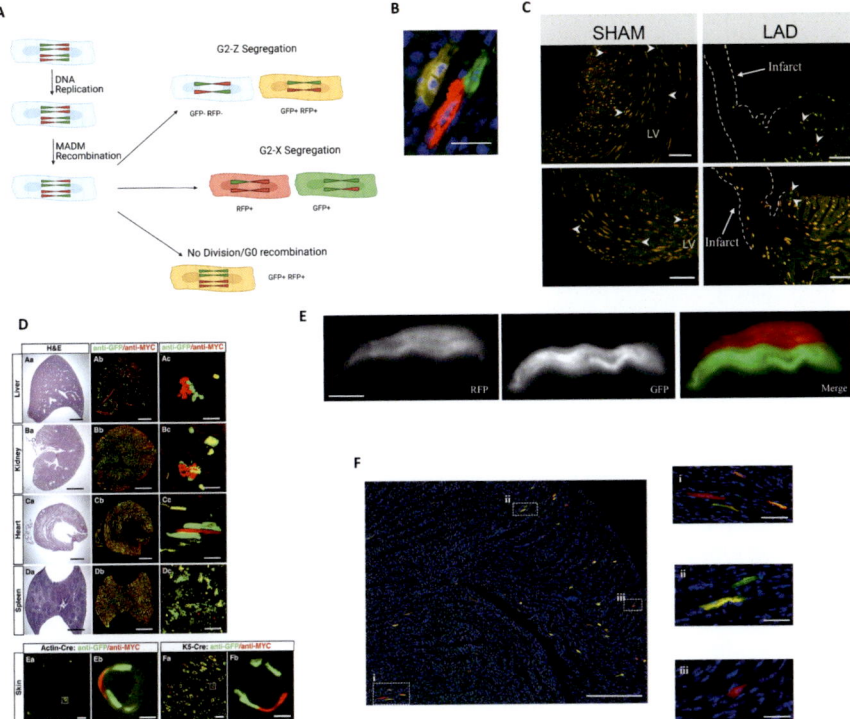

Fig. 1 Mosaic Analysis with Double Markers (MADM). (**a**) Schematic detailing how MADM works. Upon DNA replication and MADM recombination, the cell can result in three different situations. Upon G2-X segregation, two single-labeled daughter cells will arise, with one being GFP+, the other RFP+. G2-Z segregation results in one double-labeled (GFP+RFP+) cell and one unlabeled (GFP−RFP−) cell. G0 recombination with no division results in one double-labeled cell (GFP+RFP+). (**b**) Two distinctly, single-labeled, sibling cardiomyocytes exhibiting intimate end-on contact (Scale bar, 10 μm) [13]. (**c**) Myocardial infarction (MI) was induced by Left Anterior Descending Artery (LAD) ligation. The number of cardiomyocytes in Sham versus MI is similar, suggesting that MI does not result in any appreciable increase of cardiomyocyte proliferation in the left ventricle (LV). White arrowheads point to single-labeled cells (Scale bars, 120 μm) [13]. (**d**) The utility of MADM in post-mitotic and mitotic tissues. Hprt-Cre was used to generate MADM labeled cells in post-mitotic tissues (liver, kidney, heart, spleen). Actin-Cre was used to label cells in the epidermis while keratin5-Cre was used to label keratinocytes in the epidermis (Scale bars: (Aa)-(Da) and (Ab)–(Db), 2 mm; (Ac)–(Dc), 50 μm; (Ea) and (Fa), 100 μm; (Eb) and (Fb), 10 μm) [12]. (**e**) Two single-labeled daughter cardiomyocytes generated by a MADM recombination/division event in the Myh6/MADM model (Scale bar, 15 μm) [13]. (**f**) Section from a P12 Myh6/MADM heart revealing the presence of sparse single-labeled and double-labeled cells (Scale bar, 100 μm). (i) GFP+ single-labeled cardiomyocyte with its sibling RFP+ single-labeled cell, and double-labeled cells. (ii) GFP+ single-labeled cell with a double-labeled cell. The GFP+ cell contains visible sarcomeric elements, demonstrating the fidelity of this model in marking cardiomyocytes. (iii) RFP+ single-labeled cardiomyocyte (Scale bars (i-iii), 10 μm) [13]

mitosis is required to generate single-labeled cells, the MADM system unambiguously labels divided cells [11, 12]. Figure 1a details a schematic for how MADM works.

In addition, MADM allows for asymmetric labeling of daughter cells so that a relationship between precursor-progeny lineages can be established [14]. This is particularly important in studies involving stem cell differentiation, organ development and tissue regeneration in response to injury. Since MADM couples mitosis with labeling, MADM can be used to identify the progenitor cell or single cells that can be tracked for clonal expansion. The two differentially labeled MADM daughter cells can be retrospectively analyzed to investigate clonal analysis and patterns of cell division. MADM has been successfully used to track stem cell division [12, 15, 16]. In the case of an asymmetric stem cell division where a daughter stem cell and a differentiated cell are generated, each daughter cell is differentially labeled. If the differentiated cell continues to proliferate, then a clone of single-labeled cells can be identified. This is particularly useful for in vivo stem cell tracking, since MADM events are rare, hence labeling a small number of cells and their progeny. This allows for retrospective tracing of cellular expansion through easily identifiable sparse clones. In addition, multilineage cell potential can be studied by tracking whether certain clones generate multiple different cell types [14].

The labeling efficiency of MADM is partly dependent on the Cre line used. Obviously, ubiquitous Cre lines result in more doublelabeled cells, which is due to the increase in post-mitotic cells, which undergo G0 recombination [12]. Tissue-specific inducible Cre lines are routinely used for two main reasons: 1) to control the extent of recombination events, and 2) to control the timing of recombination events according to the study design. When experimental planning requires clonal analysis at specific time points (i.e. at precise stages of development or after injury), temporal control of MADM labeling can be achieved using transiently expressed Cre lines [12]. While the amount of tamoxifen induces a desired recombination event, the timing of tamoxifen administration allows for temporal control of recombination and fluorescent labeling of cells.

2.2 Labeling Rare Populations and Lineage Tracing

The first study to use MADM for clonal analysis in the heart was reported by Ali et al. [13]. First, they used an HprtCre$^{+/-}$/MADM-11$^{GT/TG}$ model, which can label any cell, and observed distinct clusters of RFP$^+$ and GFP$^+$ cells in the heart. In addition, Tasic et al. generated another transgenic mouse model where they inserted the reconstituted GFP gene into one Hipp11 locus and the reconstituted RFP gene into the other Hipp11 locus of each cell [17]. In this model, MADM-11$^{GG/TT}$, they observed only double-labeled cells and found no evidence of single or unlabeled cells. This further validated the use of MADM in the heart because there was no observed silencing of the MADM transgene, thereby ensuring no false negative results. After confirming the validity of MADM in the heart, Ali and colleagues created an inducible Myh6CreERT2;MADM-11$^{GT/TG}$ mouse model in which upon tamoxifen induction, Myh6 dividing cells may undergo homologous recombination, resulting in two distinctly labeled daughter cells (Fig. 1b). Newborn pups were given tamoxifen and their hearts were analyzed at P12. They observed that approximately 11% of labeled cells were single-labeled, which were progeny of Myh6-expressing cardiomyocytes. These cells all expressed alpha-actinin and contained sarcomeric elements, demonstrating that dividing cardiomyocytes gave rise to new cardiomyocytes. In addition, they observed equivalent frequencies of GFP$^+$ and RFP$^+$ labeled cells, demonstrating that cardiomyocytes divide to generate further cardiomyocytes in a symmetric fashion. In many cases, the cardiomyocyte clones were noncontiguous, separating from each other after division.

They next induced MADM recombination at E13.5 and analyzed the labeling of Myh6 expressing cells during development. They found that a majority of the cells were single-labeled, confirming that a majority of these cells are mitotically active. In contrast, they observed a significant number of double-labeled cardiomyocytes after birth, likely arising from G0 inter-chromosomal recombination, indicating that after birth, cardiomyocytes are not mitotically active. While their work showed that cardiomyocytes were the source of proliferating cells during development, Ali et al. demonstrated that only a very small portion of cardiomyocytes divide in the adult heart. Their lineage tracing experiments confirmed that pre-existing cardiomyocytes generate cardiomyocytes in adults at a low rate after birth.

Next, they used a β-ActinCreER/MADM-11$^{GT/TG}$ model, which permits MADM recombination in any cell-type. The use of the β-ActinCreER/MADM model allowed the investigators to determine if there is a stem/progenitor cell source for cardiomyocytes, and their findings clearly and unambiguously argued against the existence of a multipotent progenitor cell in the adult heart akin to canonical stem cells in other tissues. To investigate the proliferative behavior of cardiomyocytes after injury, MI was induced by ligation of the left anterior descending artery (LAD) at 8 weeks of age, with tamoxifen administration for 2 weeks. They analyzed the hearts 4 weeks after MI and found similar frequencies of single-labeled cells in both sham and MI, suggesting that injury in mice does not necessarily induce cardiomyocyte proliferation above the basal level (Fig. 1c). An alternative explanation could be the inefficient inter-chromosomal recombination in the setting of induced Cre recombinase in the adult cardiomyocytes. Further iteration of MADM for more efficient recombination would improve our understanding of adult cardiomyocyte proliferation in homeostasis and after injuries. The study by Ali et al. demonstrated for the first time how cardiomyocytes can be labeled through MADM and how their fate can be tracked both in response to injury and through several stages of development.

MADM has also been used to study cardiomyocyte division stimulated by cell-cycle gene induction. Mohamed and colleagues used a combination of four cell cycle regulators (CDK1:CCNB and CDK4:CCND complexes) to induce cardiomyocyte proliferation and growth in vitro and in vivo [18]. Particularly, they used MADM lineage tracing and demonstrated that adult cardiomyocyte division could be induced in vivo at an efficiency of at least 15%.

2.3 Utility of MADM in Other Organ Systems

Zong et al. first reported the development of the MADM system [12]. In their groundbreaking report, they demonstrated that Cre-dependent inter-chromosomal recombination can be induced efficiently in vivo in mitotic and post-mitotic cells (Fig. 1d). They used this system for conditional gene knockout, lineage analysis, and neural connection tracing. To illustrate the utility of MADM, they studied the fate of granule cell progenitors in the cerebellar cortex. Using the MADM system, they identified 26 distinct subclusters of granule cells in the cerebellar cortex. They showed that these granule cell clusters exhibit limited dispersion, that there was a low frequency of generating these clusters, and that there was a small chance that clusters were generated from two separate clonal lineages. They reasoned that each cluster was likely the progeny of a single-labeled clone. Thus, Zong et al. successfully used the MADM system for the first time to demonstrate that granule cell progenitors are fated to give rise to adult granule cells which distinctly localize and project axons to specific sublayers of the cerebellar cortex.

MADM has been used extensively for developmental studies in the field of neuroscience. Mihalas and Hevner used MADM to study the differentiation of early intermediate progenitors (IP) and their role in the developing cerebral cortex [16]. IP cells are derived from radial glial progenitor cells and give rise to pyramidal projection neurons in the cerebral cortex. Their data suggested three main models for IP cell differentiation. Their analysis revealed that IP cells can have asymmetric fates and generate multilayered clones, or undergo rapid or delayed terminal differentiation to produce either upper or lower-layer cortical neurons, respectively. In all of the suggested processes, the authors observed asymmetric cell death.

MADM has since been used in other systems, particularly cancer. Liu et al. used a MADM-based model for glioma to lineage trace neural stem cells (NSC) and distinguish between cell-of-mutation and cell-of-origin [15]. The lineage tracing feature of MADM allowed this group to track clones throughout the process of tumorigenesis. The mutant cells were labeled GFP+, whereas the wild-type cells, which served as internal controls, were labeled RFP+. Importantly, the MADM system was utilized to trace the cells at pre-malignant stages. They induced p53 and NF1 mutations in NSCs and observed that oligodendrocyte progenitor cells exhibited higher proliferative capacity, thus pointing to these as the cell-of-origin in their glioma model. By using MADM to differentiate between mutant and wild type cells, Liu et al. were able to conduct a detailed analysis of the precise physiological changes that occur during tumor formation and identify a cell-of-origin in their glioma model.

The MADM system has also been used in a variety of genetic imprinting studies. A particular advantage is that MADM can be used to study uniparental disomy (UPD) by indelibly and unambiguously labeling either unimaternal or unipaternal disomic cells. Hippenmeyer et al. studied the effects of genomic imprinting in chromosome 7 and 12 and explored chromosome and cell-type specific imprinting [19]. In addition, Laukoter et al. found that UPD in the neocortex results in highly cell-type specific genome-wide changes [20]. They also used MADM to reveal differences in paternal dominant and maternal dominant UPD within cortical astrocytes.

2.4   Limitations/Future Directions

The major limitation of traditional systems for conditional gene knockout has been the difficulty to achieve strict coupling of knockout and labeling. Since there is a single chromosomal exchange event in MADM, generation of homozygous cells and labeling is coupled, hence leaving little chance for ambiguity. However, there are several limitations with MADM. First, the efficiency of interchromosomal recombination is much less than intrachromosomal recombination used in traditional knockout systems. Although this could be a desirable feature for analyzing single-cell autonomous gene function, it may become a problem where high frequency of gene knockout is desired. Also, since MADM is based on the availability of a pair of MADM knock-ins between the gene of interest and the centromere, there is a need to generate knock-in cassettes for other chromosomes, an effort that has been realized in recent publications [12, 15, 19, 20].

For lineage tracing and clonal analysis studies, a potential limitation of the MADM system is performing event quantification. A binucleated daughter cell resulting from G2-Z segregation without cytokinesis would be double-labeled, an observation that is frequently encountered when analyzing heart tissue during development (Fig. 1e). However, if cytokinesis occurs, G2-Z segregation could lead to one double-labeled daughter cell while its sibling would be unlabeled (Fig. 1f). This ultimately could cause underestimation of the number of proliferated cells. Another limitation is the fact that MADM cannot be used as effectively in postmitotic cells as it depends on mitosis in order to label cells [12]. Terminally differential cells in G0 can also undergo recombination without cytokinesis, leading to the presence of a double-labeled cell.

3   Rainbow Reporter

The rainbow reporter system is a novel stochastic four-color Cre-dependent reporter system that has been used for clonal analysis studies. The advantage of this system lies in its ability to randomly assign different fluorescent labels to cells of interest, allowing for retrospective tracing of their progeny with easily distinguishable clones in vivo. When used in combination with an inducible tissue-specific Cre mouse line, recombination events can be controlled in a spatiotemporal manner. Random recombination events in proliferating cells will result in clones of cells that retain the same fluorescent label as the parental cell. When rare recombination events occur (i.e. as a result of limited amount of tamoxifen administration), sparse clones are direct evidence of cell proliferation, whereas the size of clones is suggestive of their proliferative capacity through a specific time window.

3.1 Methodology

The Rainbow system relies on Cre-dependent recombination to induce indelible labeling of one of three random fluorescent markers. Rainbow mice carry a cassette of four fluorescent genes (GFP, mCerulean, mOrange, and mCherry) inserted in the Rosa26 (R26) locus. Without Cre-mediated recombination, all tissues express the default GFP reporter label. When the Rainbow mouse is crossed with a tissue-specific inducible Cre line, tamoxifen administration leads to random excision of a pair of the mutated LoxP sites, resulting in permanent and exclusive expression of one of the three fluorescent proteins. Rainbow can be used to retrospectively trace cell lineages by identifying and counting same-colored clones that arise from a common progenitor. When crossed with a mouse line that expresses Cre/CreER under the promoter of a certain marker gene, this model can be used to label and track the fate of a population of cells of interest. The permanent labeling feature also allows for analysis over a long period of time in processes such as cellular dynamics and symmetry of cell divisions in a single cell lineage [21]. Figure 2a illustrates a schematic of how the Rainbow reporter works.

We previously reported the utility of the Rainbow system to retrospectively identify the source of new cardiomyocytes during fetal and neonatal development, as well as in adult hearts after injury [23]. Through 3D clonal analysis of cardiovascular progenitors and cardiomyocytes, we demonstrated that cardiac progenitors are the main source of cardiomyocytes during murine cardiac development. The lineage tracing experiments revealed that immature cardiomyocytes maintain their proliferative potential throughout embryonic development, however, there is a decline in their proliferation as they progress to more mature cardiomyocytes. In this study, several inducible mouse models were used in combination with the Rainbow system which allowed for distinguishable reporter expression in the heart as opposed to a mosaic pattern generated by a non-inducible Cre model. Clones of cells were permanently labeled with one of the three fluorescent proteins and further staining for α-sarcomeric actinin confirmed their cardiomyocyte identity. In order to determine the size of the generated clonal clusters, the cell counter tool on ImageJ software was used to quantify the number of cells per clone (Fig. 2b). Additionally, for a more detailed three-dimensional clone volume analysis and anatomical localization, a modified CLARITY technique was used in which the heart was transformed into an optically translucent but structurally preserved organ. The cleared hearts were subsequently imaged by confocal and light-sheet fluorescence microscopy. This advanced imaging modality facilitated an accurate measurement of clone volumes at different time points during heart development.

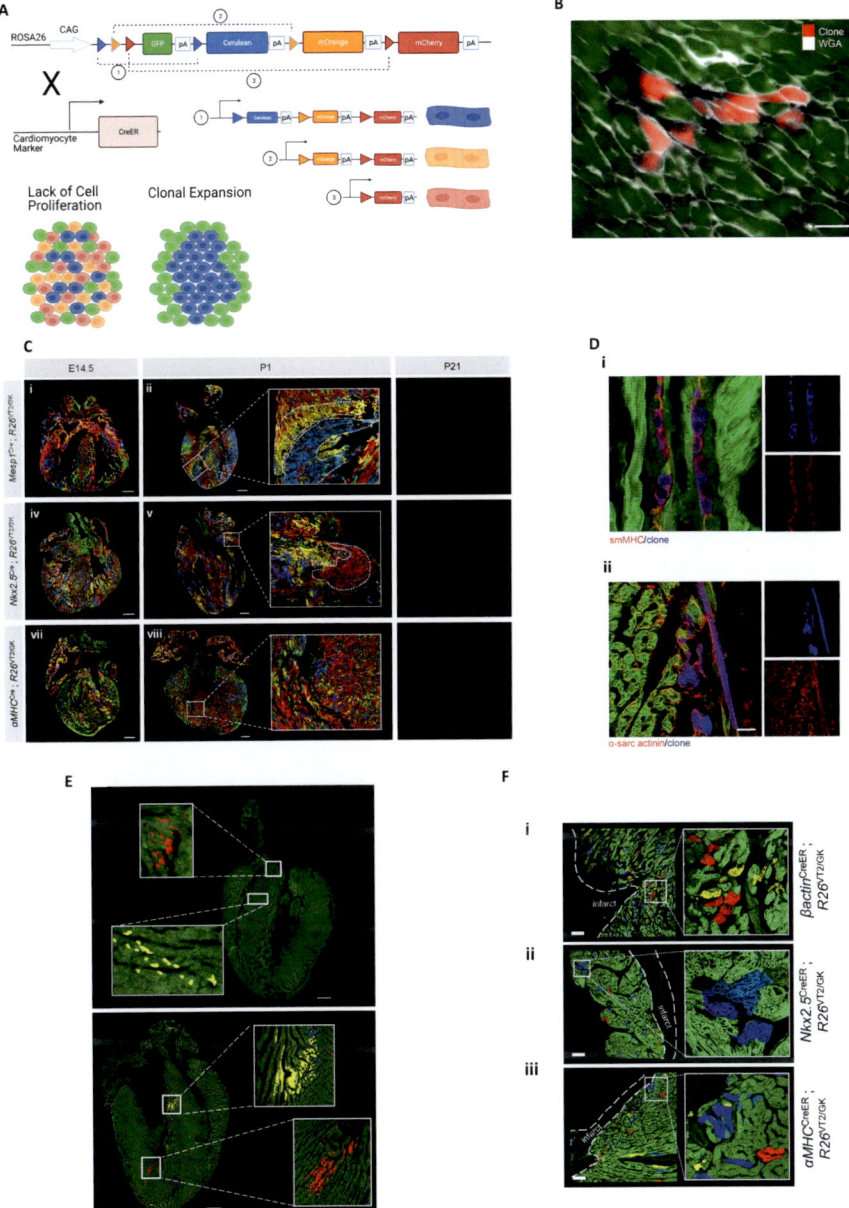

Fig. 2 Rainbow Reporter: (**a**) Schematic of how Rainbow works. When a Rosa26 Rainbow mouse is crossed with a tissue-specific CreER mouse, expression of the tissue-specific marker results in Cre expression, which allows for the excision of a random pair of loxP sites and results in expression of one of three fluorescent proteins (mCerulean, mOrange, or mCherry). Cells that do not undergo recombination express GFP. In the absence of cell proliferation, the tissue gives a mosaic appearance. When cell proliferation occurs, a clear abundance of same-colored cells is visible. (**b**)

3.2 Labeling of Rare Cell Populations and Lineage Tracing

Sereti et al. were the first to use Rainbow mice to study the regenerative capacity of the heart [22]. In order to first demonstrate the utility of Rainbow in the heart, they utilized 3 different transgenic Cre mouse models, under the control of either the cardiovascular progenitor genes Mesp1 and Nkx2.5 or the more mature cardiomyocyte marker αMHC. Analysis of hearts at embryonic day 14.5 (E14.5), postnatal day 1 (P1) and P21 revealed that the hearts marked by progenitor markers formed clonal clusters while those marked by αMHC showed a mosaic pattern of singletons (Fig. 2c). The use of constitutively active Cre lines, although very informative, normally produces high levels of recombination, and one cannot exclude the possibility that the observed single-color cell clusters are the results of random recombination events. On the other hand, a tamoxifen-inducible Cre line permits spatiotemporal control of recombination events. By administering a limiting amount of tamoxifen, one can achieve a handful of labeled cells in an organ and follow their fate retrospectively.

The authors next crossed Rainbow mice with mice harboring an inducible Cre under the control of a βactin, Nkx2.5, or αMHC promoter. This approach allowed for distinction between non-cardiomyocyte-derived clonal expansion (βactin^{CreER}; R26$^{VT2/GK}$), cardiac progenitor-derived clonal expansion (Nkx2.5^{CreER}; R26$^{VT2/GK}$), or mature cardiomyocyte (αMHC^{CreER}; R26$^{VT2/GK}$) clonal expansion. When a limiting amount of tamoxifen was administered at E9.5 or E12.5 to βactin^{CreER}; R26$^{VT2/GK}$ mice, postnatal heart analysis revealed clear clones of cardiomyocytes, fibroblasts, endothelial and vascular smooth muscle cells (Fig. 2d). Clonal analysis of Nkx2.5^{CreER}; R26$^{VT2/GK}$ mice labeled at E9.5 or E12.5, and analyzed postnatally, also revealed a similar pattern of clonal expansion with comparable clone size and volume to those observed in βactin^{CreER}; R26$^{VT2/GK}$ mice. These findings supported the

---

Fig. 2 (continued) Marking a cardiomyocyte clone for counting purposes. Cells are pseudo-colored red and WGA staining is used to mark cell boundaries (Scale bar, 100 μm) [22]. (**c**) Longitudinal sections from E14.5, P1 and P21 transgenic mice under the control of the progenitor markers Mesp1 (i-iii) and Nkx2.5 (iv-vi) or the adult CM marker αMHC (vii-ix). Fluorescent microscope images from hearts under the control of progenitor markers (ii, v) revealed the presence of clear clonal clusters. Images from hearts under the control of αMHC (viii) revealed a mosaic pattern of singletons, with no definite clonal clusters (Scale bar (iii, vi, ix), 500 μm; all others 50 μm) [22]. (**d**) The utility of the βactin^{CreER}; R26$^{VT2/GK}$ model in marking other cell types in the heart. (i) shows a close up confocal image of a clone (blue) containing vascular smooth muscle cells in a P7 heart stained for smooth muscle Myosin Heavy Chain (smMHC). (ii) shows a confocal image of cardiomyocyte clones in a P15 Nkx2.5^{CreER}; R26$^{VT2/GK}$ heart stained for α-sarcomeric actinin (Scale bar (i), 50 μm; (ii), 100 μm) [22]. (**e**) Limited tamoxifen administration allows for rare recombination events. Section of a βactin^{CreER}; R26$^{VT2/GK}$ adult heart, in which recombination was induced at E9.5, shows the presence of sparse single-colored clones labeled with mOrange and mCherry (Scale bar, 500 μm) [22]. (**f**) Representative confocal images of sections from βactin^{CreER}; R26$^{VT2/GK}$ (i), Nkx2.5^{CreER}; R26$^{VT2/GK}$ (ii) and αMHC^{CreER}; R26$^{VT2/GK}$ (iii) neonatal mice that received LAD ligation at P0. Clonal analysis performed 21 days after MI reveals the presence of sparse single-labeled clones, suggesting that neonatal mice undergo cardiomyocyte regeneration in response to injury [22]

proliferative capacity of progenitor cells to generate cardiomyocytes during early fetal development (Fig. 2e).

In order to explore the proliferative capacity of cardiomyocytes during fetal development, recombination was induced in cardiomyocytes at different embryonic time points using αMHC^{CreER}; R26$^{VT2/GK}$ mice. When tamoxifen was administered at E12.5, postnatal analysis revealed mostly singleton cardiomyocytes with few small size clones. However, interrogation of αMHC^{CreER}; R26$^{VT2/GK}$ mice labeled at E9.5 revealed similar size cardiomyocyte clones compared to βactin^{CreER}; R26$^{VT2/GK}$ or Nkx2.5^{CreER}; R26$^{VT2/GK}$ mice. These data suggest that αMHC-expressing cardiomyocytes at E9.5 retain the ability to proliferate and that this capacity is significantly diminished by E12.5.

They went on to perform single cell transcriptional analysis of αMHC-expressing cardiomyocytes at E9.5, E12.5 and P1. Their investigation demonstrated the existence of a heterogeneous population of cardiomyocytes within the early stages of cardiac development and their transition into a mature, less proliferative, and homogenous population by the early postnatal period. Overall, the use of the Rainbow model revealed that clonal dominance of differentiating progenitors mediates cardiac development, while a distinct subpopulation of cardiomyocytes may have the potential for limited proliferation during late fetal and early postnatal life. Such precise analyses at a single cell resolution would be challenging and prone to inaccurate interpretations if traditional lineage tracing experiments were utilized. It would be important to incorporate new technology in future, eg. DNA barcoding, for high-throughput analysis of a large number of individual cardiomyocytes and their progenies in developing hearts.

Sereti et al. also used Rainbow to study clonal expansion in neonatal and adult cardiomyocytes in response to cardiac injury. Newborn (P0) αMHC^{CreER}; R26$^{VT2/GK}$, βactin^{CreER}; R26$^{VT2/GK}$ and Nkx2.5^{CreER}; R26$^{VT2/GK}$ mice received tamoxifen followed by LAD ligation or Sham operation at P1. At 21 days after MI, hearts were analyzed and frequent clones of cardiomyocytes were observed in the infarct and border zone areas of αMHC^{CreER}; R26$^{VT2/GK}$ mice (Fig. 2f). Similar observations were made with βactin^{CreER}; R26$^{VT2/GK}$ and Nkx2.5^{CreER}; R26$^{VT2/GK}$ mice after injury. This suggested that regeneration of the heart after injury in neonates was largely due to cardiomyocyte proliferation, which was confirmed by recent fate mapping studies of cardiomyocytes and non-cardiomyocytes in neonates [24]. In adults, LAD ligation was performed at 8 weeks of age and analysis at 21 days post-MI revealed the presence of sparse single-labeled clones in the infarct and border zone areas. These observations suggested that injury induces cardiomyocyte proliferation in the neonatal heart but not in the adult heart.

Wang et al. used the Rainbow system to study the clonal expansion of smooth muscle cells (SMC) in atherosclerosis [25]. Myh11^{CreERT2}; R26$^{VT2/GK}$; ApoE$^{-/-}$ mice were fed a high-fat Western diet to induce atherogenesis. The authors observed that during early atherogenesis, there was a distinct subpopulation of SMCs that dedifferentiated and upregulated Sca1 (a stem cell marker). Sca1 staining was most intense within the core of the dominant clone and Sca1$^+$ cells appeared to colocalize near the necrotic core of the atherosclerotic plaque. Single cell RNA sequencing analysis of this Sca1$^+$ population revealed that these cells down-regulated SMC marker genes and upregulated genes relating to inflammation and the complement cascade (complement C3 was one of the most significantly up-regulated factors). These findings translated to human specimens as well. Histological analysis of post-mortem carotid artery specimens revealed localization of C3 expression to the necrotic core of human atherosclerotic plaque, similar to mice. Furthermore, this Sca1$^+$ SMC signature was found in humans and was associated with coronary artery disease, with an enriched expression of inflammatory genes (including C3). Production of C3 by the Sca1$^+$ SMC population may play a role in triggering further SMC proliferation and vascular inflammation, which could explain the rapid expansion of SMCs during early atherogenesis. Overall, Wang et. al's study reveals the potential of the Rainbow system to identify rapidly proliferating and differentiating populations of cells to gain insight into the pathophysiology of complex diseases such as atherosclerosis.

3.3   Utility of Rainbow in Other Organ Systems

Rinkevich et al. used an inducible lineage tracing mouse model (βactin^{CreER}; R26$^{VT2/GK}$) to perform lineage tracing and clonal analysis of individual cells of mouse hind limb tissue during regeneration of the digit tip, cutaneous wound healing, and normal maintenance [26]. They removed nerve supply and observed clonal expansion, revealing that cellular regeneration remains largely intact in the absence of nerve supply. In a study of the kidney, Rinkevich et al. used the Rainbow mouse model for clonal analysis and lineage tracing of cells that contribute to the development, maintenance and regeneration of the kidney [27]. In all three processes, they found that cells generating distinct parts of the nephron (i.e. Proximal, Distal tubules or collecting duct) were fate-restricted and stayed within their lineage. Furthermore, they used an Axin2^{CreER}; R26$^{VT2/GK3}$ mouse line to track Wnt pathway responsive cells (WRC). They showed that WRCs increased their proliferative capacity and that their clones were restricted to either a proximal tubule or collecting duct fate.

Recent studies using dual recombinases and Confetti reporter (also random labeling by one of the three fluorescences) demonstrated that bronchioalveolar stem cells residing in the bronchioalveolar duct junction could clonally expand to form bronchial epithelial cells and/or alveolar type I and II cells during lung repair and regeneration [28, 29]. This demonstrates the utility of clonal analysis by rainbow/confetti reporters for resolving the uni- or bi-differentiation potential of stem cells in tissue regeneration.

Interestingly, Rainbow has also been used in cancer models. For example, Corey et al. lineage traced endothelial cells within the tumor microenvironment and concluded that clonal expansion within the microvasculature is crucial to an invasive melanoma phenotype [30]. Particularly, their studies showed that there is a diminishing of founder clones to produce subclones, with tumor blood vessels upregulating genes associated with angiogenesis and a downregulation of lymphocyte adhesion molecules. Thus, the Rainbow mouse model allowed lineage tracing of single cells and their progeny to conclude that clonal evolution within melanoma can induce changes within the microvasculature to confer cancer cells an advantage.

3.4   Limitations/Future Directions

The ability of Rainbow to precisely track the fate of certain cells across multiple organ systems makes it a strong candidate for clonal analysis and lineage tracing within the heart. This model allows researchers to differentiate between proliferating and quiescent cardiomyocytes, a phenomenon that remains controversial in the field of cardiac development and regeneration. However, it is not without limitations. A major consideration when using the Rainbow mouse model is the efficiency of recombination events that partially depends on the Cre line used. While in many mouse lines, Rainbow faithfully induces equal expression of fluorescent proteins in labeled cells, some Cre lines have been reported to show uneven expression of the markers. Additionally, the model's dependence on tamoxifen could leave the possibility that a smaller dose does not induce enough recombination to mark all possible clones, thus underestimating the number of clones [22]. Furthermore, since in clonal expansion studies, only a small number of cells are labeled and their fate is monitored retrospectively, it is possible that rare populations of cells cannot be labeled by this strategy. The Rainbow system can be used for clonal analysis studies and the presence of clusters of single-colored cells confidently supports the existence of proliferating cells. However, the absence of an observation does not confirm its lack of existence.

References

1. Kretzschmar, K., Watt, F.M.: Lineage tracing. Cell. **148**(1-2), 33–45 (2012). https://doi.org/10.1016/j.cell.2012.01.002
2. Bergmann, O., Bhardwaj, R.D., Bernard, S., Zdunek, S., Barnabé-Heider, F., Walsh, S., et al.: Evidence for cardiomyocyte renewal in humans. Science. **324**(5923), 98–102 (2009). https://doi.org/10.1126/science.1164680
3. Bergmann, O., Zdunek, S., Felker, A., Salehpour, M., Alkass, K., Bernard, S., et al.: Dynamics of cell generation and turnover in the human heart. Cell. **161**(7), 1566–1575 (2015). https://doi.org/10.1016/j.cell.2015.05.026
4. Cai, C.L., Molkentin, J.D.: The elusive progenitor cell in cardiac regeneration: slip Slidin' away. Circ. Res. **120**(2), 400–406 (2017). https://doi.org/10.1161/CIRCRESAHA.116.309710
5. Eschenhagen, T., Bolli, R., Braun, T., Field, L.J., Fleischmann, B.K., Frisén, J., et al.: Cardiomyocyte regeneration: a consensus statement. Circulation. **136**(7), 680–686 (2017). https://doi.org/10.1161/CIRCULATIONAHA.117.029343
6. Vagnozzi, R.J., Molkentin, J.D., Houser, S.R.: New myocyte formation in the adult heart: endogenous sources and therapeutic implications. Circ. Res. **123**(2), 159–176 (2018). https://doi.org/10.1161/CIRCRESAHA.118.311208

7. Wei, K., Serpooshan, V., Hurtado, C., Diez-Cuñado, M., Zhao, M., Maruyama, S., et al.: Epicardial FSTL1 reconstitution regenerates the adult mammalian heart. Nature. **525**(7570), 479–485 (2015). https://doi.org/10.1038/nature15372

8. Yu, W., Huang, X., Tian, X., Zhang, H., He, L., Wang, Y., et al.: GATA4 regulates Fgf16 to promote heart repair after injury. Development. **143**(6), 936–949 (2016). https://doi.org/10.1242/dev.130971

9. Laflamme, M.A., Murry, C.E.: Heart regeneration. Nature. **473**(7347), 326–335 (2011). https://doi.org/10.1038/nature10147

10. Soonpaa, M.H., Kim, K.K., Pajak, L., Franklin, M., Field, L.J.: Cardiomyocyte DNA synthesis and binucleation during murine development. Am. J. Phys. **271**(5 Pt 2), H2183–H2189 (1996). https://doi.org/10.1152/ajpheart.1996.271.5.H2183

11. Soonpaa, M.H., Field, L.J.: Survey of studies examining mammalian cardiomyocyte DNA synthesis. Circ. Res. **83**(1), 15–26 (1998). https://doi.org/10.1161/01.res.83.1.15

12. Zong, H., Espinosa, J.S., Su, H.H., Muzumdar, M.D., Luo, L.: Mosaic analysis with double markers in mice. Cell. **121**(3), 479–492 (2005). https://doi.org/10.1016/j.cell.2005.02.012

13. Ali, S.R., Hippenmeyer, S., Saadat, L.V., Luo, L., Weissman, I.L., Ardehali, R.: Existing cardiomyocytes generate cardiomyocytes at a low rate after birth in mice. Proc. Natl. Acad. Sci. U. S. A. **111**(24), 8850–8855 (2014). https://doi.org/10.1073/pnas.1408233111

14. Weissman, I.L.: Stem cells: units of development, units of regeneration, and units in evolution. Cell. **100**(1), 157–168 (2000). https://doi.org/10.1016/s0092-8674(00)81692-x

15. Liu, C., Sage, J.C., Miller, M.R., Verhaak, R.G., Hippenmeyer, S., Vogel, H., et al.: Mosaic analysis with double markers reveals tumor cell of origin in glioma. Cell. **146**(2), 209–221 (2011). https://doi.org/10.1016/j.cell.2011.06.014

16. Mihalas, A.B., Hevner, R.F.: Clonal analysis reveals laminar fate multipotency and daughter cell apoptosis of mouse cortical intermediate progenitors. Development. **145**(17) (2018). https://doi.org/10.1242/dev.164335

17. Tasic, B., Miyamichi, K., Hippenmeyer, S., Dani, V.S., Zeng, H., Joo, W., et al.: Extensions of MADM (mosaic analysis with double markers) in mice. PLoS One. **7**(3), e33332 (2012). https://doi.org/10.1371/journal.pone.0033332

18. Mohamed, T.M.A., Ang, Y.S., Radzinsky, E., Zhou, P., Huang, Y., Elfenbein, A., et al.: Regulation of cell cycle to stimulate adult cardiomyocyte proliferation and cardiac regeneration. Cell. **173**(1), 104–116 (2018). https://doi.org/10.1016/j.cell.2018.02.014

19. Hippenmeyer, S., Johnson, R.L., Luo, L.: Mosaic analysis with double markers reveals cell-type-specific paternal growth dominance. Cell Rep. **3**(3), 960–967 (2013). https://doi.org/10.1016/j.celrep.2013.02.002

20. Laukoter, S., Pauler, F.M., Beattie, R., Amberg, N., Hansen, A.H., Streicher, C., et al.: Cell-type specificity of genomic imprinting in cerebral cortex. Neuron. **107**(6), 1160–1179 (2020). https://doi.org/10.1016/j.neuron.2020.06.031

21. He, L., Nguyen, N.B., Ardehali, R., Zhou, B.: Heart regeneration by endogenous stem cells and cardiomyocyte proliferation: controversy, fallacy, and Progress. Circulation. **142**(3), 275–291 (2020). https://doi.org/10.1161/CIRCULATIONAHA.119.045566

22. Sereti, K.I., Nguyen, N.B., Kamran, P., Zhao, P., Ranjbarvaziri, S., Park, S., et al.: Analysis of cardiomyocyte clonal expansion during mouse heart development and injury. Nat. Commun. **9**(1), 754 (2018). https://doi.org/10.1038/s41467-018-02891-z

23. Nguyen, N., Fernandez, E., Ding, Y., Hsiai, T., Ardehali, R.: In vivo clonal analysis of cardiomyocytes. In: Poss, K., Kühn, B. (eds.) Cardiac Regeneration: Methods and Protocols. Methods in Molecular Biology, vol. 2158, pp. 243–256. Springer (2021)

24. Li, Y., Lv, Z., He, L., Huang, X., Zhang, S., Zhao, H., et al.: Genetic tracing identifies early segregation of the cardiomyocyte and nonmyocyte lineages. Circ. Res. **125**(3), 343–355 (2019). https://doi.org/10.1161/CIRCRESAHA.119.315280

25. Wang, Y., Nanda, V., Direnzo, D., Ye, J., Xiao, S., Kojima, Y., et al.: Clonally expanding smooth muscle cells promote atherosclerosis by escaping efferocytosis and activating the

complement cascade. Proc. Natl. Acad. Sci. U. S. A. **117**(27), 15818–15826 (2020). https://doi.org/10.1073/pnas.2006348117

26. Rinkevich, Y., Montoro, D.T., Muhonen, E., Walmsley, G.G., Lo, D., Hasegawa, M., et al.: Clonal analysis reveals nerve-dependent and independent roles on mammalian hind limb tissue maintenance and regeneration. Proc. Natl. Acad. Sci. U. S. A. **111**(27), 9846–9851 (2014). https://doi.org/10.1073/pnas.1410097111

27. Rinkevich, Y., Montoro, D.T., Contreras-Trujillo, H., Harari-Steinberg, O., Newman, A.M., Tsai, J.M., et al.: In vivo clonal analysis reveals lineage-restricted progenitor characteristics in mammalian kidney development, maintenance, and regeneration. Cell Rep. **7**(4), 1270–1283 (2014). https://doi.org/10.1016/j.celrep.2014.04.018

28. Liu, Q., Liu, K., Cui, G., Huang, X., Yao, S., Guo, W., et al.: Lung regeneration by multipotent stem cells residing at the bronchioalveolar-duct junction. Nat. Genet. **51**(4), 728–738 (2019). https://doi.org/10.1038/s41588-019-0346-6

29. Liu, J., Cao, R., Xu, M., Wang, X., Zhang, H., Hu, H., et al.: Hydroxychloroquine, a less toxic derivative of chloroquine, is effective in inhibiting SARS-CoV-2 infection in vitro. Cell Discov. **6**, 16 (2020). https://doi.org/10.1038/s41421-020-0156-0

30. Corey, D.M., Rinkevich, Y., Weissman, I.L.: Dynamic patterns of clonal evolution in tumor vasculature underlie alterations in lymphocyte-endothelial recognition to Foster tumor immune escape. Cancer Res. **76**(6), 1348–1353 (2016). https://doi.org/10.1158/0008-5472.CAN-15-1150

Mechanisms that Govern Endothelial Lineage Development and Vasculogenesis

Daniel J. Garry and Javier E. Sierra-Pagan

Abbreviations

AGM	Aorta-gonad-mesonephros
Cas9	CRISPR-associated protein 9
CD31	Cluster of differentiation 31
CD41	Integrin alpha chain 2b
CD44	Cluster of differentiation 44
CD45	Leukocyte common antigen
Cdh5	Vascular endothelial cadherin
CRISPR	Clustered regularly interspaced short palindromic repeats
ES/EB	Embryonic stem cells/embryoid bodies
ETV2	Ets variant transcription factor 2
FLK1	Fetal liver kinase 1
Gata4	GATA transcription factor 4
HE	hematoendothelial
hiPSC	Human induced pluripotent stem cell

D. J. Garry (✉)
Department of Medicine, University of Minnesota, Minneapolis, MN, USA

Developmental Biology Center, University of Minnesota, Minneapolis, MN, USA

Lillehei Heart Institute, University of Minnesota, Minneapolis, MN, USA

Masonic Cancer Center, University of Minnesota, Minneapolis, MN, USA

Stem Cell Institute, University of Minnesota, Minneapolis, MN, USA

Paul and Sheila Wellstone Muscular Dystrophy Center, University of Minnesota, Minneapolis, MN, USA
e-mail: garry@umn.edu

J. E. Sierra-Pagan
Department of Medicine, University of Minnesota, Minneapolis, MN, USA

Lillehei Heart Institute, University of Minnesota, Minneapolis, MN, USA

© The Author(s), under exclusive license to Springer Nature
Switzerland AG 2022
J. Zhang, V. Serpooshan (eds.), *Advanced Technologies in Cardiovascular Bioengineering*, https://doi.org/10.1007/978-3-030-86140-7_3

HSC	Hematopoietic stem cell
Mlc2v	Myosin light chain 2v
Myf5	Myogenic factor 5
MyHC	Myosin heavy chain
Myod	Myoblast determination protein 1
Nkx2-5	Homeobox Protein Nkx2–5
OE	Overexpression
Pdx1	Pancreatic and duodenal homeobox 1
qPCR	Quantitative polymerase chain reaction
scRNA-seq	Single cell RNA sequencing
Tie2	Endothelial cell specific receptor tyrosine kinase 2
Ve-Cad	Vascular endothelial cadherin

1 Cardiovascular Diseases Are Common and Have Considerable Morbidity and Mortality

Peripheral artery disease affects more than 10 M Americans resulting in more than 150,000 limb amputations each year in the U.S. In addition, more than 300,000 patients have coronary artery bypass grafting (surgical revascularization) [1]. These diseases collectively are amplified by the rising incidence of diabetes, obesity and cardiovascular disease. These complications result in considerable morbidity and mortality [1, 2]. Current medical therapies for vascular diseases include limb amputation and vascular bypass grafting. However, these therapeutic interventions have significant limitations. These diseases are chronic, debilitating, lethal and they warrant new and novel therapies. The definition of the molecular mechanisms that govern the endothelial lineage and vascular development will provide a platform to modulate these pathways and promote vasculogenesis as a therapeutic initiative. The overall goal for this chapter is to highlight the key regulators that govern endothelial and vascular development.

2 Master Regulators Govern Fate Decisions and Lineage Development

Loss of function and gain of function genetic studies have defined essential factors that govern cell fate and lineage development (Fig. 1a). These factors also known as master regulators occupy or sit at the top of a regulatory hierarchy [3, 4]. Perhaps a prototypic example are members of the MYOD family of transcription factors. The MYOD family consists of bHLH transcription factors that have distinct and overlapping functional roles for the regulation of the myogenic lineage [5]. These master regulators also have the capacity to convert another differentiated cell type (usually a fibroblast) to a specific lineage (i.e. skeletal muscle) using a

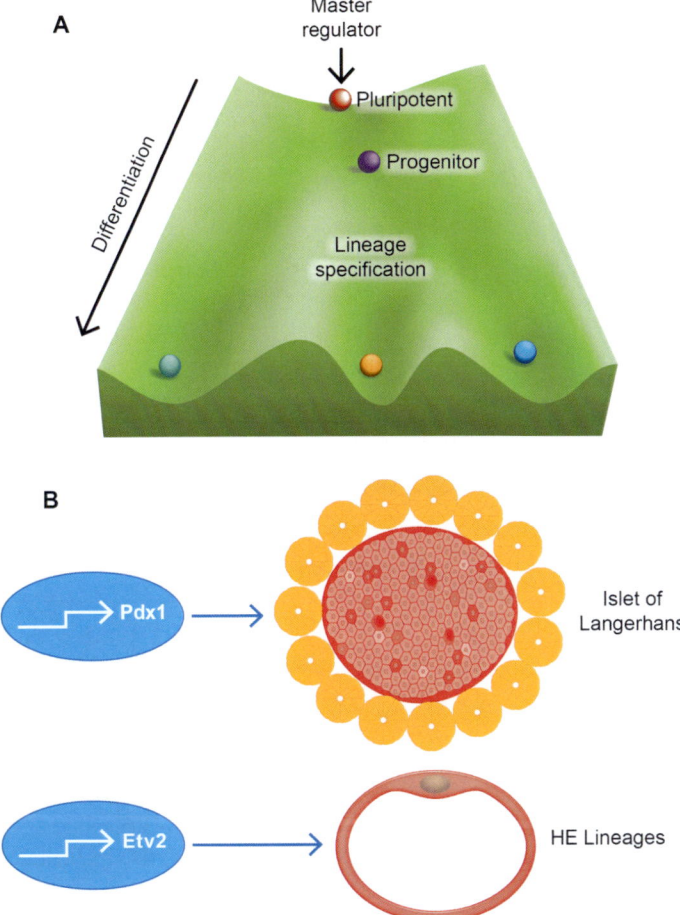

Fig. 1 Master regulators specify lineage-specific development. (**a**) Adaptation of Waddington's landscape that outlines the role of master regulators to govern fate decisions during embryogenesis. (**b**) Schematic outlining examples of master regulators for specific lineages. PDX1 is a master regulator for pancreatogenesis and ETV2 is master regulator for hematoendothelial lineages

promoter-reporter construct that demonstrates expression with lineage specific differentiation (Fig. 1) [6–9]. These assays were initially referred to as conversion assays. Using conversion assays in combination with gene disruption and transgenic technologies, hundreds of master regulators have been described [3]. In addition to the MYOD family, PDX1 (pancreas), OCT4/SOX2/NANOG (pluripotency), SCL/TAL1 (blood), HIF1 (hypoxia), and others have been identified (Fig. 1b) [6–8, 10–12]. ETV2 is a recently identified master regulator that has been shown to be essential for the specification and development of the endothelial lineage (Fig. 1b) [13, 14].

3   Developmental Milestones for Endothelial Development

The initial developmental stage for vasculogenesis occurs as specified progenitors (i.e. angioblasts) that migrate from the primitive streak in the developing mouse embryo to the yolk sac [15]. Later, these angioblasts migrate from the extraembryonic yolk sac to the embryo proper to form cord like structures, lumens (a process known as tubulogenesis) and ultimately form a vascular plexus (Fig. 2) [15]. This process continues and is associated with the onset of cardiac contractility in the E8.25 heart tube, the appearance of primitive blood cells associated with the primitive circulation (E8.5) and the establishment of a complete circulation with the propagation of blood throughout the E10-E10.25 mouse embryo [16]. This circulation is impacted by the growth of the embryo and the transition from diffusion to the circulation of blood in response to hypoxic signals (HIF1 and HIF2) and signaling pathways (VEGF1-FLK1/FLT1/FLT2, SHH-GLI1/2/3, SRC-CDH5, ANG1/2-TIE2, etc.) [16, 17]. This latter process is termed angiogenesis, which is characterized by the formation of new vessels that originate or sprout from pre-established or

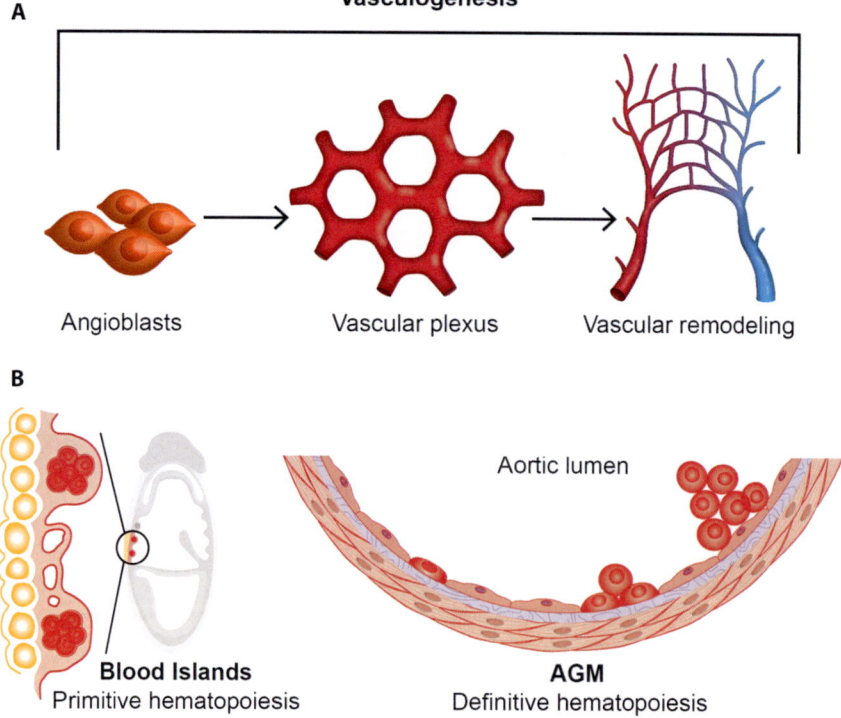

Fig. 2 Overview of hematoendothelial development. (**a**) Schematic highlighting the role of angioblasts being recruited to form a vascular plexus, tubulogenesis and sprouts to generate vascular networks and remodeling. (**b**) Schematic highlighting the role of hemogenic endothelium during primitive (yolk sac) and definitive (AGM) hematopoiesis

pre-existing vessels [17]. The vasculature continues to architecturally evolve and mature with the determination of venous and arterial endothelial fates. This maturation phase is marked by the expression of integrins, the balance of apoptosis and cell proliferation and the impact of the extracellular matrix [15]. Overall, the endothelial and vascular lineages are coordinated and responsive to microenvironmental cues and signals.

The endothelial-endocardial relationship is established early during embryogenesis. The endothelial lineage is characterized by a single cell thickness of developing vascular networks that form a tight syncytium separating the luminal space with the underlying vascular wall. This endothelial lining is contiguous with the endocardium, which lines the four-chambered heart. The ontogeny of the endocardium is distinct from the endothelial lineage and is reflected in the expression of a lineage specific molecular program [18].

4 The Common Origin of Endothelial and Hematopoietic Lineages

The endothelial and hematopoietic lineages are both derivatives of the mesodermal germ layer [19]. As both lineages develop in close proximity (i.e., blood islands of the yolk sac and the blood containing vessels) and express overlapping molecular programs, studies support a common origin for the lineages [19]. The extraembryonic yolk sac is the source for primitive hematopoiesis (Fig. 2b) [20]. Indeed, studies have demonstrated that multilineage hematopoietic stem cells are derivatives of hemogenic endothelium. Hemogenic endothelium, despite its endothelial gene expression profile, loses its endothelial potential early in development and only gives rise to blood [21]. Hemogenic endothelium are flat shaped cells that undergo endothelial-to-hematopoietic transition (EHT and marked by CD44 expression) and acquire a spherical shape characteristic of blood cells [20, 22]. These hemogenic endothelial cells are found within the allantois, the yolk sac, the endocardium and the aorta-gonad-mesonephros (AGM) of the developing embryo [20, 23]. The AGM is closely associated with the ventral wall of the dorsal aorta (~E9.5-E11.5) and has been shown to produce hematopoietic stem cells that are capable of engrafting and reconstituting the irradiated bone marrow. Therefore, the AGM is recognized for its critical role in definitive hematopoiesis (Fig. 2b) [23]. Single cell RNA-seq of the AGM in the developing mouse have identified multiple cell types and defined the expression of GATA2, RUNX1, LYL1, ERG, FLI1, LMO2 and TAL1 transcription factors, which result in the loss of endothelial gene expression and acquisition of hematopoietic gene expression [22].

5   ETV2 Is Necessary and Sufficient for Endothelial Lineage Development

ETV2 (Ets Variant Transcription Factor 2) was initially sequenced from the testis by Steve McKnight's laboratory [24]. Later, ETV2 was co-discovered as an essential factor for hematoendothelial (HE) lineage development by the Choi laboratory and the Garry laboratory [25, 26]. These independent efforts resulted in studies by the Choi laboratory, which defined a BMP/NOTCH/WNT-ETV2 axis during mouse embryogenesis and demonstrated that *Etv2* null embryos were lethal and lacked hematopoietic and vascular lineages [25]. Similarly, the Garry laboratory used a *Nkx2-5-reporter* transgenic strategy and identified *Etv2* as a putative downstream target. Further, they demonstrated that *Etv2* null embryos were nonviable at E9-E9.5 and had an absence of blood and endothelial lineages and defined that ETV2 was a direct upstream regulator of the *Tie2* gene [26]. Studies further demonstrated the essential role for ETV2 in zebrafish, xenopus, pig and human [13, 14, 27–35]. These latter results support the evolutionary conserved role for ETV2 as an essential factor for hematoendothelial development. Furthermore, forced overexpression (OE) of ETV2 using mouse embryonic stem cells/embryoid bodies (ES/EB) differentiation assays demonstrated that it was sufficient for HE development (Fig. 3) [25, 35–37]. Collectively, these studies provided a foundation for ETV2 as a master regulator for the HE lineages.

6   ETV2 Expression during Mouse Embryogenesis and the Postnatal Period

Using in situ hybridization, PCR, immunohistochemistry and the *3.9Kb Etv2-EYFP* reporter transgenic expression, ETV2 expression was restricted to the angioblasts at E6.5, hematoendothelial cells and endocardium until ~E10-E10.5 after which it was rapidly extinguished with the exception of a small subpopulation of cells associated with the dorsal aorta (Fig. 4) [25, 26, 36–41]. Similarly, using the ES/EB differentiation assay, ETV2 was robustly expressed on Days 3–4 following differentiation after which it was extinguished (Fig. 3) [25, 42]. These expression studies support the notion that ETV2 functions as a "rheostat" to tightly regulate HE lineage development similar to other master regulators for the specification of other lineages. While the mechanisms that regulate extinguished expression of ETV2 are incompletely defined, the feedback mechanisms involving FLT1 contribute, in part, to the negative regulation of *Etv2* gene expression [42]. Future studies will need to focus on the definition of these mechanisms to enhance our understanding of ETV2 expression.

ETV2 is expressed postnatally in the testis and in the HSC population (Lin-Sca1+cKit+ cells) in adult mouse bone marrow [44, 45]. ETV2 is induced and upregulated following tissue and vascular injury without sustained long-term or persistent

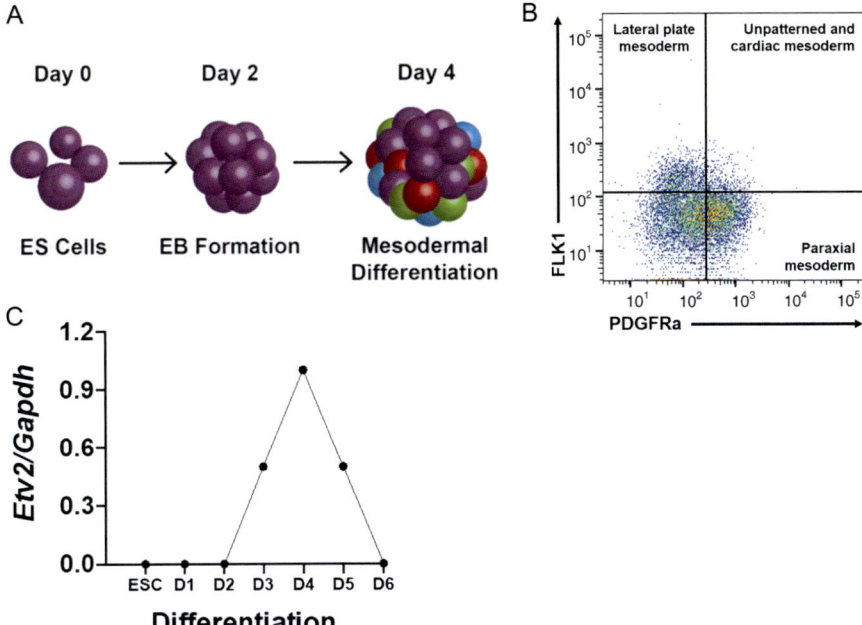

Fig. 3 Embryoid body (EBs) differentiation assays recapitulate developmental mechanisms. (**a**) Schematic highlighting ES cells differentiating to EBs and forming mesodermal derivatives including: hematoendothelial, cardiac and skeletal muscle lineages. (**b**) FACS profile of dissociated EBs stained for FLK1 and PDGFR-a demonstrate the HE (FLK1$^+$/ PDGFR-a$^-$), cardiac (FLK1$^+$/PDGFR-a$^+$) and skeletal muscle (FLK1$^-$/PDGFR-a$^+$) lineages. (**c**) Schematic of the expression profile of *Etv2* during mesodermal EB differentiation showing its transient pattern of expression

expression [46]. Furthermore, recent studies support the notion that ETV2 is expressed in tumorigenic tissues and may be associated with the angiogenic response observed with various solid tumors (as outlined below) [47–49]. These expression patterns and its role as a master regulator suggest that ETV2 may be an important target to promote or repress angiogenesis depending on the physiological context.

7 ETV2 Is Dynamically and Transcriptionally Regulated by Upstream Factors

Previous studies by our laboratory demonstrated that the 3.9 kb upstream fragment of the *Etv2* gene harbored all the modules and motifs necessary for the spatial and temporal expression pattern of endogenous ETV2 (Fig. 5) [26, 31, 36, 41, 42, 50, 51]. These studies established that the EYFP reporter expression pattern recapitulated endogenous ETV2 activity demonstrating the onset of expression and

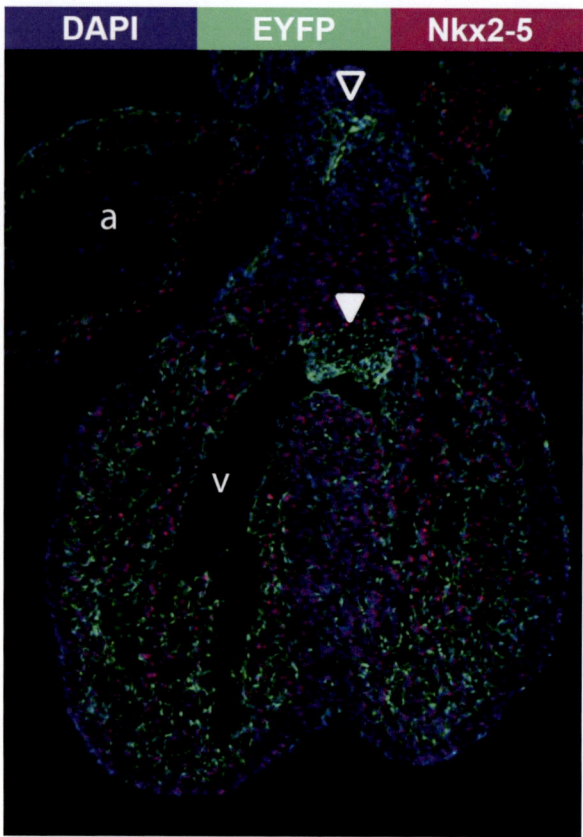

Fig. 4 ETV2 fate mapping identifies ETV2 expressing lineages and its descendants in the heart. Using the *3.9 kb Etv2-Cre* transgenic mouse model, the ETV2 contribution to embryogenesis was mapped during cardiogenesis in the developing mouse. E12.5 heart co-stained with NKX2–5 (red), DAPI (blue), and ETV2 cells/descendants (green) [43] demonstrate that every endothelial and endocardial cell is labeled with GFP including the developing valves (arrowhead) and aorta (open arrowhead) (a, atria and v, ventricle)

extinguished activity (Fig. 4) [41]. Bioinformatics analysis revealed evolutionary conservation of two modules (CRI and CRII) within the 3.9 kb fragment and the deletion of this fragment phenocopied the global deletion of *Etv2* (Fig. 5b) [42]. Furthermore, the characterization of the *Etv2* cis-regulatory modules in vitro revealed the importance of the CRI and CRII modules using luciferase assays [26, 42, 52]. Collectively, these studies confirmed that the upstream regulation of this gene utilized the binding motifs contained within the 3.9 kb fragment. Using gene disruption models, transcriptional assays, EMSAs and mutagenesis, transcriptional regulators of *Etv2* gene expression included: ETV2, GATA2, VEFGF/FLK1-Calcineurin-NFAT, CREB1, MESP1, NKX2-5 and other signaling pathways (BMP, Notch and Wnt signaling pathways) (Fig. 5c) [25, 26, 39, 42, 52, 53]. While all

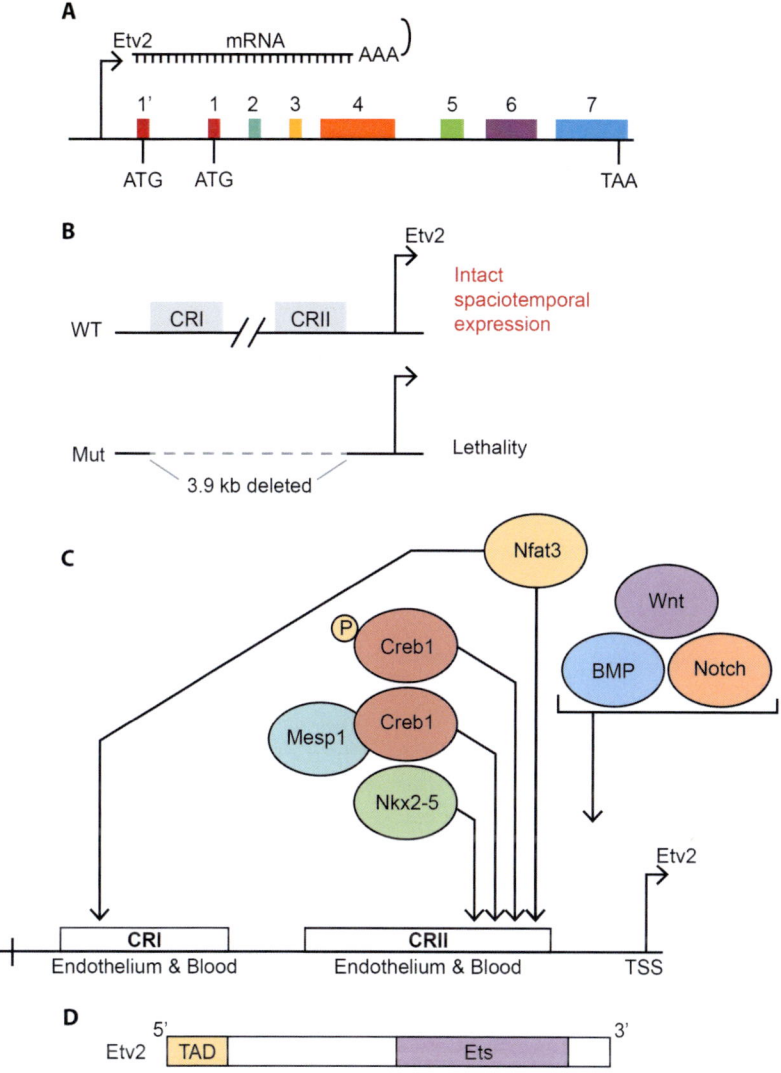

Fig. 5 The regulatory mechanisms that govern the *Etv2* gene. (**a**) Schematic highlighting the ATG start site and seven exons associated with the *Etv2* gene. (**b**) Schematic demonstrating the CRI and CRII modules that regulate *Etv2* gene expression. Deletion of the CRI and CRII modules pheno-copies the *Etv2* global gene KO with embryonic lethality and absence of blood and vasculature. (**c**) Upstream cis transcriptional regulators of the *Etv2* gene and their binding motifs are outlined. (**d**) Schematic highlighting the ETV2 protein with its transcriptional activation domain (TAD) and DNA binding or Ets domain

these factors were important regulators, it appeared that MESP1-CREB1 may initiate *Etv2* gene expression in the mesodermal lineage. Gene activation may also be context dependent, as in zebrafish, the overexpression of Nkx2-5 functions to repress *Etv2* gene activity whereas in the mouse it appears to function as a direct upstream regulator of *Etv2* in the endocardium [54].

8   Definition of Transcriptional Targets for ETV2 Mediate Distinct Developmental Events

Master regulators have a number of functions that are mediated by their respective downstream targets. ETV2 binds a canonical GGAA/T Ets motif and transactivates the hematoendothelial molecular program including: *Scl/Tal1, Lmo2, Tie2/Tek, VE-Cadherin/Cdh5, Pecam1/CD31, Gata1, Gata2, Flk1, Elk3, Fli1, Sox7, Cepd,* and others (Fig. 6) [14, 26, 34, 37, 42, 43, 53, 55–61]. These targets were identified and confirmed using CHiP-seq, transcriptional assays, EMSA, Standard ChIP and other molecular techniques [14, 55, 62]. More recent studies have demonstrated that ETV2 promotes cell proliferation by regulating *Yes1* gene expression, which interacts with the Hippo signaling pathway (Fig. 6) [63]. ETV2 also transcriptionally regulates *Rhoj* gene expression to modulate endothelial progenitor cell migration during embryogenesis (Fig. 6) [64]. Finally, ETV2 regulates microRNAs (i.e. miR130a), which govern fate determination by promoting endothelial differentiation but not hematopoietic differentiation (Fig. 6) [59, 60]. Context specific gene regulation is observed in the testis where ETV2 is a direct upstream regulator of *Sox9* gene expression. In a positive feed-back loop, SOX9 binds to the *Etv2* promoter and serves to transactivate its regulator thereby maintaining the sertoli cell phenotype [65]. ETV2-Chip-seq datasets are publicly available and are continuing to be mined to further define and explore additional targets and functional roles for this master regulator.

9   Protein-Protein Interacting Factors for ETV2 Are Important Coregulators

The *Etv2* gene harbors seven exons and encodes a protein that has carboxy terminal domain, a DNA binding domain (amino acids 316–336 which overlaps with the ETS domain that spans from 231 to 315 aa) and an amino terminal domain (amino acids 1–157) (Fig. 5d) [44]. Using an array of biochemical and molecular techniques (i.e. mass spectrometry, yeast two hybrid screening assay, FRET, etc.) interacting factors have been identified for ETV2. Early studies identified FOXC2 as an interacting factor for ETV2 and suggested that adjacent Ets-Fox binding motifs (Ets-Fox enhancer motif) in more than 20% of endothelial specific genes were

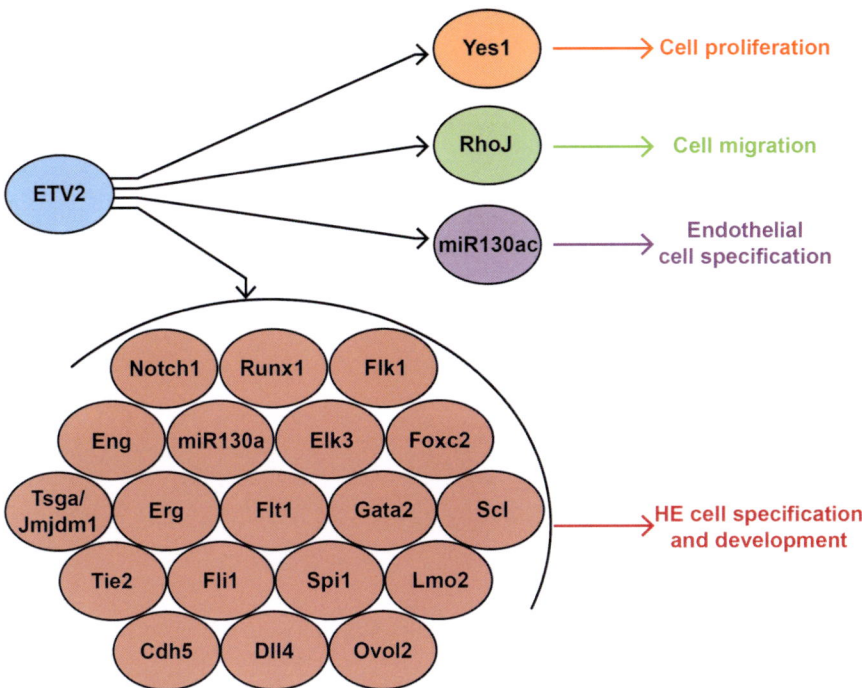

Fig. 6 ETV2 functions as an upstream regulator of gene expression. Schematic which demonstrates direct downstream targets of ETV2 and their functional role during vasculogenesis and angiogenesis

potent coactivators of gene expression [66]. Furthermore, GATA2 has been shown using multiple assays to interact with ETV2 and coactivate hematoendothelial gene expression [53]. Other factors (OVol11311) have also been shown to cooperate with ETV2 [67]. For example, forced overexpression of ETV2 and GATA2 in hiPSCs promoted a hematopoietic fate [68]. These interacting factors are important context dependent cofactors that function to amplify and modify the functional role of ETV2.

10   ETV2 Is a Master Regulator for Hematoendothelial Lineages Using Conversion Assays

Previous studies have demonstrated that forced overexpression of ETV2 promotes a hematoendothelial cell fate (Fig. 7) [9, 36, 37]. Forced overexpression of ETV2 alone or in combination with other factors (i.e. GATA2, SCL, etc.) during murine EB differentiation promotes a hematoendothelial fate (FLK1$^+$ cells) (Fig. 7) [14, 36, 37, 55, 68]. Furthermore, the delivery of ETV2 converted cell populations into

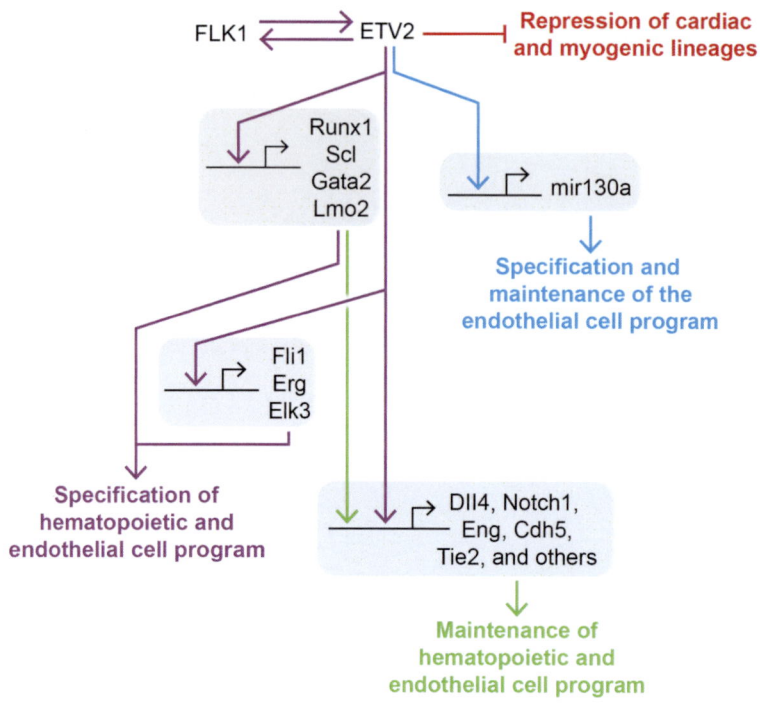

Fig. 7 The *Etv2* network regulates hematoendothelial lineage development. Upstream and downstream regulators and ETV2 effectors specify HE lineage development. ETV2 also represses non-hematoendothelial lineages

ischemic hindlimbs demonstrated that these converted cells were endothelial cells as they participated in the repair of ischemic hindlimb mouse models [9]. Additionally, the delivery of lentivirus overexpressing ETV2 promoted repair in response to ischemic injury and reduced the fibroproliferative response in mouse hearts [69]. Collectively, these preliminary studies provide an important platform for therapeutic initiatives focused on the overexpression of ETV2.

11   ETV2 Overexpression in Tumor Angiogenesis

Previous studies have demonstrated that master regulators not only promote fate determination during embryogenesis but also are overexpressed in the context of cancer [70]. Solid tumors require an enhanced vascular supply for growth, cell proliferation and metastasis. Therefore, if ETV2 is at the top of the transcriptional hierarchy for vasculogenesis/angiogenesis, then ETV2 should be expressed at distinct time-periods during tumorigenesis. Expression analysis demonstrated coexpression of ETV2 and the histone dymethylase, Junonji domain containing 2A in neuroendocrine prostate tumors [49]. In a separate study, the knockdown of *Etv2* using siRNA nanoparticles resulted in the inhibition of tumor angiogenesis and the inhibition of tumor growth [48]. Furthermore, using a zebrafish xenotransplantation model, ETV2 and FLI1b were shown to have redundant roles in promoting tumor angiogenesis [71–73]. Collectively, these initial studies support the conclusion that strategies focused on the inhibition of ETV2 may be effective therapies that target tumor growth and angiogenesis.

12 ETV2 Functions to Repress Nonhematoendothelial Fate Decisions

Master regulators function to reprogram the fate of differentiated cells [3, 4]. They also have the capacity to repress other lineages and direct progenitor cell populations down a specified pathway as outlined in the Waddington's landscape (Fig. 1) [4, 74]. In the *Etv2* global knockout mouse model, mesodermal progenitors that typically are destined for the hematoendothelial lineage are redirected to the cardiomyocyte lineage using genetic fate mapping techniques (Fig. 7) [26, 41, 75]. Similarly, the *Etv2* null zebrafish endothelial progenitors are redirected to the skeletal muscle lineage (Fig. 7) [29, 30]. These studies using a gene disruption strategy emphasize that ETV2 represses nonhematopoietic lineage formation during early stages of embryogenesis.

13 Summary

The endothelial and vascular lineages require a complex network of gene expression that governs fate determination. ETV2 is an important master regulator for the endothelial and vascular lineages. Definition of the ETV2 mediated molecular networks provide a platform for future therapeutic interventions.

Acknowledgements The authors acknowledge Cynthia Faraday for her assistance with the preparation of the figures.

References

1. Roger, V.L., Go, A.S., Lloyd-Jones, D.M., Benjamin, E.J., Berry, J.D., Borden, W.B., et al.: Executive summary: heart disease and stroke statistics–2012 update: a report from the American Heart Association. Circulation. **125**(1), 188–197 (2012). https://doi.org/10.1161/CIR.0b013e3182456d46

2. Garry, D.J., Goetsch, S.C., McGrath, A.J., Mammen, P.P.: Alternative therapies for orthotopic heart transplantation. Am J Med Sci. **330**(2), 88–101 (2005). https://doi.org/10.1097/00000441-200508000-00006

3. Chan, S.S., Kyba, M.: What is a master regulator? J. Stem Cell Res. Ther. **3** (2013). https://doi.org/10.4172/2157-7633.1000e114

4. Davis, T.L., Rebay, I.: Master regulators in development: views from the drosophila retinal determination and mammalian pluripotency gene networks. Dev. Biol. **421**(2), 93–107 (2017). https://doi.org/10.1016/j.ydbio.2016.12.005

5. Tapscott, S.J.: The circuitry of a master switch: Myod and the regulation of skeletal muscle gene transcription. Development. **132**(12), 2685–2695 (2005). https://doi.org/10.1242/dev.01874

6. Vinogradova, T.V., Sverdlov, E.D.: PDX1: a unique pancreatic master regulator constantly changes its functions during embryonic development and progression of pancreatic cancer. Biochemistry (Mosc). **82**(8), 887–893 (2017). https://doi.org/10.1134/S000629791708003X

7. Davis, R.L., Weintraub, H., Lassar, A.B.: Expression of a single transfected cDNA converts fibroblasts to myoblasts. Cell. **51**(6), 987–1000 (1987). https://doi.org/10.1016/0092-8674(87)90585-x

8. Tapscott, S.J., Davis, R.L., Thayer, M.J., Cheng, P.F., Weintraub, H., Lassar, A.B.: MyoD1: a nuclear phosphoprotein requiring a Myc homology region to convert fibroblasts to myoblasts. Science. **242**(4877), 405–411 (1988). https://doi.org/10.1126/science.3175662

9. Lee, S., Park, C., Han, J.W., Kim, J.Y., Cho, K., Kim, E.J., et al.: Direct reprogramming of human dermal fibroblasts into endothelial cells using ER71/ETV2. Circ. Res. **120**(5), 848–861 (2017). https://doi.org/10.1161/CIRCRESAHA.116.309833

10. Boyer, L.A., Lee, T.I., Cole, M.F., Johnstone, S.E., Levine, S.S., Zucker, J.P., et al.: Core transcriptional regulatory circuitry in human embryonic stem cells. Cell. **122**(6), 947–956 (2005). https://doi.org/10.1016/j.cell.2005.08.020

11. Porcher, C., Swat, W., Rockwell, K., Fujiwara, Y., Alt, F.W., Orkin, S.H.: The T cell leukemia oncoprotein SCL/tal-1 is essential for development of all hematopoietic lineages. Cell. **86**(1), 47–57 (1996). https://doi.org/10.1016/s0092-8674(00)80076-8

12. Semenza, G.L.: Hypoxia-inducible factor 1: master regulator of O2 homeostasis. Curr. Opin. Genet. Dev. **8**(5), 588–594 (1998). https://doi.org/10.1016/s0959-437x(98)80016-6

13. Garry, D.J.: Etv2 IS a MASTER REGULATOR OF HEMATOENDOTHELIAL LINEAGES. Trans. Am. Clin. Climatol. Assoc. **127**, 212–223 (2016)

14. Koyano-Nakagawa, N., Garry, D.J.: Etv2 as an essential regulator of mesodermal lineage development. Cardiovasc. Res. **113**(11), 1294–1306 (2017). https://doi.org/10.1093/cvr/cvx133

15. De Val, S., Black, B.L.: Transcriptional control of endothelial cell development. Dev. Cell. **16**(2), 180–195 (2009). https://doi.org/10.1016/j.devcel.2009.01.014

16. Garcia, M.D., Larina, I.V.: Vascular development and hemodynamic force in the mouse yolk sac. Front. Physiol. **5**, 308 (2014). https://doi.org/10.3389/fphys.2014.00308

17. Krock, B.L., Skuli, N., Simon, M.C.: Hypoxia-induced angiogenesis: good and evil. Genes Cancer. **2**(12), 1117–1133 (2011). https://doi.org/10.1177/1947601911423654

18. Harris, I.S., Black, B.L.: Development of the endocardium. Pediatr. Cardiol. **31**(3), 391–399 (2010). https://doi.org/10.1007/s00246-010-9642-8

19. Huber, T.L., Kouskoff, V., Fehling, H.J., Palis, J., Keller, G.: Haemangioblast commitment is initiated in the primitive streak of the mouse embryo. Nature. **432**(7017), 625–630 (2004). https://doi.org/10.1038/nature03122

20. de Bruijn, M.F., Speck, N.A., Peeters, M.C., Dzierzak, E.: Definitive hematopoietic stem cells first develop within the major arterial regions of the mouse embryo. EMBO J. **19**(11), 2465–2474 (2000). https://doi.org/10.1093/emboj/19.11.2465

21. Choi, K.D., Vodyanik, M.A., Togarrati, P.P., Suknuntha, K., Kumar, A., Samarjeet, F., et al.: Identification of the hemogenic endothelial progenitor and its direct precursor in human pluripotent stem cell differentiation cultures. Cell Rep. **2**(3), 553–567 (2012). https://doi.org/10.1016/j.celrep.2012.08.002

22. Oatley, M., Bolukbasi, O.V., Svensson, V., Shvartsman, M., Ganter, K., Zirngibl, K., et al.: Single-cell transcriptomics identifies CD44 as a marker and regulator of endothelial to haematopoietic transition. Nat. Commun. **11**(1), 586 (2020). https://doi.org/10.1038/s41467-019-14171-5

23. Pietila, I., Vainio, S.: The embryonic aorta-gonad-mesonephros region as a generator of haematopoietic stem cells. APMIS. **113**(11–12), 804–812 (2005). https://doi.org/10.1111/j.1600-0463.2005.apm_368.x

24. Brown, T.A., McKnight, S.L.: Specificities of protein-protein and protein-DNA interaction of GABP alpha and two newly defined ets-related proteins. Genes Dev. **6**(12B), 2502–2512 (1992). https://doi.org/10.1101/gad.6.12b.2502

25. Lee, D., Park, C., Lee, H., Lugus, J.J., Kim, S.H., Arentson, E., et al.: ER71 acts downstream of BMP, Notch, and Wnt signaling in blood and vessel progenitor specification. Cell Stem Cell. **2**(5), 497–507 (2008). https://doi.org/10.1016/j.stem.2008.03.008

26. Ferdous, A., Caprioli, A., Iacovino, M., Martin, C.M., Morris, J., Richardson, J.A., et al.: Nkx2-5 transactivates the Ets-related protein 71 gene and specifies an endothelial/endocardial fate in the developing embryo. Proc. Natl. Acad. Sci. U. S. A. **106**(3), 814–819 (2009). https://doi.org/10.1073/pnas.0807583106

27. Casie Chetty, S., Rost, M.S., Enriquez, J.R., Schumacher, J.A., Baltrunaite, K., Rossi, A., et al.: Vegf signaling promotes vascular endothelial differentiation by modulating etv2 expression. Dev. Biol. **424**(2), 147–161 (2017). https://doi.org/10.1016/j.ydbio.2017.03.005

28. Casie Chetty, S., Sumanas, S.: Ets1 functions partially redundantly with Etv2 to promote embryonic vasculogenesis and angiogenesis in zebrafish. Dev. Biol. **465**(1), 11–22 (2020). https://doi.org/10.1016/j.ydbio.2020.06.007

29. Chestnut, B., Casie Chetty, S., Koenig, A.L., Sumanas, S.: Single-cell transcriptomic analysis identifies the conversion of zebrafish Etv2-deficient vascular progenitors into skeletal muscle. Nat. Commun. **11**(1), 2796 (2020). https://doi.org/10.1038/s41467-020-16515-y

30. Chestnut, B., Sumanas, S.: Zebrafish etv2 knock-in line labels vascular endothelial and blood progenitor cells. Dev. Dyn. **249**(2), 245–261 (2020). https://doi.org/10.1002/dvdy.130

31. Das, S., Koyano-Nakagawa, N., Gafni, O., Maeng, G., Singh, B.N., Rasmussen, T., et al.: Generation of human endothelium in pig embryos deficient in ETV2. Nat. Biotechnol. **38**(3), 297–302 (2020). https://doi.org/10.1038/s41587-019-0373-y

32. Lammerts van Bueren, K., Black, B.L.: Regulation of endothelial and hematopoietic development by the ETS transcription factor Etv2. Curr. Opin. Hematol. **19**(3), 199–205 (2012). https://doi.org/10.1097/MOH.0b013e3283523e07

33. Salanga, M.C., Meadows, S.M., Myers, C.T., Krieg, P.A.: ETS family protein ETV2 is required for initiation of the endothelial lineage but not the hematopoietic lineage in the Xenopus embryo. Dev. Dyn. **239**(4), 1178–1187 (2010). https://doi.org/10.1002/dvdy.22277

34. Sumanas, S., Choi, K.: ETS transcription factor ETV2/ER71/Etsrp in hematopoietic and vascular development. Curr. Top. Dev. Biol. **118**, 77–111 (2016). https://doi.org/10.1016/bs.ctdb.2016.01.005

35. Ginsberg, M., Schachterle, W., Shido, K., Rafii, S.: Direct conversion of human amniotic cells into endothelial cells without transitioning through a pluripotent state. Nat. Protoc. **10**(12), 1975–1985 (2015). https://doi.org/10.1038/nprot.2015.126

36. Kataoka, H., Hayashi, M., Nakagawa, R., Tanaka, Y., Izumi, N., Nishikawa, S., et al.: Etv2/ER71 induces vascular mesoderm from Flk1+PDGFRalpha+ primitive mesoderm. Blood. **118**(26), 6975–6986 (2011). https://doi.org/10.1182/blood-2011-05-352658

37. Koyano-Nakagawa, N., Kweon, J., Iacovino, M., Shi, X., Rasmussen, T.L., Borges, L., et al.: Etv2 is expressed in the yolk sac hematopoietic and endothelial progenitors and regulates Lmo2 gene expression. Stem Cells. **30**(8), 1611–1623 (2012). https://doi.org/10.1002/stem.1131

38. Kobayashi, K., Ding, G., Nishikawa, S., Kataoka, H.: Role of Etv2-positive cells in the remodeling morphogenesis during vascular development. Genes Cells. **18**(8), 704–721 (2013). https://doi.org/10.1111/gtc.12070

39. Rasmussen, T.L., Shi, X., Wallis, A., Kweon, J., Zirbes, K.M., Koyano-Nakagawa, N., et al.: VEGF/Flk1 signaling cascade transactivates Etv2 gene expression. PLoS One. **7**(11), e50103 (2012). https://doi.org/10.1371/journal.pone.0050103

40. Wareing, S., Eliades, A., Lacaud, G., Kouskoff, V.: ETV2 expression marks blood and endothelium precursors, including hemogenic endothelium, at the onset of blood development. Dev. Dyn. **241**(9), 1454–1464 (2012). https://doi.org/10.1002/dvdy.23825

41. Rasmussen, T.L., Kweon, J., Diekmann, M.A., Belema-Bedada, F., Song, Q., Bowlin, K., et al.: ER71 directs mesodermal fate decisions during embryogenesis. Development. **138**(21), 4801–4812 (2011). https://doi.org/10.1242/dev.070912

42. Koyano-Nakagawa, N., Shi, X., Rasmussen, T.L., Das, S., Walter, C.A., Garry, D.J.: Feedback mechanisms regulate Ets variant 2 (Etv2) gene expression and Hematoendothelial lineages. J. Biol. Chem. **290**(47), 28107–28119 (2015). https://doi.org/10.1074/jbc.M115.662197

43. Kim, J.Y., Lee, D.H., Kim, J.K., Choi, H.S., Dwivedi, B., Rupji, M., et al.: ETV2/ER71 regulates the generation of FLK1(+) cells from mouse embryonic stem cells through miR-126-MAPK signaling. Stem Cell Res. Ther. **10**(1), 328 (2019). https://doi.org/10.1186/s13287-019-1466-8

44. De Haro, L., Janknecht, R.: Functional analysis of the transcription factor ER71 and its activation of the matrix metalloproteinase-1 promoter. Nucleic Acids Res. **30**(13), 2972–2979 (2002). https://doi.org/10.1093/nar/gkf390

45. Lee, D., Kim, T., Lim, D.S.: The Er71 is an important regulator of hematopoietic stem cells in adult mice. Stem Cells. **29**(3), 539–548 (2011). https://doi.org/10.1002/stem.597

46. Park, C., Lee, T.J., Bhang, S.H., Liu, F., Nakamura, R., Oladipupo, S.S., et al.: Injury-mediated vascular regeneration requires endothelial ER71/ETV2. Arterioscler. Thromb. Vasc. Biol. **36**(1), 86–96 (2016). https://doi.org/10.1161/ATVBAHA.115.306430

47. Zhao, C., Gomez, G.A., Zhao, Y., Yang, Y., Cao, D., Lu, J., et al.: ETV2 mediates endothelial transdifferentiation of glioblastoma. Signal Transduct. Target. Ther. **3**, 4 (2018). https://doi.org/10.1038/s41392-018-0007-8

48. Kabir, A.U., Lee, T.J., Pan, H., Berry, J.C., Krchma, K., Wu, J., et al.: Requisite endothelial reactivation and effective siRNA nanoparticle targeting of Etv2/Er71 in tumor angiogenesis. JCI Insight. **3**(8) (2018). https://doi.org/10.1172/jci.insight.97349

49. Li, X., Moon, G., Shin, S., Zhang, B., Janknecht, R.: Cooperation between ETS variant 2 and Jumonji domaincontaining 2 histone demethylases. Mol. Med. Rep. **17**(4), 5518–5527 (2018). https://doi.org/10.3892/mmr.2018.8507

50. De Haro, L., Janknecht, R.: Cloning of the murine ER71 gene (Etsrp71) and initial characterization of its promoter. Genomics. **85**(4), 493–502 (2005). https://doi.org/10.1016/j.ygeno.2004.12.003

51. Gong, W., Rasmussen, T.L., Singh, B.N., Koyano-Nakagawa, N., Pan, W., Garry, D.J.: Dpath software reveals hierarchical haemato-endothelial lineages of Etv2 progenitors based on single-cell transcriptome analysis. Nat. Commun. **8**, 14362 (2017). https://doi.org/10.1038/ncomms14362

52. Shi, X., Zirbes, K.M., Rasmussen, T.L., Ferdous, A., Garry, M.G., Koyano-Nakagawa, N., et al.: The transcription factor Mesp1 interacts with cAMP-responsive element binding protein 1 (Creb1) and coactivates Ets variant 2 (Etv2) gene expression. J. Biol. Chem. **290**(15), 9614–9625 (2015). https://doi.org/10.1074/jbc.M114.614628

53. Shi, X., Richard, J., Zirbes, K.M., Gong, W., Lin, G., Kyba, M., et al.: Cooperative interaction of Etv2 and Gata2 regulates the development of endothelial and hematopoietic lineages. Dev. Biol. **389**(2), 208–218 (2014). https://doi.org/10.1016/j.ydbio.2014.02.018

54. Schupp, M.O., Waas, M., Chun, C.Z., Ramchandran, R.: Transcriptional inhibition of etv2 expression is essential for embryonic cardiac development. Dev. Biol. **393**(1), 71–83 (2014). https://doi.org/10.1016/j.ydbio.2014.06.019

55. Abedin, M.J., Nguyen, A., Jiang, N., Perry, C.E., Shelton, J.M., Watson, D.K., et al.: Fli1 acts downstream of Etv2 to govern cell survival and vascular homeostasis via positive autoregulation. Circ. Res. **114**(11), 1690–1699 (2014). https://doi.org/10.1161/circresaha.1134303145

56. Choi, K.: ETS transcription factor ETV2/ER71/Etsrp in haematopoietic regeneration. Curr. Opin. Hematol. **25**(4), 253–258 (2018). https://doi.org/10.1097/MOH.0000000000000430
57. Davis, J.A., Koenig, A.L., Lubert, A., Chestnut, B., Liu, F., Palencia Desai, S., et al.: ETS transcription factor Etsrp/Etv2 is required for lymphangiogenesis and directly regulates vegfr3/flt4 expression. Dev. Biol. **440**(1), 40–52 (2018). https://doi.org/10.1016/j.ydbio.2018.05.003
58. Rasmussen, T.L., Martin, C.M., Walter, C.A., Shi, X., Perlingeiro, R., Koyano-Nakagawa, N., et al.: Etv2 rescues Flk1 mutant embryoid bodies. Genesis. **51**(7), 471–480 (2013). https://doi.org/10.1002/dvg.22396
59. Singh, B.N., Kawakami, Y., Akiyama, R., Rasmussen, T.L., Garry, M.G., Gong, W., et al.: The Etv2-miR-130a network regulates mesodermal specification. Cell Rep. **13**(5), 915–923 (2015). https://doi.org/10.1016/j.celrep.2015.09.060
60. Singh, B.N., Tahara, N., Kawakami, Y., Das, S., Koyano-Nakagawa, N., Gong, W., et al.: Etv2-miR-130a-Jarid2 cascade regulates vascular patterning during embryogenesis. PLoS One. **12**(12), e0189010 (2017). https://doi.org/10.1371/journal.pone.0189010
61. Sumanas, S., Gomez, G., Zhao, Y., Park, C., Choi, K., Lin, S.: Interplay among Etsrp/ER71, Scl, and Alk8 signaling controls endothelial and myeloid cell formation. Blood. **111**(9), 4500–4510 (2008). https://doi.org/10.1182/blood-2007-09-110569
62. Liu, F., Li, D., Yu, Y.Y., Kang, I., Cha, M.J., Kim, J.Y., et al.: Induction of hematopoietic and endothelial cell program orchestrated by ETS transcription factor ER71/ETV2. EMBO Rep. **16**(5), 654–669 (2015). https://doi.org/10.15252/embr.201439939
63. Singh, B.N., Gong, W., Das, S., Theisen, J.W.M., Sierra-Pagan, J.E., Yannopoulos, D., et al.: Etv2 transcriptionally regulates Yes1 and promotes cell proliferation during embryogenesis. Sci. Rep. **9**(1), 9736 (2019). https://doi.org/10.1038/s41598-019-45841-5
64. Singh, B.N., Sierra-Pagan, J.E., Gong, W., Das, S., Theisen, J.W.M., Skie, E., et al.: ETV2 (Ets variant transcription factor 2)-Rhoj Cascade regulates endothelial progenitor cell migration during embryogenesis. Arterioscler. Thromb. Vasc. Biol. **40**(12), 2875–2890 (2020). https://doi.org/10.1161/ATVBAHA.120.314488
65. DiTacchio, L., Bowles, J., Shin, S., Lim, D.S., Koopman, P., Janknecht, R.: Transcription factors ER71/ETV2 and SOX9 participate in a positive feedback loop in fetal and adult mouse testis. J. Biol. Chem. **287**(28), 23657–23666 (2012). https://doi.org/10.1074/jbc.M111.320101
66. De Val, S., Chi, N.C., Meadows, S.M., Minovitsky, S., Anderson, J.P., Harris, I.S., et al.: Combinatorial regulation of endothelial gene expression by ets and forkhead transcription factors. Cell. **135**(6), 1053–1064 (2008). https://doi.org/10.1016/j.cell.2008.10.049
67. Kim, J.Y., Lee, R.H., Kim, T.M., Kim, D.W., Jeon, Y.J., Huh, S.H., et al.: OVOL2 is a critical regulator of ER71/ETV2 in generating FLK1+, hematopoietic, and endothelial cells from embryonic stem cells. Blood. **124**(19), 2948–2952 (2014). https://doi.org/10.1182/blood-2014-03-556332
68. Elcheva, I., Brok-Volchanskaya, V., Kumar, A., Liu, P., Lee, J.H., Tong, L., et al.: Direct induction of haematoendothelial programs in human pluripotent stem cells by transcriptional regulators. Nat. Commun. **5**, 4372 (2014). https://doi.org/10.1038/ncomms5372
69. Lee, S., Lee, D.H., Park, B.W., Kim, R., Hoang, A.D., Woo, S.K., et al.: In vivo transduction of ETV2 improves cardiac function and induces vascular regeneration following myocardial infarction. Exp. Mol. Med. **51**(2), 1–14 (2019). https://doi.org/10.1038/s12276-019-0206-6
70. Hepburn, A.C., Steele, R.E., Veeratterapillay, R., Wilson, L., Kounatidou, E.E., Barnard, A., et al.: The induction of core pluripotency master regulators in cancers defines poor clinical outcomes and treatment resistance. Oncogene. **38**(22), 4412–4424 (2019). https://doi.org/10.1038/s41388-019-0712-y
71. Craig, M.P., Grajevskaja, V., Liao, H.K., Balciuniene, J., Ekker, S.C., Park, J.S., et al.: Etv2 and fli1b function together as key regulators of vasculogenesis and angiogenesis. Arterioscler. Thromb. Vasc. Biol. **35**(4), 865–876 (2015). https://doi.org/10.1161/atvbaha.114.304768
72. Li, Y., Luo, H., Liu, T., Zacksenhaus, E., Ben-David, Y.: The ets transcription factor Fli-1 in development, cancer and disease. Oncogene. **34**(16), 2022–2031 (2015). https://doi.org/10.1038/onc.2014.162

73. Baltrunaite, K., Craig, M.P., Palencia Desai, S., Chaturvedi, P., Pandey, R.N., Hegde, R.S., et al.: ETS transcription factors Etv2 and Fli1b are required for tumor angiogenesis. Angiogenesis. **20**(3), 307–323 (2017). https://doi.org/10.1007/s10456-017-9539-8
74. Waddington, C.H.: The Strategy of the Genes. Routledge (2014)
75. Liu, F., Kang, I., Park, C., Chang, L.W., Wang, W., Lee, D., et al.: ER71 specifies Flk-1+ hemangiogenic mesoderm by inhibiting cardiac mesoderm and Wnt signaling. Blood. **119**(14), 3295–3305 (2012). https://doi.org/10.1182/blood-2012-01-403766

Part II
Cellular Approaches to Cardiac Repair and Regeneration

Remuscularization of Ventricular Infarcts Using the Existing Cardiac Cells

Yang Zhou and Jianyi Zhang

1 Introduction

As the leading cause of death for both men and women, heart disease accounts for one in every four death in the United States [1]. In worldwide, more than 17.8 million deaths were attributed to heart diseases, producing numerous health and economic burdens for years [1]. The high mortality has been largely driven by the most common type of heart disease, coronary artery disease, which can cause myocardial infarction, also known as heart attack. The myocardial infarction occurs when the myocardial blood flow is suddenly or severely reduced or blocked, leading to the permanent loss of cardiomyocytes and subsequent adverse remodeling. The devastating damage of cardiomyocytes and the compensatory scar formation result in compromised cardiac function and eventually heart failure, or incapability of pumping adequate blood. To date, although the advanced surgical intervention and modern medicines have emerged to treat the patients, limited therapeutic options are available to directly solve the underlying cause of heart failure and replenish the cardiomyocyte loss. Holding the promise to treat heart failure, the idea of cardiac regeneration or to rebuild the myocardial is one of the key objectives in the cardiovascular biomedical research.

Y. Zhou (✉)
Department of Biomedical Engineering, School of Medicine and School of Engineering, University of Alabama at Birmingham, Birmingham, AL, USA
e-mail: yangzhou@uab.edu

J. Zhang
Department of Biomedical Engineering, School of Medicine and School of Engineering, University of Alabama at Birmingham, Birmingham, AL, USA

Department of Medicine/Division of Cardiovascular Disease, School of Medicine and School of Engineering, The University of Alabama at Birmingham, Birmingham, AL, USA

© The Author(s), under exclusive license to Springer Nature Switzerland AG 2022
J. Zhang, V. Serpooshan (eds.), *Advanced Technologies in Cardiovascular Bioengineering*, https://doi.org/10.1007/978-3-030-86140-7_4

51

The central basis of the insufficient cardiac regeneration is the low proliferative capacity of adult cardiomyocytes. Human cardiomyocytes have an estimated turnover rate of 0.3% to 1% each year in the first decade of life [2]. In contrast to the extremely low regeneration of adult human heart, several lower vertebrates, such as zebrafish and newts, as well as certain neonatal mammals can fully recover cardiac function after injury. In past decades, the comparative studies in animal models with different potential in cardiac regeneration and the innovation of stem cell and reprogramming technologies have provided critical insights into the process of heart regeneration, as well as new strategies to regenerate cardiomyocytes. Currently, there are three leading strategies of heart regeneration to repair left ventricle after heart injury: (1) induced proliferation of existing cardiomyocytes, (2) direct conversion of fibroblasts to cardiomyocytes, and (3) remuscularization of the myocardium via transplantation of pluripotent-stem-cell-derived cardiomyocytes (Fig. 1). As an important strategy emerging for myocardial regeneration, the remuscularization of the myocardium via exogenous cardiomyocytes generated from pluripotent stem cells has been extensively reviewed and discussed in other chapters of this book. This chapter is focused on the other two approaches, both of which leverage endogenous cardiac cells and repair mechanisms as therapeutic strategies. Here, we review the development of these two strategies, summarize the underlying molecular basis, and discuss the up-to-date breakthrough, advantages, potential limitations and future translation.

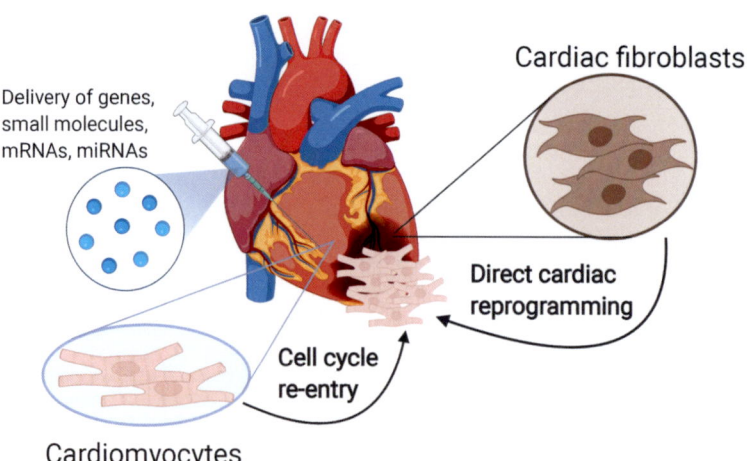

Fig. 1 Remuscularization strategies using endogenous cardiac cells. To replenish the loss of massive cardiomyocytes in the ischemic heart, two promising strategies have emerged to generate new cardiomyocytes by delivery of genes, small molecules, mRNAs and/or miRNAs into the endogenous cardiac cells. (1) Cell cycle re-entry: stimulation or re-trigger of cardiomyocyte proliferation. (2) Direct cardiac reprogramming: direct conversion of cardiac fibroblasts into functional cardiomyocytes

2 Cell Cycle Activation of Existing Cardiomyocytes

Although rare cardiomyocyte division events occur in adult mammalian hearts, natural cardiac regeneration does occur in several lower vertebrates such as adult zebrafish [3] and neonatal small and large mammals, including mice [4, 5] and pigs [6, 7]. Importantly, in-depth studies using lineage tracing and fate mapping techniques in these natural regenerative models have provided growing evidence that the cardiac regeneration is mediated by proliferation of pre-existing cardiomyocytes rather than a population of stem cells or progenitor cells [8, 9]. Therefore, understanding whether, how, and to what extent the proliferation ability of endogenous cardiomyocytes can be stimulated in mammalian hearts will help to develop a promising strategy to augment heart regeneration after injury.

2.1 Cardiac Regeneration Achieved with Cardiomyocyte Proliferation

Cardiac regeneration in zebrafish was first reported in 2002, occurring after apical resection of 20% of the single ventricle [3]. During the regenerative process, little or no fibrosis is observed. Instead, almost all the lost muscle tissue is replaced, with normal structures showing contractile function. Furthermore, fate mapping to trace the origins of newly generated cardiomyocytes showed the robust proliferation of pre-existing cardiomyocytes accompanied with dedifferentiation features, including the disassembly of sarcomere structures, detachment from one another and cell cycle gene reactivation [8]. In addition to the mechanical injury models, including apical resection and cryoinjury [10–12], myocardial regeneration can also be observed in hearts undergoing heart injury induced by genetic ablation [13]. The massive destroy of cardiomyocytes (more than 60%) expressing cmlc2 led to cardiac failure, which is eventually reversed by cell cycle re-entry of spared cardiomyocytes [13].

In mammals, cardiac regeneration has only been reported in neonatal hearts during the first few days of life. Similar apical resection of 15% of the left ventricle at P1 (1 day postpartum) mice resulted in full recovery of myocardium and heart function 21 days after resection [4]. In contrast, P7 mouse hearts, which underwent apical resection, was unable to regenerate and formed the scar tissues. It is interesting to note that loss of the endogenous cardiac regenerative potential within the first week of life coincides with cell cycle withdrawal of cardiomyocytes, and switch from hyperplasic to hypertrophic growth in rodents. Lineage tracing experiments using transgenic mice further demonstrated that the regenerative response was mediated by proliferation of preexisting cardiomyocytes [4], which is consistent with the finding in zebrafish [8].

More importantly, two independent groups recently reported that the newborn porcine hearts are capable of regenerating from injury for only the first 2 days of life and that the regeneration is associated with cardiomyocyte proliferation, presumably from single nucleated pre-existing cardiomyocytes [6, 7]. More recently, Zhang team further suggested an activated regenerative machinery initiated in postnatal day 1 pig hearts after apical resection, which was able to protect the hearts against the second injury of myocardial injury at postnatal day 28 [14]. No scar and normal cardiac function were observed in the double injured pigs at postnatal day 56. It was the first time that the cardiac regenerative window in large mammals was identified and prolonged to 1 month [6, 7, 14].

In humans, the proliferation rate in adult cardiomyocytes is limited. A low turnover rate around 1% of human cardiomyocytes at age 25 decreases to 0.45% by age 75 [15, 16]. Nevertheless, a regeneration window of human heart in postnatal stage was suggested by a case report of a newborn patient, whose heart function was indicated fully recovery from a myocardial infarction that occurred shortly after birth [17]. The regenerative potential of early postnatal human heart needs to be determined in order to provide more impactful insight to promote regenerative medicine.

The above-mentioned findings not only demonstrate the proliferation potential of cardiomyocytes, but also provide the cardiac regeneration models to study the molecular mechanisms controlling myocyte proliferation. Understanding the regulation of cell cycle, the key factors and signaling pathways involved in myocyte proliferation enables to develop approaches to achieve reactivation of cell cycle in adult cardiomyocytes.

2.2 Approaches to Stimulate Cell Cycle Re-entry of Cardiomyocytes

Given the changes of cell cycle activity in cardiomyocytes at different developmental stages in various species, considerable efforts have been made to determine the molecular signals, including cell cycle regulators, microRNAs, signaling pathways and transcription regulators that are able to stimulate cardiomyocyte re-entering cell cycle and promote cardiac regeneration after injury (Fig. 2).

Cell Cycle Regulators

Recently, a subset of cell-cycle regulators have been implicated in re-entry of postmitotic cardiomyocytes into cell cycle. Overexpression of cyclin A2, which is a cofactor of cyclin-dependent kinases (CDK) promoting G1/S and/or G2/M transition, in cardiomyocytes led to increased postnatal mitosis [18–21]. After myocardial infarction, cyclin A2 transgenic mice exhibited improved cardiac repair with a remarkable increase of cardiomyocyte mitosis [19]. The therapeutic potential of cyclin A2 has been further demonstrated using pig myocardial infarction models following   adenovirus-mediated   gene   delivery   [21].   In   addition,

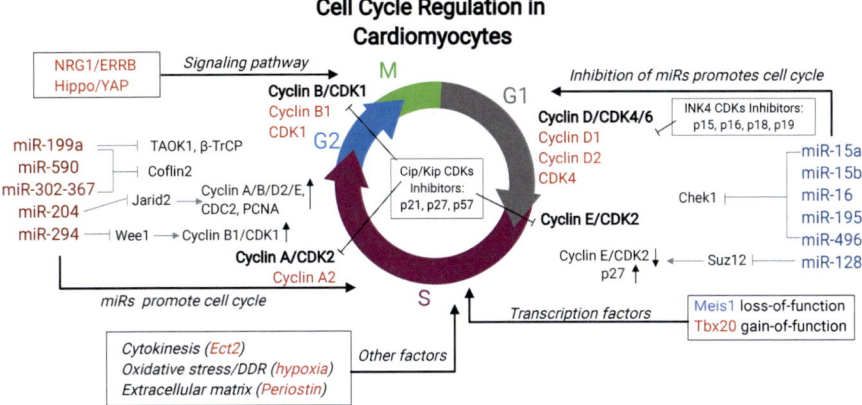

Fig. 2 Existing approaches to stimulate cell-cycle re-entry of cardiomyocytes. Overexpression of positive cell cycle regulators, including cyclins and cyclin dependent kinases (indicated in red) have been shown to promote cardiomyocyte proliferation. Cell-cycle-associated microRNAs (miRs) directly target cell cycle regulators and show the ability to activate cell cycle in adult cardiomyocytes. Among them, miR-199a, miR-590, miR-302-367, miR-204 and miR-294 positively regulate cell cycle, while miR-15 family miRs inhibit cell cycle. Besides, signaling pathways, transcription factors and other factors involved in cytokinesis, oxidative stress/DNA damage response (DDR) and extracellular matrix have been shown to regulate cell cycle in cardiomyocytes. The positive regulators are highlighted in red and the negative regulators are in blue

cardiomyocyte-specific overexpression of cyclin D2, or deletion of Rb and p130 also promotes cell cycle activity in adult mouse cardiomyocytes [22, 23]. Moreover, the cell cycle activation of transplanted cardiomyocytes has beneficial effects when used for cell therapy. Overexpression of cyclin D2 in human induced pluripotent stem cell-derived cardiomyocytes (hiPSC-CMs) engraftment to myocardial infarcted mice significantly improved cardiac function through cardiomyocyte proliferation [24]. Transplantation of cyclin D2 overexpressed hiPSC-CMs resulted in replacement of >50% of the scar over 6 months and electromechanical integration with host myocardium [25]. More recently, Mohamed et al. found that the forced expression of four cell cycle genes, CDK1/cyclin B1 and CDK4/cyclin D1, combinatorically reactivates cell division in post-mitotic mouse, rat, and human cardiomyocytes [26]. In vivo, intramyocardial injection of these four factors enhanced cardiac function via cardiomyocyte proliferation, which was evident by in vivo lineage tracing using Cre-recombinase-dependent mosaic analysis with double markers (MADM) system. They also found that chemical inhibition of TGF-β and Wee1 can replace CDK1/cyclin B1 to achieve stable cell division in adult cardiomyocytes.

In addition, microRNAs are alternative targets for cell cycle reactivation in adult cardiomyocytes. A growing number of microRNAs have been reported in regulation of cardiomyocyte cell cycle both in vitro and in animal models. Using functional screening, Eulalio et al. found that miR-199a and miR-590 are able to promote adult cardiomyocyte proliferation and stimulate remarkable cardiac regeneration after myocardial infarction in mice [27]. Further study revealed that both of the microR-NAs activate YAP translocation via directly targeting YAP inhibitory regulators, TAOK1, β-TrCP, and Coflin2 [28]. More recently, the same research group studied the therapeutic potential of miR-199a to treat myocardial infarction in a large animal model [29]. Infarcted pig hearts were injected with AAV6-mediated miR-199a and showed reduced scar size and improved heart function when compared with control virus injected hearts. Consistent with their findings in rodents, miR-199a overexpression in pig model resulted in elevated BrdU incorporation and proliferation marker expression in cardiomyocytes. However, most of the pigs treated with AAV6-miR-199a died after 6 weeks bearing arrhythmias, which emerges as a critical issue for cardiac regeneration therapy. Chen et al. reported that transgenic overexpression of miR-17-92 cluster in mouse hearts led to more proliferating cardiomyocytes and better cardiac regeneration against myocardial infarction [30]. Tian et al. identified another microRNA cluster miR-302-367, which is downregulated along embryonic to adult development, sufficient to reactivate cell proliferation in adult cardiomyocytes [31]. They found that only transient miR-302 mimic was able to improve cardiac function after myocardial infarction. Long-term overexpression of miR-302-367 in the infarcted hearts showed reduced scar size but compromised cardiac function due to the dedifferentiation of cardiomyocytes [31]. Liang and colleagues found that miR-204 promotes cell cycle activity of neonatal and adult cardiomyocytes by reactivating Cyclin A, Cyclin B, Cyclin D2, Cyclin E, CDC2 and PCNA via targeting Jarid2 [32]. Borden et al. recently identified miR-294 highly expressed in prenatal hearts and enhancing myocyte proliferation via targeting Wee1 to activate cyclin B1/CDK1 complex. The transient activation of miR-294 via AAV9-mediated delivery in myocardial infarcted mice resulted in improved LV functions, decreased infarct size together with decreased apoptosis and increased proliferation [33].

On the other hand, some microRNAs are upregulated upon cell cycle arrest in cardiomyocytes after birth, suppressing cell cycle gene expression. Thus, stimulation of cardiomyocyte proliferation can be achieved by repression of these microR-NAs as well. Porrello et al. found that knockdown of miR-15 family microRNAs (miR-15a, miR-15b, miR-16, miR-195, and miR-496) resulted in an increased number of mitotic cardiomyocytes with derepressed checkpoint kinase 1 (Chek1) expression [5, 34]. The same group further demonstrated that the long-term inhibition of miR-15 family after birth reserved the proliferative ability of cardiomyocytes in adult and protected adult heart from myocardial infarction [5]. In addition, another miR-15 family member miR-128 was also identified to be associated with cell cycle arrest during postnatal development [35]. The loss of miR-128 promotes proliferation of adult cardiomyocytes in normal and myocardial infarcted hearts [35]. MiR-128 inhibits cardiomyocyte proliferation through, at least partly, targeting the regulatory axis of SUZ12, p27, and cyclin E/CDK2 [35]. Taken together, manipulation of cell-cycle regulators has been successful in boosting cardiomyocyte proliferation in the injured small and large animals, suggesting its potential to be a therapeutic strategy for clinical application in the future.

Signaling Transduction Pathways

In addition to cell cycle regulators, an increasing number of signaling pathways have been identified that can induce cell cycle re-entry of cardiomyocytes and promote cardiac regeneration. Neuregulin1 (NRG1) is an epidermal growth factor (EGF)-like growth factor that binds a set of tyrosine kinase receptors, such as ERBB4 and ERBB2, to phosphorylate and transduce downstream signaling, involving in cardiomyocyte proliferation, differentiation and cardiac homeostasis. The recombinant NRG1 has been identified to promote cell cycle reactivation in cardiomyocytes [36], suggesting its potential as a therapeutic agent. D'Uva et al. found that the transgenic expression of NRG1's coreceptor ERBB2 prolongs the neonatal regenerative window and promotes cardiac recovery in adult mice after myocardial infarction [37]. Although the ERK/AKT pathways are required for NRG1/ERBB-mediated cardiomyocyte proliferation [37], how NRG1/ERBB signaling regulates cell cycle stimulation in cardiomyocytes is still largely unknown. Others speculate that NRG1 promotes cardiac regeneration through different mechanisms rather than pro-proliferation [38], such as anti-apoptosis [39] and pro- angiogenesis [39]. Further investigation of NRG1/ERBB signaling is necessary to dissect the detailed mechanisms and of importance for its future translation as a therapeutic strategy.

The Hippo signaling pathway, which plays an evolutionarily conserved role in organ size control during development, has also been implicated in promoting cardiomyocyte proliferation in adult mammalian hearts. In mice, Mst1/2 kinase together with regulatory protein Salv1 to phosphorylate the large tumor suppressor homolog kinases (Lats) 1/2 and their regulatory protein Mob1, which in turn phosphorylate and inhibit the downstream transcription co-activators YAP and TAZ. Dephosphorylated YAP and TAZ translocate into the nucleus to induce expression of genes associated with cell proliferation and apoptosis [40]. Heallen et al. first determined the role of Hippo in cardiomyocyte proliferation and heart regeneration via genetic inhibition of Hippo in adult cardiomyocytes [41]. They found that inducible deletion of Salv or Lats1/2 in adult mice led to cell cycle re-entry of post-mitotic cardiomyocytes. Two different injury models were used to demonstrate the inhibitory role of Hippo on heart regeneration. After apex resection at postnatal day 8, the full recovery of heart structure and function were observed in Salv deficient mouse hearts but not in normal hearts. By performing left anterior descending (LAD) coronary artery occlusion at both postnatal day 8 and adult mice, Heallen and colleagues further demonstrated that Hippo deficiency resulted in enhanced heart regeneration and proliferation in injured hearts. Recently, Leach et al. used adeno-associated virus 9 (AAV9) to deliver short hairpin RNA (shRNA) targeting Salv in infarcted myocardium and showed similar cardiac repair reported by genetic inhibition of Hippo [42]. Meanwhile, several studies genetically deleted YAP or TAZ in cardiomyocytes and found increased scar size and impaired neonatal heart regeneration [43, 44]. Conversely, transgenic mice of a constitutively active form of YAP (S112A) resulted in increased heart size, enhanced cardiac regeneration at postnatal day 7, and increased cardiomyocyte proliferation post myocardial infarction [43]. Independently, Lin and colleagues found that genetic or AAV9-mediated expression of human YAP mutant (S127A), which inhibits its cytoplasmic degradation by Hippo kinase phosphorylation, led to similar cardiac repair and cell cycle re-entry of cardiomyocytes in ischemic hearts [45]. Altogether, these studies highlight the potential of the Hippo pathway as a therapeutic target to stimulate the cell cycle re-entry of post-mitotic cardiomyocytes.

Developmental Transcription Factors

Other efforts have focused on identifying developmental transcription factors with the ability to stimulate cardiomyocyte cell cycle re-entry and dedifferentiation. Meis1 is a homeodomain transcription, required for normal embryonic heart development. Mahmoud et al. identified that Mesi1 plays a critical role to regulate cell cycle activity in postnatal cardiomyocytes [46]. Meis1 deletion resulted in a reactivation of cardiomyocyte proliferation in mice both at postnatal day 14 and 5–6 weeks. On the other hand, they showed that overexpression of Meis1 inhibits neonatal cardiomyocytes proliferation as well as the regenerative response after myocardial infarction at P1. Mechanistically, Meis1 inhibits proliferation via transcriptional activation of CDK inhibitors p15, p16 and p21 [46], suggesting Meis1's role as an antiproliferation factor.

T-box transcription factor Tbx20 is another factor identified with the ability to promote cardiomyocyte proliferation and regeneration. Tbx20 is widely expressed during heart development and required for cardiomyocyte cell proliferation in mouse and zebrafish hearts [47–50]. Tbx20 knockout mice are embryonic lethal with cardiac hypoplasia [51, 52]. While Tbx20 transgenic mice show thick myocardium and increased number of cardiomyocytes [53, 54]. Tbx20 gain-of-function in developing cardiomyocytes leads to increased proliferation and sustained fetal characteristics in adult hearts, which are partially mediated by the activation of BMP2/pSmad1/5/8 and PI3K/AKT/GSK3β/β-catenin signaling [53]. Furthermore, the induced overexpression of Tbx20 in adult mouse hearts has shown to promote cardiomyocyte proliferation without alteration of heart morphology or pathology. More importantly, induction of Tbx20 in the infarcted hearts leads to improved cardiac function, reduced scar size and repressed cardiac remodeling [55]. Multiple negative cell cycle regulators, including p21, Meis1 and Btg2, have been identified as direct downstream targets of Tbx20, contributing to Tbx20-induced cardiomyocyte proliferation [55].

2.3   Other Factors

Other factors also have been reported to contribute to proliferation regulation of cardiomyocytes. The incomplete cell cycle, missing the final cytokinesis event, occurs in postnatal cardiomyocytes and coincides with the loss of proliferation ability. The cytokinesis component epithelial cell transforming 2 (Ect2) has been shown upregulated and associated with the maintenance of mononucleated and diploid cardiomyocytes during heart regeneration in zebrafish [56]. Overexpression of a dominant negative Ect2 mutant produced a higher degree of polyploidy and prevented regeneration in zebrafish [56]. Recently, Liu et al. demonstrated that the cytokinesis failure in the early postnatal stage can be repressed by Ect2 or pharmacological inhibition of β-adrenergic receptor in patient cells and mouse model. The β-block also conferred benefit after myocardial infarction in adult mice [57].

Oxidative stress and the DNA damage response (DDR) have also been implicated in cell cycle arrest of postnatal cardiomyocytes. Puente et al. found evaluated reactive oxygen species (ROS) and DDR markers in cardiomyocytes along postnatal day 1 to day 7. The exposure of neonates (starting from E18.5) to mildly hypoxic (15% O_2) environment led to increase of proliferation markers in cardiomyocytes, while hyperoxia (100% O_2) treated hearts resulted in decreased proliferation. Meanwhile, ROS generators accelerate the cell cycle arrest, but scavenging ROS or pharmacological inhibition of DDR extend the regenerative window in postnatal mice [58]. Furthermore, using a fate-mapping system to track hypoxia cells, Kimura et al. identified a rare population of hypoxic cardiomyocytes, which can re-enter cell cycle upon ischemic injury [59]. More recently, hypoxia-induced cell cycle activation has been determined in the adult mouse heart as well. Severe hypoxia exposure after myocardial infarction led to cardiomyocyte proliferation with improved heart function [60]. These reports highlight the pivotal role of oxidative stress in regulating cardiomyocyte cell cycle and its potential to be utilized to stimulate cell cycle activity.

Periostin, an extracellular matrix (ECM) component, is mainly expressed and secreted by fibroblasts after myocardial injury. Surprisingly, it can also induce re-entry of post-mitotic cardiomyocytes into cell cycle. In myocardial infarcted hearts, activation of PI3K signaling pathway induced by recombinant Periostin led to reduced scar size and fibrosis, and improved cardiac function [61]. The delivery of Periostin in pig infarcted hearts had similar cardiac repair, but significantly increased fibrosis [62]. In a recent study, the genetic deletion of Periostin suppresses post-infarction myocardial regeneration via inhibiting the PI3K/GSK3β/cyclin D1 signaling pathway, suggesting an essential role of Periostin in heart regeneration [63]. Taken the contradictory effects of Periostin on fibrosis and cardiomyocyte proliferation, further investigation using different approaches will shed light on its role on cardiac regeneration and underlying mechanisms.

3   Direct Conversion of Cardiac Fibroblasts into Cardiomyocytes

Direct cellular reprogramming, rekindled by the generation of iPSCs [64], is generally defined as the direct conversion of one specialized cell type into another with defined factors, without passing through an intermediate progenitor or pluripotent

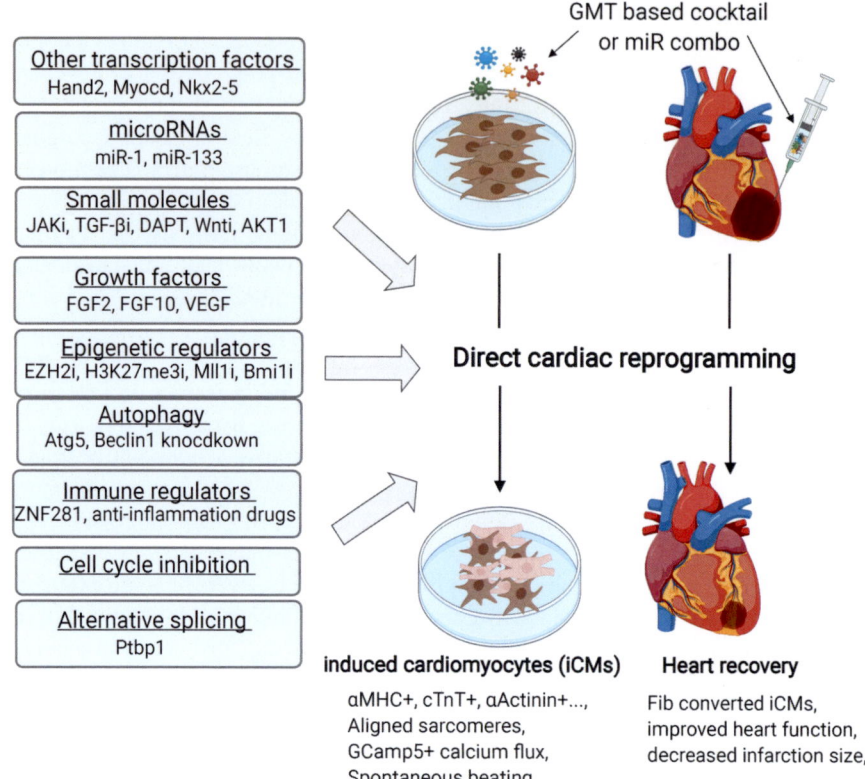

Fig. 3 Direct cardiac reprogramming serves as a promising approach to achieve heart regeneration. Direct reprogramming fibroblasts into induced cardiomyocytes (iCMs) has been reported in both in vitro and in vivo systems. The cultured iCMs exhibit gene signatures, structural and electrophysiological features of functional cardiomyocytes. The in vivo reprogramming is sufficient to reduce scar tissue and improve heart function. Several strategies for improving iCM reprogramming in vitro and in vivo have been reported and are summarized here. miR stands for microRNA. The lowercase i indicates inhibitor of the small molecules or epigenetic regulators (JAKi, TGF-β i, Wnti, EZH2i, H3K27me3i, Mll1i, Bmi1i)

stage [65]. Direct cardiac reprogramming converts fibroblasts into cardiomyocytes, holding the promise as a therapeutic strategy to treat injured hearts with dual benefits of reducing scar tissue and simultaneously generating new cardiomyocytes (Fig. 3). In 2010, Ieda et al. made the evolutionary discovery that overexpression of three factors is sufficient to convert mouse cardiac fibroblasts into functional cardiomyocyte-like cells [66]. They first found that transduction of 14 cardiac-enriched developmental transcription factors in cardiac fibroblasts derived from transgenic mice led to the expression of EGFP driven by cardiac promoter of alpha myosin heavy chain (αMHC). After a series of optimization, they showed that the minimal combination of Gata4, Mef2c and Tbx5, or GMT for short, produced 15–20% of EGFP positive cells, which are named induced cardiomyocytes (iCMs).

In addition, iCMs exhibit well-defined sarcomeric structures, cardiomyocyte-specific gene signatures as well as calcium flux and electrophysiological features. Onset of spontaneously beating occurred after 4 weeks of transduction though only in 0.01%–0.1% of cells. This pioneer study further demonstrated that no intermediate undifferentiated stage was identified by the lineage tracing using two progenitor makers, Isl1 and Mesp1, suggesting that the progenitor stage is not necessary for the cardiac cell fate induction.

Following the success of iCM conversion in vitro, the direct cardiac reprogramming was reported in vivo after myocardial infarction, highlighting its potential usefulness in future heart regenerative medicine. In 2012, Qian et al. firstly used genetic lineage-tracing to show that mouse resident cardiac fibroblasts labeled with Thy1 can be converted into iCMs in vivo by intra myocardial injection of GMT retroviruses immediately after coronary artery ligation [67]. These induced cardiomyocytes became bi-nuclear with sarcomere formation, cardiomyocytes-like gene profile and could electrically couple with viable endogenous cardiomyocytes. Three months post GMT delivery, the mouse cardiac ejection fraction and stroke volume were dramatically improved, accompanied with decreased infarct size. Furthermore, co-delivery of thymosin β4, a pro-angiogenic and fibroblast activating peptide, along with GMT resulted in further improvement of cardiac function and less scarring. Independently, Inagawa et al. also successfully converted endogenous cardiac fibroblasts into functional iCMs in vivo by retroviral delivery of GMT cocktail into infarcted mouse hearts [68]. Around 3% of virus-infected cells activated αMHC-EGFP reporter 1 week after reprogramming. These existing findings demonstrated that direct cardiac reprogramming in vivo enables the cardiomyocyte transition from resident cardiac fibroblasts, which emerges as an alternative target existing in the injured hearts for remuscularization.

3.1 Reprogramming Factors

Following the foundational studies, different combinations of cocktails have been developed to generate cardiomyocytes both in vitro and in vivo. Song et al. showed that addition of Hand2 to GMT (GHMT) enhanced reprogramming of mouse cardiac fibroblasts into iCMs with a higher reprogramming efficiency and improved cardiac repair in vivo [69]. Twelve weeks after myocardial infarction, the ejection fraction showed two-fold increase and the scar area was reduced by 50% in the reprogramming-induced mice compared to the control mice. Protze et al. utilized a read out of multiple cardiac-specific genes to identify a new combination of transcription factors, Tbx5, Mef2c, and Myocd that can upregulate a broader spectrum of cardiac genes in iCMs [70]. Using a transgenic calcium fluorescent reporter (GCamp5) driven by cardiomyocyte-specific Troponin T promoter, Addis et al. demonstrated that GMT together with Hand2 and Nkx2.5 (HNMGT for short) led to 1.5% of mouse embryonic fibroblasts (MEFs) or 5% of adult mouse cardiac fibroblasts reprogrammed into functional iCMs exhibiting GCamp5 activity [71].

Altogether, it is important to notice that these studies used different types of fibroblasts, focused on different readout to measure the reprogramming efficiency, and might generate different subtypes of iCMs. Nam et al. reported that GHMT-induced reprogramming generated diverse subtypes of cardiomyocytes, expressing markers of pacemaker, ventricular or atrial cardiomyocytes [72]. Furthermore, using Hcn4-GFP as a readout to screen the transcription factors for reprogramming specialized pacemaker cells, they found that a combination of Gata6, Tbx3, Tbx5 and Rarg or Rxra, was sufficient to activate Hcn4-GFP in 40% of transduced cells. However, this combination was insufficient to activate other sarcomeric proteins, which are required for the functional maturation of cardiomyocytes.

MicroRNAs also serve as critical reprogramming factors in cardiac direct reprogramming. The addition of miR-1 and miR-133 into GMT cocktail significantly increased the ratio of spontaneously beating iCMs from mouse adult cardiac fibroblasts compared to GMT treatment alone [73, 74]. Muraoka et al. also found that miR-133 functioned through repression of Snai1, which is associated with epithelial to mesenchymal transition [73]. In addition, Jayawarden et al. identified a combination of miR-1, miR-133, miR-208, and miR-499, termed as miR combo, was able to convert neonatal mouse fibroblasts into a cardiomyocyte-like phenotype, including cardiomyocyte-specific marker expression, sarcomeric organization and spontaneous beating [75]. Importantly, the in vivo cardiac conversion (~1%) has been demonstrated after lentiviral injection of miR combo in myocardial infarcted hearts with dual fibroblast (Fsp1-Cre/tdTomato) and cardiomyocyte (αMHC-CFP) reporters. Their findings suggest that microRNA also has the capability to convert cardiac fibroblasts into functional iCMs.

Besides the combination of reprogramming factors, the stoichiometry of reprogramming factors also significantly affects reprogramming efficiency. Since at least three reprogramming factors are required to induce cardiomyocyte reprogramming, it is important to ensure their equally transduction in individual cell. Therefore, Wang and colleagues took advantage of polycistronic gene expression, which allows expression of multiple genes split by different 2A self-cleaving peptides in one retroviral vector, to ensure the simultaneous expression of GMT factors [76]. Compared to the transduction of multiple retroviruses expressing separate factors, this approach led to enhanced reprogramming efficiency with a two-fold increase of αMHC-GFP positive cells and a ten-fold increase of cells expressing cardiac troponin T (cTnT). They also found that the order of three factors affected their relative expression levels as well as reprogramming efficiency. The desirable GMT expression for reprogramming, which is a relative high protein expression of Mef2c and low levels of Gata4 and Tbx5 expression, can be achieved by polycistronic expression of GMT factors in the order of MGT. Later on, the same group further showed that the polycistronic MGT expression also enhanced in vivo cardiac reprogramming and improved the heart repair [77]. They performed lineage tracing experiments using Periostin-Cre:R26r-lacZ mice, where the starting cardiac fibroblasts was permanently labeled with β -galactosidase. Four weeks after myocardial infarction, they found that retroviral delivery of polycistronic MGT led to a greater number of iCMs showing double positive staining for β-gal and α-Actinin compared to separate delivery.

3.2   Barriers and Boosters to Direct Cardiac Reprogramming

The cell fate conversion occurs initially with the rebuild of the transcriptional regulatory networks, but also requires the re-establishment of signaling pathways and epigenetic landscapes in the target cell type. Understanding the molecular components involved in direct reprogramming could lead to a better understanding of the reprogramming mechanisms and the overall improvement of the approach for its future clinical translation (Fig. 3).

Small Molecules and Growth Factors

Recently, many efforts have been made to enhance cardiac reprogramming by stimulating or inhibiting signaling pathways via small molecules and/or growth factors. Jayawarden et al. found that the addition of JAK inhibitor I with their miR combo led to improved reprogramming efficiency and increased spontaneously beating iCMs [75]. Ifkovits et al. found that the treatment of SB431542, an inhibitor of the TGF-β type I receptor ALK5, to the HNGMT-transduced MEFs resulted in a five-fold increase of GCamp5 activity compared to the DMSO-treated control [78]. TGF-β signaling has been known involved in fibrotic events. Zhao et al. first determined the up-down changes of TGF-β signaling and ECM expression during direct cardiac reprogramming. They further found that treatment of fibroblasts with TGF-β1 suppressed cardiac gene expression and decreased beating iCMs by 100-fold. On the other hand, a selective TGF-β inhibitor, A83–01, significantly enhanced the iCM generation from adult mouse cardiac fibroblasts. GHMT plus A83–01 also dramatically increased both fraction and yield of the beating cells [74]. In a later independent study, Mohamed and colleagues screened 5500 small molecules during MGT induced direct cardiac reprogramming [79]. Similarly, they found that the treatment of TGF-β inhibitor SB431542 enhanced the iCM generation, showing 15% of beating cells 5 weeks after transduction. The combinatorial treatment of SB431542 with XAV939 (an inhibitor of WNT signaling) led to a further increase of beating cells to around 30%. The transcriptomic analysis showed that dual inhibition of TGF-β and WNT signaling resulted in lower expression of ECM genes but higher expression of calcium and contractility genes. Together with GMT, these two compounds also enhanced in vivo reprogramming and improved the heart function repair after myocardial infarction [79].

In addition, a screen of 192 genes encoding protein kinases discovered that Akt1 activation significantly enhanced and expedited the production of iCMs from mouse adult cardiac fibroblasts via the expression of GHMT [80]. The addition of Akt1 to

GHMT led to a dramatic increase of spontaneous beating iCMs, which also exhibited a more mature phenotype, including polynucleation, hypertrophy, and response to β-adrenoreceptor pharmacologic modulation. Insulin-like growth factor 1 (IGF1)/PI3K-induced activation of Akt also promoted iCM conversion. Meanwhile, mitochondrial target of rapamycin complex 1 (mTORC1) and forkhead box o3 (Foxo3a) were identified as functional downstream of Akt1. The pharmacological inhibition of mTORC1 or gene knockdown of Foxo3a blunted the pro-reprogramming effect of Akt1 [80]. A later study from the same group showed that the treatment of DAPT, which is an inhibition of Notch signaling, also enhanced direct cardiac reprogramming driven by GHMT [81]. However, the canonical Notch signaling were not suppressed by the treatment of DAPT during direct reprogramming. Instead, DAPT promoted the binding of Mef2c on its targets, which led to the upregulation of Myh6, Tnnt2 and Actc1. Meanwhile, they found that DAPT could cooperates with Akt1, and increased cTnT+ cells up to 70% and beating cells up to 40%, suggesting a synergetic effect by manipulating both signaling pathways.

Rather than promoting the cell fate conversion, growth factors also show critical roles on functional maturation of iCMs. Mimicking the developmental signals in embryo, cardiomyocyte differentiation from pluripotent stem cells has been achieved using serum free medium supplemented with growth factors [82, 83]. Similarly, via a small scale of compound screen, Yamakawa et al. identified a defined serum free medium for culture of reprogrammed iCMs [84]. The optimized medium, supplemented with fibroblast growth factor (FGF) 2, FGF10, and vascular endothelial growth factor (VEGF), was termed FFV. Culture of reprogramming cells in FFV specifically increased spontaneously beating iCMs by 100-fold but not the percentage of cTnT or αMHC+ cells. Furthermore, they found that the FFV also promoted GHMT-induced reprogramming from tail tip fibroblasts at late stage. Mechanistically, p38 MAPK and PI3K/AKT pathways are required for FFV to boost the generation of functional contractile iCMs.

Epigenetic Regulators

During normal development, cell fate decisions are made at the transcriptional level in response to environmental cues. Reversely, for cellular reprogramming, it is required to remove the transcriptional "memory" maintained by epigenetic mechanisms and to re-establish the new epigenetic landscape to maintain the reprogrammed cell fate [85]. Several early studies have profiled the changes of histone modifications on fibroblasts signatures and cardiomyocyte genes [66, 86, 87]. The active histone marks, such as histone H3 trimethylation at lysine 4 (H3K4me3), show higher signals on cardiac gene loci, correspondence to their increased expression levels. While, the repressive histone H3 trimethylation at lysine 27 (H3K27me3) as well as DNA methylation are significantly depleted on cardiac gene loci, which occurs as early as day 3 post transduction [86]. The opposite changes of H3K4me3 and H3K27me3 are initiated later, around day 10, at the fibroblast signature gene loci. In an independent study, a global reduction of H3K27me3 was also determined and found as an essential event for miR combo-induced cardiac reprogramming [87]. The genetic knockdown of two H3K27 demethylases hindered cardiac reprogramming effect of miR combo. On the other hand, either pharmacological inhibition or knockdown of H3K27 methyltransferases led to the activation of cardiac transcription factors, Mef2c, Tbx5, Hand2 and Gata4. Similarly, inhibition of H3K27me2/3 with GSK126, a small molecule inhibitor of EZH2, which is a catalytic unit of polycomb repressive complex 2 (PRC2), enhanced cardiac reprogramming [88]. Meanwhile, they identified the most effective time window for GSK126 is the first 4 days. In the same study, Hirai et al. identified UNC0638, a potent inhibitor of G9a histone methyltransferase, which promoted reprogramming only at the late stage of reprogramming, highlighting the importance of sequential changes of epigenetic events. Taken together, these findings suggest the critical role of epigenetic remodeling during direct cardiac reprograming and provide druggable targets to facilitate iCM generation.

In an effort to identify the functional epigenetic regulators during direct cardiac reprogramming, Zhou et al. performed an unbiased loss-of-function screen of epigenetic factors [89]. They identified Bmi1 as a critical obstacle in the early stage of cellular reprogramming process. The shRNA-mediated knockdown of Bmi1 markedly enhanced efficiency of converting mouse fibroblasts into functional iCMs regardless of reprogramming cocktail used. As a key component of polycomb repressive complex 1 (PRC1), which mediates the ubiquitination of histone H2A at lysine 119 (H2AK119ub), Bmi1 plays a repressive role during cardiac reprogramming. Bmi1 inhibition removed the repressive H2AK119ub at cardiac gene loci and increased active H3K4me3 to reactivate cardiac gene expression. Since Gata4 is one of the major targets of Bmi1, knockdown of Bmi1 could replace Gata4 and achieved reprogramming together with Mef2c and Tbx5 [89]. In a later independent study, the pharmacological inhibition of Bmi1 also reported to promote direct cardiac reprogramming [90]. The pre-treatment of MEFs or adult cardiac fibroblasts with the Bmi1 inhibitor (PTC-209) 24 hours before reprogramming induction was sufficient to increase reprogramming efficiency with higher cardiac marker expression and more beating cells. Both JAK/STAT3 and MAPK/ERK1–2 pathways were downregulated upon pre-treatment of PTC-209, suggesting a potential mechanism which Bmi1 inhibition promotes reprogramming by repressing inflammatory related pathways.

In another study using a gain-of-function strategy to assess the roles of 47 epigenetic regulators during direct cardiac reprogramming, Liu et al. identified the H3K4 methyltransferase Mll1 as a key barrier [91]. Overexpression of Mll1 in MGT-transduced MEFs led to three-fold reduction of αMHC-GFP+ cells. Therefore, they targeted Mll1 via a novel Mll1 complex-specific inhibitor, MM408, and generated a four-fold increase in beating iCMs compared to the control group. It is likely that inhibition of Mll1 suppresses the ectopic expression of adipocyte lineage genes at early stage [91].

Autophagy

Autophagy is a cellular self-digestion process that double membrane-bound autophagosomes assemble, engulf cytoplasmic material and fuse with lysosomes for degradation [92, 93]. Autophagy is required for cellular homeostasis, thus contributing to development and many diseases. Wang et al. first determined the involvement of autophagy during direct cardiac reprogramming [94]. They first showed the activated autophagy during early direct reprogramming. The reprogramming efficiency could be further increased by activation of autophagy via small molecules or overexpression of an autophagy-related gene, Atg5. Reversely, the knockdown of Atg5 impaired autophagy and reduced reprogramming efficiency, suggesting that autophagy is required for iCM induction. During their study, the authors unexpectedly found that another well-known autophagy regulator Beclin 1 (Becn1) plays an opposite role during iCM reprogramming. Genetic depletion of Beclin 1 enhanced reprogramming and iCM maturation. Further investigation clarified that loss of Beclin 1 marginally affected autophagy flux but promoted reprogramming via activation of Wnt/β-catenin signaling. This autophagy independent role of Beclin 1 was found through its interaction with PI3K complex, which showed to be regulated by the upstream ULK1. Therefore, Wang and colleagues' findings support the notion that autophagy plays a crucial role in the cellular reprogramming and interestingly reveal a new regulatory mechanism associated with ULK1–PI3K complex–Wnt/β-catenin signaling network.

Immune Regulators

Among an unbiased screen of transcription factors and cytokines, Zhou et al. identified a zinc finger transcription factor 281 (ZNF281) that enhanced the cardiogenic activity of AGHMT in adult mouse fibroblasts [95]. ZNF281 was found recruited by GATA4 and facilitating the expression of cardiac genes. Concomitantly, ZNF281 also solely binds on the inflammatory enhancers and represses gene expression. Furthermore, they found that inhibition of inflammation via dexamethasone (Dex) or cyclooxygenase enzyme (COX) inhibitor nabumetone (Nab) increased the cardiac marker expression. The underlying mechanism is likely through the nucleosome remodeling and deacetylase (NuRD) chromatin remodeling complex.

In another high-throughput screening of 8400 compounds, Muraoka et al. discovered a non-steroidal anti-inflammatory drug, diclofenac, that was able to enhance GMT- or GHMT-induced cardiac reprogramming of postnatal and adult tail tip fibroblasts (TTFs) [96]. Diclofenac mainly inhibits the activity of COX2, which is an enzyme for prostanoid synthesis, and highly expressed in adult and aged fibroblasts. The COX2 specific inhibitor could also improve cardiac reprogramming. Meanwhile, the addition of prostanoid component, PGE2 blunted diclofenac-induced reprogramming enhancement through its receptor, EP4. Furthermore, the cAMP/PKA and IL-1β/IL-IR1 signaling were found suppressed by inhibition of COX-2/PGE2/EP4-induced inflammatory responses and thereby improved cardiac reprogramming [68]. This study demonstrated anti-inflammation as a target for improvement of direct reprogramming and the complexity among aging, reprogramming and inflammation.

Taken together, these studies highlighted the involvement of inflammatory pathways during cardiomyocyte fate conversion. Of note, all these findings were reported using retrovirus to deliver reprogramming factors. It is still elusive how the immune response regulates cell fate conversion and whether the same effect would be determined during reprogramming without using retroviruses. Neither is known how inflammation after myocardial infarction would affect direct cardiac reprogramming in vivo. The inflammation is suggested to play a critical role during in vivo reprogramming since both the studies by Qian et al. [67] and Song et al. [69] discovered that reprogramming efficiency and maturity of iCMs was greater in vivo, where the fibroblasts undergoing remodeling and severe inflammation response. Further understanding immune response and inflammation might provide new insights into cardiac reprogramming and in situ heart repair.

3.3   Molecular Trajectory of Cardiac Reprogramming

The cardiac reprogramming has been achieved in different studies though with the high variability in reprogramming efficiency, which indicates the heterogenous population of iCMs and asynchronized reprogramming process. To overcome these difficulties, the advanced single-cell genomics emerged for comprehensive understanding of the molecular behavior in individual cells. Liu et al. first investigated the global transcriptome changes at single cell level during early direct reprogramming in mouse cells [97]. They identified four major cell populations and reconstituted the reprogramming routes following the cell fate conversion from fibroblasts, intermediate fibroblasts, pre-iCMs and iCMs. It is interesting to find that pre-iCMs and iCMs showed less cell cycle activity when compared to fibroblasts. Furthermore, the genetic inhibition of cell cycle or cell-cycle synchronization resulted in increased reprogramming efficiency, suggesting a pre-requisite role of cell-cycle exit. This notion was further supported by a later observation using time-lapse imaging that showed enhanced cell-cycle exit could facilitate GMT-induced reprogramming [98]. The single-cell RNA-seq (scRNA-seq) analysis also first revealed the critical role of RNA splicing during direct cardiac reprogramming. It has been known that postnatal heart development is accompanied with a series of alternative splicing events, regarding functional maturation of cardiomyocytes [99]. Liu and colleagues also identified one of the key splicing factors, Ptbp1, that suppresses reprogramming. Silencing Ptbp1 changed the alternative splicing of cardiac genes as well as additional splicing factors, leading to an enhanced cardiac reprogramming. In a loss-of-function screen, Zhou et al. also reported that depletion of several RNA splicing factors, Sf3a1, Sf3b1, Zrsr2 and U2af1 affected reprogramming efficiency [100]. Taken together, both transcriptome profiling and functional assays support that direct cardiac reprogramming undergoes an early cell-cycle exit and a changed pattern of alternative splicing.

In another effort to dissect the detailed molecular dynamics during direct reprogramming, Stone et al. performed a comprehensive single-cell transcriptome analysis in 29,718 cells collected at 1 day before and 1, 2, 3 and 7 days after transduction with retroviral GMT. They confirmed the argument that initial reprogramming cells emerged as early as 48 h after transduction, showing activation of the general cardiac marker, Tnnt2, and incomplete repression of fibroblast markers. Meanwhile, the late iCMs without fibroblast signatures could be captured in a wide range of time window from day 3 to day 14, which is consistent with previous reports [97, 101]. In addition, they also identified three branches of cell trajectory containing reprogramming, proliferating and un-infected fibroblast cells, suggesting that GMT do not induce any alternative cell lineages during direct cardiac reprograming. The extensive changes of chromatin accessibility were also determined by day 2 of reprogramming via a time-course ATAC-seq analysis. Furthermore, the integrated analysis of ATAC-seq and ChIP-seq of GMT allowed the investigation of relationships between chromatin dynamics and transcription regulation. The computational modeling also predicted novel factors that exhibited GMT-bound motifs. These factors were further determined to be involved in GMT-induced cardiac fate conversion to increase accessibility and expression of cardiac genes. All the epigenomic and transcriptomic data provide a valuable source to understand the transcription and chromatin dynamics induced by combinatorial regulation of reprogramming factors and allow the prediction of novel boosters and barriers during direct cardiac reprogramming.

3.4   Direct Cardiac Reprogramming in Human

In 2013, three major publications reported the successful direct cardiac reprogramming of human fibroblast cells. Wada et al. expanded the three-factor mouse reprogramming cocktail, by adding two additional transcription factors (MESP1 and MYOCD) to successfully reprogram human neonatal and adult cardiac fibroblasts in vitro. This five-factor cocktail activated cTnT expression in about 5% of human cardiac fibroblasts. The resulting iCMs demonstrated sarcomere structure, calcium oscillations, action potentials, and spontaneous contractility in coculture with murine cardiomyocytes [102]. Similarly, Fu et al. included two additional transcription factors (ESRRG and MESP1) in GMT cocktail and showed activated cTnT expression in <10% of human primary fibroblasts. The inclusion of two additional factors, MYOCD and ZFPM2, to form a seven-factor cocktail enhanced human direct cardiac reprogramming and activated cTnT expression in approximately 13% of human primary fibroblasts. However, the cells failed to spontaneously beat even after 16 weeks in culture [103, 104]. Meanwhile, Nam et al. achieve the human direct cardiac reprogramming via a combination of transcription factors and microRNAs, including GATA4, HAND2, TBX5, MYOCD and two muscle-specific microRNAs (miR-1 and miR-133). The transcription factor-only cocktail activated cTnT expression in 13% - 17% of human foreskin fibroblasts. However, the addition of miR-1 and miR-133 increased cTnT+ cells to 34.1%. The resulting iCMs expressed cardiac specific marker genes and displayed sarcomere structure and calcium transients. Few spontaneous contractile iCMs emerged in the long-term cultured for more than 2 months [105].

Recently, Zhou et al. efficiently generated iCMs from human cardiac fibroblasts with a cocktail that contains human polycistronic transgene in the splicing order of MEF2C, GATA4, and TBX5 and miR-133 [106] (hMGT133 in short for this cocktail). They showed 40%–50% of hMGT133-transduced cells expressing cTnT/αActinin and exhibiting calcium flux. Using this robust reprogramming platform, they performed time course scRNA-seq to dissect the molecular route during human cardiac reprogramming. The scRNA-seq analysis of day 3 reprogramming cells highlighted the role of cell cycle and immune response. Consistent with the report in mouse [97, 98], cell cycle exit was found concomitant along reprogramming initiation. They also found enriched immune response genes in hMGT133-transduced cells and further demonstrated its role via the loss-of-function assay of key immune regulators, toll-like receptor 3 (TLR3), nuclear factor kappa B subunit 1 (NFKB1 or NF-κB) and prostaglandin-endoperoxide synthase 2 (PTGS2 or COX2). Knockdown of either genes reduced reprogramming efficiency, suggesting that immune response might be required for human cardiac reprogramming. It is interesting to notice that the anti-inflammation has been identified as a target for reprogramming improvement in mouse study [68, 95, 96], which suggests potential differences between mouse and human cells. Besides, they took advantage of the state-of-the-art computational algorithms to reconstitute the trajectory during reprogramming progression. They found that at the early time point of day 3, the bifurcated cell fate decision was made to convert to cardiomyocytes or regress to fibroblasts. The scRNA-seq also revealed the inefficiency on the maturation gene expressions of the human iCMs when compared to mouse iCMs [106], suggesting the incomplete acquisition of cardiomyocyte identity and function. Therefore, it highlights the importance to understand how to re-establish the whole transcriptomes of functional cardiomyocytes during reprogramming process, which will be helpful for better human iCMs production and its further application to restore heart functions in an injured heart.

In summary, unlike mouse iCM reprogramming, human iCM reprogramming so far has suffered from a longer time period and more complex cocktails along with lower efficiency and immature functions [69, 74, 78–81, 107, 108]. It is probably because the fibroblasts at different ages are used for mouse and human reprogramming; and the molecular basis of human and mouse cardiomyocytes varies. Therefore, further understanding of human cardiac cell fate conversion would be important and necessary to help generate more mature and functional human cardiomyocytes from non-myocytes.

4 Conclusions and Perspectives

To achieve the remuscularization in the injured heart, the idea of using endogenous cardiac cells to generate functional cardiomyocytes has been greatly investigated by either stimulating cell-cycle re-entry of cardiomyocytes or reprogramming iCMs from non-myocytes via different approaches. However, the efficiency of either cell cycle reactivation in cardiomyocytes or direct cardiac reprogramming is relatively low and currently insufficient to replenish the billions of cardiomyocytes loss in the injured hearts. Therefore, understanding the molecular mechanisms underlying cardiomyocyte cell-cycle stimulation and direct cardiac reprogramming might provide new insights to achieve a better efficiency and reproducibility for the future application. Meanwhile, additional issues have to be resolved before the clinical translation since increasing complexity emerges as genes or other factors need to be transferred in the specific types of cardiac cells in a mixed environment undergoing remodeling after heart injury.

Moving forward to the next step for translation, there is an emerging need to develop the clinical applicable approaches to deliver genes with a cell type specificity. Since no chemical alone approach has been reported to stimulate cardiomyocyte proliferation or direct cardiac reprogramming of cardiac fibroblasts, it is important to develop an efficient and safe way for gene delivery. The use of lentivirus or retrovirus is sufficient for gene delivery in both in vitro and in vivo systems. However, they hold the risk to disrupt the genomic integrity and cause insertional mutagenesis. So far, non-integrating adenoviruses [109], adeno-associated viruses (AAVs) [75], and Sendi viruses (SeV) [110] have been used to deliver transcription factors or miR combo for direct cardiac reprogramming. All the vectors are also feasible and efficient to induce in vivo reprogramming and cardiac repair after myocardial injection in the infarcted hearts. SeV-mediated GMT generated 100-fold more beating iCMs than retroviral-GMT, shortened the duration to induce beating cells from 30 to 10 days in mouse fibroblasts, and led to more efficient gene transfer in mouse infarct hearts [110]. More recently, synthetic modified mRNA (modRNA) has been found safe, feasible and efficacious to transfer genes in heart tissues [111], emerging as a promising gene delivery method for in vivo applications. However, how the modRNA delivery can be used to induce cardiomyocyte proliferation or direct reprograming still need to be further investigated. In addition, all of these vectors have limited ability to target the specific cardiac cells. It still needs multidisciplinary approaches such as nanoparticles to be incorporated and facilitate the specific delivery of genes. A variety of vectors are needed to provide more options for gene therapy at different context to stimulate cardiomyocyte cell-cycle re-entry and/or convert cardiac fibroblasts into cardiomyocytes.

The in vivo environment is more complicated than in vitro culture system. Although the current studies have shown that the inflammation signals influenced the cardiac reprogramming in vitro [95, 96] and macrophage deletion blunted the regenerative capacity of neonatal hearts [112], suggesting the roles of other cardiac cells and biological processes in the injured hearts. It is still largely unknow how cell-cell interaction and paracrine communication could affect the remuscularization induced by cardiomyocyte stimulation or direct reprogramming. Moreover, the role of matrix stiffness and mechanotransduction has been highlighted in the maturation of reprogrammed cardiomyocytes. The 3D fibrin-based hydrogel or soft matrix mimicking the naïve myocardium environment could improve the quantity and quality of iCMs [113, 114]. Following MI, the stiffness of cardiac tissue may change with the amount of collagen deposition in the scar area. Therefore, the ECM controlling serves as an additional target to facilitate the heart regeneration induced by both strategies.

Finally, regardless of the limitations and challenges, it is promising to witness the progresses in the field of heart regeneration. Unlike the transplantation method to regenerate the injured heart, the methods utilizing the resident cardiac cells avoid potential immune rejection issue and invasive intervention. The realization of heart remuscularization will be fulfilled in the future with a better understanding of underlying molecular mechanisms and rapid advance in technologies to facilitate the bench to beside translation.

Acknowledgments This work was supported in part by the following funding sources: NIH RO1s HL114120, HL 131017, HL138023, HL149137, HL153220 and UO1 HL134764.

References

1. Virani, S.S., Alonso, A., Benjamin, E.J., Bittencourt, M.S., Callaway, C.W., Carson, A.P., et al.: Heart disease and stroke Statistics-2020 update: a report from the American Heart Association. Circulation. **141**(9), e139–e596 (2020). https://doi.org/10.1161/CIR.0000000000000757
2. Bergmann, O., Zdunek, S., Felker, A., Salehpour, M., Alkass, K., Bernard, S., et al.: Dynamics of cell generation and turnover in the human heart. Cell. (2015). https://doi.org/10.1016/j.cell.2015.05.026
3. Poss, K.D., Wilson, L.G., Keating, M.T.: Heart regeneration in zebrafish. Science. **298**(5601), 2188–2190 (2002). https://doi.org/10.1126/science.1077857
4. Porrello, E.R., Mahmoud, A.I., Simpson, E., Hill, J.A., Richardson, J.A., Olson, E.N., et al.: Transient regenerative potential of the neonatal mouse heart. Science. **331**(6020), 1078–1080 (2011). https://doi.org/10.1126/science.1200708
5. Porrello, E.R., Mahmoud, A.I., Simpson, E., Johnson, B.A., Grinsfelder, D., Canseco, D., et al.: Regulation of neonatal and adult mammalian heart regeneration by the miR-15 family. Proc. Natl. Acad. Sci. U. S. A. **110**(1), 187–192 (2013). https://doi.org/10.1073/pnas.1208863110
6. Ye, L., D'Agostino, G., Loo, S.J., Wang, C.X., Su, L.P., Tan, S.H., et al.: Early regenerative capacity in the porcine heart. Circulation. **138**(24), 2798–2808 (2018). https://doi.org/10.1161/CIRCULATIONAHA.117.031542
7. Zhu, W., Zhang, E., Zhao, M., Chong, Z., Fan, C., Tang, Y., et al.: Regenerative potential of neonatal porcine hearts. Circulation. **138**(24), 2809–2816 (2018). https://doi.org/10.1161/CIRCULATIONAHA.118.034886
8. Jopling, C., Sleep, E., Raya, M., Marti, M., Raya, A., Izpisua Belmonte, J.C.: Zebrafish heart regeneration occurs by cardiomyocyte dedifferentiation and proliferation. Nature. **464**(7288), 606–609 (2010). https://doi.org/10.1038/nature08899

9. Lam, N.T., Sadek, H.A.: Neonatal heart regeneration: comprehensive literature review. Circulation. **138**(4), 412–423 (2018). https://doi.org/10.1161/CIRCULATIONAHA.118.033648

10. Chablais, F., Veit, J., Rainer, G., Jazwinska, A.: The zebrafish heart regenerates after cryoinjury-induced myocardial infarction. BMC Dev. Biol. **11**(1), 21 (2011). https://doi.org/10.1186/1471-213X-11-21

11. Schnabel, K., Wu, C.C., Kurth, T., Weidinger, G.: Regeneration of cryoinjury induced necrotic heart lesions in zebrafish is associated with epicardial activation and cardiomyocyte proliferation. PLoS One. **6**(4), e18503 (2011). https://doi.org/10.1371/journal.pone.0018503

12. Gonzalez-Rosa, J.M., Martin, V., Peralta, M., Torres, M., Mercader, N.: Extensive scar formation and regression during heart regeneration after cryoinjury in zebrafish. Development. **138**(9), 1663–1674 (2011). https://doi.org/10.1242/dev.060897

13. Wang, J., Panakova, D., Kikuchi, K., Holdway, J.E., Gemberling, M., Burris, J.S., et al.: The regenerative capacity of zebrafish reverses cardiac failure caused by genetic cardiomyocyte depletion. Development. **138**(16), 3421–3430 (2011). https://doi.org/10.1242/dev.068601

14. Zhao, M., Zhang, E., Wei, Y., Zhou, Y., Walcott, G.P., Zhang, J.: Apical resection prolongs the cell cycle activity and promotes myocardial regeneration after left ventricular injury in neonatal pig. Circulation. **142**(9), 913–916 (2020). https://doi.org/10.1161/CIRCULATIONAHA.119.044619

15. Bergmann, O., Zdunek, S., Frisen, J., Bernard, S., Druid, H., Jovinge, S.: Cardiomyocyte renewal in humans. Circ. Res. **110**(1), e17-8; author reply e9-21 (2012). https://doi.org/10.1161/CIRCRESAHA.111.259598

16. Bergmann, O., Bhardwaj, R.D., Bernard, S., Zdunek, S., Barnabe-Heider, F., Walsh, S., et al.: Evidence for cardiomyocyte renewal in humans. Science. **324**(5923), 98–102 (2009). https://doi.org/10.1126/science.1164680

17. Haubner, B.J., Schneider, J., Schweigmann, U., Schuetz, T., Dichtl, W., Velik-Salchner, C., et al.: Functional recovery of a human neonatal heart after severe myocardial infarction. Circ. Res. **118**(2), 216–221 (2016). https://doi.org/10.1161/CIRCRESAHA.115.307017

18. Chaudhry, H.W., Dashoush, N.H., Tang, H., Zhang, L., Wang, X., Wu, E.X., et al.: Cyclin A2 mediates cardiomyocyte mitosis in the postmitotic myocardium. J. Biol. Chem. **279**(34), 35858–35866 (2004). https://doi.org/10.1074/jbc.M404975200

19. Cheng, R.K., Asai, T., Tang, H., Dashoush, N.H., Kara, R.J., Costa, K.D., et al.: Cyclin A2 induces cardiac regeneration after myocardial infarction and prevents heart failure. Circ. Res. **100**(12), 1741–1748 (2007). https://doi.org/10.1161/CIRCRESAHA.107.153544

20. Woo, Y.J., Panlilio, C.M., Cheng, R.K., Liao, G.P., Atluri, P., Hsu, V.M., et al.: Therapeutic delivery of cyclin A2 induces myocardial regeneration and enhances cardiac function in ischemic heart failure. Circulation. **114**(1 Suppl), I206–I213 (2006). https://doi.org/10.1161/CIRCULATIONAHA.105.000455

21. Shapiro, S.D., Ranjan, A.K., Kawase, Y., Cheng, R.K., Kara, R.J., Bhattacharya, R., et al.: Cyclin A2 induces cardiac regeneration after myocardial infarction through cytokinesis of adult cardiomyocytes. Sci. Transl. Med. **6**(224), 224ra27 (2014). https://doi.org/10.1126/scitranslmed.3007668

22. Sdek, P., Zhao, P., Wang, Y., Huang, C.J., Ko, C.Y., Butler, P.C., et al.: Rb and p130 control cell cycle gene silencing to maintain the postmitotic phenotype in cardiac myocytes. J. Cell Biol. **194**(3), 407–423 (2011). https://doi.org/10.1083/jcb.201012049

23. Pasumarthi, K.B., Nakajima, H., Nakajima, H.O., Soonpaa, M.H., Field, L.J.: Targeted expression of cyclin D2 results in cardiomyocyte DNA synthesis and infarct regression in transgenic mice. Circ. Res. **96**(1), 110–118 (2005). https://doi.org/10.1161/01.RES.0000152326.91223.4F

24. Zhu, W., Zhao, M., Mattapally, S., Chen, S., Zhang, J.: CCND2 overexpression enhances the regenerative potency of human induced pluripotent stem cell-derived cardiomyocytes: Remuscularization of injured ventricle. Circ. Res. **122**(1), 88–96 (2018). https://doi.org/10.1161/CIRCRESAHA.117.311504

25. Fan, C., Fast, V.G., Tang, Y., Zhao, M., Turner, J.F., Krishnamurthy, P., et al.: Cardiomyocytes from CCND2-overexpressing human induced-pluripotent stem cells repopulate the myocardial scar in mice: a 6-month study. J. Mol. Cell. Cardiol. **137**, 25–33 (2019). https://doi.org/10.1016/j.yjmcc.2019.09.011

26. Mohamed, T.M.A., Ang, Y.S., Radzinsky, E., Zhou, P., Huang, Y., Elfenbein, A., et al.: Regulation of cell cycle to stimulate adult cardiomyocyte proliferation and cardiac regeneration. Cell. **173**(1), 104–116. e12 (2018). https://doi.org/10.1016/j.cell.2018.02.014

27. Eulalio, A., Mano, M., Dal Ferro, M., Zentilin, L., Sinagra, G., Zacchigna, S., et al.: Functional screening identifies miRNAs inducing cardiac regeneration. Nature. **492**(7429), 376–381 (2012). https://doi.org/10.1038/nature11739

28. Torrini, C., Cubero, R.J., Dirkx, E., Braga, L., Ali, H., Prosdocimo, G., et al.: Common regulatory pathways mediate activity of MicroRNAs inducing cardiomyocyte proliferation. Cell Rep. **27**(9), 2759–2771. e5 (2019). https://doi.org/10.1016/j.celrep.2019.05.005

29. Gabisonia, K., Prosdocimo, G., Aquaro, G.D., Carlucci, L., Zentilin, L., Secco, I., et al.: MicroRNA therapy stimulates uncontrolled cardiac repair after myocardial infarction in pigs. Nature. **569**(7756), 418–422 (2019). https://doi.org/10.1038/s41586-019-1191-6

30. Chen, J., Huang, Z.P., Seok, H.Y., Ding, J., Kataoka, M., Zhang, Z., et al.: Mir-17-92 cluster is required for and sufficient to induce cardiomyocyte proliferation in postnatal and adult hearts. Circ. Res. **112**(12), 1557–1566 (2013). https://doi.org/10.1161/CIRCRESAHA.112.300658

31. Tian, Y., Liu, Y., Wang, T., Zhou, N., Kong, J., Chen, L., et al.: A microRNA-Hippo pathway that promotes cardiomyocyte proliferation and cardiac regeneration in mice. Sci. Transl. Med. **7**(279), 279ra38 (2015). https://doi.org/10.1126/scitranslmed.3010841

32. Liang, D., Li, J., Wu, Y., Zhen, L., Li, C., Qi, M., et al.: miRNA-204 drives cardiomyocyte proliferation via targeting Jarid2. Int. J. Cardiol. **201**, 38–48 (2015). https://doi.org/10.1016/j.ijcard.2015.06.163

33. Borden, A., Kurian, J., Nickoloff, E., Yang, Y., Troupes, C.D., Ibetti, J., et al.: Transient introduction of miR-294 in the heart promotes cardiomyocyte cell cycle reentry after injury. Circ. Res. **125**(1), 14–25 (2019). https://doi.org/10.1161/CIRCRESAHA.118.314223

34. Porrello, E.R., Johnson, B.A., Aurora, A.B., Simpson, E., Nam, Y.J., Matkovich, S.J., et al.: MiR-15 family regulates postnatal mitotic arrest of cardiomyocytes. Circ. Res. **109**(6), 670–679 (2011). https://doi.org/10.1161/CIRCRESAHA.111.248880

35. Huang, W., Feng, Y., Liang, J., Yu, H., Wang, C., Wang, B., et al.: Loss of microRNA-128 promotes cardiomyocyte proliferation and heart regeneration. Nat. Commun. **9**(1), 700 (2018). https://doi.org/10.1038/s41467-018-03019-z

36. Bersell, K., Arab, S., Haring, B., Kuhn, B.: Neuregulin1/ErbB4 signaling induces cardiomyocyte proliferation and repair of heart injury. Cell. **138**(2), 257–270 (2009). https://doi.org/10.1016/j.cell.2009.04.060

37. D'Uva, G., Aharonov, A., Lauriola, M., Kain, D., Yahalom-Ronen, Y., Carvalho, S., et al.: ERBB2 triggers mammalian heart regeneration by promoting cardiomyocyte dedifferentiation and proliferation. Nat. Cell Biol. **17**(5), 627–638 (2015). https://doi.org/10.1038/ncb3149

38. Reuter, S., Soonpaa, M.H., Firulli, A.B., Chang, A.N., Field, L.J.: Recombinant neuregulin 1 does not activate cardiomyocyte DNA synthesis in normal or infarcted adult mice. PLoS One. **9**(12), e115871 (2014). https://doi.org/10.1371/journal.pone.0115871

39. Russell, K.S., Stern, D.F., Polverini, P.J., Bender, J.R.: Neuregulin activation of ErbB receptors in vascular endothelium leads to angiogenesis. Am. J. Phys. **277**(6), H2205–H2211 (1999). https://doi.org/10.1152/ajpheart.1999.277.6.H2205

40. Zhao, B., Tumaneng, K., Guan, K.L.: The hippo pathway in organ size control, tissue regeneration and stem cell self-renewal. Nat. Cell Biol. **13**(8), 877–883 (2011). https://doi.org/10.1038/ncb2303

41. Heallen, T., Morikawa, Y., Leach, J., Tao, G., Willerson, J.T., Johnson, R.L., et al.: Hippo signaling impedes adult heart regeneration. Development. **140**(23), 4683–4690 (2013). https://doi.org/10.1242/dev.102798

42. Leach, J.P., Heallen, T., Zhang, M., Rahmani, M., Morikawa, Y., Hill, M.C., et al.: Hippo pathway deficiency reverses systolic heart failure after infarction. Nature. **550**(7675), 260–264 (2017). https://doi.org/10.1038/nature24045

43. Xin, M., Kim, Y., Sutherland, L.B., Murakami, M., Qi, X., McAnally, J., et al.: Hippo pathway effector yap promotes cardiac regeneration. Proc. Natl. Acad. Sci. U. S. A. **110**(34), 13839–13844 (2013). https://doi.org/10.1073/pnas.1313192110

44. Del Re, D.P., Yang, Y., Nakano, N., Cho, J., Zhai, P., Yamamoto, T., et al.: Yes-associated protein isoform 1 (Yap1) promotes cardiomyocyte survival and growth to protect against myocardial ischemic injury. J. Biol. Chem. **288**(6), 3977–3988 (2013). https://doi.org/10.1074/jbc.M112.436311

45. Lin, Z., von Gise, A., Zhou, P., Gu, F., Ma, Q., Jiang, J., et al.: Cardiac-specific YAP activation improves cardiac function and survival in an experimental murine MI model. Circ. Res. **115**(3), 354–363 (2014). https://doi.org/10.1161/CIRCRESAHA.115.303632

46. Mahmoud, A.I., Kocabas, F., Muralidhar, S.A., Kimura, W., Koura, A.S., Thet, S., et al.: Meis1 regulates postnatal cardiomyocyte cell cycle arrest. Nature. **497**(7448), 249–253 (2013). https://doi.org/10.1038/nature12054

47. Lu, F., Langenbacher, A., Chen, J.N.: Tbx20 drives cardiac progenitor formation and cardiomyocyte proliferation in zebrafish. Dev. Biol. **421**(2), 139–148 (2017). https://doi.org/10.1016/j.ydbio.2016.12.009

48. Boogerd, C.J., Zhu, X., Aneas, I., Sakabe, N., Zhang, L., Sobreira, D.R., et al.: Tbx20 is required in mid-gestation cardiomyocytes and plays a central role in atrial development. Circ. Res. **123**(4), 428–442 (2018). https://doi.org/10.1161/CIRCRESAHA.118.311339

49. Szeto, D.P., Griffin, K.J.P.: Kimelman D. HrT is required for cardiovascular development in zebrafish, Development (2002)

50. Cai, C.L., Zhou, W., Yang, L., Bu, L., Qyang, Y., Zhang, X., et al.: T-box genes coordinate regional rates of proliferation and regional specification during cardiogenesis. Development. **132**(10), 2475–2487 (2005). https://doi.org/10.1242/dev.01832

51. Stennard, F.A., Costa, M.W., Elliott, D.A., Rankin, S., Haast, S.J., Lai, D., et al.: Cardiac T-box factor Tbx20 directly interacts with Nkx2-5, GATA4, and GATA5 in regulation of gene expression in the developing heart. Dev. Biol. **262**(2), 206–224 (2003). https://doi.org/10.1016/s0012-1606(03)00385-3

52. Stennard, F.A., Costa, M.W., Lai, D., Biben, C., Furtado, M.B., Solloway, M.J., et al.: Murine T-box transcription factor Tbx20 acts as a repressor during heart development, and is essential for adult heart integrity, function and adaptation. Development. **132**(10), 2451–2462 (2005). https://doi.org/10.1242/dev.01799

53. Chakraborty, S., Sengupta, A., Yutzey, K.E.: Tbx20 promotes cardiomyocyte proliferation and persistence of fetal characteristics in adult mouse hearts. J. Mol. Cell. Cardiol. **62**, 203–213 (2013). https://doi.org/10.1016/j.yjmcc.2013.05.018

54. Chakraborty, S., Yutzey, K.E.: Tbx20 regulation of cardiac cell proliferation and lineage specialization during embryonic and fetal development in vivo. Dev. Biol. **363**(1), 234–246 (2012). https://doi.org/10.1016/j.ydbio.2011.12.034

55. Xiang, F.L., Guo, M., Yutzey, K.E.: Overexpression of Tbx20 in adult cardiomyocytes promotes proliferation and improves cardiac function after myocardial infarction. Circulation. **133**(11), 1081–1092 (2016). https://doi.org/10.1161/CIRCULATIONAHA.115.019357

56. Gonzalez-Rosa, J.M., Sharpe, M., Field, D., Soonpaa, M.H., Field, L.J., Burns, C.E., et al.: Myocardial Polyploidization creates a barrier to heart regeneration in zebrafish. Dev. Cell. **44**(4), 433–446. e7 (2018). https://doi.org/10.1016/j.devcel.2018.01.021

57. Liu, H., Zhang, C.H., Ammanamanchi, N., Suresh, S., Lewarchik, C., Rao, K., et al.: Control of cytokinesis by beta-adrenergic receptors indicates an approach for regulating cardiomyocyte endowment. Sci. Transl. Med. **11**(513) (2019). https://doi.org/10.1126/scitranslmed.aaw6419

58. Puente, B.N., Kimura, W., Muralidhar, S.A., Moon, J., Amatruda, J.F., Phelps, K.L., et al.: The oxygen-rich postnatal environment induces cardiomyocyte cell-cycle arrest through DNA damage response. Cell. **157**(3), 565–579 (2014). https://doi.org/10.1016/j.cell.2014.03.032

59. Kimura, W., Xiao, F., Canseco, D.C., Muralidhar, S., Thet, S., Zhang, H.M., et al.: Hypoxia fate mapping identifies cycling cardiomyocytes in the adult heart. Nature. **523**(7559), 226–230 (2015). https://doi.org/10.1038/nature14582

60. Nakada, Y., Canseco, D.C., Thet, S., Abdisalaam, S., Asaithamby, A., Santos, C.X., et al.: Hypoxia induces heart regeneration in adult mice. Nature. **541**(7636), 222–227 (2016). https://doi.org/10.1038/nature20173

61. Kuhn, B., del Monte, F., Hajjar, R.J., Chang, Y.S., Lebeche, D., Arab, S., et al.: Periostin induces proliferation of differentiated cardiomyocytes and promotes cardiac repair. Nat. Med. **13**(8), 962–969 (2007). https://doi.org/10.1038/nm1619

62. Ladage, D., Yaniz-Galende, E., Rapti, K., Ishikawa, K., Tilemann, L., Shapiro, S., et al.: Stimulating myocardial regeneration with periostin peptide in large mammals improves function post-myocardial infarction but increases myocardial fibrosis. PLoS One. **8**(5), e59656 (2013). https://doi.org/10.1371/journal.pone.0059656

63. Chen, Z., Xie, J., Hao, H., Lin, H., Wang, L., Zhang, Y., et al.: Ablation of periostin inhibits post-infarction myocardial regeneration in neonatal mice mediated by the phosphatidylinositol 3 kinase/glycogen synthase kinase 3beta/cyclin D1 signalling pathway. Cardiovasc. Res. **113**(6), 620–632 (2017). https://doi.org/10.1093/cvr/cvx001

64. Takahashi, K., Yamanaka, S.: Induction of pluripotent stem cells from mouse embryonic and adult fibroblast cultures by defined factors. Cell. **126**(4), 663–676 (2006). https://doi.org/10.1016/j.cell.2006.07.024

65. Xu, J., Du, Y., Deng, H.: Direct lineage reprogramming: strategies, mechanisms, and applications. Cell Stem Cell. **16**(2), 119–134 (2015). https://doi.org/10.1016/j.stem.2015.01.013

66. Ieda, M., Fu, J.D., Delgado-Olguin, P., Vedantham, V., Hayashi, Y., Bruneau, B.G., et al.: Direct reprogramming of fibroblasts into functional cardiomyocytes by defined factors. Cell. **142**(3), 375–386 (2010). https://doi.org/10.1016/j.cell.2010.07.002

67. Qian, L., Huang, Y., Spencer, C.I., Foley, A., Vedantham, V., Liu, L., et al.: In vivo reprogramming of murine cardiac fibroblasts into induced cardiomyocytes. Nature. **485**(7400), 593–598 (2012). https://doi.org/10.1038/nature11044

68. Inagawa, K., Miyamoto, K., Yamakawa, H., Muraoka, N., Sadahiro, T., Umei, T., et al.: Induction of cardiomyocyte-like cells in infarct hearts by gene transfer of Gata4, Mef2c, and Tbx5. Circ. Res. **111**(9), 1147–1156 (2012). https://doi.org/10.1161/CIRCRESAHA.112.271148

69. Song, K., Nam, Y.J., Luo, X., Qi, X., Tan, W., Huang, G.N., et al.: Heart repair by reprogramming non-myocytes with cardiac transcription factors. Nature. **485**(7400), 599–604 (2012). https://doi.org/10.1038/nature11139

70. Protze, S., Khattak, S., Poulet, C., Lindemann, D., Tanaka, E.M., Ravens, U.: A new approach to transcription factor screening for reprogramming of fibroblasts to cardiomyocyte-like cells. J. Mol. Cell. Cardiol. **53**(3), 323–332 (2012). https://doi.org/10.1016/j.yjmcc.2012.04.010

71. Addis, R.C., Ifkovits, J.L., Pinto, F., Kellam, L.D., Esteso, P., Rentschler, S., et al.: Optimization of direct fibroblast reprogramming to cardiomyocytes using calcium activity as a functional measure of success. J. Mol. Cell. Cardiol. **60**, 97–106 (2013). https://doi.org/10.1016/j.yjmcc.2013.04.004

72. Y-j, N., Lubczyk, C., Bhakta, M., Zang, T., Fernandez-perez, A., McAnally, J., et al.: Induction of diverse cardiac cell types by reprogramming fibroblasts with cardiac transcription factors. Development (Cambridge, England). **2014**, 4267–4278. https://doi.org/10.1242/dev.114025

73. Muraoka, N., Yamakawa, H., Miyamoto, K., Sadahiro, T., Umei, T., Isomi, M., et al.: MiR-133 promotes cardiac reprogramming by directly repressing Snai1 and silencing fibroblast signatures. EMBO J., 1–17 (2014)

74. Zhao, Y., Londono, P., Cao, Y., Sharpe, E.J., Proenza, C., O'Rourke, R., et al.: High-efficiency reprogramming of fibroblasts into cardiomyocytes requires suppression of pro-fibrotic signalling. Nat. Commun. **6**, 8243 (2015). https://doi.org/10.1038/ncomms9243

75. Jayawardena, T.M., Egemnazarov, B., Finch, E.A., Zhang, L., Payne, J.A., Pandya, K., et al.: MicroRNA-mediated in vitro and in vivo direct reprogramming of cardiac fibroblasts to cardiomyocytes. Circ. Res. **110**(11), 1465–1473 (2012). https://doi.org/10.1161/CIRCRESAHA.112.269035

76. Wang, L., Liu, Z., Yin, C., Asfour, H., Chen, O., Li, Y., et al.: Stoichiometry of Gata4, Mef2c, and Tbx5 influences the efficiency and quality of induced cardiac myocyte reprogramming. Circ. Res. **116**(2), 237–244 (2015). https://doi.org/10.1161/CIRCRESAHA.116.305547

77. Ma, H., Wang, L., Yin, C., Liu, J., Qian, L.: In vivo cardiac reprogramming using an optimal single polycistronic construct. Cardiovasc. Res. **108**(2), 217–219 (2015). https://doi.org/10.1093/cvr/cvv223

78. Ifkovits, J.L., Addis, R.C., Epstein, J.A., Gearhart, J.D.: Inhibition of TGFbeta signaling increases direct conversion of fibroblasts to induced cardiomyocytes. PLoS One. **9**(2), e89678 (2014). https://doi.org/10.1371/journal.pone.0089678

79. Mohamed, T.M., Stone, N.R., Berry, E.C., Radzinsky, E., Huang, Y., Pratt, K., et al.: Chemical enhancement of in vitro and in vivo direct cardiac reprogramming. Circulation. **135**(10), 978–995 (2017). https://doi.org/10.1161/CIRCULATIONAHA.116.024692

80. Zhou, H., Dickson, M.E., Kim, M.S., Bassel-Duby, R., Olson, E.N.: Akt1/protein kinase B enhances transcriptional reprogramming of fibroblasts to functional cardiomyocytes. Proc. Natl. Acad. Sci. U. S. A. **112**(38), 11864–11869 (2015). https://doi.org/10.1073/pnas.1516237112

81. Abad, M., Hashimoto, H., Zhou, H., Morales, M.G., Chen, B., Bassel-Duby, R., et al.: Notch inhibition enhances cardiac reprogramming by increasing MEF2C transcriptional activity. Stem Cell Rep. **8**(3), 548–560 (2017). https://doi.org/10.1016/j.stemcr.2017.01.025

82. Kattman, S.J., Witty, A.D., Gagliardi, M., Dubois, N.C., Niapour, M., Hotta, A., et al.: Stage-specific optimization of activin/nodal and BMP signaling promotes cardiac differentiation of mouse and human pluripotent stem cell lines. Cell Stem Cell. **8**(2), 228–240 (2011). https://doi.org/10.1016/j.stem.2010.12.008

83. Burridge, P.W., Matsa, E., Shukla, P., Lin, Z.C., Churko, J.M., Ebert, A.D., et al.: Chemically defined generation of human cardiomyocytes. Nat. Methods. **11**(8), 855–860 (2014). https://doi.org/10.1038/nmeth.2999

84. Yamakawa, H., Muraoka, N., Miyamoto, K., Sadahiro, T., Isomi, M., Haginiwa, S., et al.: Fibroblast growth factors and vascular endothelial growth factor promote cardiac reprogramming under defined conditions. Stem Cell Rep. **5**(6), 1128–1142 (2015)

85. Francis, N.J., Kingston, R.E.: Mechanisms of transcriptional memory. Nat. Rev. Mol. Cell Biol. **2**(6), 409–421 (2001). https://doi.org/10.1038/35073039

86. Liu, Z., Chen, O., Zheng, M., Wang, L., Zhou, Y., Yin, C., et al.: Re-patterning of H3K27me3, H3K4me3 and DNA methylation during fibroblast conversion into induced cardiomyocytes. Stem Cell Res. **16**(2), 507–518 (2016)

87. Dal-Pra, S., Hodgkinson, C.P., Mirotsou, M., Kirste, I., Dzau, V.J.: Demethylation of H3K27 is essential for the induction of direct cardiac reprogramming by miR combo. Circ. Res. **120**(9), 1403–1413 (2017). https://doi.org/10.1161/CIRCRESAHA.116.308741

88. Hirai, H., Kikyo, N.: Inhibitors of suppressive histone modification promote direct reprogramming of fibroblasts to cardiomyocyte-like cells. Cardiovasc. Res. **102**(1), 188–190 (2014). https://doi.org/10.1093/cvr/cvu023

89. Zhou, Y., Wang, L., Vaseghi, H.R., Liu, Z., Lu, R., Alimohamadi, S., et al.: Bmi1 is a key epigenetic barrier to direct cardiac reprogramming. Cell Stem Cell. **18**(3), 382–395 (2016)

90. Testa, G., Russo, M., Di Benedetto, G., Barbato, M., Parisi, S., Pirozzi, F., et al.: Bmi1 inhibitor PTC-209 promotes chemically-induced direct cardiac reprogramming of cardiac fibroblasts into cardiomyocytes. Sci. Rep. **10**(1), 7129 (2020). https://doi.org/10.1038/s41598-020-63992-8

91. Liu, L., Lei, I., Karatas, H., Li, Y., Wang, L., Gnatovskiy, L., et al.: Targeting Mll1 H3K4 methyltransferase activity to guide cardiac lineage specific reprogramming of fibroblasts. Cell Discov. **2**, 16036 (2016). https://doi.org/10.1038/celldisc.2016.36

92. Levine, B., Kroemer, G.: Autophagy in the pathogenesis of disease. Cell. **132**(1), 27–42 (2008). https://doi.org/10.1016/j.cell.2007.12.018

93. Mizushima, N., Levine, B., Cuervo, A.M., Klionsky, D.J.: Autophagy fights disease through cellular self-digestion. Nature. **451**(7182), 1069–1075 (2008). https://doi.org/10.1038/nature06639

94. Wang, L., Ma, H., Huang, P., Xie, Y., Near, D., Wang, H., et al.: Down-regulation of Beclin1 promotes direct cardiac reprogramming. Sci. Transl. Med. **12**(566) (2020). https://doi.org/10.1126/scitranslmed.aay7856

95. Zhou, H., Morales, M.G., Hashimoto, H., Dickson, M.E., Song, K., Ye, W., et al.: ZNF281 enhances cardiac reprogramming by modulating cardiac and inflammatory gene expression. Genes Dev. **31**(17), 1770–1783 (2017). https://doi.org/10.1101/gad.305482.117

96. Muraoka, N., Nara, K., Tamura, F., Kojima, H., Yamakawa, H., Sadahiro, T., et al.: Role of cyclooxygenase-2-mediated prostaglandin E2-prostaglandin E receptor 4 signaling in cardiac reprogramming. Nat. Commun. **10**(1), 674 (2019). https://doi.org/10.1038/s41467-019-08626-y

97. Liu, Z., Wang, L., Welch, J.D., Ma, H., Zhou, Y., Vaseghi, H.R., et al.: Single-cell transcriptomics reconstructs fate conversion from fibroblast to cardiomyocyte. Nature. **551**(7678), 100–104 (2017)

98. Bektik, E., Dennis, A., Pawlowski, G., Zhou, C., Maleski, D., Takahashi, S., et al.: S-phase synchronization facilitates the early progression of induced-cardiomyocyte reprogramming through enhanced cell-cycle exit. Int. J. Mol. Sci. **19**(5), 1364 (2018). https://doi.org/10.3390/ijms19051364

99. van den Hoogenhof, M.M., Pinto, Y.M., Creemers, E.E.: RNA splicing: regulation and dysregulation in the heart. Circ. Res. **118**(3), 454–468 (2016). https://doi.org/10.1161/CIRCRESAHA.115.307872

100. Zhou, Y., Alimohamadi, S., Wang, L., Liu, Z., Wall, J.B., Yin, C., et al.: A loss of function screen of epigenetic modifiers and splicing factors during early stage of cardiac reprogramming. Stem Cells Int. **2018** (2018)

101. Sauls, K., Greco, T.M., Wang, L., Zou, M., Villasmil, M., Qian, L., et al.: Initiating events in direct cardiomyocyte reprogramming. Cell Rep. **22**(7), 1913–1922 (2018). https://doi.org/10.1016/j.celrep.2018.01.047

102. Wada, R., Muraoka, N., Inagawa, K., Yamakawa, H., Miyamoto, K., Sadahiro, T., et al.: Induction of human cardiomyocyte-like cells from fibroblasts by defined factors. Proc. Natl. Acad. Sci. U. S. A. **110**(31), 12667–12672 (2013). https://doi.org/10.1073/pnas.1304053110

103. Fu, J.D., Stone, N.R., Liu, L., Spencer, C.I., Qian, L., Hayashi, Y., et al.: Direct reprogramming of human fibroblasts toward a cardiomyocyte-like state. Stem Cell Rep. **1**(3), 235–247 (2013). https://doi.org/10.1016/j.stemcr.2013.07.005

104. Fu, J.D., Srivastava, D.: Direct reprogramming of fibroblasts into cardiomyocytes for cardiac regenerative medicine. Circ. J. Off. J. Jpn. Circul. Soc. **79**(2), 245–254 (2015). https://doi.org/10.1253/circj.CJ-14-1372

105. Nam, Y.J., Song, K., Luo, X., Daniel, E., Lambeth, K., West, K., et al.: Reprogramming of human fibroblasts toward a cardiac fate. Proc. Natl. Acad. Sci. U. S. A. **110**(14), 5588–5593 (2013). https://doi.org/10.1073/pnas.1301019110

106. Zhou, Y., Liu, Z., Welch, J.D., Gao, X., Wang, L., Garbutt, T., et al.: Single-cell transcriptomic analyses of cell fate transitions during human cardiac reprogramming. Cell Stem Cell. **25**(1), 149–164. e9 (2019). https://doi.org/10.1016/j.stem.2019.05.020

107. Muraoka, N., Yamakawa, H., Miyamoto, K., Sadahiro, T., Umei, T., Isomi, M., et al.: MiR-133 promotes cardiac reprogramming by directly repressing Snai1 and silencing fibroblast signatures. EMBO J. **33**(14), 1565–1581 (2014). https://doi.org/10.15252/embj.201387605

108. Yamakawa, H., Muraoka, N., Miyamoto, K., Sadahiro, T., Isomi, M., Haginiwa, S., et al.: Fibroblast growth factors and vascular endothelial growth factor promote cardiac reprogramming under defined conditions. Stem Cell Rep. 5(6), 1128–1142 (2015). https://doi.org/10.1016/j.stemcr.2015.10.019

109. Mathison, M., Gersch, R.P., Nasser, A., Lilo, S., Korman, M., Fourman, M., et al.: In vivo cardiac cellular reprogramming efficacy is enhanced by angiogenic preconditioning of the infarcted myocardium with vascular endothelial growth factor. J. Am. Heart Assoc. 1(6), e005652 (2012). https://doi.org/10.1161/JAHA.112.005652

110. Miyamoto, K., Akiyama, M., Tamura, F., Isomi, M., Yamakawa, H., Sadahiro, T., et al.: Direct in vivo reprogramming with Sendai virus vectors improves cardiac function after myocardial infarction. Cell Stem Cell. 22(1), 91–103. e5 (2018). https://doi.org/10.1016/j.stem.2017.11.010

111. Zangi, L., Lui, K.O., von Gise, A., Ma, Q., Ebina, W., Ptaszek, L.M., et al.: Modified mRNA directs the fate of heart progenitor cells and induces vascular regeneration after myocardial infarction. Nat. Biotechnol. 31(10), 898–907 (2013). https://doi.org/10.1038/nbt.2682

112. Aurora, A.B., Porrello, E.R., Tan, W., Mahmoud, A.I., Hill, J.A., Bassel-Duby, R., et al.: Macrophages are required for neonatal heart regeneration. J. Clin. Invest. 124(3), 1382–1392 (2014). https://doi.org/10.1172/JCI72181

113. Kurotsu, S., Sadahiro, T., Fujita, R., Tani, H., Yamakawa, H., Tamura, F., et al.: Soft matrix promotes cardiac reprogramming via inhibition of YAP/TAZ and suppression of fibroblast signatures. Stem Cell Rep. 15(3), 612–628 (2020). https://doi.org/10.1016/j.stemcr.2020.07.022

114. Li, Y., Dal-Pra, S., Mirotsou, M., Jayawardena, T.M., Hodgkinson, C.P., Bursac, N., et al.: Tissue-engineered 3-dimensional (3D) microenvironment enhances the direct reprogramming of fibroblasts into cardiomyocytes by microRNAs. Sci. Rep. 6, 38815 (2016). https://doi.org/10.1038/srep38815

Allogeneic Immunity Following Transplantation of Pluripotent Stem Cell-Derived Cardiomyocytes

Yuji Shiba

1 Introduction

Human pluripotent stem cells (PSCs), both embryonic stem (ES) [1] cells and induced pluripotent stem (iPS) cells [2], have the capacity to proliferate practically infinitely and differentiate into virtually every somatic cell type. These unique characteristics have the potential to repair injured organs caused by conditions including heart failure. Indeed, many relatively homogeneous cardiomyocytes (CMs) can be generated from human PSCs [3] and these PSC-derived CMs (PSC-CMs) have been shown to regenerate injured hearts in multiple animal models [4–7]. These preclinical studies were conducted under a xenogeneic basis, transplanting human derived cardiomyocytes into animal recipients. These xenogeneic transplantation studies have provided valuable insights, including CM engraftment and survival in injured hearts [5], electrical integration of grafted CMs with host CMs [6, 8], and mechanical restoration of injured hearts [9]; however, immune response following allogeneic or autologous transplantation, as is the case in clinics, has yet to be sufficiently examined.

Non-human primate preclinical transplantation studies are useful for evaluating the immune response following allogeneic PSC-derivatives because of the following two aspects. First, both monkey ES cells and iPS cells were established and shown to have similar characteristics to human PSCs [10, 11]. Second, the structure of the major histocompatibility complex (MHC), which plays an essential role in the immune response, is identical to human leukocyte antigen (HLA) [12]. Here, I will summarize a recent approach for controlling the immune response following

Y. Shiba (✉)
Department of Regenerative Science and Medicine, Institute for Biomedical Sciences,
Shinshu University, Matsumoto, Japan
e-mail: yshiba@shinshu-u.ac.jp

© The Author(s), under exclusive license to Springer Nature
Switzerland AG 2022
J. Zhang, V. Serpooshan (eds.), *Advanced Technologies in Cardiovascular
Bioengineering*, https://doi.org/10.1007/978-3-030-86140-7_5

PSC-CM transplantation, including an allogeneic transplantation study in non-human primates.

1.1 Immunosuppressive Therapies After Heart Transplantation

Survival after heart transplantation has steadily improved since the first heart transplant was performed in 1967 [13] essentially due to the development of an effective immunosuppression protocol. Immune reactions are known to be the highest immediately after heart transplantation and decrease gradually thereafter. In this regard, the current protocol following heart transplantation uses the highest levels of immunosuppression immediately after transplantation. Depending on the risk of hyperacute rejection, approximately half of the recipients have received induction therapy, such as anti-thymocyte globulin and basiliximab [14]. Subsequent maintenance therapy is generally a combination of three classes of medications: anti-metabolite medications, calcineurin inhibitors, and corticosteroids. Anti-metabolite medications include mycophenolate mofetil (MMF) and azathioprine (AZA), but mycophenolate mofetil is now more commonly used because of its better outcome in clinical study [15]. Calcineurin inhibitors inhibit the transport of transcription factor, nuclear factor of activated T-cells (NFAT), leading to a reduction in T-cell activation. Cyclosporine was approved in 1983 for prevention or treatment of immune rejection against transplanted organs and the use of cyclosporine dramatically improved the outcomes of organ transplantations [16]. Tacrolimus was discovered in the early 1980s and approved for the prevention of transplanted liver rejection in 1994 [17]. Since then, the use of tacrolimus has expanded quite rapidly for the prevention or treatment of immune rejection in other organs. There have been multiple randomized clinical trials comparing cyclosporine and tacrolimus as a maintenance therapy following heart transplantation; overall, there was no difference in recipient survival and graft rejection; however, tacrolimus seems to be superior to cyclosporine with respect to adverse effects such as hypertension, hyperlipidemia, and gingival hyperplasia [18]. Corticosteroids affect multiple mechanisms of action, including the innate and adaptive immune system [19]. High-dose intravenous methylprednisolone was administered immediately after transplantation followed by oral prednisolone with gradual tapering of doses.

1.2 Current Immunosuppression Therapies Following Transplantation of PSC-CMs in Clinical Trials

Currently, there are two clinical trials of PSC-derived cardiac derivatives. The Menasche group created fibrin patches including human ES cell-derived cardiovascular progenitors and delivered epicardially to the hearts of six patients with ischemic heart disease [20]. These patients received methylprednisolone, cyclosporine, and mycophenolate mofetil for up to 2 months. One patient died from unrelated comorbidities, but the other five patients showed symptomatic improvement without any adverse effects including tumor formation, post-transplant arrhythmia, and immunosuppression-related side effects. Three patients developed donor-specific antibodies, but graft survival or immune rejection was not histologically confirmed in this study. Another clinical trial was conducted by the Sawa group [21]. In this study, three patients with ischemic cardiomyopathy were enrolled and received iPS-CM sheets generated on temperature-responsive cell culture plates [22] under the treatment of corticosteroids, tacrolimus, and mycophenolate mofetil. Both transplantation studies were performed on an MHC-mismatched allogeneic basis and immunosuppressant protocols seem to be equivalent to those after heart transplantation. However, in addition to cardiomyocytes, the human heart consists of a wide variety of cells including fibroblasts, endothelial cells, smooth muscle cells, and neural cells. Therefore, optimal immunosuppression therapy should be determined separately. Recent preclinical transplantation studies in large animal models have revealed that hPSC-CMs can remuscularize only <5% of the left ventricle [7, 9, 23]. It would be impossible to evaluate immune rejection by transcatheter endomyocardial biopsy, which is a well-established technique following heart transplantation. Therefore, it is crucial to identify optical immunosuppression therapy following PSC-CM transplantation in preclinical studies.

1.3   Allogeneic Transplantation Model in Non-human Primates

Many preclinical animal studies have shown engraftment and long-term survival of transplanted PSC-CMs without significant immune rejection; however, in these studies, human PSC-CMs were transplanted into severely immunodeficient recipient animals by either genetically modified T-cell deficiency [5, 24] or a large amount of immunosuppressants [6–9, 23]. To properly evaluate the immune response following PSC-CM transplantation, an allogeneic transplantation study is required. The cynomolgus monkey (*Macaca fascicularis*) is a model organism that has advantages for PSC transplantation studies. First, cynomolgus ES cells [11] and iPS cells [10] have already been established and show characteristics similar to those of humans, including the ability to differentiate cardiomyocytes [25]. More importantly, the structure and function of cynomolgus MHC, which plays an essential role in immune rejection, were shown to be identical to those of HLA [12, 26, 27].

We established cynomolgus iPS cell lines whose MHC class I and II were homozygous on both chromosomes and generated iPSC-CMs. When we transplanted these iPSC-CMs into MHC heterozygous monkeys, in which either of the MHC haplotypes were identical to that of donor iPS cells (Fig. 1), the grafted CMs survived for 12 weeks without apparent immune rejection under the treatment of clinically relevant doses of methylprednisolone and tacrolimus [28]. Notably, when we transplanted the same batch of CMs into MHC-mismatched recipients with the

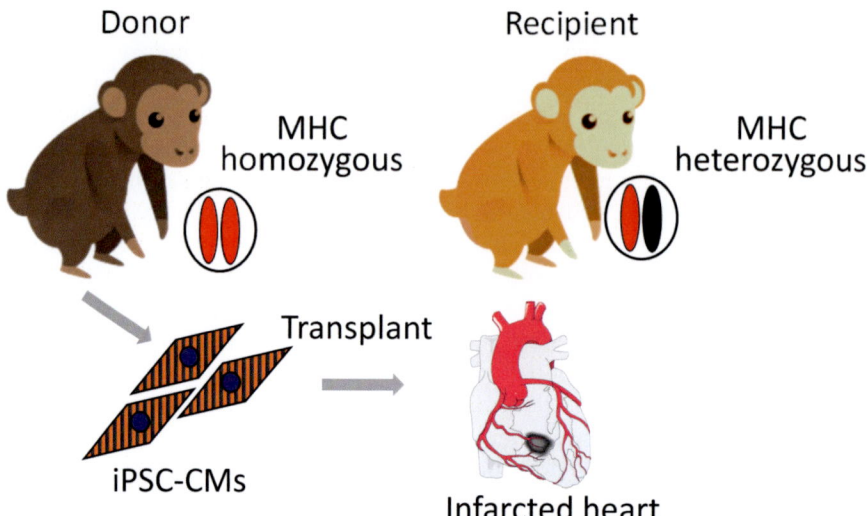

Fig. 1 MHC-matched allogeneic transplantation study of cynomolgus iPSC-CMs. iPS cells were established from MHC homozygous monkeys and differentiated into cardiomyocytes. The recipient animals, in which either of the MHC haplotypes were identical to the donor MHC, underwent induction of myocardial infarction and were transplanted with iPSC-CMs

same immunosuppressant therapy, only a small amount of grafted CMs survived with significant immune rejection at 4 weeks post-transplantation. This study is the first to show that allogeneic transplantation of PSC-CMs regenerates injured hearts; however, we have not attempted to reduce immunosuppressants to clarify the amount of immunosuppressants required for graft CM survival. In the future, it is desirable to identify the appropriate amount of immunosuppressants for graft survival in multiple conditions, including MHC-matched and MHC-mismatched transplantations in this allogeneic model.

1.4 Strategies to Reduce Immunosuppressants Following PSC-CM Transplantation

The ideal strategy to prevent immune rejection is to use autologous iPS-CMs; however, at present, it is not clear whether this strategy is feasible in clinical practice. It would be quite challenging to prepare autologous patient-specific iPSC-CMs that are sufficiently validated for clinical use at an affordable cost. In addition, it may be crucial to prepare CMs in a short period because many studies have shown the beneficial effects of PSC-CM transplantation in acute or subacute models of myocardial infarction, but not in chronic myocardial infarction models [8, 29]. Alternatively, allogeneic transplantation using off-the-shelf cell products is a viable approach, but this approach requires a strategy to control the host immune reaction. As mentioned above, MHC-matched transplantation using MHC homozygous iPS cells could reduce the reliance on immunosuppressants, but numerous iPS cell lines still need to be established to cover most of the populations in the world. To reduce the number of cell lines for covering the global population, several approaches have been proposed using gene editing techniques. HLA class I molecules are expressed on most cells in the human body and play an essential role in allogeneic rejection through the presentation of peptide antigens to CD8$^+$ T cells. Simple disruption of beta-2 microglobulin (B2M), which is a protein subunit required for surface expression of all HLA class I chains, was shown to enhance natural killer (NK) cell-dependent rejection due to lack of cell surface expression of HLA class I; this is called the "missing self" response (Fig. 2). Gornalusse et al. created universal PSCs by the overexpression of HLA-E molecules into the B2M gene locus and showed that hematopoietic cells derived from their universal PSCs were able to avoid the immune response by T cells and NK cells [30]. A similar approach was reported by Xu et al., whereby HLA-A, HLA-B, and HLA-II were deleted in HLA homozygous iPSCs; these iPSCs still retained HLA-C, E, F, and G, which can inhibit NK cell activation [31]. CD47 is a cell surface protein that can interact with receptors to inhibit phagocytosis [32]. Deuse et al. created hypoimmunogenic PSCs by overexpression of CD47 concomitant with inactivation of HLA class I and II by knocking out B2M and class II MHC transactivator (CIITA), respectively; they showed that hypoimmunogenic iPSC-derivatives evaded immune rejection in allogeneic recipients without the use of immunosuppressants [33].

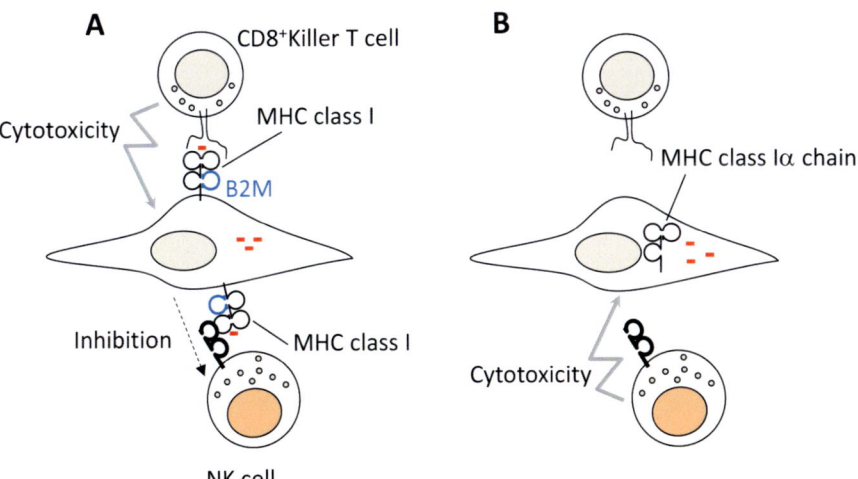

Fig. 2 Role of MHC class I molecules in immune rejection. (**a**) CD8$^+$ killer T cells recognize peptides displayed on MHC class I molecules and provoke an immune response unless the peptide or MHC molecules are recognized as self-derived. MHC class I molecules also function as ligands for inhibitory receptors of NK cells. (**b**) Disruption of beta-2 microglobulin (B2M) induces cytotoxic activity of NK cells due to the lack of cell surface HLA expression

Another promising strategy to reduce the use of immunosuppressants is to induce immune tolerance mediated by regulatory T cells (Tregs). Tregs receive antigen presentation through T cell receptors, but do not produce IL-2. Furthermore, Tregs express CD25 and CTLA4, both of which suppress T cell activation. Interestingly, recipients of a higher proportion of peripheral Tregs were more likely to be able to terminate immunosuppressants after liver transplantation [34]. The potential immunomodulation therapy was also proposed by the use of thymic epithelial cells [35] and mesenchymal stem cells [36].

Many clinical trials have already been conducted or plan to use allogeneic PSCs, but we still do not know the appropriate way to control immune rejection. More studies on allogeneic immunity are required to develop this novel therapy as a standard treatment.

2 Conclusion

Although clinical trials using human PSC-CMs have already started, little is known about host immune reactions against graft CMs. This is attributed to the lack of an appropriate model to evaluate in vivo allogeneic immune reactions. As heart transplantations have become the standard therapy for end-stage heart failure due to the development of immunosuppression protocols, it is crucial to control the immune response following PSC-CM transplantation. Allogeneic transplantation models using non-human primates are one of the few models that provide important information about host-graft immune reactions following cell transplantation.

References

1. Thomson, J.A., Itskovitz-Eldor, J., Shapiro, S.S., Waknitz, M.A., Swiergiel, J.J., Marshall, V.S., et al.: Embryonic stem cell lines derived from human blastocysts. Science. **282**(5391), 1145–1147 (1998)
2. Takahashi, K., Tanabe, K., Ohnuki, M., Narita, M., Ichisaka, T., Tomoda, K., et al.: Induction of pluripotent stem cells from adult human fibroblasts by defined factors. Cell. **131**(5), 861–872 (2007)
3. Minami, I., Yamada, K., Otsuji, T.G., Yamamoto, T., Shen, Y., Otsuka, S., et al.: A small molecule that promotes cardiac differentiation of human pluripotent stem cells under defined, cytokine- and xeno-free conditions. Cell Rep. **2**(5), 1448–1460 (2012). https://doi.org/10.1016/j.celrep.2012.09.015
4. van Laake, L.W., Passier, R., Doevendans, P.A., Mummery, C.L.: Human embryonic stem cell-derived cardiomyocytes and cardiac repair in rodents. Circ. Res. **102**(9), 1008–1010 (2008). https://doi.org/10.1161/circresaha.108.175505
5. Laflamme, M.A., Chen, K.Y., Naumova, A.V., Muskheli, V., Fugate, J.A., Dupras, S.K., et al.: Cardiomyocytes derived from human embryonic stem cells in pro-survival factors enhance function of infarcted rat hearts. Nat. Biotechnol. **25**(9), 1015–1024 (2007)

6. Shiba, Y., Fernandes, S., Zhu, W.Z., Filice, D., Muskheli, V., Kim, J., et al.: Human ES-cell-derived cardiomyocytes electrically couple and suppress arrhythmias in injured hearts. Nature. **489**(7415), 322–325 (2012). https://doi.org/10.1038/nature11317

7. Chong, J.J., Yang, X., Don, C.W., Minami, E., Liu, Y.W., Weyers, J.J., et al.: Human embryonic-stem-cell-derived cardiomyocytes regenerate non-human primate hearts. Nature. **510**(7504), 273–277 (2014). https://doi.org/10.1038/nature13233

8. Shiba, Y., Filice, D., Fernandes, S., Minami, E., Dupras, S.K., Biber, B.V., et al.: Electrical integration of human embryonic stem cell-derived cardiomyocytes in a Guinea pig chronic infarct model. J. Cardiovasc. Pharmacol. Ther. **19**(4), 368–381 (2014). https://doi.org/10.1177/1074248413520344

9. Liu, Y.W., Chen, B., Yang, X., Fugate, J.A., Kalucki, F.A., Futakuchi-Tsuchida, A., et al.: Human embryonic stem cell-derived cardiomyocytes restore function in infarcted hearts of non-human primates. Nat. Biotechnol. **36**(7), 597–605 (2018). https://doi.org/10.1038/nbt.4162

10. Okahara-Narita, J., Umeda, R., Nakamura, S., Mori, T., Noce, T., Torii, R.: Induction of pluripotent stem cells from fetal and adult cynomolgus monkey fibroblasts using four human transcription factors. Primates. **53**(2), 205–213 (2012). https://doi.org/10.1007/s10329-011-0283-1

11. Suemori, H., Tada, T., Torii, R., Hosoi, Y., Kobayashi, K., Imahie, H., et al.: Establishment of embryonic stem cell lines from cynomolgus monkey blastocysts produced by IVF or ICSI. Dev. Dyn. **222**(2), 273–279 (2001). https://doi.org/10.1002/dvdy.1191

12. Ishigaki, H., Shiina, T., Ogasawara, K.: MHC-identical and transgenic cynomolgus macaques for preclinical studies. Inflamm. Regen. **38**, 30 (2018). https://doi.org/10.1186/s41232-018-0088-3

13. Barnard, C.N.: The operation. A human cardiac transplant: an interim report of a successful operation performed at Groote Schuur Hospital, Cape Town. S. Afr. Med. J. **41**(48), 1271–1274 (1967)

14. Chang, D.H., Youn, J.-C., Dilibero, D., Patel, J.K., Kobashigawa, J.A.: Heart transplant immunosuppression strategies at cedars-Sinai medical center. Int. J. Heart Fail. **3**(1), 15–30 (2021)

15. Eisen, H.J., Kobashigawa, J., Keogh, A., Bourge, R., Renlund, D., Mentzer, R., et al.: Three-year results of a randomized, double-blind, controlled trial of mycophenolate mofetil versus azathioprine in cardiac transplant recipients. J. Heart Lung Transplant. **24**(5), 517–525 (2005). https://doi.org/10.1016/j.healun.2005.02.002

16. Kahan, B.D.: Cyclosporine. N. Engl. J. Med. **321**(25), 1725–1738 (1989). https://doi.org/10.1056/nejm198912213212507

17. A comparison of tacrolimus (FK 506) and cyclosporine for immunosuppression in liver transplantation. N. Engl. J. Med. **331**(17), 1110–1115 (1994). https://doi.org/10.1056/nejm199410273311702

18. Penninga, L., Møller, C.H., Gustafsson, F., Steinbrüchel, D.A., Gluud, C.: Tacrolimus versus cyclosporine as primary immunosuppression after heart transplantation: systematic review with meta-analyses and trial sequential analyses of randomised trials. Eur. J. Clin. Pharmacol. **66**(12), 1177–1187 (2010). https://doi.org/10.1007/s00228-010-0902-6

19. Lindenfeld, J., Miller, G.G., Shakar, S.F., Zolty, R., Lowes, B.D., Wolfel, E.E., et al.: Drug therapy in the heart transplant recipient: part II: immunosuppressive drugs. Circulation. **110**(25), 3858–3865 (2004). https://doi.org/10.1161/01.cir.0000150332.42276.69

20. Menasché, P., Vanneaux, V., Hagège, A., Bel, A., Cholley, B., Parouchev, A., et al.: Transplantation of human embryonic stem cell-derived cardiovascular progenitors for severe ischemic left ventricular dysfunction. J. Am. Coll. Cardiol. **71**(4), 429–438 (2018). https://doi.org/10.1016/j.jacc.2017.11.047

21. Cyranoski, D.: 'Reprogrammed' stem cells approved to mend human hearts for the first time. Nature. **557**(7707), 619–620 (2018). https://doi.org/10.1038/d41586-018-05278-8

22. Yamada, N., Okano, T., Sakai, H., Karikusa, F., Sawasaki, Y., Sakurai, Y.: Thermo-responsive polymeric surfaces; control of attachment and detachment of cultured cells. Die

Makromolekulare Chemie Rapid Commun. **11**(11), 571–576 (1990). https://doi.org/10.1002/marc.1990.030111109

23. Romagnuolo, R., Masoudpour, H., Porta-Sánchez, A., Qiang, B., Barry, J., Laskary, A., et al.: Human embryonic stem cell-derived cardiomyocytes regenerate the infarcted pig heart but induce ventricular Tachyarrhythmias. Stem Cell Rep. **12**(5), 967–981 (2019). https://doi.org/10.1016/j.stemcr.2019.04.005

24. Ogasawara, T., Okano, S., Ichimura, H., Kadota, S., Tanaka, Y., Minami, I., et al.: Impact of extracellular matrix on engraftment and maturation of pluripotent stem cell-derived cardiomyocytes in a rat myocardial infarct model. Sci. Rep. **7**(1), 8630 (2017). https://doi.org/10.1038/s41598-017-09217-x

25. Wunderlich, S., Haase, A., Merkert, S., Beier, J., Schwanke, K., Schambach, A., et al.: Induction of pluripotent stem cells from a cynomolgus monkey using a polycistronic simian immunodeficiency virus-based vector, differentiation toward functional cardiomyocytes, and generation of stably expressing reporter lines. Cell. Reprogram. **14**(6), 471–484 (2012). https://doi.org/10.1089/cell.2012.0041

26. Saito, Y., Naruse, T.K., Akari, H., Matano, T., Kimura, A.: Diversity of MHC class I haplotypes in cynomolgus macaques. Immunogenetics. **64**(2), 131–141 (2012). https://doi.org/10.1007/s00251-011-0568-y

27. Ling, F., Wei, L.Q., Wang, T., Wang, H.B., Zhuo, M., Du, H.L., et al.: Characterization of the major histocompatibility complex class II DOB, DPB1, and DQB1 alleles in cynomolgus macaques of Vietnamese origin. Immunogenetics. **63**(3), 155–166 (2011). https://doi.org/10.1007/s00251-010-0498-0

28. Shiba, Y., Gomibuchi, T., Seto, T., Wada, Y., Ichimura, H., Tanaka, Y., et al.: Allogeneic transplantation of iPS cell-derived cardiomyocytes regenerates primate hearts. Nature. **538**(7625), 388–391 (2016). https://doi.org/10.1038/nature19815

29. Fernandes, S., Naumova, A.V., Zhu, W.Z., Laflamme, M.A., Gold, J., Murry, C.E.: Human embryonic stem cell-derived cardiomyocytes engraft but do not alter cardiac remodeling after chronic infarction in rats. J. Mol. Cell. Cardiol. **49**(6), 941–949 (2010). https://doi.org/10.1016/j.yjmcc.2010.09.008

30. Gornalusse, G.G., Hirata, R.K., Funk, S.E., Riolobos, L., Lopes, V.S., Manske, G., et al.: HLA-E-expressing pluripotent stem cells escape allogeneic responses and lysis by NK cells. Nat. Biotechnol. **35**(8), 765–772 (2017). https://doi.org/10.1038/nbt.3860

31. Xu, H., Wang, B., Ono, M., Kagita, A., Fujii, K., Sasakawa, N., et al.: Targeted disruption of HLA genes via CRISPR-Cas9 generates iPSCs with enhanced immune compatibility. Cell Stem Cell. **24**(4), 566–78.e7 (2019). https://doi.org/10.1016/j.stem.2019.02.005

32. Jaiswal, S., Jamieson, C.H., Pang, W.W., Park, C.Y., Chao, M.P., Majeti, R., et al.: CD47 is upregulated on circulating hematopoietic stem cells and leukemia cells to avoid phagocytosis. Cell. **138**(2), 271–285 (2009). https://doi.org/10.1016/j.cell.2009.05.046

33. Deuse, T., Hu, X., Gravina, A., Wang, D., Tediashvili, G., De, C., et al.: Hypoimmunogenic derivatives of induced pluripotent stem cells evade immune rejection in fully immunocompetent allogeneic recipients. Nat. Biotechnol. **37**(3), 252–258 (2019). https://doi.org/10.1038/s41587-019-0016-3

34. Nafady-Hego, H., Li, Y., Ohe, H., Zhao, X., Satoda, N., Sakaguchi, S., et al.: The generation of donor-specific CD4+CD25++CD45RA+ naive regulatory T cells in operationally tolerant patients after pediatric living-donor liver transplantation. Transplantation. **90**(12), 1547–1555 (2010). https://doi.org/10.1097/TP.0b013e3181f9960d

35. Otsuka, R., Wada, H., Tsuji, H., Sasaki, A., Murata, T., Itoh, M., et al.: Efficient generation of thymic epithelium from induced pluripotent stem cells that prolongs allograft survival. Sci. Rep. **10**(1), 224 (2020). https://doi.org/10.1038/s41598-019-57088-1

36. Yoshida, S., Miyagawa, S., Toyofuku, T., Fukushima, S., Kawamura, T., Kawamura, A., et al.: Syngeneic mesenchymal stem cells reduce immune rejection after induced pluripotent stem cell-derived allogeneic cardiomyocyte transplantation. Sci. Rep. **10**(1), 4593 (2020). https://doi.org/10.1038/s41598-020-58126-z

Vascular Regeneration with Induced Pluripotent Stem Cell-Derived Endothelial Cells and Reprogrammed Endothelial Cells

Sangho Lee and Young-sup Yoon

1 Introduction

Extensive research and clinical trials have been performed over the last several decades to find better treatment for patients suffering from cardiovascular diseases (CVD) such as myocardial infarction (MI) and peripheral artery disease (PAD). Owing to these efforts, the treatment options have improved, and the mortality rate has been reduced. However, for chronic, severe conditions, treatment options are still limited For example, critical limb ischemia (CLI), a severe form of PAD, can lead to a 50% risk of amputation [1] with high incidences of second leg loss and mortality within 2 ~ 5 years after the first amputation. Pathophysiologically, the major cause of these clinical entities is loss or dysfunction of blood vessels, of which the major component is endothelial cells (ECs). Thus, to treat these diseases, therapeutic neovascularization, which includes both angiogenesis and vasculogenesis, has been proposed as an attractive approach to re-establish or re-enforce functional vasculature that can support proper blood perfusion and tissue repair [2]. Among several available modalities, cell-based therapy has drawn great interest as it can generate new blood vessels.

S. Lee
Division of Cardiology, Department of Medicine, Emory University School of Medicine, Atlanta, GA, USA
e-mail: sangho.lee@emory.edu

Y.-s. Yoon (✉)
Division of Cardiology, Department of Medicine, Emory University School of Medicine, Atlanta, GA, USA

Severance Biomedical Science Institute, Yonsei University College of Medicine, Seoul, South Korea
e-mail: yyoon5@emory.edu

J. Zhang, V. Serpooshan (eds.), *Advanced Technologies in Cardiovascular Bioengineering*, https://doi.org/10.1007/978-3-030-86140-7_6

For cell therapy, various cell types including mesenchymal stem cells (MSCs), mononuclear cells from bone marrow (BM-MNCs) or peripheral blood (PB-MNCs), or endothelial progenitor cells (EPCs) have been attempted, with limited success. Studies using these adult stem or progenitor cells demonstrated that transdifferentiation of these adult cells to ECs was very limited and the modest therapeutic effects of these adult cells were mainly ascribed to the paracrine effects by soluble factors secreted by these cells [3–6]. More recently, human pluripotent stem cells (PSCs), which include embryonic stem cells (ESCs) and induced pluripotent stem cells (iPSCs), were used to generate ECs. iPSCs were generated by overexpression of pluripotent transcription factors (TFs) in the somatic cells through a process called cellular reprogramming. Despite therapeutic effects and vessel-forming potential of PSC-derived ECs, there are potential concerns for tumorigenicity and aberrant tissue formation and, for ESCs, limitations for clinical use due to ethical issues [7]. Clinical use of PSCs also includes issues such as lengthy and complex differentiation processes, low differentiation efficiency, and difficulties in maintaining the phenotype [8, 9].

More recently, another cell type referred to as induced or directly reprogrammed cells emerged as a new source for cell therapy. These somatic cells (target cells), such as ECs, neurons, cardiomyocytes, or hepatocytes, are generated directly from another type of somatic cells (source cells) via direct reprogramming (or direct conversion or transdifferentiation) using lineage or cell type-specific TFs or micro RNAs (miRNAs), without first de-differentiating into a pluripotent state (reviewed in [10]).

This approach has received notable attention and is considered the third-generation modality for cell therapy and regenerative medicine, as it can reduce not only the time and cost of target cell generation, but also the potential side effects and inefficiency associated with the use of adult or pluripotent stem cells mentioned above. Thus, in this chapter, we will address the progress in the field of iPSC-derived ECs and directly reprogrammed ECs.

2   Endothelial Cells Generated from Human Pluripotent Stem Cells (hPSC-ECs)

2.1   Derivation of Endothelial Cells from Human Pluripotent Stem Cells

To generate ECs, human PSCs (hPSCs), which include both hESCs and hiPSCs, must be differentiated into ECs. Two EC differentiation systems have been developed: three-dimensional (3D) embryoid body (EB)-mediated differentiation and two-dimensional (2D) monolayer-directed differentiation. In EB-mediated differentiation methods, hPSCs are permitted to differentiate spontaneously into a mass (EBs) of multiple lineage cells by suspension culture, and thus the yield of ECs is generally low and inconsistent [11]. Monolayer 2D systems allow more homogeneous exposure of cells to the differentiation medium, generating higher and more consistent yields of ECs. These systems also employ stepwise differentiation methods following the embryonic developmental stages of the vascular system, enabling modular control of the yield at each stage and enhancing the final efficiency of EC generation [12–15].

In 2D systems, PSCs are first differentiated into the mesodermal lineage. Combinations of BMP4, FGF2, and small molecule inhibitors of GSK-3β are generally used [12–14]. In the next stage, mesodermal cells are differentiated into vascular progenitor cells or endothelial cells by adding angiogenic factors and small molecules [12, 13, 15]. Studies demonstrated that in combinations with VEGFA, treatment with a small molecular inhibitor of TGF-β (SB431542) or with forskolin showed higher expression of CDH5 (VE-Cadherin) in hPSC-ECs [13, 15]. Even higher expression of CDH5 was achieved when DLL4, a Notch ligand, was added [12]. The final stage was to select EC lineage cells via sorting with antibodies against EC surface markers. KDR or CD34 was used for selecting progenitor stage ECs [14, 16], and PECAM1 [11], CDH5 [12, 13], or VWF [17] was used for isolating more mature or committed ECs.

The criteria to determine the identity of hPSC-ECs include EC-specific gene and protein expression, cell biological characteristics, and EC generation in vivo, with EC generation in vivo being the ultimate assay to prove the identity and function of hPSC-ECs. However, only a few studies performed in vivo experiments together with histological studies to confirm the genuine incorporation of transplanted hPSC-ECs into vascular walls [12, 17]. Short-term cell survival after in vivo transplantation was also a problem for hPSC-ECs. Recently, transplantation of hPSC-ECs encapsulated with biomaterials was shown to prolong cell survival and to induce sustained incorporation of hPSC-ECs within the vessel wall [12]. Surprisingly, this study showed that many hPSC-ECs were incorporated into the vascular wall after 6 months, suggesting the importance of long-term survival of the transplanted cells for vasculogenesis.

2.2   Remaining Questions and Challenges for hPSC-ECs

hPSC-ECs are a promising source for cell therapy. However, several hurdles must be addressed before clinical application. First, guidelines should be established to verify the characteristics of hPSC-ECs. Many protocols to differentiate hPSC-ECs have been developed and there is no consensus on the criteria to define the EC characteristics to be used for clinical application. Second, scalability of hPSC-ECs is still a concern. For clinical application, a large number of cells are required. To date, no studies have reported large scale generation of hPSC-ECs. More studies are needed to develop methods to proliferate hPSC-ECs in culture while maintaining the phenotype. This culturability or scalability of hPSC-ECs is an important bottleneck even for an autologous approach, not to mention an allogeneic one. Third, the genomic stability of differentiated hPSC-ECs during this expansion should be addressed during the scale-up as well as during the differentiation processes. Fourth, long-term safety of transplanted hPSC-ECs must be verified using animal models. Thus far, in cardiovascular cell therapy, this has not been a serious issue because adult cells were used, which were easily washed away in the host tissues. However, hPSCs have the inborn capacity for tumorigenicity and multi-lineage differentiation [7]. Thus, extensive studies are needed to resolve this concern. Finally, when considering autologous cell therapy, the effects of cardiovascular risk factors on hPSCs and hPSC-EC must be also counted.

Additionally, many biological questions remain in this field. For example, the maturity of hPSC-ECs needs to be addressed together with their biological or therapeutic potency. Type-specific EC generation, particularly arterial versus venous specification was not investigated. There are still gaps between the therapeutic effects and mechanisms, particularly direct vessel-forming or vasculogenic effects of hPSC-ECs. The important applications of stem cells such as disease modeling and drug testing with hPSC-ECs lag relative to other stem cell-derived cells such as hPSC-derived cardiomyocytes in the cardiovascular field.

3 Direct Reprogramming of Somatic Cells into Endothelial Cells

3.1 Reprogramming into Endothelial Cells by Pluripotent Transcription Factors

Early attempts to reprogram somatic cells into ECs used pluripotent transcription factors (TFs). Margariti et al. reported generation of ECs by reprogramming human fibroblasts (HFs) using iPSC-inducing transcription factors OCT4, SOX2, KLF4, and c-MYC (OSKM) with EC differentiation conditions [18]. Overexpression of OSKM factors for 4 days altered the plasticity of HFs to a de-differentiated state, partial-iPSCs (PiPSs), which were then differentiated into ECs under defined culture conditions. These ECs were referred to as PiPS-ECs. In addition, Kurian and colleagues overexpressed OSKM factors as well as other pluripotency inducing factors in human fetal and adult fibroblasts for 8 days to convert their cellular states into the intermediate de-differentiated or partial pluripotent state. These cells were in turn cultured in mesodermal induction medium to generate angioblast-like progenitor cells [19]. Another group showed the generation of functional ECs from human neonatal fibroblasts by using only two pluripotent TFs, OCT4 and KLF4, and defined culture conditions for endothelial differentiation similar to the previous two studies [20]. In this strategy, unlike for iPSCs, somatic cells were not fully reprogrammed to pluripotency, but directed or induced to form endothelial cells by culture conditions at the stage of partial pluripotency, which basically mimics the differentiation process from iPSCs. Therefore, these methods are not true direct reprogramming, although this strategy is still considered one of direct reprogramming methods. While these methods are valuable to expedite the process of generating ECs, there remain concerns about tumorigenesis due to the use of iPSC-inducing factors. In addition, other concerns related to hPSC-derived ECs mentioned above still remain.

3.2   Direct Reprogramming by EC Specific Transcription Factors

To accomplish true "direct reprogramming" of somatic cells into ECs without going through stem-cell-like stages, investigators developed a new approach using lineage specific TFs. Dr. Rafii's group pioneered approaches in this category, demonstrating that expandable ECs could be produced from human amniotic cells (ACs) by overexpression of ETS TFs, ETV2, FLI1, and ERG1 [8]. The successful reprogramming into ECs required transient expression of ETV2 in the early phase of the reprogramming process along with treatment with SB431542, a TGF-β receptor I inhibitor, and continuous expression of FLI1 and ERG1. These reprogrammed cells, called reprogrammed AC vascular endothelial cells (rAC-VECs), formed tubular structures in Matrigel in vitro and in vivo, and showed transcriptome profiles similar to HUVECs. rAC-VECs were also found engrafted into the new blood vessels in regenerating liver, indicating the capability to establish functional blood vessels. The authors acknowledged that ETV2 expression alone was insufficient to turn on all EC genes, and an effort to use human postnatal fibroblasts as source cells was unfruitful. Additionally, due to the origin of the source cells, it is not clear if the source cells were fully differentiated and if there was potential contamination with stem or progenitor cells. Nevertheless, this study was the first to demonstrate the feasibility of direct cellular reprogramming into ECs with lineage specific TFs. Still, this study could only allow an allogeneic approach due to limitations of obtaining the source cells. Two years later, direct reprogramming of mouse skin fibroblasts into ECs was reported by Han et al. using EC/hematopoietic lineage TFs [21]. Using adult skin fibroblasts isolated from Tie2-GFP reporter mice, they found Etv2 alone was insufficient to drive successful reprogramming and all five factors (Etv2, Foxo1, Klf2, Lmo2, and Tal1) were required for efficient reprogramming into functional induced ECs (iECs) (Tie2-GFP+ 4%).

More recently, studies have shown successful reprogramming of postnatal human fibroblasts into ECs [22, 23]. Both groups reported that transduction of single TF ETV2 using a doxycycline-inducible lentiviral system was sufficient to directly convert fibroblasts into functional ECs. Morita et al. demonstrated that PECAM1^{high} cells sorted after 15 days of ETV2 transduction, called ETVECs, displayed EC-like cellular phenotypes in vitro. They also observed that ETVECs implanted within Matrigel improved blood flow in ischemic hindlimbs of NOD SCID mice, and ETVECs were engrafted into blood vessels in vivo [22]. However, during the culture period of more than 50 days, ETVECs maintained high expression of ETV2, which should be minimally expressed in any mammalian postnatal ECs. Therefore, it can be argued that ETVECs are not reprogrammed or induced ECs but rather selected cells displaying the ectopic expression of CD31 (PECAM1), one of the direct targets of ETV2 [24]. Our group more convincingly demonstrated successful direct reprogramming of human postnatal fibroblasts into functional ECs with ETV2 alone [23, 25]. In this study, fibroblasts were converted to reprogrammed ECs (rECs) through two stages. At an early stage of reprogramming, KDR$^+$ cells were sorted at day 7 after ETV2 transduction, and were referred to as early rECs. These cells displayed less mature but enriched EC characteristics in vitro. When early rECs were implanted into ischemic hindlimbs of mice, they were incorporated into the functional vessels, enhanced neovascularization, and repaired tissue ischemia, suggesting their functional and therapeutic potential. When these early rECs were further cultured for another 2 months after transient re-induction of ETV2, they formed late rECs. Late rECs exhibited increased expression of PECAM1 and other EC markers such as CDH5, KDR, and VWF, whereas the level of ETV2 expression was diminished as in mature ECs. These late rECs showed a transcriptome profile similar to mature EC's and produced nitric oxide, indicating maturation of rECs at a later stage through further culture. They also were incorporated into functional vessels in vivo, like early rECs. Notably, treatment with VPA, a histone deacetylase (HDAC) inhibitor, during the ETV2 re-induction enhanced the reprogramming process in this late stage [23]. Thus, these studies clearly demonstrated that ETV2 alone can directly reprogram human fibroblasts into ECs, which have several phenotypes, and can be applied for cell therapy.

3.3 Direct Reprogramming Using Biological Molecules

Another approach to directly reprogram fibroblasts into ECs, which did not involve overexpression of any transcription factors, modulated signaling pathways of innate immunity to trigger cellular reprogramming. A study reported that the activation of Toll-like receptor (TLR3) by poly I:C (polyinosinic: polycytidylic acid) induced global changes in the expression and activity of epigenetic modifiers, increasing cellular plasticity favorable for reprogramming [26]. Based on these results, the same research group demonstrated direct reprogramming of human fibroblasts into ECs by employing the activation of TLR3 with poly I:C and EC differentiation culture medium [27]. The resulting ECs, referred to as iECs, improved blood flow recovery and increased capillary density in mouse hindlimb ischemia. However, vascular incorporation of iECs was not observed, suggesting that the therapeutic potential of iECs is mainly attributed to paracrine effects. Although this method has the clear advantage of a transgene-free strategy, there are shortcomings to be resolved such as lack of vasculogenesis and low reprogramming efficiency.

Interestingly, in this study, in addition to poly I:C, authors used 8-Br-cAMP and SB431542 to increase the reprogramming efficiency [27]. A water-soluble cAMP analog, 8-Br-cAMP is an activator of cAMP-dependent protein kinase and is known to enhance endothelial specification and decrease cell proliferation [28]. For the cell reprogramming study, 8-Br-cAMP was first shown to increase the reprogramming efficiency of human foreskin fibroblasts into iPSCs [29] and Li et al. also used

8-Br-cAMP to increase the efficiency of reprogramming of human neonatal fibroblasts into endothelial cells through partial pluripotency [20]. SB431542 is an inhibitor of a TGF-β/activin/NODAL pathway, which inhibits ALK5, ALK4, ALK7, and TGF-β RI kinases. It has been used to enhance efficiency of iPSC generation [30, 31], direct reprogramming of neurons [32], and cardiomyocyte reprogramming by suppression of pro-fibrotic signals [33]. SB431542 has also been shown to enhance the growth, integrity, and maintenance of ESC-derived endothelial cells [9, 34]. Of note, it was also used in the first study of direct reprogramming into endothelial cells by Ginsberg et al. [8]. One of the components in the reprogramming media in this study is BMP4, a member of TGF-β superfamily. BMP4 has been used for the differentiation of mesodermal lineage cells such as cardiomyocyte and endothelial cells. Thus, BMP4 was used as a component of reprogramming media not only in this study but also in the study involving the partial pluripotency state by Li et al. [20]. Epigenetic modifiers such as DNMT inhibitors or HDAC inhibitors have been widely used for cellular reprogramming to enhance the efficiency. However, only one study used VPA, a HDAC inhibitor, for EC direct reprogramming [23].

3.4   Remaining Questions and Challenges for rECs

Clinically Compatibility

This direct reprogramming approach for EC generation is still at an early stage, leaving many questions and challenges. First of all, clinically acceptable gene delivery methods are required as one major goal of direct reprogramming. However, most of the studies for EC direct reprogramming have used lentiviral [8, 18, 20–23, 35] or retroviral [19] vectors which cause integration of ectopic genes into host genome to overexpress EC-specific TFs. Although the insertional mutation induced by these viral vectors is not problematic for proof-of-concept studies, it is a major hindrance for clinical application. Thus, clinically compatible gene delivery methods must be employed to generate reprogrammed ECs (rECs). For example, Sendai viral vector, a cytoplasmic RNA vector, has been able to deliver pluripotency genes efficiently for iPSC generation without integrating into the host cell genome [36–38]. Also, using this vector, induced cardiomyocytes [39] have been successfully generated indicating that the Sendai viral vector is a promising candidate. An advantage is that the Sendai viral genome diminishes and disappears after a dozen passages, which is beneficial for iPSC generation. However, it is unclear if the viral genome is retained in the reprogrammed cells, which have a fewer than 12 passages in culture. Although Sendai viruses containing pluripotency genes are commercially available and ready for use, Sendai viral vector is not available for cloning of lineage specific TFs such as ETV2. Other candidates are non-integrating DNA viral vectors, such as adenoviruses and adeno-associated viruses (AAVs) which have already been approved and utilized for gene therapy in clinical settings. To date, no study demonstrated the use of these viral vectors for direct reprogramming, possibly due to the lack of sustained expression of ectopic genes in these vectors. However, as addressed in the previous section, the requirement for transient expression of ETV2 fits the nature of these viral vectors, although more precise control of gene expression may be desired. Of note, there have been recent warnings of liver cancer risks by AAV in a number of animal studies of gene therapy [40–44]. Thus, a long-term monitoring for the development of cancer in AAV treated patients is warranted.

Another appealing method to overexpress TFs is to use modified mRNA. In early studies, modified mRNAs were used to replace viral vectors for ectopic overexpression of TFs to generate iPSCs [45–47]. In these studies, modified nucleosides and 5'-cap structure were used for in vitro transcription (IVT) of mRNAs to reduce the innate immunity and to improve the efficiency of translation and the stability of mRNAs. Modified mRNAs have several advantages: they are non-integrative, non-viral, and short-lived, and therefore clinically compatible and safe and able to keep host genome integrity. In addition, compared to viral vectors, using modified mRNAs makes it easier to control gene overexpression temporally and stoichiometrically. Modified mRNAs have also been employed for direct reprogramming of human fibroblasts into myoblast-like cells [48] and hepatocyte-like cells [49]. However, there are several drawbacks to the use of modified mRNAs. Due to the short half-life of mRNAs, repeated transfection is required to maintain expression of TFs. The capability of mRNAs to initiate direct reprogramming seems weak, while a strong driving force is desired. Therefore, there is a great need for the development of efficient and safe transfection reagents or delivery methods which have low cytotoxicity to maintain healthy cells during repeated transfections and to sustain protein expression, such as self-assembled mRNA nanoparticles (mRNA-NPs) [50] and synthetic self-replicative RNAs [51].

The CRISPR system of genome editing has become the state-of-the-art technology. Recently, a mutant form of CRISPR-associated protein 9, Cas9 (dCas9), which has nuclease activity deactivated and fused with the activation domain of TFs has been utilized for endogenous gene activation. Using this CRISPR-dCas9 system, Black et al. were able to activate endogenous Myod1 and successfully converted mouse embryonic fibroblasts into skeletal myocytes [52]. Another study demonstrated that mouse embryonic fibroblasts were reprogrammed into induced neuronal cells by direct activation of endogenous Brn2, Ascl1, and Myt1l genes with CRISPR-dCas9-based transcriptional activator [53]. More recently, the CRISPR-dCas9 system was used to activate Oct4 and Sox2 loci to induce mouse iPSCs [54]. These studies suggest that transcriptional activation and epigenetic remodeling of endogenous master TFs are sufficient for direct reprogramming. However, all these studies involved mouse cells, which are known to be more easily reprogrammed than human cells. Thus, we must determine whether these technologies are sufficient for reprogramming human cells. In addition, the same problems of the clinical compatibility of delivery methods for single guided RNA (sgRNA) and dCAS9-activator gene still remain.

Clinical Applicability

For clinical application of rECs, practicality is another important issue: particularly, the availability of source cells and the reprogramming efficiency needs to be considered. For EC direct reprogramming studies, fetal fibroblasts and neonatal foreskin fibroblasts from human have been used. Interestingly, in a recent study, Hong et al. used human vascular smooth muscle cells (human umbilical artery smooth muscle cells, UASMC) as source cells for direct reprogramming through partial pluripotency [35]. Like vascular smooth muscle cells, blood cells or bone marrow cells are close to ECs in the development lineage, so that they can be more readily reprogrammed. Thus, the plasticity or readiness for reprogramming of source cells, which is contingent on age and lineage, is an important factor in the choice of source cells.

Reprogramming efficiency is another barrier to overcome. Although the underlying concepts of iPSC generation and direct reprogramming are the same, there is a fundamental difference in the process. Reprogramming efficiency for iPSC generation is less important since a large number of iPSCs for the differentiation process can be produced by continuous proliferation from a healthy colony. In contrast, direct reprogramming is the process of generating a type of somatic cells, which have limited proliferation capacity. Thus, reprogramming efficiency is much more important in the direct reprogramming process than in iPSC generation. The efficiency of direct reprogramming of human somatic cells into endothelial cells to date is still less than 20% [19, 22]. To enhance the reprogramming efficiency, the use of additional or more powerful factors including transcription factors, epigenetic modulators, and non-coding RNAs would be desired. It is noteworthy that these additional factors can also further specify subtypes of ECs, arterial, venous and lymphatic ECs, whose existence in heterogenous population of previous studies was speculated by the expression of specific genes of all three subtypes [19, 22, 23, 27]. These results suggest that all three subtypes of EC can be induced with the current strategies of direct reprogramming, and additional measures specific to each subtype could be included for further specification of EC subtypes.

4 Future Perspectives

Although human iPSC-ECs and rECs/iECs are becoming more propitious, there are several impediments to overcome to progress to clinical application One is the scalability of hiPSC-ECs. Another important issue is immunogenicity, which obviously would be less problematic for an autologous approach. However, for commercial development, an allogeneic approach is much favored. Researchers have been seeking to establish a cell bank for iPSCs as well as iPSC-derived cells from various individuals with different human leukocyte antigens (HLA) types. However, this cell bank would hold only major HLA types among the population. Interestingly, recent advances in gene editing technology, such as CRISPR technology, can personalize cell therapy by allowing universal donors or by integrating and customizing HLA proteins for matching [55–57].

Directly reprogrammed or induced ECs also have several problems to be resolved for clinical compatibility and practicality. The use of lenti- or retro-viral vectors for transduction of transcription factors hinders clinical application. This problem can be resolved by employing alternative gene delivery methods or expression systems. The source cells can be diversified to ensure practicality, reliability, and convenience. Low efficiency of reprogramming requires improvement of reprogramming protocols and discovery of new transcription factors and modulators of critical signaling pathways. Deep sequencing of RNAs and bioinformatics will help identify such additional factors that can facilitate and augment reprogramming. Especially, due to the nature of direct reprogramming to induce heterogenous populations, single cell sequencing would be a great method to decipher the mechanisms of reprogramming and provide valuable information for identifying additional factors. These additional factors could also be used to further specify EC subtypes of rECs among the heterogenous population.

After generation of iPSC-ECs and rECs in a clinically compatible manner, we should consider successful application in patients. It is well known that the main hurdle for optimal vascular regeneration by any type of cell therapy is poor engraftment and survival of transplanted cells in the ischemic tissues [58–62]. Studies have shown that injected or transplanted cells stayed at the site of treatment only for a short duration, leading to reduced therapeutic efficacy of the transplanted cells [63–66]. Due to the inflammatory environment and the lack of oxygen and nutrients in the ischemic area, the injected or transplanted cells easily die or are washed away. To prolong cell retention and improve cell survival, several classes of biomaterials including natural and synthetic hydrogels have been successfully employed for encapsulation to serve as carriers for the cells. These biomaterials can provide a matrix to support cell adhesion, and function as a barrier against inflammatory cell infiltration. For example, a chitosan-based hydrogel showed high cell survival and minimal cytotoxicity in vitro, and robust cell retention and enhanced therapeutic effects in a mouse hindlimb model [67]. As briefly addressed in the iPSC-EC section, encapsulation of human iPSC-ECs in an injectable self-assembled nanomatrix gel consisting of a synthetic biomaterial and peptide amphiphile (PA) enhanced cell retention, prolonged cell survival, and manifested authentic vasculogenic effects [12]. These studies suggest that optimization of biomaterial-mediated cell delivery and tissue engineering would be necessary. For example, small molecules modulating immune responses can be incorporated into biomaterials to reduce immune reactions.

5 Conclusion

Cellular reprogramming is a promising strategy for cell therapy. Using pluripotency transcription factors, human iPSCs have been successfully generated, and were further differentiated into various cell types. Great progress made in these processes brings us closer to clinical trials. Using cell type- or lineage-specific transcription factors, researchers have demonstrated that somatic cells can be directly reprogrammed into another type of somatic cells. Among these various cells, ECs are a key player for therapeutic neovascularization to treat advanced ischemic cardiovascular diseases.

Two recent advances, ECs derived from human iPSCs and ECs directly reprogrammed from somatic cells, have emerged as promising options for cell-based revascularization. Particularly, notable progress has been made in the generation of hiPSC-ECs and the exploration of their therapeutic utility and mechanisms. Thus, hiPSC-ECs are getting closer to clinical translation. Also, important discoveries have been made in the reprogramming of human somatic cells into ECs. Overexpression of several transcription factors and small molecules were found to induce the conversion of somatic cells into ECs. Notably, ETV2 alone is capable of inducing reprogramming of human fibroblasts into ECs. These rECs or iECs manifested a wide range of characteristics of ECs from immature to mature phenotypes and have therapeutic effects in ischemic animal models.

Together, these two sources of ECs will advance our understanding of EC biology and can become a new therapeutic option for treating ischemic cardiovascular diseases as summarized in Fig. 1. While not covered in our review, the utility of iPSC-ECs and rECs is not limited to cell therapy but can also be a good platform for disease investigation, drug screening, and precision medicine.

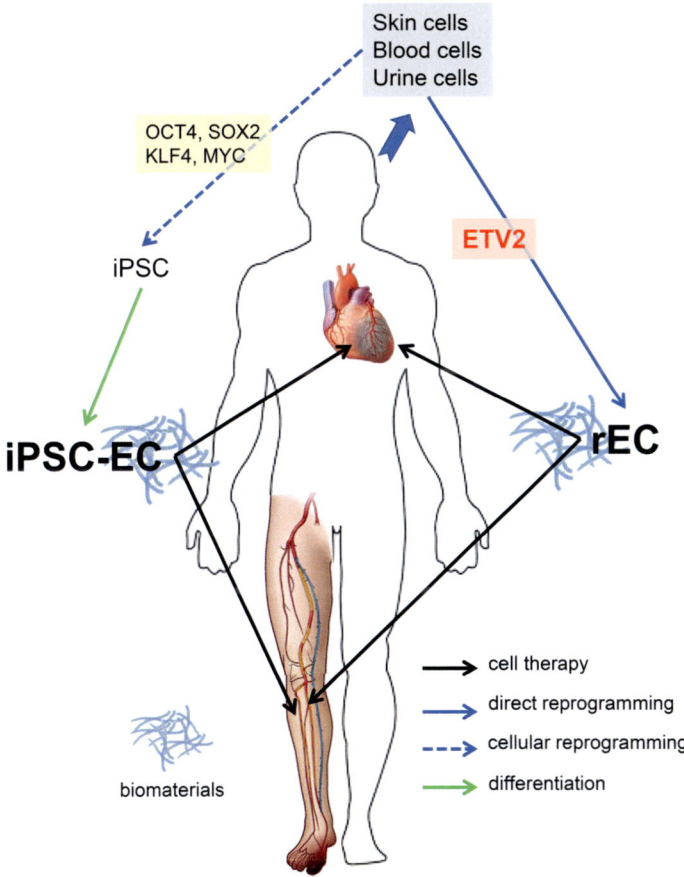

Fig. 1 Therapeutic options using iPSC-EC and rECs

References

1. Datta, D., Atkinson, G.: Amputation, rehabilitation and prosthetic developments. In: Beard, J.D., Gaines, P.A. (eds.) Vascular and Endovascular Surgery, 4th edn, pp. 97–109. Elsevier, Edinburgh (2009)
2. Losordo, D.W., Dimmeler, S.: Therapeutic angiogenesis and vasculogenesis for ischemic disease: part II: cell-based therapies. Circulation. **109**(22), 2692–2697 (2004). https://doi.org/10.1161/01.CIR.0000128596.49339.05
3. Janssens, S., Dubois, C., Bogaert, J., Theunissen, K., Deroose, C., Desmet, W., et al.: Autologous bone marrow-derived stem-cell transfer in patients with ST-segment elevation myocardial infarction: double-blind, randomised controlled trial. Lancet. **367**(9505), 113–121 (2006). https://doi.org/10.1016/S0140-6736(05)67861-0
4. Lee, S., Yoon, Y.S.: Revisiting cardiovascular regeneration with bone marrow-derived angiogenic and vasculogenic cells. Br. J. Pharmacol. **169**(2), 290–303 (2013). https://doi.org/10.1111/j.1476-5381.2012.01857.x
5. Lunde, K., Solheim, S., Aakhus, S., Arnesen, H., Abdelnoor, M., Egeland, T., et al.: Intracoronary injection of mononuclear bone marrow cells in acute myocardial infarction. N. Engl. J. Med. **355**(12), 1199–1209 (2006). https://doi.org/10.1056/NEJMoa055706
6. Ziegelhoeffer, T., Fernandez, B., Kostin, S., Heil, M., Voswinckel, R., Helisch, A., et al.: Bone marrow-derived cells do not incorporate into the adult growing vasculature. Circ. Res. **94**(2), 230–238 (2004). https://doi.org/10.1161/01.RES.0000110419.50982.1C
7. Yamanaka, S.: A fresh look at iPS cells. Cell. **137**(1), 13–17 (2009). https://doi.org/10.1016/j.cell.2009.03.034
8. Ginsberg, M., James, D., Ding, B.S., Nolan, D., Geng, F., Butler, J.M., et al.: Efficient direct reprogramming of mature amniotic cells into endothelial cells by ETS factors and TGFbeta suppression. Cell. **151**(3), 559–575 (2012). https://doi.org/10.1016/j.cell.2012.09.032
9. James, D., Nam, H.S., Seandel, M., Nolan, D., Janovitz, T., Tomishima, M., et al.: Expansion and maintenance of human embryonic stem cell-derived endothelial cells by TGFbeta inhibition is Id1 dependent. Nat. Biotechnol. **28**(2), 161–166 (2010). https://doi.org/10.1038/nbt.1605
10. Xu, J., Du, Y., Deng, H.: Direct lineage reprogramming: strategies, mechanisms, and applications. Cell Stem Cell. **16**(2), 119–134 (2015). https://doi.org/10.1016/j.stem.2015.01.013
11. Levenberg, S., Golub, J.S., Amit, M., Itskovitz-Eldor, J., Langer, R.: Endothelial cells derived from human embryonic stem cells. Proc. Natl. Acad. Sci. U. S. A. **99**(7), 4391–4396 (2002). https://doi.org/10.1073/pnas.032074999
12. Lee, S.J., Sohn, Y.D., Andukuri, A., Kim, S., Byun, J., Han, J.W., et al.: Enhanced therapeutic and long-term dynamic vascularization effects of human pluripotent stem cell-derived endothelial cells encapsulated in a Nanomatrix gel. Circulation. **136**(20), 1939–1954 (2017). https://doi.org/10.1161/CIRCULATIONAHA.116.026329
13. Patsch, C., Challet-Meylan, L., Thoma, E.C., Urich, E., Heckel, T., O'Sullivan, J.F., et al.: Generation of vascular endothelial and smooth muscle cells from human pluripotent stem cells. Nat. Cell Biol. **17**(8), 994–1003 (2015). https://doi.org/10.1038/ncb3205
14. Park, S.W., Jun Koh, Y., Jeon, J., Cho, Y.H., Jang, M.J., Kang, Y., et al.: Efficient differentiation of human pluripotent stem cells into functional CD34+ progenitor cells by combined modulation of the MEK/ERK and BMP4 signaling pathways. Blood. **116**(25), 5762–5772 (2010). https://doi.org/10.1182/blood-2010-04-280719
15. Orlova, V.V., van den Hil, F.E., Petrus-Reurer, S., Drabsch, Y., Ten Dijke, P., Mummery, C.L.: Generation, expansion and functional analysis of endothelial cells and pericytes derived from human pluripotent stem cells. Nat. Protoc. **9**(6), 1514–1531 (2014). https://doi.org/10.1038/nprot.2014.102
16. White, M.P., Rufaihah, A.J., Liu, L., Ghebremariam, Y.T., Ivey, K.N., Cooke, J.P., et al.: Limited gene expression variation in human embryonic stem cell and induced pluripotent

stem cell-derived endothelial cells. Stem Cells. **31**(1), 92–103 (2013). https://doi.org/10.1002/stem.1267

17. Cho, S.W., Moon, S.H., Lee, S.H., Kang, S.W., Kim, J., Lim, J.M., et al.: Improvement of postnatal neovascularization by human embryonic stem cell derived endothelial-like cell transplantation in a mouse model of hindlimb ischemia. Circulation. **116**(21), 2409–2419 (2007). https://doi.org/10.1161/CIRCULATIONAHA.106.687038

18. Margariti, A., Winkler, B., Karamariti, E., Zampetaki, A., Tsai, T.N., Baban, D., et al.: Direct reprogramming of fibroblasts into endothelial cells capable of angiogenesis and reendothelialization in tissue-engineered vessels. Proc. Natl. Acad. Sci. U. S. A. **109**(34), 13793–13798 (2012). https://doi.org/10.1073/pnas.1205526109

19. Kurian, L., Sancho-Martinez, I., Nivet, E., Aguirre, A., Moon, K., Pendaries, C., et al.: Conversion of human fibroblasts to angioblast-like progenitor cells. Nat. Methods. **10**(1), 77–83 (2013). https://doi.org/10.1038/nmeth.2255

20. Li, J., Huang, N.F., Zou, J., Laurent, T.J., Lee, J.C., Okogbaa, J., et al.: Conversion of human fibroblasts to functional endothelial cells by defined factors. Arterioscler. Thromb. Vasc. Biol. **33**(6), 1366–1375 (2013). https://doi.org/10.1161/ATVBAHA.112.301167

21. Han, J.K., Chang, S.H., Cho, H.J., Choi, S.B., Ahn, H.S., Lee, J., et al.: Direct conversion of adult skin fibroblasts to endothelial cells by defined factors. Circulation. **130**(14), 1168–1178 (2014). https://doi.org/10.1161/CIRCULATIONAHA.113.007727

22. Morita, R., Suzuki, M., Kasahara, H., Shimizu, N., Shichita, T., Sekiya, T., et al.: ETS transcription factor ETV2 directly converts human fibroblasts into functional endothelial cells. Proc. Natl. Acad. Sci. U. S. A. **112**(1), 160–165 (2015). https://doi.org/10.1073/pnas.1413234112

23. Lee, S., Park, C., Han, J.W., Kim, J.Y., Cho, K., Kim, E.J., et al.: Direct reprogramming of human dermal fibroblasts into endothelial cells using ER71/ETV2. Circ. Res. **120**(5), 848–861 (2017). https://doi.org/10.1161/CIRCRESAHA.116.309833

24. Kohler, E.E., Wary, K.K., Li, F., Chatterjee, I., Urao, N., Toth, P.T., et al.: Flk1+ and VE-cadherin+ endothelial cells derived from iPSCs recapitulates vascular development during differentiation and display similar angiogenic potential as ESC-derived cells. PLoS One. **8**(12), e85549 (2013). https://doi.org/10.1371/journal.pone.0085549

25. Lee, S., Park, C., Han, J.W., Kim, J.Y., Cho, K., Kim, E.J., et al.: Direct reprogramming of human dermal fibroblasts into endothelial cells using a single transcription factor. Circulation. **130**(Suppl 2), A18205 (2014)

26. Lee, J., Sayed, N., Hunter, A., Au, K.F., Wong, W.H., Mocarski, E.S., et al.: Activation of innate immunity is required for efficient nuclear reprogramming. Cell. **151**(3), 547–558 (2012). https://doi.org/10.1016/j.cell.2012.09.034

27. Sayed, N., Wong, W.T., Ospino, F., Meng, S., Lee, J., Jha, A., et al.: Transdifferentiation of human fibroblasts to endothelial cells: role of innate immunity. Circulation. **131**(3), 300–309 (2015). https://doi.org/10.1161/CIRCULATIONAHA.113.007394

28. Yamamizu, K., Kawasaki, K., Katayama, S., Watabe, T., Yamashita, J.K.: Enhancement of vascular progenitor potential by protein kinase a through dual induction of Flk-1 and Neuropilin-1. Blood. **114**(17), 3707–3716 (2009). https://doi.org/10.1182/blood-2008-12-195750

29. Wang, Y., Adjaye, J.: A cyclic AMP analog, 8-Br-cAMP, enhances the induction of pluripotency in human fibroblast cells. Stem Cell Rev. **7**(2), 331–341 (2011). https://doi.org/10.1007/s12015-010-9209-3

30. Ichida, J.K., Blanchard, J., Lam, K., Son, E.Y., Chung, J.E., Egli, D., et al.: A small-molecule inhibitor of tgf-Beta signaling replaces sox2 in reprogramming by inducing nanog. Cell Stem Cell. **5**(5), 491–503 (2009). https://doi.org/10.1016/j.stem.2009.09.012

31. Lin, T., Ambasudhan, R., Yuan, X., Li, W., Hilcove, S., Abujarour, R., et al.: A chemical platform for improved induction of human iPSCs. Nat. Methods. **6**(11), 805–808 (2009). https://doi.org/10.1038/nmeth.1393

32. Li, X., Zuo, X., Jing, J., Ma, Y., Wang, J., Liu, D., et al.: Small-molecule-driven direct reprogramming of mouse fibroblasts into functional neurons. Cell Stem Cell. **17**(2), 195–203 (2015). https://doi.org/10.1016/j.stem.2015.06.003

33. Zhao, Y., Londono, P., Cao, Y., Sharpe, E.J., Proenza, C., O'Rourke, R., et al.: High-efficiency reprogramming of fibroblasts into cardiomyocytes requires suppression of pro-fibrotic signalling. Nat. Commun. **6**, 8243 (2015). https://doi.org/10.1038/ncomms9243

34. Watabe, T., Nishihara, A., Mishima, K., Yamashita, J., Shimizu, K., Miyazawa, K., et al.: TGF-beta receptor kinase inhibitor enhances growth and integrity of embryonic stem cell-derived endothelial cells. J. Cell Biol. **163**(6), 1303–1311 (2003). https://doi.org/10.1083/jcb.200305147

35. Hong, X., Margariti, A., Le Bras, A., Jacquet, L., Kong, W., Hu, Y., et al.: Transdifferentiated human vascular smooth muscle cells are a new potential cell source for endothelial regeneration. Sci. Rep. **7**(1), 5590 (2017). https://doi.org/10.1038/s41598-017-05665-7

36. Fusaki, N., Ban, H., Nishiyama, A., Saeki, K., Hasegawa, M.: Efficient induction of transgene-free human pluripotent stem cells using a vector based on Sendai virus, an RNA virus that does not integrate into the host genome. Proc. Jpn. Acad. Ser. B Phys. Biol. Sci. **85**(8), 348–362 (2009)

37. Seki, T., Yuasa, S., Oda, M., Egashira, T., Yae, K., Kusumoto, D., et al.: Generation of induced pluripotent stem cells from human terminally differentiated circulating T cells. Cell Stem Cell. **7**(1), 11–14 (2010). https://doi.org/10.1016/j.stem.2010.06.003

38. Ban, H., Nishishita, N., Fusaki, N., Tabata, T., Saeki, K., Shikamura, M., et al.: Efficient generation of transgene-free human induced pluripotent stem cells (iPSCs) by temperature-sensitive Sendai virus vectors. Proc. Natl. Acad. Sci. U. S. A. **108**(34), 14234–14239 (2011). https://doi.org/10.1073/pnas.1103509108

39. Miyamoto, K., Akiyama, M., Tamura, F., Isomi, M., Yamakawa, H., Sadahiro, T., et al.: Direct in vivo reprogramming with Sendai virus vectors improves cardiac function after myocardial infarction. Cell Stem Cell. **22**(1), 91–103. e5 (2018). https://doi.org/10.1016/j.stem.2017.11.010

40. Chandler, R.J., Sands, M.S., Venditti, C.P.: Recombinant adeno-associated viral integration and genotoxicity: insights from animal models. Hum. Gene Ther. **28**(4), 314–322 (2017). https://doi.org/10.1089/hum.2017.009

41. Dalwadi, D.A., Torrens, L., Abril-Fornaguera, J., Pinyol, R., Willoughby, C., Posey, J., et al.: Liver injury increases the incidence of HCC following AAV gene therapy in mice. Mol. Ther. **29**(2), 680–690 (2021). https://doi.org/10.1016/j.ymthe.2020.10.018

42. Li, Y., Miller, C.A., Shea, L.K., Jiang, X., Guzman, M.A., Chandler, R.J., et al.: Enhanced efficacy and increased long-term toxicity of CNS-directed, AAV-based combination therapy for Krabbe disease. Mol. Ther. **29**(2), 691–701 (2021). https://doi.org/10.1016/j.ymthe.2020.12.031

43. Logan, G.J., Dane, A.P., Hallwirth, C.V., Smyth, C.M., Wilkie, E.E., Amaya, A.K., et al.: Identification of liver-specific enhancer-promoter activity in the 3′ untranslated region of the wild-type AAV2 genome. Nat. Genet. **49**(8), 1267–1273 (2017). https://doi.org/10.1038/ng.3893

44. Nguyen, G.N., Everett, J.K., Kafle, S., Roche, A.M., Raymond, H.E., Leiby, J., et al.: A long-term study of AAV gene therapy in dogs with hemophilia A identifies clonal expansions of transduced liver cells. Nat. Biotechnol. **39**(1), 47–55 (2021). https://doi.org/10.1038/s41587-020-0741-7

45. Plews, J.R., Li, J., Jones, M., Moore, H.D., Mason, C., Andrews, P.W., et al.: Activation of pluripotency genes in human fibroblast cells by a novel mRNA based approach. PLoS One. **5**(12), e14397 (2010). https://doi.org/10.1371/journal.pone.0014397

46. Warren L, Manos PD, Ahfeldt T, Loh YH, Li H, Lau F, et al. Highly efficient reprogramming to pluripotency and directed differentiation of human cells with synthetic modified mRNA. Cell Stem Cell 2010;7(5):618–630. S1934-5909(10)00434-0 [pii] https://doi.org/10.1016/j.stem.2010.08.012

47. Yakubov E, Rechavi G, Rozenblatt S, Givol D. Reprogramming of human fibroblasts to pluripotent stem cells using mRNA of four transcription factors. Biochem. Biophys. Res.

Commun. 2010;394(1):189–193. S0006-291X(10)00384-0 [pii] https://doi.org/10.1016/j.bbrc.2010.02.150

48. Preskey, D., Allison, T.F., Jones, M., Mamchaoui, K., Unger, C.: Synthetically modified mRNA for efficient and fast human iPS cell generation and direct transdifferentiation to myoblasts. Biochem. Biophys. Res. Commun. **473**(3), 743–751 (2016). https://doi.org/10.1016/j.bbrc.2015.09.102

49. Simeonov, K.P., Uppal, H.: Direct reprogramming of human fibroblasts to hepatocyte-like cells by synthetic modified mRNAs. PLoS One. **9**(6), e100134 (2014). https://doi.org/10.1371/journal.pone.0100134

50. Kim, H., Park, Y., Lee, J.B.: Self-assembled messenger RNA nanoparticles (mRNA-NPs) for efficient gene expression. Sci. Rep. **5**, 12737 (2015). https://doi.org/10.1038/srep12737

51. Yoshioka, N., Gros, E., Li, H.R., Kumar, S., Deacon, D.C., Maron, C., et al.: Efficient generation of human iPSCs by a synthetic self-replicative RNA. Cell Stem Cell. **13**(2), 246–254 (2013). https://doi.org/10.1016/j.stem.2013.06.001

52. Chakraborty, S., Ji, H., Kabadi, A.M., Gersbach, C.A., Christoforou, N., Leong, K.W.: A CRISPR/Cas9-based system for reprogramming cell lineage specification. Stem Cell Rep. **3**(6), 940–947 (2014). https://doi.org/10.1016/j.stemcr.2014.09.013

53. Black, J.B., Adler, A.F., Wang, H.G., D'Ippolito, A.M., Hutchinson, H.A., Reddy, T.E., et al.: Targeted epigenetic Remodeling of endogenous loci by CRISPR/Cas9-based transcriptional activators directly converts fibroblasts to neuronal cells. Cell Stem Cell. **19**(3), 406–414 (2016). https://doi.org/10.1016/j.stem.2016.07.001

54. Liu, P., Chen, M., Liu, Y., Qi, L.S., Ding, S.: CRISPR-based chromatin Remodeling of the endogenous Oct4 or Sox2 locus enables reprogramming to pluripotency. Cell Stem Cell. **22**(2), 252–261. e4 (2018). https://doi.org/10.1016/j.stem.2017.12.001

55. Lee, J., Sheen, J.H., Lim, O., Lee, Y., Ryu, J., Shin, D., et al.: Abrogation of HLA surface expression using CRISPR/Cas9 genome editing: a step toward universal T cell therapy. Sci. Rep. **10**(1), 17753 (2020). https://doi.org/10.1038/s41598-020-74772-9

56. Xu, H., Wang, B., Ono, M., Kagita, A., Fujii, K., Sasakawa, N., et al.: Targeted disruption of HLA genes via CRISPR-Cas9 generates iPSCs with enhanced immune compatibility. Cell Stem Cell. **24**(4), 566–578. e7 (2019). https://doi.org/10.1016/j.stem.2019.02.005

57. Yin, Y., Reed, E.F., Zhang, Q.: Integrate CRISPR/Cas9 for protein expression of HLA-B*38:68Q via precise gene editing. Sci. Rep. **9**(1), 8067 (2019). https://doi.org/10.1038/s41598-019-44336-7

58. Hoffmann, J., Glassford, A.J., Doyle, T.C., Robbins, R.C., Schrepfer, S., Pelletier, M.P.: Angiogenic effects despite limited cell survival of bone marrow-derived mesenchymal stem cells under ischemia. Thorac. Cardiovasc. Surg. **58**(3), 136–142 (2010). https://doi.org/10.1055/s-0029-1240758

59. Huang, N.F., Niiyama, H., Peter, C., De, A., Natkunam, Y., Fleissner, F., et al.: Embryonic stem cell-derived endothelial cells engraft into the ischemic hindlimb and restore perfusion. Arterioscler. Thromb. Vasc. Biol. **30**(5), 984–991 (2010). https://doi.org/10.1161/ATVBAHA.110.202796

60. Huang, N.F., Okogbaa, J., Babakhanyan, A., Cooke, J.P.: Bioluminescence imaging of stem cell-based therapeutics for vascular regeneration. Theranostics. **2**(4), 346–354 (2012). https://doi.org/10.7150/thno.3694

61. Laurila, J.P., Laatikainen, L., Castellone, M.D., Trivedi, P., Heikkila, J., Hinkkanen, A., et al.: Human embryonic stem cell-derived mesenchymal stromal cell transplantation in a rat hind limb injury model. Cytotherapy. **11**(6), 726–737 (2009). https://doi.org/10.3109/14653240903067299

62. Rufaihah, A.J., Huang, N.F., Jame, S., Lee, J.C., Nguyen, H.N., Byers, B., et al.: Endothelial cells derived from human iPSCS increase capillary density and improve perfusion in a mouse model of peripheral arterial disease. Arterioscler. Thromb. Vasc. Biol. **31**(11), e72–e79 (2011). https://doi.org/10.1161/ATVBAHA.111.230938

63. Wollert, K.C., Drexler, H.: Clinical applications of stem cells for the heart. Circ. Res. **96**(2), 151–163 (2005). https://doi.org/10.1161/01.RES.0000155333.69009.63
64. Menasche, P.: Skeletal myoblasts as a therapeutic agent. Prog. Cardiovasc. Dis. **50**(1), 7–17 (2007). https://doi.org/10.1016/j.pcad.2007.02.002
65. Hofmann, M., Wollert, K.C., Meyer, G.P., Menke, A., Arseniev, L., Hertenstein, B., et al.: Monitoring of bone marrow cell homing into the infarcted human myocardium. Circulation. **111**(17), 2198–2202 (2005). https://doi.org/10.1161/01.CIR.0000163546.27639.AA
66. Qian, H., Yang, Y., Huang, J., Gao, R., Dou, K., Yang, G., et al.: Intracoronary delivery of autologous bone marrow mononuclear cells radiolabeled by 18F-fluoro-deoxy-glucose: tissue distribution and impact on post-infarct swine hearts. J. Cell. Biochem. **102**(1), 64–74 (2007). https://doi.org/10.1002/jcb.21277
67. Lee, S., Valmikinathan, C.M., Byun, J., Kim, S., Lee, G., Mokarram, N., et al.: Enhanced therapeutic neovascularization by CD31-expressing cells and embryonic stem cell-derived endothelial cells engineered with chitosan hydrogel containing VEGF-releasing microtubes. Biomaterials. **63**, 158–167 (2015). https://doi.org/10.1016/j.biomaterials.2015.06.009

The Guinea Pig Model in Cardiac Regeneration Research; Current Tissue Engineering Approaches and Future Directions

Tim Stüdemann and Florian Weinberger

1 Introduction

The human heart contains about three billion cardiomyocytes [1]. Up to 25% of them can be lost during an acute myocardial infarction. Moreover, acute myocardial infarction is regularly followed by a remodeling process involving fibrosis, cardiomyocyte apoptosis, and myocardial hypertrophy [2, 3]. The loss of cardiomyocytes is not compensated for by de-novo cardiomyogenesis and regularly results in a decline of left-ventricular function [4]. This loss of cardiomyocytes leads to chronic heart failure (CHF) a major disease burden that affects millions of patients worldwide [5]. Current pharmacotherapy aims to inhibit neurohumoral activation during CHF but is not able to remuscularize the injured heart [6]. Thus, heart transplantation is the only curative therapeutic option for CHF. However, heart transplantation comes with serious immunological problems [7] and the donor shortage restricts its use to only the most severely sick patients. Currently, about 2500 heart transplantations are conducted in the US per year [8].

Hence, new therapeutic options are urgently needed to counter the growing number of heart failure patients and donor organ shortage [6, 8]. Replacement of lost cardiomyocytes by new cardiomyocytes to improve left-ventricular function and reduce disease burden seems to be a straightforward approach (Fig. 1). Over the past decade, diverse strategies to replenish cardiomyocyte numbers have emerged and step-by-step brought us closer to the goal of regenerating the human heart [9]. For now, cardiac transplantation studies are mostly in late preclinical stages but first clinical trials are underway. In this chapter, we will review tissue engineering

T. Stüdemann · F. Weinberger (✉)
Institute of Experimental Pharmacology and Toxicology, University Medical Center Hamburg-Eppendorf, Hamburg, Germany

German Center for Cardiovascular Research (DZHK), Hamburg/Kiel/Lübeck, Germany
e-mail: f.weinberger@uke.de

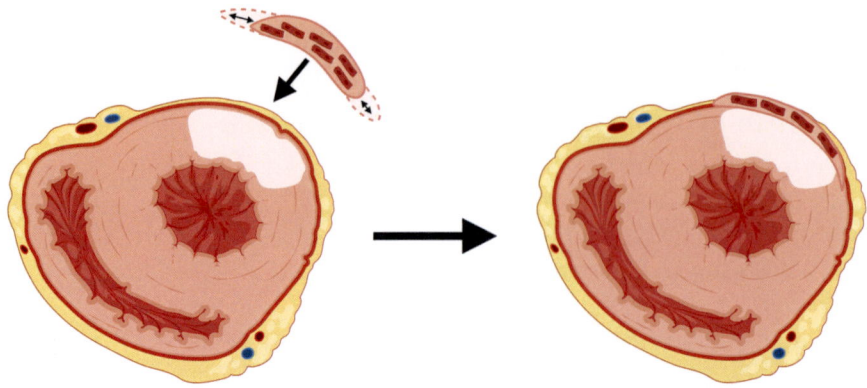

Replenishment of cardiomyocytes

Fig. 1 Conceptual overview of a tissue engineering approach to remuscularize the injured heart. Schematic short axis view of the damaged heart. This strategy aims to replace lost cardiomyocytes in the scar (pale area) by delivering new myocytes that actively contribute to the heart function. Created with biorender

approaches and discuss the value of the cryoinjury guinea pig model for preclinical transplantation studies.

2 Tissue Engineered Heart Repair. Cardiac Tissue Engineering Based Regeneration

2.1 How Much Maturation Do We Need?

Different tissue engineering strategies have been developed for heart repair [10]. Simplified these can be divided into two groups i) tissues that contain no or only non-contractile cells and ii) contractile engineered tissue constructs. The first option gained attention mainly because of its simplicity, which led to early clinical application. These patches can be loaded with growth-factors or cytokines but primarily aim for mechanical stabilization. Preclinical studies demonstrated improvement in left-ventricular function [11, 12] but effects in clinical studies remained modest [13].

Contractile engineered tissue constructs aim for bonafide remuscularization, a hallmark of cardiac regeneration [14]. This strategy intends to deliver cardiomyocytes that actively participate in the contractile function of the heart (Fig. 1). Most tissue engineering approaches use hydrogels that are either based on collagen [15] or fibrin [16, 17] (Table 1). Other strategies include cell sheets, preformed matrices, or 3D-printed constructs (reviewed in [18, 19], Table 1). Cardiac tissue engineering

Table 1 Tissue engineering based strategies for cardiac regeneration

Animal model	Tissue Engineering Approach	Cell Types	Main Findings	Source
Rat, LAD ligation	Bioprinting of spheroids in a patch without biomaterials	hiPSC-CMs, hCFs and HUVECs in spheroids	Reduced scar area Increased vascularization of scar area	[21]
Rat	Collagen hydrogel	PSC-CMs	Grafting achieved	[15]
Guinea pig, cryoinjury	Fibrin hydrogel	hiPSC-CMs	Improved left-ventricular function No arrythmia discovered	[22]
Rat, LAD ligation	Macroporous 3D iron oxide scaffold	hiPSC CMs + human MSCs	Increased LV-function Infarct size decreased	[23]
Rat, LAD ligation	Biomaterial-free net mold or collagen patch	HiPSC-CMs, HCFs, and HUVEC in spheroids	Increased LV-function	[24]
Mouse, LAD ligation	3D printing of gelatin methacrylate scaffold	hciPSC-CMs, -SMCs, and -ECs	LV-function improved Reduced infarct size Increased wall thickness	[25]
Rat, LAD ligation	Patterend collagen matrix to preform microvessels	hESC-CM and-ECs	Improved vascularization Enhanced CMs engraftment	[26]
Rat, LAD ligation	Fibrin hydrogels, CM and microvessel patch > bi-layer formation	hiPSC-CMs, BOECs and PCs	Bi-layer increased wall thickness, regardless of orientation	[27]
Mouse, LAD ligation	Spheroids embedded in fibrin patch	hiPSC-CMs in spheroids	Improved LV-function Infarct size reduced Increased vascularization	[28]
Rat, LAD ligation	Electrospun aligned nanofibers	hiPSC-CMs	Increased vascularization Increased EF and FS	[29]
Rat, LAD ligation	Polyglactin 910 knitted mesh	hiPSC-CMs and human neonatal fibroblasts	Increased LV function	[30]
Rat, LAD ligation	Biomaterial-free bioprinting of spheroids	Cardiospheres from hiPSC-CMs, fibroblasts and ECs	Minimal engrafting	[31]
Rat, LAD ligation	Collagen mesh casting	hiPSC-CMs, -ECs and -MCs	Increased LV function Decreased scar size	[32]
Rat, LAD ligation	Fibronectin or gelatin coating of individual cells and layering	hiPSC-CMs, human CFs, ECs	CMs engraftment and vascularization	[33]

(continued)

Table 1 (continued)

Animal model	Tissue Engineering Approach	Cell Types	Main Findings	Source
Mouse, LAD ligation	Fibrin patch and nanoparticles containing FGF1 and CHIR99021	hiPSC-CMs	Increased graft size and survival of grafted cells Increased LV-function Increased CMs proliferation and vascularization	[34]
Rat, LAD ligation	Collagen and fibrin net with alginate microspheres releasing VEGF	hiPSC-CMs,	Increased LV function Improved vascularization	[35]
Rat, LAD ligation	Fibrin patch with Laminin-221	hiPSC-CMs	Improved LV function Improved vascularization	[36]
Guinea pig, cryoinjury	Fibrin patch	hiPSC-CMs	FAC increased Vascularization increased	[37]
Rat	Fibrin patch	hiPSC-CMs	CMs engraftment No electrical coupling	[38]

improved continuously over the years, but general principles remained. Support systems that allow cardiomyocytes to perform isometric or auxotonic contractions are required for tissue development and result in cardiomyocyte maturation, which can be further stimulated e.g. by dynamic culture conditions [17] or electrical pacing [20].

Strategies to mature cardiomyocytes have been extensively studied because disease modeling and drug toxicity screening aim for systems that most closely mirror the mature adult heart. Irrespective of the differentiation protocol that is used pluripotent stem cell-derived cardiomyocytes (PSC-CMs) display an immature, fetal phenotype (comprehensively reviewed in [39]). Overall, PSC-CMs are smaller compared to adult CMs and have less organized myofibrils [40–42] attributed to a reduced expression of distinct sarcomeric isoforms (e.g. myosin heavy chain, titin, and troponin I isoforms [43–45]). PSC-CMs express the hyperpolarization-activated cyclic nucleotide-gated channel 4, show differences in calcium handling [46–48], and most importantly beat spontaneously. They mainly use glucose as an energy substrate whereas fatty acid utilization [49, 50] and OXPHOS activity, accompanied by increasing mitochondria number and organization, are signs of maturation [50–52]. Moreover, PSC-CMs are still in the cell cycle [42].

Besides tissue engineering, strategies to enhance maturation include long-term culture (up to one year) [53], hormonal supplementation, e.g. with triiodothyronine [54] or optimization of extracellular matrix combinations during cell culture [55]. Metabolism has emerged as an important factor in PSC-CM maturation [39, 56] and fatty acid supplementation matured PSC-CMs morphologically, metabolically, and Electrophysiologically [57, 58]. Whereas enhanced maturation intuitively is favorable for in vitro assays, the situation is more complex for regenerative medicine.

The first transplantation studies of heart tissue were performed in 1978. Bader and Oberpriller minced heart tissue of newts and transferred it back to the resection site, remarkably observing remuscularization of the injury site [59]. Newts and other non-mammals such as zebrafish possess a regenerative capacity also in adult stages of life [60, 61], indicating greater cardiomyocyte plasticity that likely facilitated transplantation success. Mammalian cardiomyocytes, however, withdraw from the cell cycle early after birth [62] and mammals show only a brief window of postnatal heart regeneration [63–65]. With increasing age, the basic turnover of mammalian cardiomyocytes declines and full functional recovery from cardiac injury is rarely observed [1, 4, 66]. Yet, transplantation of syngeneic neonatal cardiomyocytes resulted in structurally engrafted cardiomyocytes [67], showing for the first time that mammalian cardiomyocytes can also be transplanted and thereby allowed a first glimpse into potential regeneration approaches in the future [67, 68]. The use of neonatal cells indicated that the state of maturation might be critical for transplantation success. Indeed, adult mature mammalian cardiomyocytes did not survive transplantation whereas embryonic and neonatal ones engrafted [69].

Based on these rodent experiments less mature cells intuitively are more favorable. Yet, there is some evidence that a certain degree of maturation can enhance the engraftment of PSC-CMs. Transplantation experiments showed that temporal maturation by culturing iPSC-CMs for three weeks resulted in better engraftment compared to younger CMs. However, too much maturation turned out to have a negative impact [70]. Engineered heart tissues (EHTs) in our lab are cultivated for three weeks prior to transplantation. CMs mature within the culture period as shown by a higher MLC2v expression, lower circularity index, a shift to oxidative metabolism, and higher mitochondrial mass [50]. Contraction force increases with time and the action potentials resemble those of the human adult myocardium [37]. Whether this maturation state reflects the optimal time point for transplantation is unknown at present.

How can we assess PSC-CMs maturation? Full metabolic or expression panels are not useful before transplantation. We currently use contractility as a surrogate parameter for cardiomyocyte maturation (and more generally to control the quality of tissue development prior to transplantation) but to evaluate the optimal timepoint for transplantation other strategies could be of help. Chanthra et al. have developed a fluorescent reporter system, based on increased Myomesin2-RFP expression upon maturation. They identified extracellular matrix components to drive maturation [71]. So far, this system was applied only in mouse CMs. Poon et al. found the cell surface marker CD36 (a fatty acid translocase) to be a marker for PSC-CM maturity. Early (day 0) CMs do not express CD36 but the $CD36^+$ proportion increased with prolonged culture to almost 60% after 60 days. $CD36^{high}$ CMs showed more binucleation, sarcomere organization, and higher MLC2V to MLC2A ratio compared to $CD36^{low}$ CMs. Besides, electrical properties and mitochondrial function of $CD36^{high}$ CMs matched a more mature state [72].

2.2 Engineered Heart Tissue Transplantation in Small Animal Models

EHT transplantation studies with EHTs from neonatal rat heart cells laid the foundation for the current work with stem cell-derived cardiomyocytes. Epicardial transplantation of collagen-based EHTs resulted in stable graft formation in the healthy [73] and injured heart [74]. Transplantation after myocardial infarction improved left-ventricular function and provided the first evidence that transplanted constructs can electrically couple to the host myocardium [74]. Since these early studies, EHT fabrication has been modified. In our studies, we use EHTs in which collagen was replaced by fibrinogen, because gelation is faster, resulting in a more homogenous cell distribution [75].

Whereas cardiac tissue engineering for disease modeling and toxicity screens aims towards smaller (and thereby cheaper) constructs that can be used in high-throughput [49], for regenerative medicine the opposite direction is necessary. The concept of remuscularization requires the replacement of many cardiomyocytes and for clinical applications tissue constructs eventually must be upscaled to human scale in size and cell number. Different strategies were followed to achieve this goal: i) the fusion of single engineered tissues prior to transplantation [15], ii) transplantation of several constructs [76], or iii) construction of larger engineered patches [38].

Transplantation of two collagen-based engineered constructs (2.5×10^6 cells) remuscularized the injured heart in an ischemia-reperfusion rat model [77]. However, this was not accompanied by an improvement of left-ventricular function compared to the control group that received irradiated constructs. Evidence for stable engraftment was also provided by a study in which engineered patches (0.5×10^6 cells) were transplanted epicardially onto healthy rat hearts [38]. In our work, we upscaled the classical EHTs which are usually generated in a 24-well format with 1×10^6 cells per construct [78] for a proof-of-concept study. For this study, EHT strips containing 7×10^6 cells (5×10^6 CMs and 2×10^6 endothelial cells) were generated in a 6-well format. Like their smaller counterparts, EHT strips coherently started to beat after 7 to 10 days in culture. CMs aligned along the force lines and showed signs of maturation such as elongation, advanced sarcomeric structure, and increased expression of the ventricular myosin light chain isoform. A guinea pig cryoinjury model was chosen for this study and two strip format EHTs per animal were transplanted one week after myocardial injury. EHT transplantation remuscularized ~12% of the scar (~25% of the left ventricle) and improved left-ventricular function compared to non-contractile tissue constructs containing either endothelial cells or no cells [76].

2.3   Upscaling Towards a Clinical Application

Moving towards clinical application mainly two upscaling ideas seem feasible. The fusion of many smaller constructs or the generation of larger engineered patches. However, it is unknown if simply enlarged constructs fulfill the same requirements as the earlier small versions. Upscaling of cardiac patches at dimensions of 15x15 mm with $2x10^6$ cells and 36x36 mm with $5x10^6$ cells have been described recently [38]. These patches morphologically and physiologically resembled smaller ones and demonstrated advanced electrophysiological properties such as high conduction velocity, indicating the feasibility of this approach. To translate our EHT strips approach further towards a clinical trial, different strategies were combined. We developed mesh structured EHT patches to completely cover the injury site in small animal models but also enabling upscaling for large animal studies and eventually patients. Regular EHT patches have dimensions of 15x25 mm, containing up to $15x10^6$ cells, and are attached to 8 silicone posts, allowing culture under continuous auxotonic stress. The EHT patches remodeled over time and cardiomyocytes aligned along the force lines that were created by the mesh structure. In contrast to the upscaled strip format EHTs, in which the cardiomyocytes mainly aligned along the outer surface, cell distribution was more homogenous, most likely resulting from an improved oxygen and nutrients supply. Electrophysiologically these EHT patches resembled native myocardium. The patch geometry allowed complete scar coverage in a guinea pig model but importantly enabled upscaling to human scale with dimensions of 5x7 cm containing ~450 x 10^6 CMs [37]. In a guinea pig study, this net structure improved graft remuscularized the injured heart (Fig. 2) in a dose-dependent manner. Transplantation of high-dose EHTs improved left-ventricular function, whereas EHTs with lower cell numbers showed no beneficial effect [37]. Although transplanted epicardially onto the scar, the grafted cardiomyocytes came in close proximity to the host myocardium (Fig. 2).

2.4   Cell Composition – Do We Need More Than Just Cardiomyocytes?

A major discrepancy between tissue engineering with neonatal heart cells and stem cell-derived cardiomyocytes is the non-myocyte fraction [79, 80]. Whereas some tissue engineering strategies require the inclusion of a stromal cell population [15, 81] others (including our group) found that the addition of a stromal cell population is not necessary for proper tissue development [80, 82]. From a regenerative medicine perspective, engineered heart constructs need to deliver cardiomyocytes to achieve remuscularization. Non-myocytes do not directly participate in the formation of engrafted myocardium but might enhance cardiomyocyte engraftment and thereby improve functional outcomes. Endothelial cells were included in many studies, including the proof-of-concept study from our group [33, 76, 83, 84].

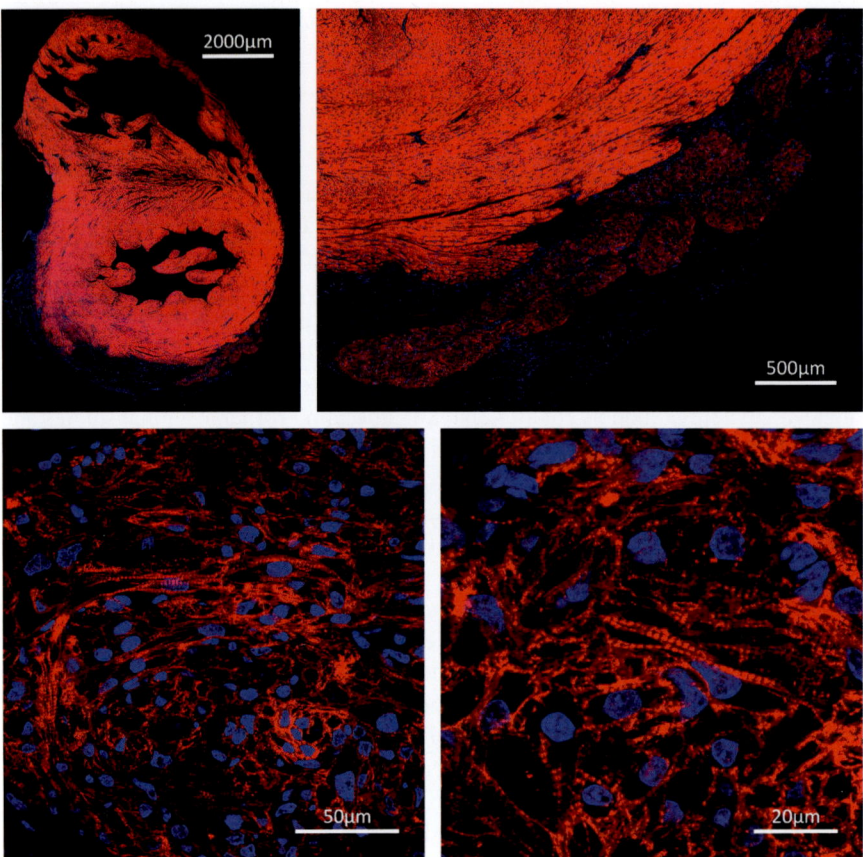

Fig. 2 Short axis view of a guinea pig heart 4 weeks after EHT patch transplantation. Troponin is shown in red and nuclei in blue. Higher magnifications show engrafted human cardiomyocytes

Conceptually this seems reasonable, as there is a post-transplantation phase during which the transplanted CMs are poorly supplied with oxygen and nutrients. Host-derived vessels were observed as early as 12 hours after EHT transplantation in our studies. There is some evidence that the co-transplantation of prevascularized constructs facilitates cardiomyocyte survival [26, 85]. However, the endothelial cells in our study did not survive transplantation and omission of endothelial cells in our recent studies did not negatively impact engraftment [76]. Lately, the co-transplantation of preformed vessels with cardiomyocytes increased vascular density in the grafts and enhanced cardiomyocyte engraftment [86]. Vascularization might become a more pressing issue when transplanting larger and thicker constructs in larger animal models, or eventually patients. Whether the incorporation of preformed vessels represents an opportunity for larger constructs remains to be seen. On the contrary, including only a single cell type will facilitate clinical translation.

3   Guinea Pig Model

3.1   Guniea Pig Model in Cardiac Research

Next to classical models such as mice and rats especially guinea pigs were employed for transplantation studies with hiPSC or hESC-derived cardiomyocytes [76, 87, 88]. The main reason for us to choose the guinea pig model is the closer similarity to human cardiac (electro)physiology. Mice and rats have high heart rates (~600 bpm and ~ 400 bpm respectively) [89] that differ substantially from human CMs with a basic frequency of ~80 bpm and the capability to be paced up to 300 bpm [76] (Table 2). The guinea pig has a heart rate of 200-270 bpm in a conscious state [90, 91]. Additionally, the electrophysiological properties of guinea pig myocardium match human cardiac physiology more closely (Table 2). Action potential (AP) shape and duration are similar to human APs while rats and mice differ in this regard [92–95]. Ca^{2+}-handling in guinea pig myocardium is also more similar to human myocardium [92, 96] than myocardium of mice and rats. These characteristics make it a useful model for predictive pharmacology studies [97–100].

However, there are also drawbacks to the guinea pig model. Ligation (temporary or permanent) of the left ascending coronary artery (LAD) is the classical method to induce cardiac injury in rodents. LAD-ligations have also been described in guinea pigs [101]. Yet, the guinea pigs´ extensive coronary collateralization impedes to reproducibly create myocardial injuries of similar size [102] (Table 2). Hence, we (and other groups) have used a cryoinjury model in which a liquid nitrogen cooled metal stamp is used to injure the left ventricle [76, 88, 103]. While coronary artery anatomy has intraspecies differences (Table 1), the stamp has a consistent size. Therefore, the main advantage of this method is the high reproducibility of the injury size, which allows to better compare remodeling, repair, and regeneration [104, 105]. Disadvantageous is the unphysiological mode of injury. Cryoinjury propagates from the epicardium to the subendocardial layers. In contrast, ischemic injuries predominantly affect the inner layers first. Another difference can either be regarded as an advantage or disadvantage of the model. Ischemic injuries do not uniformly kill all myocytes in the affected region, but myocardial islands in the core

Table 2 Species used for cardiomyocyte transplantation studies

Species	Mouse [89, 92, 94]	Rat [89, 92–94, 102, 106]	Guinea Pig [89–92, 95, 106]
AP shape	No plateau phase	No plateau phase	Similar but no notch
AP duration (ventricular)	< 100 ms	< 100 ms	~ 200 ms
LAD collateralization	LCA diverse	Variety between strains but LAD is present Little collateralization	Extensive
Heart rate	600-800	400	250

zone are spared. In contrast, cryoinjury results in uniform destruction of the ventricular wall with surviving myocardium only endocardially. We have seen spared host myocardium within the scar only in a few (< 5) cases in over 150 hearts analyzed. This can be an advantage because it facilitates the detection of new transplanted CMs but it is disadvantageous for functional coupling. We rather consider this a disadvantage of the model because there are antibodies that unequivocally distinguish between human and guinea pig cells. Nevertheless, the closest similarity to the human electrophysiology outweighs the drawbacks of the guinea pig model in our opinion. In our experience, cryoinjury (metal stamp diameter: 5 mm) results in an injury size of 20-25% (in rare cases up to 40%) of the left-ventricle and a substantial reduction of left ventricular function (basal fractional area change (FAC): ~45%, 7 days after injury: ~35%) [76, 103]. Other studies reported a scar area of ~15% resulting in a reduction of left-ventricular function (fractional shortening (FS)) from ~45% to ~35% [87] seven days after injury or ~ 20% four weeks after injury [88].

3.2   Lessons That Cannot Be Learned from Small Animal Models

Two methods to deliver CMs to the injured heart dominate: transplantation of preformed tissue constructs and direct injection of single-cell suspension into the myocardium. Injected cells reach deep into the scar while EHTs are applied epicardially, covering the injured myocardium. Our proof-of-concept study provided evidence that the transplanted cells can electrically couple to the host myocardium [76]. However, two other studies did not find any evidence for electrical coupling after epicardial transplantation [38, 107]. In our experience, grafted EHTs are physically separated from the host myocardium by a fibrous layer but close contacts are regularly seen (Fig. 2). Coupling of immature spontaneously beating cardiomyocytes to the host myocardium intuitively could induce arrhythmias. Yet, guinea pig studies not only demonstrated safety but hinted towards a reduced arrhythmogenic burden after CM transplantation [22, 87]. Eventually, though large animal studies revealed that CM transplantation can induce ventricular tachycardia [108–110], which was not recognized in small animal models, most likely because of the small heart size and the fast heart frequency. Whether the same holds true for tissue engineering strategies remains to be seen.

The cardiac regeneration field moves rapidly towards first clinical trials (e.g. BioVAT-HF, HEAL-CHF [111]) raising the question of whether a small animal model is still of use. Problems towards clinical application are mainly the automaticity of the immature cells that can cause ventricular tachycardia and the immunogenicity of the transplanted cells. Immunosuppression comes with clear risks, e.g. higher risk for melanomas and infections but also adverse drug effects, such as nephrotoxicity in calcineurin-inhibitors [7]. Hypoimmunogenic cell lines have been developed and might circumvent this problem [112–114]. However, none of the hurdles that remain towards clinical application will be addressable in the guinea pig model because it is i) of low predictive value for the occurrence of ventricular arrhythmias, ii) as a xenogeneic model, requires clinically not tolerable immuno-suppressive regiments and is not suited to assess transplant immunology.

3.3   Future Use for the Guinea Pig Model?

Although certain limitations of the guinea pig model are recognized, we see the future of the model in basic science studies. These include i) evaluation of the mode of action, ii) assessment of strategies to improve transplantation success.

Evaluation of the Mode of Action

Even though it is an intuitive assumption that transplanted cardiomyocytes actively contribute to the heart's function, this has not been directly demonstrated and the exact mechanism of action of EHT transplantation has not been defined yet.

There are possibly three major mechanisms by which cell transplantation can improve cardiac function after injury: i) replenishment of the scar with viable car-diomyocytes, ii) paracrine mechanisms such as enhanced vascularization, antifi-brotic effects, and reduction of cardiomyocyte apoptosis in the border zone and iii) mechanical stabilization. So far there is only indirect evidence that the beneficial effect depends on the active participation of the transplanted cardiomyocytes. This hypothesis is supported by results that showed that only the transplantation of con-tractile EHTs improved left-ventricular function, whereas EHTs that contained no or only non-contractile endothelial cells did not improve left ventricular function [76, 87]. Further support comes from studies in which larger grafts correlated to a greater improvement of left-ventricular function [81, 86]. In these studies, co-transplantation of either CMs and preformed vessels [86] or CMs and epicardial cells [81] resulted in larger grafts than the transplantation of CMs only, which even-tually resulted in a greater improvement of left-ventricular function.

Paracrine mechanisms are supported by studies that demonstrated an improve-ment in left ventricular function with only minimal or even no cell engraftment [30, 115]. There is also some evidence that transplanted constructs stabilize the injured heart thereby improving left-ventricular function. However, the interpretation of these results is difficult because they are derived from different animal models and first clinical studies [11–13]. A better understanding of the exact mode of action is not only desirable from an academic point of view but also to further improve trans-plantation success and translation.

Improving Transplantation Success

This brings us to the second field in which we foresee a future for the guinea pig model. To substantially support the heart's function, a large number of CMs have to engraft [9]. Employing small animal models we can evaluate strategies to boost EHTs transplantation, e.g. there is growing evidence that CM proliferation participates in the final graft development [76, 110, 116] and can be stimulated to increase graft size and improve transplantation success. First attempts used genetically modified PSCs to stimulate CM proliferation and demonstrated that graft size can be enlarged with this strategy [117, 118]. Additionally, engineered tissue constructs can be loaded with small molecular substances that will be released over time to exert a prolonged effect after transplantation. This strategy has successfully been applied by targeting Notch-signaling [119], the Wnt pathway [34] but also to stimulate angiogenesis [35]. Stimulation of CM proliferation resulted in larger grafts and greater improvement in cardiac function in each of these studies. Several strategies to induce cardiomyocyte proliferation have been described (reviewed in [120]) that prospectively can be used to enhance transplantation results. Additionally, as mentioned above CM maturation is crucial for transplantation success and the guinea pig model can help define an optimal maturation state for engraftment.

Eventually, there is one more question that can be addressed in small animal models. So far, most cell transplantation studies focus on (sub)acute injury models but in most patients with advanced heart failure, the disease has progressed over many years. Myocardial scars structurally change over time [121] and the chronically injured heart is a more difficult target for regenerative strategies [88]. The exact mechanisms are not well understood but fibroblast populations, extracellular matrix protein composition, and inflammatory cells most likely are crucial factors and the guinea pig model is suitable to evaluate strategies to enhance transplantation success for the chronically injured heart.

4   Conclusion

EHT transplantation, and more generally pluripotent stem cell-based strategies, are currently translated to first clinical trials, which will teach us more about the feasibility and safety of this approach. Eventually, these trials will hopefully answer the question of whether heart failure patients can benefit from EHT transplantation. At the same time, we also see a need for further studies that address mechanistic questions and aim to improve this therapeutic strategy.

References

1. Bergmann, O., Zdunek, S., Felker, A., Salehpour, M., Alkass, K., Bernard, S., et al.: Dynamics of cell generation and turnover in the human heart. Cell. **161**(7), 1566–1575 (2015). https://doi.org/10.1016/j.cell.2015.05.026
2. Nakamura, M., Sadoshima, J.: Mechanisms of physiological and pathological cardiac hypertrophy. Nat. Rev. Cardiol. **15**(7), 387–407 (2018). https://doi.org/10.1038/s41569-018-0007-y
3. Rockey, D.C., Bell, P.D., Hill, J.A.: Fibrosis--a common pathway to organ injury and failure. N. Engl. J. Med. **373**(1), 96 (2015). https://doi.org/10.1056/NEJMc1504848
4. Eschenhagen, T., Bolli, R., Braun, T., Field, L.J., Fleischmann, B.K., Frisen, J., et al.: Cardiomyocyte regeneration: a consensus statement. Circulation. **136**(7), 680–686 (2017). https://doi.org/10.1161/CIRCULATIONAHA.117.029343
5. Ponikowski, P., Anker, S.D., AlHabib, K.F., Cowie, M.R., Force, T.L., Hu, S., et al.: Heart failure: preventing disease and death worldwide. ESC Heart Failure. **1**, 4–25 (2014). https://doi.org/10.1002/ehf2.12005
6. Murphy, S.P., Ibrahim, N.E., Januzzi Jr., J.L.: Heart failure with reduced ejection fraction: a review. JAMA. **324**(5), 488–504 (2020). https://doi.org/10.1001/jama.2020.10262
7. Soderlund, C., Radegran, G.: Immunosuppressive therapies after heart transplantation--the balance between under- and over-immunosuppression. Transplant. Rev. (Orlando). **29**(3), 181–189 (2015). https://doi.org/10.1016/j.trre.2015.02.005
8. Miller, L., Birks, E., Guglin, M., Lamba, H., Frazier, O.H.: Use of ventricular assist devices and heart transplantation for advanced heart failure. Circ. Res. **124**(11), 1658–1678 (2019). https://doi.org/10.1161/CIRCRESAHA.119.313574
9. Weinberger, F., Eschenhagen, T.: Cardiac regeneration: new Hope for an old dream. Annu. Rev. Physiol. **83**:annurev-physiol-031120-103629 (2021). https://doi.org/10.1146/annurev-physiol-031120-103629
10. Weinberger, F., Mannhardt, I., Eschenhagen, T.: Engineering cardiac muscle tissue: a maturating Field of research. Circ. Res. **120**, 1487–1500 (2017). https://doi.org/10.1161/CIRCRESAHA.117.310738
11. Sabbah, H.N., Wang, M., Gupta, R.C., Rastogi, S., Ilsar, I., Sabbah, M.S., et al.: Augmentation of left ventricular wall thickness with alginate hydrogel implants improves left ventricular function and prevents progressive remodeling in dogs with chronic heart failure. JACC: Heart Failure. **1**, 252–258 (2013). https://doi.org/10.1016/j.jchf.2013.02.006
12. Christman, K.L., Vardanian, A.J., Fang, Q., Sievers, R.E., Fok, H.H., Lee, R.J.: Injectable fibrin scaffold improves cell transplant survival, reduces infarct expansion, and induces neovasculature formation in ischemic myocardium. J. Am. Coll. Cardiol. **44**(3), 654–660 (2004). https://doi.org/10.1016/j.jacc.2004.04.040
13. Anker, S.D., Coats, A.J., Cristian, G., Dragomir, D., Pusineri, E., Piredda, M., et al.: A prospective comparison of alginate-hydrogel with standard medical therapy to determine impact on functional capacity and clinical outcomes in patients with advanced heart failure (AUGMENT-HF trial). Eur. Heart J. **36**(34), 2297–2309 (2015). https://doi.org/10.1093/eurheartj/ehv259
14. Bertero, A., Murry, C.E.: Hallmarks of cardiac regeneration. Nat. Rev. Cardiol. **15**(10), 579–580 (2018). https://doi.org/10.1038/s41569-018-0079-8
15. Tiburcy, M., Hudson, J.E., Balfanz, P., Schlick, S., Meyer, T., Chang Liao, M.-L., et al.: Defined engineered human myocardium with advanced maturation for applications in heart failure modeling and repair. Circulation. **135**, 1832–1847 (2017). https://doi.org/10.1161/CIRCULATIONAHA.116.024145
16. Schaaf, S., Shibamiya, A., Mewe, M., Eder, A., Stöhr, A., Hirt, M.N., et al.: Human engineered heart tissue as a versatile tool in basic research and preclinical toxicology. PLoS One. **6**, e26397 (2011). https://doi.org/10.1371/journal.pone.0026397

17. Jackman, C.P., Carlson, A.L., Bursac, N.: Dynamic culture yields engineered myocardium with near-adult functional output. Biomaterials. **111**, 66–79 (2016). https://doi.org/10.1016/j. biomaterials.2016.09.024

18. Alonzo, M., AnilKumar, S., Roman, B., Tasnim, N., Joddar, B.: 3D bioprinting of cardiac tissue and cardiac stem cell therapy. Transl. Res. (2019). https://doi.org/10.1016/j. trsl.2019.04.004

19. Pomeroy JE, Helfer A, Bursac N. Biomaterializing the promise of cardiac tissue engineering. Biotechnol. Adv. 2020;42:0-1. doi: https://doi.org/10.1016/j.biotechadv.2019.02.009

20. Hirt, M.N., Boeddinghaus, J., Mitchell, A., Schaaf, S., Börnchen, C., Müller, C., et al.: Functional improvement and maturation of rat and human engineered heart tissue by chronic electrical stimulation. J. Mol. Cell. Cardiol. **74**, 151–161 (2014). https://doi.org/10.1016/j. yjmcc.2014.05.009

21. Yeung, E., Fukunishi, T., Bai, Y., Bedja, D., Pitaktong, I., Mattson, G., et al.: Cardiac regeneration using human-induced pluripotent stem cell-derived biomaterial-free 3D-bioprinted cardiac patch in vivo. J. Tissue Eng. Regen. Med. **13**, 2031–2039 (2019). https://doi. org/10.1002/term.2954

22. Pecha, S., Yorgan, K., Röhl, M., Geertz, B., Hansen, A., Weinberger, F., et al.: Human iPS cell-derived engineered heart tissue does not affect ventricular arrhythmias in a Guinea pig cryo-injury model. Sci. Rep. **9**, 9831 (2019). https://doi.org/10.1038/s41598-019-46409-z

23. Yang, H., Wei, L., Liu, C., Zhong, W., Li, B., Chen, Y., et al.: Engineering human ventricular heart tissue based on macroporous iron oxide scaffolds. Acta Biomater. **88**, 540–553 (2019). https://doi.org/10.1016/j.actbio.2019.02.024

24. Yang, B., Lui, C., Yeung, E., Matsushita, H., Jeyaram, A., Pitaktong, I., et al.: A net Mold-based method of biomaterial-free three-dimensional cardiac tissue creation. Tissue Eng. Part C: Methods. **25**, 243–252 (2019). https://doi.org/10.1089/ten.tec.2019.0003

25. Gao, L., Kupfer, M.E., Jung, J.P., Yang, L., Zhang, P., Da Sie, Y., et al.: Myocardial tissue engineering with cells derived from human-induced pluripotent stem cells and a native-like, high-resolution, 3-dimensionally printed scaffold. Circ. Res. **120**, 1318–1325 (2017). https:// doi.org/10.1161/CIRCRESAHA.116.310277

26. Redd, M.A., Zeinstra, N., Qin, W., Wei, W., Martinson, A., Wang, Y., et al.: Patterned human microvascular grafts enable rapid vascularization and increase perfusion in infarcted rat hearts. Nat. Commun. **10**(1), 584 (2019). https://doi.org/10.1038/s41467-019-08388-7

27. Schaefer, J.A., Guzman, P.A., Riemenschneider, S.B., Kamp, T.J., Tranquillo, R.T.: A cardiac patch from aligned microvessel and cardiomyocyte patches HHS public access. J. Tissue Eng. Regen. Med. **12**, 546–556 (2018). https://doi.org/10.1002/term.2568

28. Mattapally, S., Zhu, W., Fast, V.G., Gao, L., Worley, C., Kannappan, R., et al.: Spheroids of cardiomyocytes derived from human-induced pluripotent stem cells improve recovery from myocardial injury in mice. Am. J. Physiol. Heart Circ. Physiol. **315**, 327–339 (2018). https:// doi.org/10.1152/ajpheart.00688.2017.-The

29. Li, J., Minami, I., Shiozaki, M., Yu, L., Yajima, S., Miyagawa, S., et al.: Human pluripotent stem cell-derived cardiac tissue-like constructs for repairing the infarcted myocardium. Stem Cell Reports. **9**, 1546–1559 (2017). https://doi.org/10.1016/j.stemcr.2017.09.007

30. Lancaster, J.J., Sanchez, P., Repetti, G.G., Juneman, E., Pandey, A.C., Chinyere, I.R., et al.: Human induced pluripotent stem cell-derived cardiomyocyte patch in rats with heart failure. Ann. Thorac. Surg. **108**(4), 1169–1177 (2019). https://doi.org/10.1016/j. athoracsur.2019.03.099

31. Ong, C.S., Fukunishi, T., Zhang, H., Huang, C.Y., Nashed, A., Blazeski, A., et al.: Biomaterial-free three-dimensional bioprinting of cardiac tissue using human induced pluripotent stem cell derived cardiomyocytes. Sci. Rep. **7**, 2–12 (2017). https://doi.org/10.1038/ s41598-017-05018-4

32. Nakane, T., Masumoto, H., Tinney, J.P., Yuan, F., Kowalski, W.J., Ye, F., et al.: Impact of cell composition and geometry on human induced pluripotent stem cells-derived engineered cardiac tissue. Sci. Rep. **7**, 1–13 (2017). https://doi.org/10.1038/srep45641

33. Narita, H., Shima, F., Yokoyama, J., Miyagawa, S., Tsukamoto, Y., Takamura, Y., et al.: Engraftment and morphological development of vascularized human iPS cell-derived 3D-cardiomyocyte tissue after xenotransplantation. Sci. Rep. **7**, 1–9 (2017). https://doi.org/10.1038/s41598-017-14053-0

34. Fan, C., Tang, Y., Zhao, M., Lou, X., Pretorius, D., Menasche, P., et al.: CHIR99021 and fibroblast growth factor 1 enhance the regenerative potency of human cardiac muscle patch after myocardial infarction in mice. J. Mol. Cell. Cardiol. **141**, 1–10 (2020). https://doi.org/10.1016/j.yjmcc.2020.03.003

35. Munarin, F., Kant, R.J., Rupert, C.E., Khoo, A., Coulombe, K.L.K.: Engineered human myocardium with local release of angiogenic proteins improves vascularization and cardiac function in injured rat hearts. Biomaterials. **251**, 120033 (2020). https://doi.org/10.1016/j.biomaterials.2020.120033

36. Samura, T., Miyagawa, S., Kawamura, T., Fukushima, S., Yokoyama, J.Y., Takeda, M., et al.: Laminin-221 enhances therapeutic effects of human-induced pluripotent stem cell-derived 3-dimensional engineered cardiac tissue transplantation in a rat ischemic cardiomyopathy model. J. Am. Heart Assoc. **9**, e015841 (2020). https://doi.org/10.1161/JAHA.119.015841

37. Querdel, E., Reinsch, M., Castro, L., Köse, D., Bähr, A., Reich, S., et al.: Human engineered heart tissue patches Remuscularize the injured heart in a dose-dependent manner. **0**(0). https://doi.org/10.1161/CIRCULATIONAHA.120.047904

38. Shadrin, I.Y., Allen, B.W., Qian, Y., Jackman, C.P., Carlson, A.L., Juhas, M.E., et al.: Cardiopatch platform enables maturation and scale-up of human pluripotent stem cell-derived engineered heart tissues. Nat. Commun. **8**, 1825 (2017). https://doi.org/10.1038/s41467-017-01946-x

39. Karbassi, E., Fenix, A., Marchiano, S., Muraoka, N., Nakamura, K., Yang, X., et al.: Cardiomyocyte maturation: advances in knowledge and implications for regenerative medicine. Nat. Rev. Cardiol. **17**(6), 341–359 (2020). https://doi.org/10.1038/s41569-019-0331-x

40. Zhang, J., Wilson, G.F., Soerens, A.G., Koonce, C.H., Yu, J., Palecek, S.P., et al.: Functional cardiomyocytes derived from human induced pluripotent stem cells. Circ. Res. **104**(4), e30–e41 (2009). https://doi.org/10.1161/CIRCRESAHA.108.192237

41. Mummery, C., Ward-van Oostwaard, D., Doevendans, P., Spijker, R., Van den Brink, S., Hassink, R., et al.: Differentiation of human embryonic stem cells to cardiomyocytes: role of coculture with visceral endoderm-like cells. Circulation. **107**, 2733–2740 (2003). https://doi.org/10.1161/01.cir.0000068356.38592.68

42. Snir, M., Kehat, I., Gepstein, A., Coleman, R., Itskovitz-Eldor, J., Livne, E., et al.: Assessment of the ultrastructural and proliferative properties of human embryonic stem cell-derived cardiomyocytes. Am. J. Physiol. Heart Circ. Physiol. **285** (2003). https://doi.org/10.1152/ajpheart.00020.2003

43. Lundy, S.D., Zhu, W.Z., Regnier, M., Laflamme, M.A.: Structural and functional maturation of cardiomyocytes derived from human pluripotent stem cells. Stem Cells Dev. **22**(14), 1991–2002 (2013). https://doi.org/10.1089/scd.2012.0490

44. Bedada, F.B., Chan, S.S., Metzger, S.K., Zhang, L., Zhang, J., Garry, D.J., et al.: Acquisition of a quantitative, stoichiometrically conserved ratiometric marker of maturation status in stem cell-derived cardiac myocytes. Stem Cell Rep. **3**(4), 594–605 (2014). https://doi.org/10.1016/j.stemcr.2014.07.012

45. Iorga, B., Schwanke, K., Weber, N., Wendland, M., Greten, S., Piep, B., et al.: Differences in contractile function of myofibrils within human embryonic stem cell-derived cardiomyocytes vs. adult ventricular myofibrils are related to distinct sarcomeric protein isoforms. Front. Physiol. **8** (2018). https://doi.org/10.3389/fphys.2017.01111

46. Ma, J., Guo, L., Fiene, S.J., Anson, B.D., Thomson, J.A., Kamp, T.J., et al.: High purity human-induced pluripotent stem cell-derived cardiomyocytes: electrophysiological properties of action potentials and ionic currents. Am. J. Physiol. Heart Circ. Physiol. **301**, H2006 (2011). https://doi.org/10.1152/ajpheart.00694.2011

47. Satin, J., Kehat, I., Caspi, O., Huber, I., Arbel, G., Itzhaki, I., et al.: Mechanism of spontaneous excitability in human embryonic stem cell derived cardiomyocytes. J. Physiol. **559**, 479–496 (2004). https://doi.org/10.1113/jphysiol.2004.068213

48. Weisbrod, D., Peretz, A., Ziskind, A., Menaker, N., Oz, S., Barad, L., et al.: SK4 Ca2+ activated K+ channel is a critical player in cardiac pacemaker derived from human embryonic stem cells. Proc. Natl. Acad. Sci. U. S. A. **110** (2013). https://doi.org/10.1073/pnas.1221022110

49. Mills, R.J., Titmarsh, D.M., Koenig, X., Parker, B.L., Ryall, J.G., Quaife-Ryan, G.A., et al.: Functional screening in human cardiac organoids reveals a metabolic mechanism for cardiomyocyte cell cycle arrest. Proc. Natl. Acad. Sci. U. S. A. **114**, E8372–E8E81 (2017). https://doi.org/10.1073/pnas.1707316114

50. Ulmer, B.M., Stoehr, A., Schulze, M.L., Patel, S., Gucek, M., Mannhardt, I., et al.: Contractile work contributes to maturation of energy metabolism in hiPSC-derived cardiomyocytes. Stem Cell Reports. **10**, 834–847 (2018). https://doi.org/10.1016/j.stemcr.2018.01.039

51. Correia, C., Koshkin, A., Duarte, P., Hu, D., Teixeira, A., Domian, I., et al.: Distinct carbon sources affect structural and functional maturation of cardiomyocytes derived from human pluripotent stem cells. Sci. Rep. **7** (2017). https://doi.org/10.1038/s41598-017-08713-4

52. Dai, D.F., Danoviz, M.E., Wiczer, B., Laflamme, M.A., Tian, R.: Mitochondrial maturation in human pluripotent stem cell derived cardiomyocytes. Stem Cells Int. **2017**, 5153625 (2017). https://doi.org/10.1155/2017/5153625

53. Kamakura, T., Makiyama, T., Sasaki, K., Yoshida, Y., Wuriyanghai, Y., Chen, J., et al.: Ultrastructural maturation of human-induced pluripotent stem cell-derived cardiomyocytes in a long-term culture. Circ. J. **77**, 1307–1314 (2013). https://doi.org/10.1253/circj.CJ-12-0987

54. Yang, X., Rodriguez, M., Pabon, L., Fischer, K.A., Reinecke, H., Regnier, M., et al.: Triiodo-l-thyronine promotes the maturation of human cardiomyocytes-derived from induced pluripotent stem cells. J. Mol. Cell. Cardiol. **72**, 296–304 (2014). https://doi.org/10.1016/j.yjmcc.2014.04.005

55. Herron, T.J., Rocha, A.M., Campbell, K.F., Ponce-Balbuena, D., Willis, B.C., Guerrero-Serna, G., et al.: Extracellular matrix-mediated maturation of human pluripotent stem cell-derived cardiac monolayer structure and electrophysiological function. Circ. Arrhythm. Electrophysiol. **9**(4), e003638 (2016). https://doi.org/10.1161/CIRCEP.113.003638

56. Marchianò, S., Bertero, A., Murry, C.E.: Learn from your elders: developmental biology lessons to guide maturation of stem cell-derived cardiomyocytes. Pediatr. Cardiol. **1**, 3. https://doi.org/10.1007/s00246-019-02165-5

57. Horikoshi, Y., Yan, Y., Terashvili, M., Wells, C., Horikoshi, H., Fujita, S., et al.: Fatty acid-treated induced pluripotent stem cell-derived human cardiomyocytes exhibit adult cardiomyocyte-like energy metabolism phenotypes. Cell. **8**, 1095 (2019). https://doi.org/10.3390/cells8091095

58. Yang, X., Rodriguez, M.L., Leonard, A., Sun, L., Fischer, K.A., Wang, Y., et al.: Fatty acids enhance the maturation of cardiomyocytes derived from human pluripotent stem cells. Stem Cell Reports. **13**(4), 657–668 (2019). https://doi.org/10.1016/j.stemcr.2019.08.013

59. Bader, D., Oberpriller, J.O.: Repair and reorganization of minced cardiac muscle in the adult newt (Notophthalmus viridescens). J. Morphol. **155**, 349–357 (1978). https://doi.org/10.1002/jmor.1051550307

60. Oberpriller, J.O., Oberpriller, J.C.: Response of the adult newt ventricle to injury. J. Exp. Zool. **187**, 249–259 (1974). https://doi.org/10.1002/jez.1401870208

61. Poss, K.D., Wilson, L.G., Keating, M.T.: Heart Regeneration in Zebrafish. Science. **298**, 2188–2190 (2002)

62. Soonpaa, M.H., Kim, K.K., Pajak, L., Franklin, M., Field, L.J.: Cardiomyocyte DNA synthesis and binucleation during murine development. Am. J. Phys. **271**, H2183–H21H9 (1996)

63. Porrello, E.R., Mahmoud, A.I., Simpson, E., Hill, J.A., Richardson, J.A., Olson, E.N., et al.: Transient regenerative potential of the neonatal mouse heart. Science (New York, N.Y.). **331**(6020), 1078–1080 (2011). https://doi.org/10.1126/science.1200708

64. Ye, L., D'Agostino, G., Loo, S.J., Wang, C.X., Su, L.P., Tan, S.H., et al.: Early regenerative capacity in the porcine heart. Circulation. **138**, 2798–2808 (2018). https://doi.org/10.1161/CIRCULATIONAHA.117.031542
65. Haubner, B.J., Schneider, J., Schweigmann, U., Schuetz, T., Dichtl, W., Velik-Salchner, C., et al.: Functional recovery of a human neonatal heart after severe myocardial infarction. Circ. Res. **118**(2), 216–221 (2016). https://doi.org/10.1161/CIRCRESAHA.115.307017
66. Bergmann, O., Bhardwaj, R.D., Bernard, S., Zdunek, S., Barnabé-Heider, F., Walsh, S., et al.: Evidence for cardiomyocyte renewal in humans. Science (New York, N.Y.). **324**, 98–102 (2009). https://doi.org/10.1126/science.1164680
67. Soonpaa, M.H., Koh, G.Y., Klug, M.G., Field, L.J.: Formation of nascent intercalated disks between grafted fetal cardiomyocytes and host myocardium. Science. **264**, 98–101 (1994). https://doi.org/10.1126/science.8140423
68. Koh, G.Y., Soonpaa, M.H., Klug, M.G., Field, L.J.: Long-term survival of AT-1 cardiomyocyte grafts in syngeneic myocardium. Am. J. Phys. **264**, H1727–H1733 (1993). https://doi.org/10.1152/ajpheart.1993.264.5.H1727
69. Reinecke, H., Zhang, M., Bartosek, T., Murry, C.E.: Survival, integration, and differentiation of cardiomyocyte grafts. Circulation. **100**, 193–202 (1999). https://doi.org/10.1161/01.CIR.100.2.193
70. Funakoshi, S., Miki, K., Takaki, T., Okubo, C., Hatani, T., Chonabayashi, K., et al.: Enhanced engraftment, proliferation, and therapeutic potential in heart using optimized human iPSC-derived cardiomyocytes. Sci. Rep. **6**, 1–14 (2016). https://doi.org/10.1038/srep19111
71. Chanthra, N., Abe, T., Miyamoto, M., Sekiguchi, K., Kwon, C., Hanazono, Y., et al.: A novel fluorescent reporter system identifies Laminin-511/521 as potent regulators of cardiomyocyte maturation. Sci. Rep. **10** (2020). https://doi.org/10.1038/s41598-020-61163-3
72. Poon ENY, Luo Xl, Webb SE, Yan B, Zhao R, Wu SCM, et al. The cell surface marker CD36 selectively identifies matured, mitochondria-rich hPSC-cardiomyocytes. Cell Research: Springer Nature; 2020. p. 1–4
73. Zimmermann, W.-H., Didié, M., Wasmeier, G.H., Nixdorff, U., Hess, A., Melnychenko, I., et al.: Cardiac grafting of engineered heart tissue in Syngenic rats. (2002). https://doi.org/10.1161/01.cir.0000032876.55215.10
74. Zimmermann, W.-H., Melnychenko, I., Wasmeier, G., Didié, M., Naito, H., Nixdorff, U., et al.: Engineered heart tissue grafts improve systolic and diastolic function in infarcted rat hearts. Nat. Med. **12**, 452–458 (2006). https://doi.org/10.1038/nm1394
75. Hansen, A., Eder, A., Bönstrup, M., Flato, M., Mewe, M., Schaaf, S., et al.: Development of a drug screening platform based on engineered heart tissue. Circ. Res. **107**, 35–44 (2010). https://doi.org/10.1161/CIRCRESAHA.109.211458
76. Weinberger, F., Breckwoldt, K., Pecha, S., Kelly, A., Geertz, B., Starbatty, J., et al.: Cardiac repair in Guinea pigs with human engineered heart tissue from induced pluripotent stem cells. Sci. Transl. Med. **8**, 363ra148 (2016). https://doi.org/10.1126/scitranslmed.aaf8781
77. Riegler, J., Tiburcy, M., Ebert, A., Tzatzalos, E., Raaz, U., Abilez, O.J., et al.: Human engineered heart muscles engraft and survive long term in a rodent myocardial infarction model. Circ. Res. **117**, 720–730 (2015). https://doi.org/10.1161/CIRCRESAHA.115.306985
78. Mannhardt, I., Saleem, U., Mosqueira, D., Loos, M.F., Ulmer, B.M., Lemoine, M.D., et al.: Comparison of 10 control hPSC lines for drug screening in an engineered heart tissue format. Stem Cell Reports. **15**, 983–998 (2020). https://doi.org/10.1016/j.stemcr.2020.09.002
79. Naito, H., Melnychenko, I., Didié, M., Schneiderbanger, K., Schubert, P., Rosenkranz, S., et al.: Optimizing engineered heart tissue for therapeutic applications as surrogate heart muscle. Circulation. **114**, I72–I78 (2006). https://doi.org/10.1161/CIRCULATIONAHA.105.001560
80. Breckwoldt, K., Letuffe-Brenière, D., Mannhardt, I., Schulze, T., Ulmer, B., Werner, T., et al.: Differentiation of cardiomyocytes and generation of human engineered heart tissue. Nat. Protoc. **12**, 1177–1197 (2017). https://doi.org/10.1038/nprot.2017.033

81. Bargehr, J., Ong, L.P., Colzani, M., Davaapil, H., Hofsteen, P., Bhandari, S., et al.: Epicardial cells derived from human embryonic stem cells augment cardiomyocyte-driven heart regeneration. Nat. Biotechnol. **37**, 895–906 (2019). https://doi.org/10.1038/s41587-019-0197-9

82. Mannhardt, I., Breckwoldt, K., Letuffe-Brenière, D., Schaaf, S., Schulz, H., Neuber, C., et al.: Human engineered heart tissue - analysis of contractile force. Stem Cell Reports. (2016) in press

83. Tulloch, N.L., Muskheli, V., Razumova, M.V., Korte, F.S., Regnier, M., Hauch, K.D., et al.: Growth of engineered human myocardium with mechanical loading and vascular coculture. Circ. Res. **109**, 47–59 (2011). https://doi.org/10.1161/CIRCRESAHA.110.237206

84. Lesman, A., Habib, M., Caspi, O., Gepstein, A., Arbel, G., Levenberg, S., et al.: Transplantation of a tissue-engineered human vascularized cardiac muscle. Tissue Eng. Part A. **16**, 115–125 (2010). https://doi.org/10.1089/ten.TEA.2009.0130

85. Stevens, K.R., Kreutziger, K.L., Dupras, S.K., Korte, F.S., Regnier, M., Muskheli, V., et al.: Physiological function and transplantation of scaffold-free and vascularized human cardiac muscle tissue. Proc. Natl. Acad. Sci. U. S. A. **106**, 16568–16573 (2009). https://doi.org/10.1073/pnas.0908381106

86. Sun X, Wu J, Qiang B, Romagnuolo R, Gagliardi M, Keller G, et al. Transplanted microvessels improve pluripotent stem cell – derived cardiomyocyte engraftment and cardiac function after infarction in rats. 2020:1-12

87. Shiba, Y., Fernandes, S., Zhu, W.-Z., Filice, D., Muskheli, V., Kim, J., et al.: Human ES-cell-derived cardiomyocytes electrically couple and suppress arrhythmias in injured hearts. Nature. **489**(7415), 322–325 (2012). https://doi.org/10.1038/nature11317

88. Shiba, Y., Filice, D., Fernandes, S., Minami, E., Dupras, S.K., Biber, B.V., et al.: Electrical integration of human embryonic stem cell-derived cardiomyocytes in a Guinea pig chronic infarct model. J. Cardiovasc. Pharmacol. Ther. **19**(4), 368–381 (2014). https://doi.org/10.1177/1074248413520344

89. Heatley, J.J.: Cardiovascular anatomy, physiology, and disease of rodents and small exotic mammals. Veterinary Clin. North Am. Exotic Animal Practice. **12**, 99–113 (2009). https://doi.org/10.1016/j.cvex.2008.08.006

90. De Silva, M., Mihailovic, A., Baron, T.M.: Two-dimensional, M-mode, and Doppler echocardiography in 22 conscious and apparently healthy pet Guinea pigs. J. Vet. Cardiol. **27**, 54–61 (2020). https://doi.org/10.1016/j.jvc.2020.01.004

91. Shiotani, M., Harada, T., Abe, J., Hamada, Y., Horii, I.: Methodological validation of an existing telemetry system for QT evaluation in conscious Guinea pigs. J. Pharmacol. Toxicol. Methods. **55**, 27–34 (2007). https://doi.org/10.1016/j.vascn.2006.04.008

92. Clauss, S., Bleyer, C., Schüttler, D., Tomsits, P., Renner, S., Klymiuk, N., et al.: Animal models of arrhythmia: classic electrophysiology to genetically modified large animals. Nat. Rev. Cardiol. **16**(8), 457–475 (2019). https://doi.org/10.1038/s41569-019-0179-0

93. Kääb, S., Nä, M.: Diversity of ion channel expression in health and disease. Eur. Heart J. Suppl., 31–40 (2001)

94. Schotten, U., Verheule, S., Kirchhof, P., Goette, A.: Pathophysiological mechanisms of atrial fibrillation: a translational appraisal. Physiol. Rev. Am. Physiol. Soc., 265–325 (2011)

95. Hume, J.R., Ueharat, A.: Configurations of single Guinea-pig atrial and ventricular myocytes. J. Physiol., 525–544 (1985)

96. Rajamohan, D., Matsa, E., Kalra, S., Crutchley, J., Patel, A., George, V., et al.: Current status of drug screening and disease modelling in human pluripotent stem cells. BioEssays. **35**, 281–298 (2013). https://doi.org/10.1002/bies.201200053

97. Himmel, H.M., Bussek, A., Hoffmann, M., Beckmann, R., Lohmann, H., Schmidt, M., et al.: Field and action potential recordings in heart slices: correlation with established in vitro and in vivo models. Br. J. Pharmacol. **166**(1), 276–296 (2012). https://doi.org/10.1111/j.1476-5381.2011.01775.x

98. Takahara, A., Sasaki, R., Nakamura, M., Sendo, A., Sakurai, Y., Namekata, I., et al.: Clobutinol delays ventricular repolarization in the Guinea pig heart: comparison with cardiac effects of hERG K+ channel inhibitor E-4031. J. Cardiovasc. Pharmacol. **54**(6) (2009)

99. Marks, L., Borland, S., Philp, K., Ewart, L., Lainée, P., Skinner, M., et al.: The role of the anaesthetised Guinea-pig in the preclinical cardiac safety evaluation of drug candidate compounds. Toxicol. Appl. Pharmacol. **263**, 171–183 (2012). https://doi.org/10.1016/j.taap.2012.06.007

100. Kågström, J., Sjögren, E.L., Ericson, A.C.: Evaluation of the Guinea pig monophasic action potential (MAP) assay in predicting drug-induced delay of ventricular repolarisation using 12 clinically documented drugs. J. Pharmacol. Toxicol. Methods. **56**, 186–193 (2007). https://doi.org/10.1016/j.vascn.2007.03.003

101. Dasagrandhi, D., Muthuswamy, A., Lennox, A.M., Jayavelu, T., Devanathan, V., et al.: Ischemia/reperfusion injury in male Guinea pigs: an efficient model to investigate myocardial damage in cardiovascular complications. Biomed. Pharmacother. **99**, 469–479 (2018). https://doi.org/10.1016/j.biopha.2018.01.087

102. Maxwell, M.P., Hearse, D.J., Yellon, D.M.: Species variation in the coronary collateral circulation during regional myocardial ischaemia: a critical determinant of the rate of evolution and extent of myocardial infarction. Cardiovasc. Res. **21**(10), 737–746 (1987). https://doi.org/10.1093/cvr/21.10.737

103. Castro, L., Geertz, B., Reinsch, M., Aksehirlioglu, B., Hansen, A., Eschenhagen, T., et al.: Implantation of hiPSC-derived cardiac-muscle patches after myocardial injury in a Guinea pig model. J. Visualized Exp. JoVE. (2019). https://doi.org/10.3791/58810

104. Lindsey, M.L., Bolli, R., Canty, J.M., Du, X.J., Frangogiannis, N.G., Frantz, S., et al.: Guidelines for experimental models of myocardial ischemia and infarction. Am. J. Physiol. Heart Circulatory Physiol. Am. Physiol. Soc., H812–H38 (2018)

105. van den Bos, E.J., Mees, B.M.E., de Waard, M.C., de Crom, R., Duncker, D.J.: A novel model of cryoinjury-induced myocardial infarction in the mouse: a comparison with coronary artery ligation. Am. J. Phys. Heart Circ. Phys. **289**(3), H1291–HH300 (2005). https://doi.org/10.1152/ajpheart.00111.2005

106. Varró, A., Lathrop, D.A., Hester, S.B., Nánási, P.P., Papp, J.G.Y.: Ionic currents and action potentials in rabbit, rat, and guinea pig ventricular myocytes. Basic Res. Cardiol. **88**, 93–102 (1993). https://doi.org/10.1007/BF00798257

107. Gerbin, K.A., Yang, X., Murry, C.E., Coulombe, K.L.K.: Enhanced electrical integration of engineered human myocardium via Intramyocardial versus Epicardial delivery in infarcted rat hearts. PLoS One. **10**(7), e0131446 (2015). https://doi.org/10.1371/journal.pone.0131446

108. Chong, J.J.H., Yang, X., Don, C.W., Minami, E., Liu, Y.-W., Weyers, J.J., et al.: Human embryonic-stem-cell-derived cardiomyocytes regenerate non-human primate hearts. Nature. **510**, 273–277 (2014). https://doi.org/10.1038/nature13233

109. Shiba, Y., Gomibuchi, T., Seto, T., Wada, Y., Ichimura, H., Tanaka, Y., et al.: Allogeneic transplantation of iPS cell-derived cardiomyocytes regenerates primate hearts. Nature. **538**, 388–391 (2016). https://doi.org/10.1038/nature19815

110. Romagnuolo, R., Masoudpour, H., Porta-Sánchez, A., Qiang, B., Barry, J., Laskary, A., et al.: Human embryonic stem cell-derived cardiomyocytes regenerate the infarcted pig heart but induce ventricular Tachyarrhythmias. Stem Cell Rep. **12**, 967–981 (2019)

111. Mallapaty, S.: Revealed: two men in China were first to receive pioneering stem-cell treatment for heart disease. Nature. **581**, 249–250 (2020). https://doi.org/10.1038/d41586-020-01285-w

112. Deuse, T., Hu, X., Gravina, A., Wang, D., Tediashvili, G., De, C., et al.: Hypoimmunogenic derivatives of induced pluripotent stem cells evade immune rejection in fully immunocompetent allogeneic recipients. Nat. Biotechnol. **37**, 252–258 (2019). https://doi.org/10.1038/s41587-019-0016-3

113. Gornalusse, G.G., Hirata, R.K., Funk, S.E., Riolobos, L., Lopes, V.S., Manske, G., et al.: HLA-E-expressing pluripotent stem cells escape allogeneic responses and lysis by NK cells. Nat. Biotechnol. **35**(8), 765–772 (2017). https://doi.org/10.1038/nbt.3860

114. Mattapally, S., Pawlik, K.M., Fast, V.G., Zumaquero, E., Lund, F.E., Randall, T.D., et al.: Human leukocyte antigen class I and II knockout human induced pluripotent stem cell–derived cells: universal donor for cell therapy. J. Am. Heart Assoc. **7** (2018). https://doi.org/10.1161/JAHA.118.010239

115. Tachibana, A., Santoso, M.R., Mahmoudi, M., Shukla, P., Wang, L., Bennett, M., et al.: Paracrine effects of the pluripotent stem cell-derived cardiac myocytes salvage the injured myocardium. Circ. Res. **121**(6), e22–e36 (2017). https://doi.org/10.1161/CIRCRESAHA.117.310803

116. Liu, Y.-W., Chen, B., Yang, X., Fugate, J.A., Kalucki, F.A., Futakuchi-Tsuchida, A., et al.: Human embryonic stem cell–derived cardiomyocytes restore function in infarcted hearts of non-human primates. Nat. Biotechnol. **36**, 597–605 (2018). https://doi.org/10.1038/nbt.4162

117. Zhu, W., Zhao, M., Mattapally, S., Chen, S., Zhang, J.: CCND2 overexpression enhances the regenerative potency of human induced pluripotent stem cell-derived cardiomyocytes: Remuscularization of injured ventricle. Circ. Res. **122**, 88–96 (2018). https://doi.org/10.1161/CIRCRESAHA.117.311504

118. Fan, C., Fast, V.G., Tang, Y., Zhao, M., Turner, J.F., Krishnamurthy, P., et al.: Cardiomyocytes from CCND2-overexpressing human induced-pluripotent stem cells repopulate the myocardial scar in mice: a 6-month study. J. Mol. Cell. Cardiol. **137**, 25–33 (2019). https://doi.org/10.1016/j.yjmcc.2019.09.011

119. Gerbin, K.A., Mitzelfelt, K.A., Guan, X., Martinson, A.M., Murry, C.E.: Delta-1 functionalized hydrogel promotes hESC-cardiomyocyte graft proliferation and maintains heart function post-injury. Mol. Therapy Methods Clin. Dev. **17**, 986–998 (2020). https://doi.org/10.1016/j.omtm.2020.04.011

120. Giacca, M.: Cardiac regeneration after myocardial infarction: an approachable goal. Curr. Cardiol. Rep. **22**(10), 122 (2020). https://doi.org/10.1007/s11886-020-01361-7

121. Fu, X., Khalil, H., Kanisicak, O., Boyer, J.G., Vagnozzi, R.J., Maliken, B.D., et al.: Specialized fibroblast differentiated states underlie scar formation in the infarcted mouse heart. J. Clin. Investig. **128**, 2127–2143 (2018). https://doi.org/10.1172/JCI98215

Part III
Genetic Approaches to Study Cardiac Differentiation and Repair

Analysing Genetic Programs of Cell Differentiation to Study Cardiac Cell Diversification

Zhixuan Wu, Sophie Shen, Yuliangzi Sun, Tessa Werner, Stephen T. Bradford, and Nathan J. Palpant

1 Introduction

Cellular differentiation involves sequential, coordination of gene expression changes which instruct cells toward a functional identity that contributes to organ and animal physiology. Uncovering mechanisms governing regulation of this process is key to understanding developmental programs and how they fail in the context of disease. Historically, this has been studied using experimental strategies designed to describe the role of genes or biological process in model organisms, but capabilities to study cell differentiation using human pluripotent stem cells (hPSCs) have opened a new, scalable platform to study mechanisms of human cell differentiation.

Current applications of stem cells are focusing primarily on evaluating differentiation of stem cells into mature cell types to understand developmental biology and disease progression, growing stem cells into cells or tissues for organ regeneration, and using stem cells or their derivatives to test the safety of drugs. These opportunities are linked with increasing investment in regenerative biology internationally. The market size for progenitor and stem cell therapies reflects this, with a projected value of $237 M USD by 2025, making up a small fraction of the total regenerative medicine market, with a forecasted value of ~$18B USD by 2025 [1]. The largest market segment is in Asia Pacific, though strong growth is

Z. Wu · S. Shen · Y. Sun · T. Werner · N. J. Palpant (✉)
Institute for Molecular Bioscience, The University of Queensland, Brisbane, Australia
e-mail: n.palpant@uq.edu.au

S. T. Bradford
Uniquest, Brisbane, Australia

J. Zhang, V. Serpooshan (eds.), *Advanced Technologies in Cardiovascular Bioengineering*, https://doi.org/10.1007/978-3-030-86140-7_8

predicted in Europe with a compound annual growth rate of 39.9% [1]. With recent advances in cellular genomics, stem cell research is positioned for breakthrough in fundamental developmental studies as well as therapeutic applications in drug discovery, disease modelling, and regenerative medicine. Despite the overwhelming attention and increasing financial investment that stem cell research has received, few products are currently approved by the FDA for use in patients. A growing number of clinical trials are beginning to emerge, however, particularly in the context of cardiovascular applications where iPSCs are being used to study disease processes and as a next-generation therapeutic for heart regeneration (Table 1).

While the promise of pluripotent stem cells for research, drug screening, and cell therapies has been promoted in recent decades, current stem cell differentiation protocols generate a growing but still limited portfolio of cell types that remain functionally immature compared to *in vivo* cell types. This highlights a lack in fundamental understanding of the genetic, signalling, and physiological mechanisms controlling cell differentiation. With the advent of innovative technologies in gene editing, sequencing and cell biology, the opportunity for hPSCs to facilitate applications in research, industry, and clinical sectors continues to grow. Advances in three-dimensional (3D) organoid culture are also enabling a better understanding of cell-cell interactions and improving the ability to model more complex physiology and diseases [2, 3]. Furthermore, recent technological advances in genome-wide single-cell transcriptomic assays have enabled gene expression analysis of millions of individual cells navigating differentiation. These sequencing, gene perturbation, and complex cell organoid technologies that draw on the versatility and scalability of hPSCs are providing new perspectives and approaches to study gene expression changes occurring throughout differentiation. Collectively, these methods are accelerating our understanding of cell autonomous mechanisms and cell-cell interactions required for development and maturation of diverse cell types, enabling derivation of new cell types from iPSCs, and facilitating development of multidisciplinary techniques to model and study human biology.

This chapter will detail the current understanding of fundamental mechanisms of cardiac cell differentiation and lineage segregations in early development. We begin with a detailed summary of embryogenesis and gastrulation as a basis to understand germ layer specification which forms the underpinning principles of multilineage stem cell differentiation. This is followed by a discussion of new capabilities in single-cell genomics in both *in vivo* developmental embryogenesis and *in vitro* mesendoderm stem cell differentiation that are emerging as powerful opportunities to reveal mechanisms, timing, and regulation of multilineage cell differentiation.

Table 1 Registered clinical trials involving human pluripotent stem cells for cardiovascular applications

iPSC reprogramming for disease modelling or *in vitro* differentiation/molecular characterisation of diseased iPSCs		
NCT number	Conditions	Sponsors/Collaborators
NCT03696628	Familial Cardiomyopathy	University Hospital, Montpellier
NCT01517425	Coronary Artery Disease	Scripps Translational Science Institute
NCT01734356	Inherited Arrhythmias and Valvulopathies	Nantes University Hospital
NCT03252613	Cardiomyopathies, Sepsis, Septic Shock	Samuel Brown, Euan Ashley
NCT02417311	Hypertrophic Cardiomyopathy Dilated Cardiomyopathy	Universitätsklinikum Hamburg-Eppendorf
NCT02413450	Inherited Cardiac Arrythmias, Long QT Syndrome, Brugada Syndrome, Catecholaminergic Polymorphic Ventricular Tachycardia, Early Repolarization Syndrome, Arrhythmogenic Cardiomyopathy, Hypertrophic Cardiomyopathy, Dilated Cardiomyopathy, Muscular Dystrophies (Duchenne, Becker, Myotonic Dystrophy), Normal Control Subjects	Johns Hopkins University, National Heart, Lung, and Blood Institute
NCT00243776	Congenital Heart Disease, Tetralogy of Fallot	Emory University, National Heart, Lung, and Blood Institute
NCT03387072	Sudden Cardiac Death	Nantes University Hospital
NCT03509441	Cardiomyopathies, Insulin Resistance, Non-ischemic Cardiomyopathy, Cardiac Fibrosis, Diabetes	Stanford University
iPSC derivatives transplantation for *in vivo* organ regeneration or cell therapy		
NCT number	Conditions	Sponsors/Collaborators
NCT04396899	Heart Failure	University Medical Center Goettingen Deutsches Zentrum für Herz-Kreislauf-Forschung University Medical Center Freiburg
NCT03763136	Heart Failure	Help Therapeutics The Affiliated Nanjing Drum Tower Hospital of Nanjing University Medical School The First Affiliated Hospital with Nanjing Medical University
NCT00629018	Dilated Cardiomyopathy	University Medical Centre LjubljanaBlood Transfusion Centre of SloveniaStanford University

2   Designing Stem Cell Biology to Recapitulate Developmental Biology

The fundamental mechanisms of iPSC differentiation are grounded in the biology of organ morphogenesis. Understanding how cell diversity is established from an undifferentiated embryo *in vivo* forms the underlying rationale for how *in vitro* protocols are designed. During the initial phases of embryonic development, the zygote forms a blastula containing an outer single layer of cells comprised of the trophoblast, which will further develop into the placenta, and the inner cell mass, which gives rise to cell lineages of the developing embryo. *In vitro* pluripotent stem cells are derived from or reprogrammed to a cell state recapitulating the cells of this embryonic inner cell mass.

Differentiation into any of the three primary germ layers, namely mesoderm, endoderm, and ectoderm, occurs within the first 2–4 days of *in vitro* differentiation. This is the first stage of cell lineage restriction to a specified set of potential cell types derived by each of these germ layers. The cell patterning processes underlying body axis formation and cell diversification is controlled by carefully orchestrated signalling pathways such as BMP4, Nodal, and Wnt/β-catenin signalling that instruct gene regulatory networks in cells to instruct stages of cell differentiation and morphogenesis [20, 21]. During gastrulation *in vivo*, epiblast cells lose cell-cell adhesion, undergo epithelial-mesenchymal-transition (EMT), elongate anteriorly to the distal end of the embryo by progressively recruiting lateral epiblast cells, and begin to ingress at the primitive streak (PS) [22]. This process of EMT is similarly recapitulated during *in vitro* differentiation [23, 24]. In the developing embryo, the first group of cells that migrate away from the posterior PS contribute to extraembryonic mesoderm [25]. Subsequently, cells exiting the middle region of the PS form the lateral plate mesoderm, intermediate mesoderm, and paraxial mesoderm [26]. Lastly, the definitive endoderm lineage emerges from the anterior primitive streak. Cells continue to migrate through the anterior-most end of primitive streak to form the notochord, which is a group of mesodermal cells that defines the embryonic midline. Finally, after the ingression stops, cells that remain in the epiblast differentiate into ectoderm cells. Decades of studies in developmental biology using lineage tracing and transcriptional analysis have identified numerous cell types and lineage-specific genetic markers that have help guide stem cell differentiation protocol development (Table 2).

3   Mesendoderm Patterning and Germ Layer Derivatives

Definitive endoderm arises from the anterior primitive streak and is specified into anterior and posterior definitive endoderm by a Wnt/FGF/BMP signalling gradient from the mesoderm and neurectoderm [17, 27–29]. TGFβ signalling in the anterior endoderm allows for pharyngeal endoderm and anterior foregut fates to arise, after

Table 2 Marker genes demarcating selected mesendoderm lineages and cell types

Gene symbol	Exit from pluripotency				Limb				Heart											Kidney					Somites					Gastrointestinal tract					
	Pluripotency	Anterior PS	Posterior PS	Pan-PS	Lateral plate mes	Limb mesoderm	Hindlimb	Forelimb	Cardiac mes	First heart field	Cardiomyocytes	Ventricular CMs	Atrial CMs	Nodal CMs	Second heart field	Endocardium	Endothelium	Epicardium	Smooth muscle cells	AIM	PIM	AIM derivatives	PIM derivatives	MM	Paraxial mes	Seg. boundary	Early somites	Dermomyotome	Sclerotome	Definitive Endoderm	Anterior foregut	Posterior foregut	Mid/Hindgut	Liver bud	Pancreas bud
SOX2	[4–6]																														[5]	[5]			
NANOG	[5, 7]																																		
HHEX		[5]																												[5]					
LHX1		[5]																		[8]		[8]	[9]												
FOXA2		[5]																												[5]					
MIXL1		[4]		[7]																					[4]										
GSC		[5]																																	
T		[4, 5]	[6]	[5, 7]																															
EOMES		[5]																																	
MESP1			[5]	[7]	[6]				[10]																										
EVX1			[5]																														[5]		
MSGN1																									[4, 7]										
HES7																									[4, 7]										

(continued)

Table 2 (continued)

Gene symbol	Exit from pluripotency				Limb				Heart												Kidney					Somites					Gastrointestinal tract					
	Pluripotency	Anterior PS	Posterior PS	Pan-PS	Lateral plate mes	Limb mesoderm	Hindlimb	Forelimb	Cardiac mes	First heart field	Cardiomyocytes	Ventricular CMs	Atrial CMs	Nodal CMs	Second heart field	Endocardium	Endothelium	Epicardium	Smooth muscle cells	MM	AIM	PIM	AIM derivatives	PIM derivatives	MM	Paraxial mes	Seg. boundary	Early somites	Dermomyotome	Sclerotome	Definitive Endoderm	Anterior foregut	Posterior foregut	Mid/Hindgut	Liver bud	Pancreas bud
TBX6				[6, 9]																						[4, 7]										
CDX2				[4]																						[4]							[5]	[5]		
PDGFRA				[4]	[4]																									[4]						
HAND1					[4]					[10]																										
PRRX1						[4]																							[7]							
HOXB5						[4]																														
PITX1							[11]																													
TBX4							[11]	[11]																												
HOXB4								[11]																												
TBX5								[11]	[4, 6]	[10]			[12]																							
NKX2-5									[4, 6]																											
TBX20									[4]																											
LRRC32									[4, 13]																											
IRX4										[10]		[14]																								
MYL2												[12]																								

Gene symbol	Pluripotency	Anterior PS	Posterior PS	Pan-PS	Lateral plate mes	Limb mesoderm	Hindlimb	Forelimb	Cardiac mes	First heart field	Cardiomyocytes	Ventricular CMs	Atrial CMs	Nodal CMs	Second heart field	Endocardium	Endothelium	Epicardium	Smooth muscle cells	AIM	PIM	AIM derivatives	PIM derivatives	MM	Paraxial mes	Seg. boundary	Early somites	Dermomyotome	Sclerotome	Definitive Endoderm	Anterior foregut	Posterior foregut	Mid/Hindgut	Liver bud	Pancreas bud
GJA5													[12, 15]																						
MEF2C										[10]																									
HCN4										[10]				[15, 16]																					
TBX3														[15, 16]																					
TNNI1											[10]																								
MYH7											[10]																								
ACTN2											[15]																								
TBX1															[10]																				
SIX2															[10]								[9]	[9]											
IRX3															[10]									[9]							[5]				
IRX5															[10]																				
FGF10															[10]																				
FGF8															[10]								[9]												
MSX1																[17]																			

(continued)

Table 2 (continued)

| Gene symbol | Exit from pluripotency | | | | Limb | | | | Heart | | | | | | | | | | | Kidney | | | | | Somites | | | | | Gastrointestinal tract | | | | | |
	Pluripotency	Anterior PS	Posterior PS	Pan-PS	Lateral plate mes	Limb mesoderm	Hindlimb	Forelimb	Cardiac mes	First heart field	Cardiomyocytes	Ventricular CMS	Atrial CMs	Nodal CMS	Second heart field	Endocardium	Endothelium	Epicardium	Smooth muscle cells	AIM	PIM	AIM derivatives	PIM derivatives	MM	Paraxial mes	Seg. boundary	Early somites	Dermomyotome	Sclerotome	Definitive Endoderm	Anterior foregut	Posterior foregut	Mid/Hindgut	Liver bud	Pancreas bud
NFATC1																[17]																			
CDH5																	[18]																		
PECAM1																	[18]																		
HOPX																	[18]									[4]									
WT1																		[19]						[9]											
TBX18																		[19]																	
TAGLN																			[19]																
CNN1																			[19]																
ACTA2																			[19]																
PAX2																				[8]	[9]			[9]											
OSR1																				[8]	[8, 9]			[9]											
ITGA8																						[8]													
PAX8																					[8]		[8, 9]	[8]											
HOXB4																							[8]												
EYA1																							[8]												
BMP7																							[8]												
SIX1																							[8]												
LRP2																							[8]												
CDH1																															[5]				

Gene symbol	Exit from pluripotency				Limb				Heart											Kidney					Somites					Gastrointestinal tract					
	Pluripotency	Anterior PS	Posterior PS	Pan-PS	Lateral plate mes	Limb mesoderm	Hindlimb	Forelimb	Cardiac mes	First heart field	Cardiomyocytes	Ventricular CMs	Atrial CMs	Nodal CMs	Second heart field	Endocardium	Endothelium	Epicardium	Smooth muscle cells	AIM	PIM	AIM derivatives	PIM derivatives	MM	Paraxial mes	Seg. boundary	Early somites	Dermomyotome	Sclerotome	Definitive Endoderm	Anterior foregut	Posterior foregut	Mid/Hindgut	Liver bud	Pancreas bud
DLL1																									[4, 13]										
DLL3																									[4]										
DUSP6																									[7]										
RSPO3																									[7]										
MESP2																										[4]									
RIPPLY2																										[4]									
FOXC2																										[4]	[4]		[4, 7]						
PAX3																											[4]	[4]							
TCF15																											[4, 7]								
MEOX1																											[4, 7]								
SOX9																													[4]						
PAX1																													[4, 7]						
TWIST1																												[7]	[4]						
PAX7																												[7]							
ALX4																												[7]							

(continued)

Table 2 (continued)

Gene symbol	Exit from pluripotency				Limb				Heart											Kidney					Somites					Gastrointestinal tract					
	Pluripotency	Anterior PS	Posterior PS	Pan-PS	Lateral plate mes	Limb mesoderm	Hindlimb	Forelimb	Cardiac mes	First heart field	Cardiomyocytes	Ventricular CMs	Atrial CMs	Nodal CMs	Second heart field	Endocardium	Endothelium	Epicardium	Smooth muscle cells	AIM	PIM	AIM derivatives	PIM derivatives	MM	Paraxial mes	Seg. boundary	Early somites	Dermomyotome	Sclerotome	Definitive Endoderm	Anterior foregut	Posterior foregut	Mid/Hindgut	Liver bud	Pancreas bud
PAX9																													[7]				[5]		
CER1																														[5]					
SOX17																														[5]					
FOXA1																														[5]					
SHISA2																											[4]			[5]	[5]				
OTX2																														[5]	[5]				
ONECUT1																																[5]			
HNF1B																																[5]	[5]		
AFP																																[5]	[5]	[5]	
TTR																																[5]	[5]	[5]	
PDX1																																[5]	[5]		[5]
NKX2-2																																[5]			[5]

Lineages are displayed in different colours and the numbers indicate relevant literature references. *PS* primitive streak, *mes* mesoderm, *CMs* cardiomyocytes, *AIM* anterior intermediate mesoderm, *PIM* posterior intermediate mesoderm, *MM* metanephric mesenchyme

which retinoic acid (RA), BMP4, and Wnt signals from the cardiopharyngeal mesoderm promote lung and thyroid fate specification [30–35]. BMP inhibition and RA signalling pattern the posterior foregut, where reciprocally repressive TGFβ/BMP and FGF/MAPK signalling promotes pancreatic and hepatic specification respectively [17, 36]. Finally, Wnt, BMP4 and FGF signalling from the tailbud mesoderm adjacent to the most posterior endoderm, paired with hedgehog (HH) signalling, drives and maintains its specification into the mid- and hindgut lineages [29, 35, 37, 38].

The mesoderm arises from cells ingressing at the posterior primitive streak, where they are exposed to high levels of BMP4, Wnt, and FGF signalling. This causes the cells to express pan-mesodermal genes including Eomesodermin (*EOMES*) and Brachyury (*T*) as they migrate anteriorly and laterally [39–43].

At the most anterior, nodal region of the primitive streak, lower levels of Wnt signalling allow for progenitors of the notochord, prechordal plate and head process mesoderm to arise [44, 45]. The notochord is an important signalling structure that provides HH signalling cues to pattern the gut endoderm, neural tube, and paraxial mesoderm, as well as antagonising Wnt, BMP and TGFβ signalling [46–48].

Mesodermal cells that remain closest to the primitive streak are exposed to sustained Wnt and FGF signalling and give rise to paraxial, or presomitic mesoderm [4, 6, 39]. Paraxial mesoderm is exposed to Noggin signalling from the notochord, ectoderm, and neural tube, which inhibit BMP signals from the nearby intermediate and lateral plate mesoderm [4, 46, 49]. As the paraxial mesoderm undergoes segmentation to form the somites, HH signalling from the notochord induces specification of bone and cartilage lineages, while Wnt signalling from the endoderm promotes dermal and skeletal muscle fates [4, 7].

Intermediate mesoderm (IM) arises from the posterior primitive streak and is exposed to moderate levels of Nodal and BMP signalling as it migrates laterally to situate as two bilateral stripes between the lateral plate mesoderm and nascent paraxial mesoderm [49–51]. Wnt and FGF signalling along the anteroposterior axis of the IM guide specification into anterior and posterior IM, giving rise to uretic bud and kidney lineages respectively [8, 52].

BMP/Smad signals at the mid-primitive streak region promote cell migration laterally away from the streak to form lateral plate mesoderm (LPM) [4, 39, 53]. The anterior LPM, exposed to BMP, FGF, noncanonical Wnt, and SHH signalling, gives rise to cardiac mesoderm, while the Wnt-exposed posterior LPM is specified into limb bud mesenchyme, which segregates into forelimb and hindlimb buds based on RA signalling [4, 11, 54–57].

Early expression of Brachyury and Eomesodermin in the gastrulating mesoderm promote expression of *MESP1*, which is considered the first known marker of cardiac specification in the mesoderm [58–60]. *MESP1* is expressed sequentially in two distinct populations within the cardiac mesoderm: the first and second heart fields (FHF & SHF) [61, 62].

The FHF expresses *MESP1* earlier, which drives mesoderm cells to migrate towards the anterior lateral plate where it is exposed to the endoderm which secretes Wnt inhibitors, non-canonical Wnt signalling, and BMP4 signalling. These in turn promote myocardial differentiation through inducing expression of cardiac transcription factors and chromatin remodellers including *MEF2C*, *GATA4*, *NKX2.5*, *HAND1*, *TBX20* and *TBX5*. These cells then migrate to form the cardiac crescent, which later fuses to become the contractile linear heart tube, that remains the primary blood pumping force as it continues to remodel and septate to become the adult heart [63–65].

The SHF meanwhile lies in the pharyngeal mesoderm anterior and dorsal to the FHF, where its cells are exposed to sustained Wnt signalling from the ectoderm, and moderate levels of BMP inhibition from the notochord. This results in activation of downstream SHF genes including *ISL1, BMP4, FGF10, SHH, TBX1,* and *WNT11*, through which the SHF maintains a proliferative and multipotent differentiation capacity [10, 66–70]. Eventually these cells are recruited ventrally to the poles of the heart tube, via endoderm-secreted FGF signalling, where they differentiate into myocardium, fibroblasts, endothelial, and smooth muscle cells contributing to the right ventricle, atria, inflow and outflow tracts, as well as surrounding pulmonary vasculature and endocardium [17, 35, 68, 71–76].

4   Translating Developmental Biology into *In Vitro* Stem Cell Differentiation Protocols

Developmental studies *in vivo* using model organisms have revealed the timing and dosing of key signalling pathways involved in early embryogenesis, which provide the framework for deriving *in vitro* differentiation protocols. Understanding how developmental regulators mediate cell differentiation is essential to guide cells towards specific lineages with purity and efficiency. Current differentiation protocols enable derivation of a multitude of cell types that are built from knowledge gained through decades of studies in developmental biology. Indeed, developmental biology has informed mechanisms of stem cell differentiation and provided benchmarking of hPSC-derived cell types against organotypic cellular functions, laying the intellectual and commercial groundwork for clinical and industry implementation [77]. Developmental biology has guided hPSC-derived models of the entire gastrointestinal tract [3, 78–81], lung [82], haematopoietic system [83, 84], kidney [85–88], brain [89], and sensory organs [90] among other tissue types. Knowledge of signalling pathways that control rostral-caudal and dorsal-ventral axis formation as well as germ layer specification have enabled the field to identify protocols for derivation of sub-lineages within the same germ layer (such as kidney and heart which are both derived from mesoderm) and subtypes of the same cell lineage (such as atrial and ventricular cardiomyocytes). What is common across all approaches, as is now becoming more and more evident with single-cell transcriptional profiling, is the embryonic nature of the tissue models being produced, with functional maturation being a major challenge.

In the context of cardiovascular lineages, developmental studies revealing the temporal and concentration-dependent mechanisms of Wnt signalling as they influence cell differentiation have formed the basis for current protocols guiding pluripotent cells into cardiomyocytes. β-catenin, the second messenger molecule of the Wnt pathway, acts as a transcriptional coactivator in the context of active signalling via the canonical Wnt pathway. β-catenin binds to transcription factors including T-cell factor (Tcf) and lymphoid enhancer factor (Lef) to initiate the transcription of Wnt target gene expression programs [91]. In the absence of Wnt ligand, β-catenin binds to a destruction complex containing the Adenomatous Polyposis Coli (APC), Axin and Glycogen synthase kinase 3β (GSK3β) components. GSK3β phosphorylate the amino terminal region of β-catenin and leads to its subsequent ubiquitination and proteasome-mediated degradation [92]. On the contrary, GSK3β inhibitors such as CHIR99021 and SB-216763 dissociate β-catenin from binding with the protein complex.

GSK3β-mediated Wnt/β-catenin signalling is required for the exit from pluripotency and mesendoderm differentiation. Mice embryos subjected to Wnt depletion are unable to initiate and maintain the specification of primitive streak, mesoderm, and endoderm [93–95]. Reciprocally, activation of Wnt signalling leads to loss of self-renewal ability. OCT4 (POU5F1), a key pluripotency factor, plays an important role in the maintenance of undifferentiated human embryonic stem cells (hESCs) by functionally repressing endogenous Wnt signalling [96]. Activation of Wnt signalling results in expression of mesoderm lineage markers and diverse DNA regulatory factors required for orchestrating cell lineage differentiation decisions [97].

Wnt signalling acts in a time-specific and biphasic manner in the process of cardiac myogenesis [98, 99] as well as in a dosage-dependent manner in the context of mesendodermal specification. Lower levels of Wnt signalling produce SOX17-positive definitive endoderm, whereas higher doses of Wnt inhibits cardiac mesoderm generation and upregulates the expression of *CDX1* and *CDX2*, directing differentiation toward paraxial/presomitic mesoderm lineage [6, 100]. As with all signalling pathway perturbations, the optimal timing and dosage of activation and/or inhibition are key determinants for deriving high purity cell types from *in vitro* differentiation protocols. Moreover, early activation of Wnt signalling promotes mesoderm commitment while late signalling, after gastrulation, represses cardiac specification and redirects cardiovascular progenitor cells to haematopoietic lineages [98]. Studies in zebrafish used heat shock transgenic models to convincingly demonstrate the temporal requirement for Wnt inhibition in cardiac differentiation *in vivo*, which was further demonstrated as critical for deriving high purity cardiomyocytes *in vitro* [101].

5 Protocol Development for Cardiovascular Lineage Differentiation in 2D and 3D Systems

5.1 Monolayer and EB-Based Differentiation

Knowledge gained from developmental biology has informed protocols for derivation of different cardiovascular lineages derived from hPSCs *in vitro*. Current state-of-the-art *in vitro* differentiation approaches generate cardiac subtypes using monolayer or embryoid body (EB) based cultures. Efficient generation of cardiomyocytes has been accomplished through manipulation of signalling cues instructing cardiac mesoderm specification including BMP, Wnt, FGF, and Activin-Nodal. For example, Kattman *et al.* optimised the cardiac differentiation by fine-tuning levels of Activin-Nodal and BMP signalling [102]. More recently, Lian *et al.* [103, 104] and Burridge *et al.* [105] described a small molecule-based, growth factor-free method for efficient cardiomyocyte differentiation with high purity (>90%), in which they solely modulate Wnt signalling by applying Wnt regulators at specific time and dosage. EB-based suspension culture has gained popularity for large-scale differentiation of hPSCs. EBs are 3D aggregates formed by hPSCs in suspension culture. Several culture platforms are available such as spinner flasks and stirred bioreactors. The first EB-based culture was employed to generate CMs with only 10% purity [106]. More recently, under appropriate control of Wnt concentration, EB aggregates size, and agitation rate, Chen *et al.* were able to scale up the yield of highly pure cardiomyocytes (>90%) to 1 litre with spinner flasks in a suspension culture system [107].

Most methods produce a mixture of non-contractile cells and cardiomyocytes including stromal cells, ventricular-like, atrial-like, and nodal-like cardiomyocytes. Thus, studies have focused on differentiating hPSCs toward specific cardiac cell subtypes. Lee and colleagues revealed that derivation of ventricular-like or atrial-like cardiomyocytes is initiated from different mesoderm populations where ventricular-like cardiomyocytes are derived from $CD235^+$ mesoderm and atrial cardiomyocyte generation requires RA treatment with $RALDH2^+$ mesoderm [12]. Indeed, addition of RA during hPSC differentiation results in atrial-like cardiomyocytes that closely resemble *in vivo* atrial cardiomyocytes [108]. Protze *et al.* successfully directed hPSCs towards sinoatrial node-like pacemaker cells by manipulating BMP and RA signalling pathways [109].

In addition to cardiomyocytes, other cell types including cardiac fibroblasts, endothelial cells, endocardial cells, and epicardial cells are essential for heart structure and function [110]. Several studies have generated endothelial cells from hPSCs [17, 38, 111]. In addition, Witty *et al.* described the derivation of $WT1^+/TBX18^+$ epicardial cell lineage from hPSCs-derived $PDGFRA^+$ mesoderm population. This method was achieved by the manipulation of the BMP and Wnt signalling. Recently, a more simplified approach for epicardial cell differentiation has been developed via Wnt and RA signalling control under defined, albumin-free culture medium [112]. Additionally, the generation of cardiac fibroblasts has been reported by Zhang *et al.* from ISL1$^+$/CXCR4$^+$ SHF progenitor cells [113]. Table 3 summarises a selection of differentiation protocols that are widely used for the generation of distinct cardiovascular lineages from PSCs.

The strategies described above enable derivation of specific cardiac cell subtypes with high purity and yield. However, most protocols often result in immature cell population that do not fully recapitulate the function and structure of *in vivo* counterparts [114].

5.2   3D and Tissue engineering Based In Vitro Differentiation Protocols

Considering the practical application of *in vitro*-derived heart tissues, 3D engineered cardiac tissues outperform cardiomyocytes acquired from 2D differentiation, with regards to their structural, electrophysiological, and contractile properties, at least in part because the cellular and environmental conditions more closely resembling *in vivo* biology. Current available 3D culture systems can be grouped into two broad categories: scaffold-based and scaffold-free models.

Spheroids are self-assembled multicellular aggregates formed under a scaffold-free system. Giacomelli *et al.* developed a scaffold-free method, in which they generated cardiac spheroids with enhanced gene expression associated with ion channels and calcium handling by co-culturing hPSCs-derived cardiomyocytes and endothelial cells [122]. Campostrini *et al.* reported a simplified protocol to form mature 3D microtissues by combining hPSCs-derived cardiomyocytes, cardiac fibroblasts, and cardiac endothelial cells [123]. The cardiac microtissues generated in this study show enhanced sarcomere structure. Spheroid formation is simple and cost-effective, however, their efficiency in generating large-scale cardiomyocytes relies on size and seeding density. More recently, with the advance of 3D printing techniques, several strategies were developed to form 3D bio-printed cardiac tissues using needle array as a scaffold-like support [124, 125]. Most recently, Xu and colleagues developed an embryo-like construct that is structural similar to neurula-stage mouse embryos and is able to form germ layer derivatives [126]. This has been achieved by scaffold-free co-culture of mouse ESCs with aggregates providing morphogen signalling. Cardiac organoid is another 3D model that been well established for *in vitro* cardiac differentiation [127–129]. Alternatively, development of engineered heart tissue (EHT) uses a scaffold to improve mechanical maturation of *in vitro* PSCs-derived cardiac tissues. Current 3D scaffold-based cardiac tissue formation falls into a variety of different formats. The implementation of 3D patch culture allows maturation of cardiac tissue with robust contractility and conduction velocities [130]. Cell sheet-based tissue engineering with octagonal column wrapping has been used to assemble tubular cardiac tissues [131]. Following this, Zhao *et al.* [132] and Goldfracht *et al.* [133] were able to generate chamber-specific cardiac tissues adapting Biowire II platform and ring-shaped constructs, respectively.

Table 3 Monolayer and EB-based differentiation protocols for cardiovascular lineages

Source	Cell subtypes		Models	Days	Conditions	Efficiency	Ref.
hPSCs	CMs	Mostly ventricular-like	Monolayer ECM-coated	15	Growth factor-free Small molecules defined	90%	[103]
hiPSCs		Mostly ventricular-like		15	Recombinant human albumin Small molecules defined	90%	[105]
hPSCs		Mostly ventricular-like		15	Albumin-free Chemically defined	90%	[104]
hPSCs		Nodal, atrial, and ventricular		10	Albumin-free Wnt signalling control	90%	[115]
hPSCs		CMs and endothelial cells		14	Wnt, Activin A and BMP4	90%	[111]
hiPSCs		CMs	Multilayer culture plates	10	With serum	80%	[116]
hESCs/ hiPSCs		CMs	EB-based protocol	7	BMP4 and activin	60%	[102]
hPSCs		Nodal, atrial, and ventricular		16	Matrix-free, serum-free Spinner flask	90%	[107]
hPSCs		Atrial and ventricular		20	RA signalling		[12]
hPSCs		Mature compact ventricular		32	Maturation media		[117]
hPSCs		Atrial		15	RA signalling		[108]
hiPSCs		Sinoatrial		20	BMP and RA	80% after sorting	[109]
hPSCs	Pre-valvular endocardial cells		Mouse embryonic fibroblasts + fibronectin-coated	10	MESP1+ sorted cells		[17]
hPSCs	Endocardial cells		EB-based + monolayer culture	9	BMP10 signalling	40%	[38]
hPSCs	Pro-epicardial cells Vascular smooth muscle cells (+14d) Cardiac fibroblasts (+14d)		Monolayer vitronectin coated	14	Wnt and RA signalling	90%	[112]

Table 3 (continued)

Source	Cell subtypes	Models	Days	Conditions	Efficiency	Ref.
hPSCs	Pro-epicardial cells Vascular smooth muscle cells (+6d) Epicardial cells (+4d) Cardiac fibroblasts (+6d)	Monolayer gelatin or Synthemax coated	12	Wnt signalling Albumin/ xeno-free	85%	[118]
hiPSCs	Epicardial cells	EB-based protocol	9	BMP4 and RA	60% ECs	[119]
hPSCs	Pro-epicardial cells Epicardial cells (+9d) Vascular smooth muscle cells (+17d) Cardiac fibroblasts (+17d)	EB-based + monolayer culture	15	BMP and Wnt signalling	85%	[19]
hPSCs	Epicardial cells Epicardium-derived smooth muscle cells (+12d) Epicardium-derived cardiac fibroblasts (+12d)	Monolayer ECM-coated	15	Wnt, BMP and RA signalling		[120]
hPSCs	Cardiac fibroblasts		20	Wnt and FGF signalling	70%	[113]
hPSCs	Cardiac fibroblasts		18	Replated iPSC-epicardial cells	90%	[121]

CMs cardiomyocytes, *ECM* extra cellular matrix, *EB* embryoid body, *RA* retinoic acid.

The goal of 3D culture models is to enhance the maturation of *in vitro* hPSCs-derived cardiomyocytes that more accurately represent *in vivo* counterparts, facilitating further applications in drug testing and regenerative medicine.

6   Single Cell Technologies Accelerating Discovery into Developmental Biology and Cell Differentiation

Drawing on this growing list of cell differentiation methods, the anticipated impact of the stem cell sector remains dependent on precision control of cell differentiation into cell types that model human physiology. However, efforts to dissect cell differentiation mechanisms and recapitulate human development using iPSCs have encountered the following major challenges: 1) we still have a limited understanding of human developmental biology, 2) we lack sufficient scale of data mapping gene expression changes controlling cell processes over time, 3) among the thousands of genes expressed in cells, we lack the ability to efficiently identify genes (especially non-transcription factors) responsible for guiding specific cell differentiation processes, and 4) we do not understand how and when to effectively perturb these specialised gene programs to customise cell differentiation decisions or functions. To address this, technologies in sequencing cell transcriptomes at single cell resolution have emerged, providing a new scale of information for dissecting cell differentiation [140].

Single-cell RNA sequencing (scRNA-seq) technology brings together computational analysis with biological interpretation, to reveal the heterogeneity present in tissues or cell populations. Multiple scRNA-seq strategies and platforms are available for isolating transcriptomes from single cells including Smart-seq2 [141], Drop-seq [142], CEL-seq2 [143], and 10X Chromium [144] with variable advantages and disadvantages regarding throughput, read-depth, and transcript coverage. For example, Smart-seq is used to sequence full-length transripts, allowing for deeper sequencing but captures a lower number of cells in comparison to other methods that sequence only the 3′ end of each transcript, but capture many more cells. To increase the number of cells that can be captured in a single experiment and reduce the cost of high-throughput scRNA-seq, sample multiplexing strategies using virally delivered genomic barcoding [145] or using cell lines with different genotypes [146] as well as external sample barcoding methods such as cell hashing antibodies [147] have been established, along with development of computational demultiplexing tools [147]. With the growing ease and reduced cost of deriving data by these methods, genomic data generation is outpacing many of the major big data disciplines and creating significant challenges in extracting meaningful and accurate insights from these unprecedented volumes of data. This has led to a requirement for mathematical techniques such as dimensionality reduction, normalisation, and clustering that are scalable across data size and cell diversity. This intersection is a frontier of science with many ongoing challenges and opportunities. Understanding strategies to implement these technological and computational approaches is critical to enable accurate interpretation of single-cell data.

The utility of scRNA-seq has been demonstrated in various contexts from constructing atlases of gene regulatory landscapes and cell subpopulations in healthy, diseased, or otherwise perturbed biological systems, to chronicle developmental processes through sequential 'snapshots' over time [148–155]. The field surrounding cellular genomics has exploded in recent years, with additional capabilities for full length-transcript sequencing, and adaptations of cell isolation protocols to profile whole genomes and epigenomes at single-cell resolution, and will likely continue to expand, in particular with an eye for single-cell spatial transcriptomics and proteomics in the near future [156–164]. Furthermore, capturing millions of genomic data points in sequencing studies can provide powerful insights into cell biology [165] with single-cell methods emerging as a powerful new genomic technology with the potential to greatly influence studies in cellular differentiation [166]. As single-cell genomics technologies have matured and diversified in the past decade, so too have the means to analyse their output and infer biological meaning in a high throughput, unsupervised manner. New computational methods are harnessing consortium-scale data to identify genetic mechanisms controlling cell decisions, infer intracellular communication events, and predict differentiation trajectories from single-cell transcriptomic data. Undoubtedly, breakthroughs across the wet and dry lab pipeline have the potential to transform stem cell biology by uncovering fundamental mechanisms driving stem cell differentiation and facilitating rapid protocol development and benchmarking.

7   Cardiac Developmental Single-Cell RNA-Seq Data Analysis

Single-cell genomics methods applied to stem cell biology are advancing fundamental studies of hPSC differentiation, demonstrated by recent work detailing the subpopulation heterogeneity and complex regulation throughout stages of *in vitro* differentiation and reprogramming [18, 167–171]. In addition to this, single-cell genomics have the capability to take hPSC protocol development to scale, through high-throughput assessment of the accuracy with which *in vitro*-derived cells recapitulate *in vivo* development [172–175], evaluation of differentiation efficiency, and identification of sources of technical variation to aid protocol reproducibility and standardisation [176–178]. Emerging multiplexed experimental designs and co-culture approaches aim to further upscale discovery in stem cell biology and redefine hPSC protocols by extending their capabilities to study multilineage perturbations and population statistical genetics [144, 179 (preprint), 180, 181].

7.1   In Vivo *Cardiac Single Cell Analysis*

scRNA-seq has been used to study mechanisms of early lineage restriction and regional segregation of the heart at the gastrulation stage of mouse and human (Fig. 1 and Table 5). Initial studies focused on murine development given the ease of access to embryonic tissues. Current scRNA-seq data sets characterised the transcriptional features of mouse heart development across a wide range of developmental time course spanning from mouse embryonic day E6.5 to postanal day 21. In a study conducted by DeLaughter *et al.*, the authors studied mouse embryonic development from E9.5 to postnatal day 21 and defined cardiomyocytes that are spatiotemporally distinctive [149]. Moreover, Lescroart *et al.* provided a molecular map of early cardiovascular lineage segregation by sequencing cardiac cells collected from mouse E6.5 to E7.5. Similarly, de Soysa *et al.* identified the transcriptional features of cardiac cell specification during E7.75 to E9.25 of mouse embryonic development [155]. Xiao and colleagues revealed an essential role of Hippo signalling during cardiac fibroblast development by sequencing embryonic hearts with or without *Lat1/2* conditional knockout [182]. More recently, Tyser *et al.* [76] and Jia *et al.* [183] performed single cell transcriptomics analyses of

Table 4 Engineered 3D *in vitro* differentiation protocol for cardiovascular lineages

Cell types	Model	Products	Ref.
PSCs	Spheroid	FHF progenitors SHF progenitors	[10]
hPSCs-CMs/ECs	Spheroid	Cardiac microtissues with CMs and ECs	[122]
hPSCs	3D microtissues	Cardiac microtissues composed of hiPSC-CMs, hiPSC-ECs, and hiPSC-CFs	[123]
hPSCs-CMs/ECs	3D printing	Whole hearts with major blood vessels	[124]
hiPSCs-CMs	3D printing	Tubular heart constructs	[125]
Mouse ESCs	Embryonic-like entities	Embryoid that are structural similar to neurula-stage embryo	[126]
hPSCs	Organoids	Cardiac chamber-like structures	[127]
hPSCs	Cardiac organoids	Cardiac microchambers	[128]
hPSCs	Heart forming organoids	Myocardial layer lined by endocardial-like cells and surrounded by septum-transversum-like anlagen	[129]
hESC-CMs	Fibrin-based cardiac patch culture	Augmented conduction velocities Longer sarcomeres Enhanced cardiac contraction	[130]
hPSCs-CM and human dermal fibroblast sheets	Cell sheet-based tissue engineering Fibrin and collagen gels Octagonal column	Tubular cardiac tissues	[131]
hPSCs-CM + hCFs	Biowire II platform Hydrogel matrix	Chamber-specific cardiac tissues Commercially available	[132]
hPSCs-CM	Circular casting molds Collagen-based hydrogel	Ring-shaped chamber-specific cardiac tissues	[133]
hiPSCs	Spheroid using microfabricated vessels	CMs with high purity and maturation levels	[134]
hiPSCs	Casting molds PDMS posts Fibrin gel	Fibrin-based engineered heart tissues	[135]
hPSCs-CM + Human foreskin fibroblasts	Casting molds Bovine collagen	Macroscale human myocardium Adult-like stage	[136]
hESC-CM + hESC-stromal cells	Cardiac organoid PDMS molds	Heart-dyno culture Mature CMs	[137]
hiPSC-CMs/ECs/CFs	Replica molding	Cardiac microtissues	[138]
hPSCs	2D expansion +3D bioreactor culture	High purity cardiac spheres Large scale	[139]

CMs cardiomyocytes, *ECs* endothelial cells, *CFs* cardiac fibroblasts, *FHF* first heart field, *SHF* second heart field, *PDMS* polydimethylsiloxane

cardiac progenitor cells from mouse embryos. They both characterised the transcriptional profile of different cardiac progenitor cell types. Recently, a single cell transcriptomic map of developing cardiac out flow tract has also been reported, providing a thorough understanding of the cell states transitions in developing out flow tract [184]. For example, the authors identified that vascular smooth muscle cells can derived from either myocardial or mesenchymal cells.

More recent studies have used scRNA-seq to analyse human embryos. Cui *et al.* profiled 4000 anatomically informed cardiac cells from 18 human embryos that range from week 5 to week 25 of gestation to investigate gene expression across developmental patterns of human heart [185]. The study identified four cardiac cell types, cardiomyocytes, cardiac fibroblasts, endothelial cells, and valve interstitial cells, and analysed stepwise developmental changes in gene expression in cardiomyocytes and fibroblasts [185]. Moreover, by comparing cardiac cell subtypes between human and mouse single-cell data they suggest that cardiomyocytes are the most transcriptionally similar among those four cell types between human and mouse [185]. Another transcriptomic study conducted by Sahara *et al.* sequenced cardiac cells from human embryonic hearts that ranging from week 4.5 to week 10 of gestation and identified *LGR5* as a key regulator promoting cono-ventriculogenesis in humans. In the same study, the authors also performed another scRNA-seq as well as bulk RNA-seq using *in vitro* hESCs-derived cardiac cells to build up a comprehensive gene profile on cardiogenesis. An additional two single-cell studies focused on post-conception week 6.5–7 embryos [152] and analysis of hearts from foetal to adult stages [186]. Suryawanshi *et al.* compared the transcriptional changes in healthy human foetal hearts to foetal hearts with congenital heart block (CHB) using scRNA-seq [187]. They identified a profound interferon response in CHB foetal hearts. Beyond human and mouse, single cell data analyses of cardiopharyngeal lineages are emerging, enabling cross-species comparisons and insights into evolutionary mechanisms of cell differentiation (Fig. 1). In summary, applications of scRNA-seq have so far provide us a comprehensive molecular atlas of *in vivo* heart development, which not only help us to understand the key molecular pathways involved in heart development, but also enable investigation of treatment cues for heart diseases.

7.2   In Vitro *Cardiac Single Cell Analysis*

By comparison to *in vivo* cardiac developmental biology in humans and mice, parallel analysis has been performed to study human stem cell differentiation to identify methods for more accurately differentiating cells and/or identifying methods for enhancing cell maturation (Fig. 2 and Table 6). Friedman *et al.* performed single-cell transcriptome analysis of over 40,000 hiPSC-derived cells during cardiac cell differentiation at day 0 (pluripotency), day 2 (gastrulation), day 5 (progenitor), day 15 (committed), and day 30 (definitive) stages of differentiation [18]. By assessing gene programs during cardiac differentiation, they show that the non-DNA

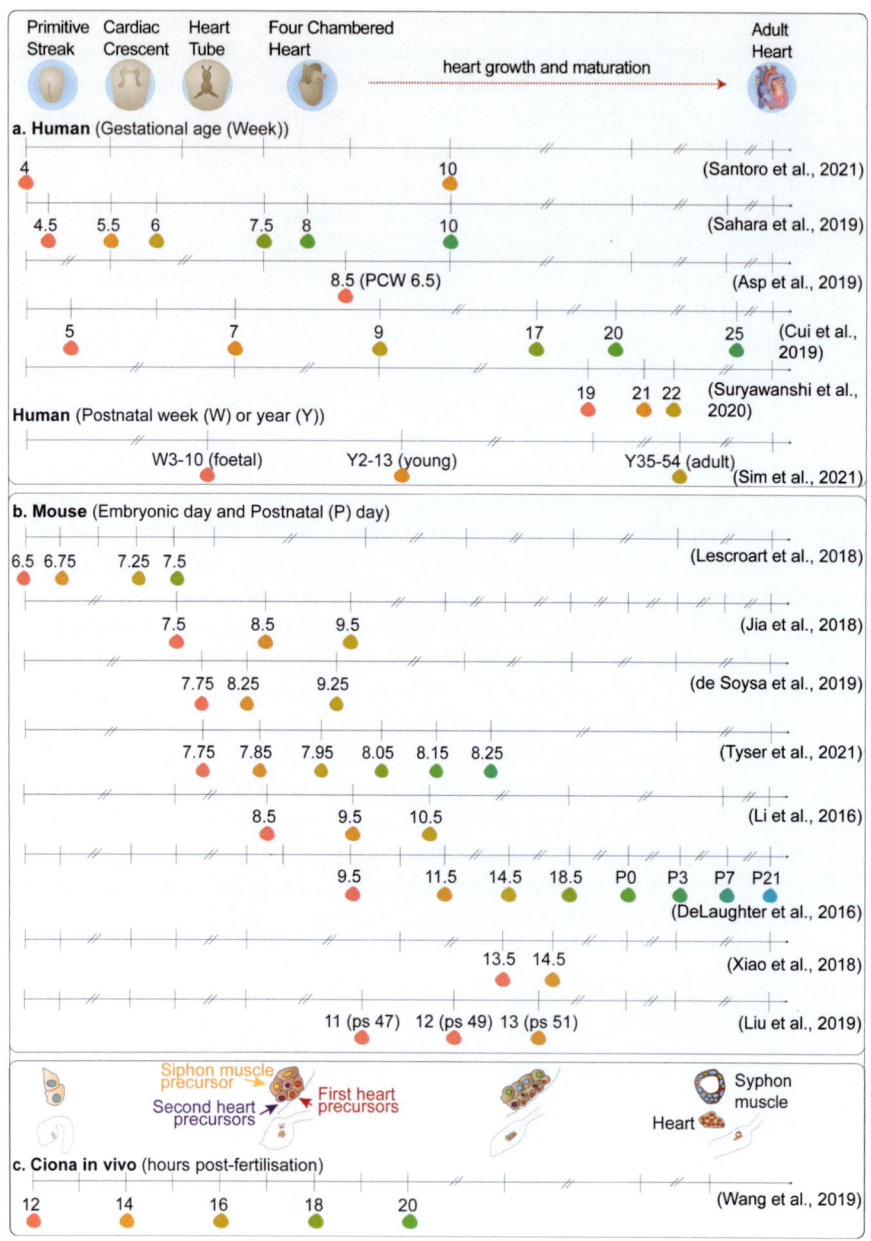

Fig. 1 Single-cell RNA-seq data sets analysing *in vivo* cardiovascular cell types from (**a**) humans, (**b**) mice, and (**c**) ciona

binding homeodomain protein HOPX plays a vital role in cell hypertrophy, and dysregulation of HOPX during differentiation leads to a persistently immature state of hiPSC-derived cardiomyocytes *in vitro* [18]. In a parallel study, Churko *et al.* sequenced differentiating hiPSCs at days 0, 5, 14, and 45 of cardiac differentiation. Focusing on cell subtype transcriptional regulators, they identified hiPSC-derived cardiomyocytes with expression of *COUP-TF2* and *TBX5* consisted of primarily immature and atrial-like cells, whereas expression of *HEY2, IRX4,* and *MYL2* enrich for ventricular-like hiPSC-derived cardiomyocytes [190]. In conjunction with chromatin immunoprecipitation sequencing (ChIP-seq) analysis, they provided evidence to elucidate distinct regulatory roles of *NR2F2* and *HEY2* in promoting atrial *vs.* ventricular cell states, respectively [190]. In addition to cardiomyocytes, hPSCs-derived endothelial cells serve as an essential model to study cardiovascular diseases *in vitro*. Paik *et al.* uncovered the heterogeneity of hiPSC-derived endothelial cells by sequencing differentiated cells on day 8 and day 12 [191]. Endothelial cell populations derived from hiPSCs are composed of 4 major subpopulations marked by enhanced expression of *CLDN5, APLNR, GJA5,* and *ESM 1*.

Apart from above mentioned time course scRNA-seq analysis of cardiovascular development, scRNA-seq has also been applied to dissect the transcriptional differences in different cardiomyocytes. For example, Lam *et al.* performed scRNA-seq of cardiomyocytes derived from either healthy individuals or patients with pulmonary atresia with intact ventricular septum (PAIVS) [192]. Combined with cardiac tissue engineering, they identified reduced contractility in cardiomyocytes generated from PAIVS-derived hiPSCs. More recently, Funakoshi *et al.* use scRNA-seq as a tool to characterise the composition of hPSCs-derived day 20 ventricular cardiomyocytes, day 32 mature cardiomyocytes, and day 32 untreated immature cardiomyocytes [117]. Similarly, Pezhouman and colleagues carried out scRNA-seq analysis comparing FHF and SHF-like cardiomyocytes on day 20 of differentiation in order to characterise the specific molecular markers for further generation of chamber specific cardiomyocytes [193].

Given the current understanding of mesendodermal differentiation and the signalling pathways involved in guiding this process, stem cells and their use in modelling these processes, and the current era of cellular genomics, there remain significant fundamental gaps and opportunities in these fields. The use of single-cell RNA-sequencing paired with the range of available multiplexing strategies has seen uptake through highly multiplexed "atlas" datasets describing cellular differentiation *in vivo*. Though cell types generated from *in vitro* stem cell differentiation are known to be physiologically immature compared to their adult counterparts *in vivo*, the causes of these differences, and how and when the phenotype-driving gene programs *in vitro* diverge from those *in vivo* are poorly characterised. This is due in part to a lack of comprehensive datasets that allow for benchmarking and comparing the cell types generated in the two systems. Finally, current experimental and computational methods for analysing the sequential transitions in differentiation in single-cell genomics data and identifying the key genetic and signalling switches governing these transitions are plentiful, but the field is still lacking strategies to remove housekeeping noise to identify key biological mechanism from single-cell

Table 5 Single cell data sets of *in vivo* development

Species	Time points	Cell types	Key question	Sequencing	Source
Human [150]	Week 5–7 of gestation Week 9–17 of gestation Week 20–25 of gestation	CMs Fibroblast-like cells Endothelial cells Valvar cells	Characterise human foetal heart development	STRT-seq	GSE106118
Human [186]	Foetal stage Young stage Adult stage	CMs Fibroblasts Endothelial cells Epicardial cells Immune cells Smooth muscle cells Neural cells Erythroid cells	Reveal the sex-dependent transcriptional changes across different cell types during heart development from foetal to adult stage	NovaSeq 6000 SnRNA-seq	GSE156707
Human [188]	Week 4.5–5.5 of gestation Week 6–7.5 of gestation Week 8–10 of gestation	Cono-ventricular progenitors Free-wall ventricular progenitors Intermediates of OFT Intermediates of ventricleEarly atria/PMs Late atria CMs expressing extracellular matrix genes Late cardiac mesenchymal cells Cono-ventricular muscle cells Free-wall ventricular muscle cells	Characterise the transcriptional dynamics of human cardiac lineage specification	SMART-seq2	PRJNA510181

Species	Sample	Cell types	Platform	Aim	Accession
Human [152]	6.5 PCW (~46 days)	Capillary endothelium; Ventricular CMs; Fibroblast-like (cardiac skeleton connective tissue); Epicardium-derived cells; Fibroblast-like (smaller vascular development); Smooth muscle cells; Atrial CMs; Fibroblast-like (larger vascular development); Epicardial cells; Endothelium/pericytes/adventia; ErythrocytesMyoz2-enriched CMs; Immune cells; Cardiac neural crest cells/Schwann progenitor cells	10x genomics	Dissect cell type heterogeneity of the heart, create organ-wide cell atlas of the human heart	EGAS00001003996
Human [187]	Healthy heart: Week 19–22 of gestation; Congenital heart block: 1 of gestation	Erythroblasts; Mesothelial cells; CMs; Endothelial cells; Stromal cells; Immune cells	10x genomics	Generate an atlas of human heart development and congenital heart diseases	PRJNA576243
Mouse [184]	OFT ps47 OFT ps49 OFT ps51	Mesenchymal; Myocardial; Vascular smooth muscle cells; Endocardial; Epicardial; Endocardial; Macrophage	10x genomics	Characterise the cell heterogeneity during OFT development and show vascular smooth muscle cells derived from myocardial and mesenchymal cells	PRJNA489304

(continued)

Table 5 (continued)

Species	Time points	Cell types	Key question	Sequencing	Source
Mouse [76]	E7.75 E7.85 E7.95 E8.05 E8.15 E8.25	Most differentiated CMs Differentiating cardiac progenitors Least differentiated CMs Endothelial mesoderm Blood mesoderm Definitive endoderm Yok sac endoderm Surface/amnion ectoderm Neuroectoderm	A transcriptional definition of cardiac progenitor cell types	SMART-seq2	E-MTAB-7403, PRJEB14363
Mouse [154]	E6.5 E6.75 E7.25 E7.5	Epiblast cells Endothelial cells Anterior SHF or pharyngeal mesoderm Cardiomyocytes or mesenchymal Notochord or definitive endoderm Posterior SHF or somatic mesoderm Others	Investigate the molecular and cellular basis of the earliest stages of cardiovascular lineage segregation	SMART-seq2	GSE100471 http://singlecell.stemcells.cam.ac.uk/mesp1
Mouse [148]	E8.5 E9.5 E10.5	Cardiomyocytes Endothelial cells Mesenchymal cells Epicardial cells	Characterise chamber-specific genes in the embryonic mouse heart	Fluidigm C1	GSE76118
Mouse [183]	E7.5 E8.5 E9.5	Early cardiac progenitor cells Intermediate cardiac progenitor cells CMs Endothelial cells	Characterise transcriptional regulations during cardiac progenitor cell fate decisions	Fluidigm C1	PRJEB23303
Mouse [149]	E9.5 E11.5 E14.5 E18.5 P0P3P7 P21	CMs Endothelial cells Fibroblast cells	Temporal gene expression during heart development	Fluidigm C1	Request from author

Mouse [182]	E13.5 E14.5	Macrophages Arterial endothelial cells Vascular endothelial cells Endocardial valve cells Endocardial cells Proliferating endocardial cells Atrioventricular canal CMs Trabecular CMs CMs Proliferating CMs Epicardial cells Valve mesenchymal cells Atrioventricular cushion mesenchymal cells Proliferating atrioventricular cushion mesenchymal cells Smooth muscle cells Fibroblasts	Characterise the role of hippo signalling in epicardial diversification	Drop-seq	GSE100861
Mouse [155]	E7.75 E8.25 E9.25	Endoderm_E7.75_1Endoderm_E7.75_2Endoderm_E7.75_3Endoderm_E7.75_4Endoderm_E7.75_5endocardial/endothelialHematoendothelialEndothelial mesenchymal transitionAnterior heart fieldPosterior SHF Branchiomeric muscle progenitorsEmbryonic myocardium progenitorsVentricleSinus venosusAtrialOFTAtrioventricular canalLeft ventricleEarly RVRight ventricle	Identify transcriptional features of cardiac cell specification and morphogenesis	10x genomics	UCSC cell browser: https://mouse-cardiac.cells.ucsc.edu
Ciona [189]	12 hpf 14 hpf 16 hpf 18 hpf 20 hpf	Trunk ventral cell First heart precursorsSecond trunk ventral cellsSecond heart precursors Atrial siphon muscle founder cellsInner atrial siphon muscle precursor Outer atrial siphon muscle precursor	Understand molecular mechanisms of cardio pharyngeal fate choices	SMART-seq2	GSE99846

CMs cardiomyocytes, *OFT* outflow tract, *SHF* second heart field, *hpf* hours post-fertilisation, *PCW* post-conception week, *ps* pairs of somites, *E* embryonic day, *P* postnatal day

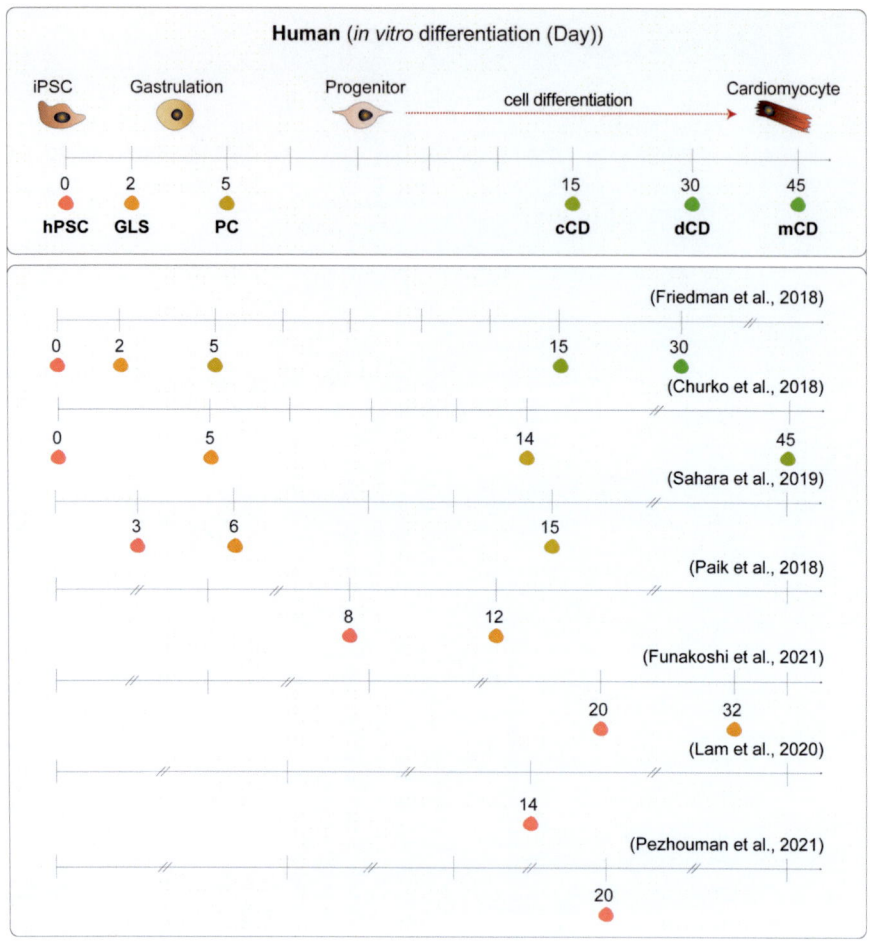

Fig. 2 Single-cell RNA-seq data sets analysing *in vitro* cardiovascular cell types differentiated from iPSCs and hESCs hPSC, human pluripotent stem cell; GLS, germ layer specification; PC, progenitor cell; cCD, committed cardiac derivative; dCD, definitive cardiac derivative; mCD, mature cardiac derivative

genomics data, particularly in the realm of cell type identification, classification, and lineage analysis.

8   Concluding Remarks

Pluripotent stem cells can self-renew and differentiate into all three germ layers to give rise to all cells or tissues of the body. They are therefore excellent cell models to study human development and pathophysiology *in vitro*. To date, PSCs have demonstrated promise in a range of applications including disease modelling, drug

Table 6 Single cell data sets of *in vitro* development

Source	Time Points	Cell types	Key findings	Source	Ref.
hiPSCs cardiac differentiation	Day 0 Day 2 Day 5 Day 15 Day 30	Definitive endoderm mesoderm Mesendoderm CM precursors Cardiovascular progenitor Non-contractile CMs Committed CMs Definitive CMs	Reveal cellular heterogeneity	E-MTAB-6268	[18]
hiPSCs cardiac differentiation	Day 0 Day 5 Day 14 Day 45	Pluripotent stem cells Definitive endoderm Mesoderm Ectoderm Stromal\Neural crest Endothelial cells Early CM progenitors Mid CM progenitors Late CM progenitors	Identify key transcription factors involved in different subpopulations of hiPSCs-CMs	GSE81585	[190]
hiPSCs EC differentiation	Day 8 Day 12	Arterial-like ECs Inflammation-responsive ECs Activated ECs Metabolically active ECs	Characterise the transcriptional heterogeneity of hiPSC-ECs	GSE116555	[191]
hESCs cardiac differentiation	Day 3 Day 6 Day 15	Immature cells Mesodermal precursors/multipotent cardiac progenitor Cardiac progenitors/intermediates Mature CMs Non-CMs	*LGR5* closely associated with SHF-related genes	PRJNA510181	[188]

(continued)

Table 6 (continued)

Source	Time Points	Cell types	Key findings	Source	Ref.
CMs derived from healthy individuals or patients with PAIVS	Day 14	Different CMs	CMs derived from PAVIS patients-derived hiPSC lines show developmental abnormalities	GSE157157	[192]
hESCs-CMs	Day 20	FHF-like CMs SHF-like CMs	Identify surface markers of FHF CMs for the generation of chamber specific CMs	GSE114373	[193]
Ventricular CMs	Day 20	Ventricular CMs Compact CMs Trabecular CMs Non-CMs	Characterise the maturation of CMs	GSE152589	[117]
Mature CMs Untreated immature CMs	Day 32	Mature CMs Immature CMs Fibroblast Smooth muscle cells Endoderm			

CMs cardiomyocytes, *ECs* endothelial cells, *PAIVS* pulmonary atresia with intact ventricular septum, *FHF* first heart field, *SHF* second heart field

discovery and screening, and regenerative medicine. The rapid development of technologies such as gene editing and tissue engineering will promote the translation of current iPSC-based therapies in clinical medicine and industry pipelines.

While the field is advancing with significant global growth anticipated, the current limitations in understanding mechanisms of cell development and cell state transitions represent a bottleneck impeding optimisation and development of *in vitro*-directed differentiation of PSCs towards specific cell types. Furthermore, derivation of cell subtypes is critical based on the unique function cells of the same organ play in healthy and disease processes, including variable responses to drugs in the context of iPSC drug screening platforms. Future studies must continue to develop knowledge into mechanisms of cell-fate decisions during differentiation, including the relationship between cell types and their environmental cues that guide genetic programs underpinning cell diversification and function. With advances in fields of genomics, gene editing, and cell biology, multidisciplinary teams with expertise in bioinformatics, gene modification techniques, and biological processes governing cell differentiation are required to collaboratively enable co-design of techniques and facilitate in their resulting interpretation.

References

1. Markets and Markets: Regenerative Medicine Market by Product (Cell Therapies (Autologous, Allogenic), Stemcell Therapy, Tissue-engineering, Gene Therapy), Application (Wound Care, Musculoskeletal, Oncology, Dental, Ocular. https://www.marketsandmarkets.com/PressReleases/regenerative-medicine.asp (2020). Accessed
2. Takebe, T., Sekine, K., Enomura, M., Koike, H., Kimura, M., Ogaeri, T., et al.: Vascularized and functional human liver from an iPSC-derived organ bud transplant. Nature. **499**(7459), 481–484 (2013). https://doi.org/10.1038/nature12271
3. McCracken, K.W., Catá, E.M., Crawford, C.M., Sinagoga, K.L., Schumacher, M., Rockich, B.E., et al.: Modelling human development and disease in pluripotent stem-cell-derived gastric organoids. Nature. **516**(7531), 400–404 (2014). https://doi.org/10.1038/nature13863
4. Loh, K.M., Chen, A., Koh, P.W., Deng, T.Z., Sinha, R., Tsai, J.M., et al.: Mapping the pairwise choices leading from pluripotency to human bone, heart, and other mesoderm cell types. Cell. **166**(2), 451–467 (2016). https://doi.org/10.1016/j.cell.2016.06.011
5. Loh, K.M., Ang, L.T., Zhang, J., Kumar, V., Ang, J., Auyeong, J.Q., et al.: Efficient endoderm induction from human pluripotent stem cells by logically directing signals controlling lineage bifurcations. Cell Stem Cell. **14**(2), 237–252 (2014). https://doi.org/10.1016/j.stem.2013.12.007
6. Mendjan, S., Mascetti, V.L., Ortmann, D., Ortiz, M., Karjosukarso, D.W., Ng, Y., et al.: NANOG and CDX2 pattern distinct subtypes of human mesoderm during exit from pluripotency. Cell Stem Cell. **15**(3), 310–325 (2014). https://doi.org/10.1016/j.stem.2014.06.006
7. Matsuda, M., Yamanaka, Y., Uemura, M., Osawa, M., Saito, M.K., Nagahashi, A., et al.: Recapitulating the human segmentation clock with pluripotent stem cells. Nature. **580**(7801), 124–129 (2020). https://doi.org/10.1038/s41586-020-2144-9
8. Khoshdel Rad, N., Aghdami, N., Moghadasali, R.: Cellular and molecular mechanisms of kidney development: from the embryo to the kidney organoid. Frontiers in Cell and Developmental Biology. **8**(183) (2020). https://doi.org/10.3389/fcell.2020.00183

9. Morizane, R., Miyoshi, T., Bonventre, J.V.: Concise review: kidney generation with human pluripotent stem cells. Stem Cells. **35**(11), 2209–2217 (2017). https://doi.org/10.1002/stem.2699

10. Andersen, P., Tampakakis, E., Jimenez, D.V., Kannan, S., Miyamoto, M., Shin, H.K., et al.: Precardiac organoids form two heart fields via Bmp/Wnt signaling. Nat. Commun. **9**(1), 3140 (2018). https://doi.org/10.1038/s41467-018-05604-8

11. Mori, S., Sakakura, E., Tsunekawa, Y., Hagiwara, M., Suzuki, T., Eiraku, M.: Self-organized formation of developing appendages from murine pluripotent stem cells. Nat. Commun. **10**(1), 3802 (2019). https://doi.org/10.1038/s41467-019-11702-y

12. Lee, J.H., Protze, S.I., Laksman, Z., Backx, P.H., Keller, G.M.: Human pluripotent stem cell-derived atrial and ventricular cardiomyocytes develop from distinct mesoderm populations. Cell Stem Cell. **21**(2), 179–94.e4 (2017). https://doi.org/10.1016/j.stem.2017.07.003

13. Koh, P.W., Sinha, R., Barkal, A.A., Morganti, R.M., Chen, A., Weissman, I.L., et al.: An atlas of transcriptional, chromatin accessibility, and surface marker changes in human mesoderm development. Scientific data. **3**, 160109 (2016). https://doi.org/10.1038/sdata.2016.109

14. Nelson, D.O., Jin, D.X., Downs, K.M., Kamp, T.J., Lyons, G.E.: Irx4 identifies a chamber-specific cell population that contributes to ventricular myocardium development. Developmental dynamics: an official publication of the American Association of Anatomists. **243**(3), 381–392 (2014). https://doi.org/10.1002/dvdy.24078

15. Goodyer, W.R., Beyersdorf, B.M., Paik, D.T., Tian, L., Li, G., Buikema, J.W., et al.: Transcriptomic profiling of the developing cardiac conduction system at single-cell resolution. Circ. Res. **125**(4), 379–397 (2019). https://doi.org/10.1161/circresaha.118.314578

16. van Eif, V.W.W., Stefanovic, S., van Duijvenboden, K., Bakker, M., Wakker, V., de Gier-de Vries, C., et al.: Transcriptome analysis of mouse and human sinoatrial node cells reveals a conserved genetic program. Development. **146**(8) (2019). https://doi.org/10.1242/dev.173161

17. Neri, T., Hiriart, E., van Vliet, P.P., Faure, E., Norris, R.A., Farhat, B., et al.: Human pre-valvular endocardial cells derived from pluripotent stem cells recapitulate cardiac patho-physiological valvulogenesis. Nat. Commun. **10**(1), 1929 (2019). https://doi.org/10.1038/s41467-019-09459-5

18. Friedman, C.E., Nguyen, Q., Lukowski, S.W., Helfer, A., Chiu, H.S., Miklas, J., et al.: Single-cell transcriptomic analysis of cardiac differentiation from human PSCs reveals HOPX-dependent cardiomyocyte maturation. Cell Stem Cell. **23**(4), 586–98.e8 (2018). https://doi.org/10.1016/j.stem.2018.09.009

19. Witty, A.D., Mihic, A., Tam, R.Y., Fisher, S.A., Mikryukov, A., Shoichet, M.S., et al.: Generation of the epicardial lineage from human pluripotent stem cells. Nat. Biotechnol. **32**(10), 1026–1035 (2014). https://doi.org/10.1038/nbt.3002

20. Marikawa, Y.: Wnt/beta-catenin signaling and body plan formation in mouse embryos. Semin. Cell Dev. Biol. **17**(2), 175–184 (2006). https://doi.org/10.1016/j.semcdb.2006.04.003

21. Arkell, R.M., Fossat, N., Tam, P.P.: Wnt signalling in mouse gastrulation and anterior development: new players in the pathway and signal output. Curr. Opin. Genet. Dev. **23**(4), 454–460 (2013). https://doi.org/10.1016/j.gde.2013.03.001

22. Williams, M., Burdsal, C., Periasamy, A., Lewandoski, M., Sutherland, A.: Mouse primitive streak forms in situ by initiation of epithelial to mesenchymal transition without migration of a cell population. Developmental dynamics: an official publication of the American Association of Anatomists. **241**(2), 270–283 (2012). https://doi.org/10.1002/dvdy.23711

23. Schauer, A., Heisenberg, C.P.: Reassembling gastrulation. Dev. Biol. **474**, 71–81 (2021). https://doi.org/10.1016/j.ydbio.2020.12.014

24. Qiao, Y., Yang, X., Jing, N.: Epigenetic regulation of early neural fate commitment. Cell. Mol. Life Sci. **73**(7), 1399–1411 (2016). https://doi.org/10.1007/s00018-015-2125-6

25. Parameswaran, M., Tam, P.P.: Regionalisation of cell fate and morphogenetic movement of the mesoderm during mouse gastrulation. Dev. Genet. **17**(1), 16–28 (1995). https://doi.org/10.1002/dvg.1020170104

26. Kinder, S.J., Tsang, T.E., Quinlan, G.A., Hadjantonakis, A.K., Nagy, A., Tam, P.P.: The orderly allocation of mesodermal cells to the extraembryonic structures and the anteroposterior axis during gastrulation of the mouse embryo. Development. **126**(21), 4691–4701 (1999)

27. Peng, G., Suo, S., Cui, G., Yu, F., Wang, R., Chen, J., et al.: Molecular architecture of lineage allocation and tissue organization in early mouse embryo. Nature. **572**(7770), 528–532 (2019). https://doi.org/10.1038/s41586-019-1469-8

28. Vandernoot, I., Haerlingen, B., Gillotay, P., Trubiroha, A., Janssens, V., Opitz, R., et al.: Enhanced canonical Wnt Signaling during early zebrafish development perturbs the interaction of cardiac mesoderm and pharyngeal endoderm and causes thyroid specification defects. Thyroid. (2020). https://doi.org/10.1089/thy.2019.0828

29. Li, L.-C., Wang, X., Xu, Z.-R., Wang, Y.-C., Feng, Y., Yang, L., et al.: Single-cell patterning and axis characterization in the murine and human definitive endoderm. Cell Res. **31**(3), 326–344 (2021). https://doi.org/10.1038/s41422-020-00426-0

30. Barlow, L.A.: Specification of pharyngeal endoderm is dependent on early signals from axial mesoderm. Development. **128**(22), 4573 (2001)

31. Marvin, M.J., Di Rocco, G., Gardiner, A., Bush, S.M., Lassar, A.B.: Inhibition of Wnt activity induces heart formation from posterior mesoderm. Genes Dev. **15**(3), 316–327 (2001). https://doi.org/10.1101/gad.855501

32. Kishimoto, K., Furukawa, K.T., Luz-Madrigal, A., Yamaoka, A., Matsuoka, C., Habu, M., et al.: Bidirectional Wnt signaling between endoderm and mesoderm confers tracheal identity in mouse and human cells. Nat. Commun. **11**(1), 4159 (2020). https://doi.org/10.1038/s41467-020-17969-w

33. Steimle, J.D., Rankin, S.A., Slagle, C.E., Bekeny, J., Rydeen, A.B., Chan, S.S.-K., et al.: Evolutionarily conserved Tbx5–Wnt2/2b pathway orchestrates cardiopulmonary development. Proc. Natl. Acad. Sci. **115**(45), E10615 (2018). https://doi.org/10.1073/pnas.1811624115

34. Rankin, S.A., Han, L., McCracken, K.W., Kenny, A.P., Anglin, C.T., Grigg, E.A., et al.: A retinoic acid-hedgehog cascade coordinates mesoderm-inducing signals and endoderm competence during lung specification. Cell Rep. **16**(1), 66–78 (2016). https://doi.org/10.1016/j.celrep.2016.05.060

35. Han, L., Chaturvedi, P., Kishimoto, K., Koike, H., Nasr, T., Iwasawa, K., et al.: Single cell transcriptomics identifies a signaling network coordinating endoderm and mesoderm diversification during foregut organogenesis. Nat. Commun. **11**(1), 4158 (2020). https://doi.org/10.1038/s41467-020-17968-x

36. Kumar, M., Jordan, N., Melton, D., Grapin-Botton, A.: Signals from lateral plate mesoderm instruct endoderm toward a pancreatic fate. Dev. Biol. **259**(1), 109–122 (2003). https://doi.org/10.1016/S0012-1606(03)00183-0

37. Davenport, C., Diekmann, U., Budde, I., Detering, N., Naujok, O.: Anterior–posterior patterning of definitive endoderm generated from human embryonic stem cells depends on the differential signaling of retinoic acid, Wnt-, and BMP-signaling. Stem Cells. **34**(11), 2635–2647 (2016). https://doi.org/10.1002/stem.2428

38. Mikryukov, A.A., Mazine, A., Wei, B., Yang, D., Miao, Y., Gu, M., et al.: BMP10 signaling promotes the development of endocardial cells from human pluripotent stem cell-derived cardiovascular progenitors. Cell Stem Cell. **28**(1), 96–111.e7 (2021). https://doi.org/10.1016/j.stem.2020.10.003

39. Nandkishore, N., Vyas, B., Javali, A., Ghosh, S., Sambasivan, R.: Divergent early mesoderm specification underlies distinct head and trunk muscle programmes in vertebrates. Development. **145**(18), dev160945 (2018). https://doi.org/10.1242/dev.160945

40. Dunn, N.R., Vincent, S.D., Oxburgh, L., Robertson, E.J., Bikoff, E.K.: Combinatorial activities of Smad2 and Smad3 regulate mesoderm formation and patterning in the mouse embryo. Development. **131**(8), 1717–1728 (2004). https://doi.org/10.1242/dev.01072

41. Kelly, O.G., Pinson, K.I., Skarnes, W.C.: The Wnt co-receptors Lrp5 and Lrp6 are essential for gastrulation in mice. Development. **131**(12), 2803–2815 (2004). https://doi.org/10.1242/dev.01137

42. Brennan, J., Lu, C.C., Norris, D.P., Rodriguez, T.A., Beddington, R.S., Robertson, E.J.: Nodal signalling in the epiblast patterns the early mouse embryo. Nature. **411**(6840), 965–969 (2001). https://doi.org/10.1038/35082103

43. van Boxtel, A.L., Economou, A.D., Heliot, C., Hill, C.S.: Long-range signaling activation and local inhibition separate the mesoderm and endoderm lineages. Developmental Cell. **44**(2), 179–91.e5 (2018). https://doi.org/10.1016/j.devcel.2017.11.021

44. Colombier, P., Halgand, B., Chédeville, C., Chariau, C., François-Campion, V., Kilens, S., et al.: NOTO transcription factor directs human induced pluripotent stem cell-derived mesendoderm progenitors to a notochordal fate. Cell. **9**(2), 509 (2020). https://doi.org/10.3390/cells9020509

45. Balmer, S., Nowotschin, S., Hadjantonakis, A.-K.: Notochord morphogenesis in mice: current understanding & open questions. Dev. Dyn. **245**(5), 547–557 (2016). https://doi.org/10.1002/dvdy.24392

46. Camus, A., Davidson, B.P., Billiards, S., Khoo, P., Rivera-Perez, J.A., Wakamiya, M., et al.: The morphogenetic role of midline mesendoderm and ectoderm in the development of the forebrain and the midbrain of the mouse embryo. Development. **127**(9), 1799 (2000)

47. Robb, L., Tam, P.P.L.: Gastrula organiser and embryonic patterning in the mouse. Semin. Cell Dev. Biol. **15**(5), 543–554 (2004). https://doi.org/10.1016/j.semcdb.2004.04.005

48. Sharon, N., Mor, I., Golan-lev, T., Fainsod, A., Benvenisty, N.: Molecular and functional characterizations of gastrula organizer cells derived from human embryonic stem cells. Stem Cells. **29**(4), 600–608 (2011). https://doi.org/10.1002/stem.621

49. Tani, S., Chung, U.-I., Ohba, S., Hojo, H.: Understanding paraxial mesoderm development and sclerotome specification for skeletal repair. Exp. Mol. Med. **52**(8), 1166–1177 (2020). https://doi.org/10.1038/s12276-020-0482-1

50. Perens, E.A., Garavito-Aguilar, Z.V., Guio-Vega, G.P., Peña, K.T., Schindler, Y.L., Yelon, D.: Hand2 inhibits kidney specification while promoting vein formation within the posterior mesoderm. elife. **5**, e19941 (2016). https://doi.org/10.7554/eLife.19941

51. Fleming, B.M., Yelin, R., James, R.G., Schultheiss, T.M.: A role for Vg1/nodal signaling in specification of the intermediate mesoderm. Development. **140**(8), 1819 (2013). https://doi.org/10.1242/dev.093740

52. Davidson, A.J., Lewis, P., Przepiorski, A., Sander, V.: Turning mesoderm into kidney. Semin. Cell Dev. Biol. **91**, 86–93 (2019). https://doi.org/10.1016/j.semcdb.2018.08.016

53. Prummel, K.D., Hess, C., Nieuwenhuize, S., Parker, H.J., Rogers, K.W., Kozmikova, I., et al.: A conserved regulatory program initiates lateral plate mesoderm emergence across chordates. Nat. Commun. **10**(1), 3857 (2019). https://doi.org/10.1038/s41467-019-11561-7

54. Schultheiss, T.M., Burch, J.B., Lassar, A.B.: A role for bone morphogenetic proteins in the induction of cardiac myogenesis. Genes Dev. **11**(4), 451–462 (1997)

55. Reifers, F., Walsh, E.C., Leger, S., Stainier, D.Y., Brand, M.: Induction and differentiation of the zebrafish heart requires fibroblast growth factor 8 (fgf8/acerebellar). Development. **127**(2), 225–235 (2000)

56. Pandur, P., Läsche, M., Eisenberg, L.M., Kühl, M.: Wnt-11 activation of a non-canonical Wnt signalling pathway is required for cardiogenesis. Nature. **418**, 636 (2002). https://doi.org/10.1038/nature00921

57. Dyer, L.A., Kirby, M.L.: Sonic hedgehog maintains proliferation in secondary heart field progenitors and is required for normal arterial pole formation. Dev. Biol. **330**(2), 305–317 (2009). https://doi.org/10.1016/j.ydbio.2009.03.028

58. Costello, I., Pimeisl, I.-M., Dräger, S., Bikoff, E.K., Robertson, E.J., Arnold, S.J.: The T-box transcription factor Eomesodermin acts upstream of Mesp1 to specify cardiac mesoderm during mouse gastrulation. Nat. Cell Biol. **13**, 1084 (2011). https://doi.org/10.1038/ncb2304. https://www.nature.com/articles/ncb2304#supplementary-information

59. Bondue, A., Lapouge, G., Paulissen, C., Semeraro, C., Iacovino, M., Kyba, M., et al.: Mesp1 acts as a master regulator of multipotent cardiovascular progenitor specification. Cell Stem Cell. **3**(1), 69–84 (2008). https://doi.org/10.1016/j.stem.2008.06.009

60. Saga, Y., Miyagawa-Tomita, S., Takagi, A., Kitajima, S., Miyazaki, J., Inoue, T.: MesP1 is expressed in the heart precursor cells and required for the formation of a single heart tube. Development. **126**(15), 3437–3447 (1999)

61. Devine, W.P., Wythe, J.D., George, M., Koshiba-Takeuchi, K., Bruneau, B.G.: Early patterning and specification of cardiac progenitors in gastrulating mesoderm. elife. **3**, e03848 (2014). https://doi.org/10.7554/eLife.03848

62. Lescroart, F., Chabab, S., Lin, X., Rulands, S., Paulissen, C., Rodolosse, A., et al.: Early lineage restriction in temporally distinct populations of Mesp1 progenitors during mammalian heart development. Nat. Cell Biol. **16**(9), 829–840 (2014). https://doi.org/10.1038/ncb3024

63. Schultheiss, T.M., Xydas, S., Lassar, A.B.: Induction of avian cardiac myogenesis by anterior endoderm. Development. **121**(12), 4203–4214 (1995)

64. Goss, C.M.: The first contractions of the heart in rat embryos. Anat. Rec. **70**(5), 505–524 (1938). https://doi.org/10.1002/ar.1090700502

65. Harvey, R.P.: Patterning the vertebrate heart. Nat. Rev. Genet. **3**, 544 (2002). https://doi.org/10.1038/nrg843

66. Kwon, C., Arnold, J., Hsiao, E.C., Taketo, M.M., Conklin, B.R., Srivastava, D.: Canonical Wnt signaling is a positive regulator of mammalian cardiac progenitors. Proc. Natl. Acad. Sci. U. S. A. **104**(26), 10894–10899 (2007). https://doi.org/10.1073/pnas.0704044104

67. Kwon, C., Qian, L., Cheng, P., Nigam, V., Arnold, J., Srivastava, D.: A regulatory pathway involving Notch1/β-catenin/Isl1 determines cardiac progenitor cell fate. Nat. Cell Biol. **11**, 951 (2009). https://doi.org/10.1038/ncb1906. https://www.nature.com/articles/ncb1906#supplementary-information

68. Prall, O.W., Menon, M.K., Solloway, M.J., Watanabe, Y., Zaffran, S., Bajolle, F., et al.: An Nkx2-5/Bmp2/Smad1 negative feedback loop controls heart progenitor specification and proliferation. Cell. **128**(5), 947–959 (2007). https://doi.org/10.1016/j.cell.2007.01.042

69. Paige, S.L., Plonowska, K., Xu, A., Wu, S.M.: Molecular regulation of cardiomyocyte differentiation. Circ. Res. **116**(2), 341–353 (2015). https://doi.org/10.1161/CIRCRESAHA.116.302752

70. Tzahor, E., Lassar, A.B.: Wnt signals from the neural tube block ectopic cardiogenesis. Genes Dev. **15**(3), 255–260 (2001)

71. Barron, M., Gao, M., Lough, J.: Requirement for BMP and FGF signaling during cardiogenic induction in non-precardiac mesoderm is specific, transient, and cooperative. Dev. Dyn. **218**(2), 383–393 (2000). https://doi.org/10.1002/(SICI)1097-0177(200006)218:2<383::AID-DVDY11>3.0.CO;2-P

72. Klaus, A., Müller, M., Schulz, H., Saga, Y., Martin, J.F., Birchmeier, W.: Wnt/β-catenin and Bmp signals control distinct sets of transcription factors in cardiac progenitor cells. Proc. Natl. Acad. Sci. U. S. A. **109**(27), 10921–10926 (2012). https://doi.org/10.1073/pnas.1121236109

73. Meilhac, S.M., Esner, M., Kerszberg, M., Moss, J.E., Buckingham, M.E.: Oriented clonal cell growth in the developing mouse myocardium underlies cardiac morphogenesis. J. Cell Biol. **164**(1), 97 (2004). https://doi.org/10.1083/jcb.200309160

74. Moretti, A., Caron, L., Nakano, A., Lam, J.T., Bernshausen, A., Chen, Y., et al.: Multipotent embryonic isl1+ progenitor cells lead to cardiac, smooth muscle, and endothelial cell diversification. Cell. **127**(6), 1151–1165 (2006). https://doi.org/10.1016/j.cell.2006.10.029

75. Spater, D., Abramczuk, M.K., Buac, K., Zangi, L., Stachel, M.W., Clarke, J., et al.: A HCN4+ cardiomyogenic progenitor derived from the first heart field and human pluripotent stem cells. Nat. Cell Biol. **15**(9), 1098–1106 (2013). https://doi.org/10.1038/ncb2824

76. Tyser, R.C.V., Ibarra-Soria, X., McDole, K., Arcot Jayaram, S., Godwin, J., van den Brand, T.A.H., et al.: Characterization of a common progenitor pool of the epicardium and myocardium. Science. **371**(6533), eabb2986 (2021). https://doi.org/10.1126/science.abb2986

77. Murry, C.E., Keller, G.: Differentiation of embryonic stem cells to clinically relevant populations: lessons from embryonic development. Cell. **132**(4), 661–680 (2008) doi: S0092-8674(08)00216-X [pii]; 10.1016/j.cell.2008.02.008

78. Múnera, J.O., Sundaram, N., Rankin, S.A., Hill, D., Watson, C., Mahe, M., et al.: Differentiation of human pluripotent stem cells into colonic organoids via transient activation of BMP signaling. Cell Stem Cell. **21**, 51 (2017). https://doi.org/10.1016/j.stem.2017.05.020

79. Singh, A., Poling, H.M., Spence, J.R., Wells, J.M., Helmrath, M.A.: Gastrointestinal organoids: a next-generation tool for modeling human development. Am. J. Physiol. Gastrointest. Liver Physiol. **319**(3), G375–Gg81 (2020). https://doi.org/10.1152/ajpgi.00199.2020

80. Spence, J.R., Mayhew, C.N., Rankin, S.A., Kuhar, M.F., Vallance, J.E., Tolle, K., et al.: Directed differentiation of human pluripotent stem cells into intestinal tissue in vitro. Nature. **470**(7332), 105–109 (2011). https://doi.org/10.1038/nature09691

81. Trisno, S.L., Philo, K.E.D., McCracken, K.W., Catá, E.M., Ruiz-Torres, S., Rankin, S.A., et al.: Esophageal organoids from human pluripotent stem cells delineate Sox2 functions during esophageal specification. Cell Stem Cell. 2018;23(4):501–15.e7. https://doi.org/10.1016/j.stem.2018.08.008

82. Miller, A.J., Dye, B.R., Ferrer-Torres, D., Hill, D.R., Overeem, A.W., Shea, L.D., et al.: Generation of lung organoids from human pluripotent stem cells in vitro. Nat. Protoc. **14**(2), 518–540 (2019). https://doi.org/10.1038/s41596-018-0104-8

83. Ng, E.S., Azzola, L., Bruveris, F.F., Calvanese, V., Phipson, B., Vlahos, K., et al.: Differentiation of human embryonic stem cells to HOXA+ hemogenic vasculature that resembles the aorta-gonad-mesonephros. Nat. Biotechnol. **34**(11), 1168–1179 (2016). https://doi.org/10.1038/nbt.3702

84. Sturgeon, C.M., Ditadi, A., Awong, G., Kennedy, M., Keller, G.: Wnt signaling controls the specification of definitive and primitive hematopoiesis from human pluripotent stem cells. Nat. Biotechnol. **32**(6), 554–561 (2014). https://doi.org/10.1038/nbt.2915

85. Freedman, B.S., Brooks, C.R., Lam, A.Q., Fu, H., Morizane, R., Agrawal, V., et al.: Modelling kidney disease with CRISPR-mutant kidney organoids derived from human pluripotent epiblast spheroids. Nat. Commun. **6**(1), 8715 (2015). https://doi.org/10.1038/ncomms9715

86. Morizane, R., Lam, A.Q., Freedman, B.S., Kishi, S., Valerius, M.T., Bonventre, J.V.: Nephron organoids derived from human pluripotent stem cells model kidney development and injury. Nat. Biotechnol. **33**(11), 1193–1200 (2015). https://doi.org/10.1038/nbt.3392

87. Taguchi, A., Kaku, Y., Ohmori, T., Sharmin, S., Ogawa, M., Sasaki, H., et al.: Redefining the in vivo origin of metanephric nephron progenitors enables generation of complex kidney structures from pluripotent stem cells. Cell Stem Cell. **14**(1), 53–67 (2014). https://doi.org/10.1016/j.stem.2013.11.010

88. Takasato, M., Er, P.X., Chiu, H.S., Maier, B., Baillie, G.J., Ferguson, C., et al.: Kidney organoids from human iPS cells contain multiple lineages and model human nephrogenesis. Nature. **526**(7574), 564–568 (2015). https://doi.org/10.1038/nature15695

89. Lancaster, M.A., Renner, M., Martin, C.-A., Wenzel, D., Bicknell, L.S., Hurles, M.E., et al.: Cerebral organoids model human brain development and microcephaly. Nature. **501**(7467), 373–379 (2013). https://doi.org/10.1038/nature12517

90. Koehler, K.R., Nie, J., Longworth-Mills, E., Liu, X.-P., Lee, J., Holt, J.R., et al.: Generation of inner ear organoids containing functional hair cells from human pluripotent stem cells. Nat. Biotechnol. **35**(6), 583–589 (2017). https://doi.org/10.1038/nbt.3840

91. Clevers, H.: Wnt/beta-catenin signaling in development and disease. Cell. **127**(3), 469–480 (2006). https://doi.org/10.1016/j.cell.2006.10.018

92. Clevers, H., Nusse, R.: Wnt/β-catenin signaling and disease. Cell. **149**(6), 1192–1205 (2012). https://doi.org/10.1016/j.cell.2012.05.012

93. Tortelote, G.G., Hernández-Hernández, J.M., Quaresma, A.J., Nickerson, J.A., Imbalzano, A.N., Rivera-Pérez, J.A.: Wnt3 function in the epiblast is required for the maintenance but not the initiation of gastrulation in mice. Dev. Biol. **374**(1), 164–173 (2013). https://doi.org/10.1016/j.ydbio.2012.10.013

94. Barrow, J.R., Howell, W.D., Rule, M., Hayashi, S., Thomas, K.R., Capecchi, M.R., et al.: Wnt3 signaling in the epiblast is required for proper orientation of the anteroposterior axis. Dev. Biol. **312**(1), 312–320 (2007). https://doi.org/10.1016/j.ydbio.2007.09.030

95. Liu, P., Wakamiya, M., Shea, M.J., Albrecht, U., Behringer, R.R., Bradley, A.: Requirement for Wnt3 in vertebrate axis formation. Nat. Genet. **22**(4), 361–365 (1999). https://doi.org/10.1038/11932

96. Davidson, K.C., Adams, A.M., Goodson, J.M., McDonald, C.E., Potter, J.C., Berndt, J.D., et al.: Wnt/β-catenin signaling promotes differentiation, not self-renewal, of human embryonic stem cells and is repressed by Oct4. Proc. Natl. Acad. Sci. U. S. A. **109**(12), 4485–4490 (2012). https://doi.org/10.1073/pnas.1118777109

97. Tan, J.Y., Sriram, G., Rufaihah, A.J., Neoh, K.G., Cao, T.: Efficient derivation of lateral plate and paraxial mesoderm subtypes from human embryonic stem cells through GSKi-mediated differentiation. Stem Cells Dev. **22**(13), 1893–1906 (2013). https://doi.org/10.1089/scd.2012.0590

98. Ueno, S., Weidinger, G., Osugi, T., Kohn, A.D., Golob, J.L., Pabon, L., et al.: Biphasic role for Wnt/beta-catenin signaling in cardiac specification in zebrafish and embryonic stem cells. Proc. Natl. Acad. Sci. U. S. A. **104**(23), 9685–9690 (2007) doi: 0702859104 [pii]; 10.1073/pnas.0702859104

99. Naito, A.T., Shiojima, I., Akazawa, H., Hidaka, K., Morisaki, T., Kikuchi, A., et al.: Developmental stage-specific biphasic roles of Wnt/beta-catenin signaling in cardiomyogenesis and hematopoiesis. Proc. Natl. Acad. Sci. U. S. A. **103**(52), 19812–19817 (2006). https://doi.org/10.1073/pnas.0605768103

100. Zhao, M., Tang, Y., Zhou, Y., Zhang, J.: Deciphering role of Wnt signalling in cardiac mesoderm and cardiomyocyte differentiation from human iPSCs: four-dimensional control of Wnt pathway for hiPSC-CMs differentiation. Sci. Rep. **9**(1), 19389 (2019). https://doi.org/10.1038/s41598-019-55620-x

101. Dohn, T.E., Waxman, J.S.: Distinct phases of Wnt/β-catenin signaling direct cardiomyocyte formation in zebrafish. Dev. Biol. **361**(2), 364–376 (2012). https://doi.org/10.1016/j.ydbio.2011.10.032

102. Kattman, S.J., Witty, A.D., Gagliardi, M., Dubois, N.C., Niapour, M., Hotta, A., et al.: Stage-specific optimization of activin/nodal and BMP signaling promotes cardiac differentiation of mouse and human pluripotent stem cell lines. Cell Stem Cell. **8**(2), 228–240 (2011). https://doi.org/10.1016/j.stem.2010.12.008

103. Lian, X., Hsiao, C., Wilson, G., Zhu, K., Hazeltine, L.B., Azarin, S.M., et al.: Robust cardiomyocyte differentiation from human pluripotent stem cells via temporal modulation of canonical Wnt signaling. Proc. Natl. Acad. Sci. U. S. A. 2012;109(27):E1848–E1857. doi: 1200250109 [pii]; https://doi.org/10.1073/pnas.1200250109

104. Lian, X., Bao, X., Zilberter, M., Westman, M., Fisahn, A., Hsiao, C., et al.: Chemically defined, albumin-free human cardiomyocyte generation. Nat. Methods. **12**(7), 595–596 (2015). https://doi.org/10.1038/nmeth.3448

105. Burridge, P.W., Matsa, E., Shukla, P., Lin, Z.C., Churko, J.M., Ebert, A.D., et al.: Chemically defined generation of human cardiomyocytes. Nat. Methods. **11**(8), 855–860 (2014). https://doi.org/10.1038/nmeth.2999

106. Kehat, I., Kenyagin-Karsenti, D., Snir, M., Segev, H., Amit, M., Gepstein, A., et al.: Human embryonic stem cells can differentiate into myocytes with structural and functional properties of cardiomyocytes. J. Clin. Invest. **108**(3), 407–414 (2001). https://doi.org/10.1172/jci12131

107. Chen, V.C., Ye, J., Shukla, P., Hua, G., Chen, D., Lin, Z., et al.: Development of a scalable suspension culture for cardiac differentiation from human pluripotent stem cells. Stem Cell Res. **15**(2), 365–375 (2015). https://doi.org/10.1016/j.scr.2015.08.002

108. Devalla, H.D., Schwach, V., Ford, J.W., Milnes, J.T., El-Haou, S., Jackson, C., et al.: Atrial-like cardiomyocytes from human pluripotent stem cells are a robust preclinical model for assessing atrial-selective pharmacology. EMBO Molecular Medicine. **7**(4), 394–410 (2015). https://doi.org/10.15252/emmm.201404757

109. Protze, S.I., Liu, J., Nussinovitch, U., Ohana, L., Backx, P.H., Gepstein, L., et al.: Sinoatrial node cardiomyocytes derived from human pluripotent cells function as a biological pacemaker. Nat. Biotechnol. **35**(1), 56–68 (2017). https://doi.org/10.1038/nbt.3745

110. Litviňuková, M., Talavera-López, C., Maatz, H., Reichart, D., Worth, C.L., Lindberg, E.L., et al.: Cells of the adult human heart. Nature. **588**(7838), 466–472 (2020). https://doi.org/10.1038/s41586-020-2797-4

111. Palpant, N.J., Pabon, L., Friedman, C.E., Roberts, M., Hadland, B., Zaunbrecher, R.J., et al.: Generating high-purity cardiac and endothelial derivatives from patterned mesoderm using human pluripotent stem cells. Nat. Protoc. **12**(1), 15–31 (2017). https://doi.org/10.1038/nprot.2016.153

112. Zhao, J., Cao, H., Tian, L., Huo, W., Zhai, K., Wang, P., et al.: Efficient differentiation of TBX18(+)/WT1(+) epicardial-like cells from human pluripotent stem cells using small molecular compounds. Stem Cells Dev. **26**(7), 528–540 (2017). https://doi.org/10.1089/scd.2016.0208

113. Zhang, J., Tao, R., Campbell, K.F., Carvalho, J.L., Ruiz, E.C., Kim, G.C., et al.: Functional cardiac fibroblasts derived from human pluripotent stem cells via second heart field progenitors. Nat. Commun. **10**(1), 2238 (2019). https://doi.org/10.1038/s41467-019-09831-5

114. Koivumäki, J.T., Naumenko, N., Tuomainen, T., Takalo, J., Oksanen, M., Puttonen, K.A., et al.: Structural immaturity of human iPSC-derived cardiomyocytes: in silico investigation of effects on function and disease modeling. Front. Physiol. **9**, 80 (2018). https://doi.org/10.3389/fphys.2018.00080

115. Lin, Y., Linask, K.L., Mallon, B., Johnson, K., Klein, M., Beers, J., et al.: Heparin promotes cardiac differentiation of human pluripotent stem cells in chemically defined albumin-free medium, enabling consistent manufacture of cardiomyocytes. Stem Cells Transl. Med. **6**(2), 527–538 (2017). https://doi.org/10.5966/sctm.2015-0428

116. Tohyama, S., Fujita, J., Fujita, C., Yamaguchi, M., Kanaami, S., Ohno, R., et al.: Efficient large-scale 2D culture system for human induced pluripotent stem cells and differentiated cardiomyocytes. Stem cell reports. **9**(5), 1406–1414 (2017). https://doi.org/10.1016/j.stemcr.2017.08.025

117. Funakoshi, S., Fernandes, I., Mastikhina, O., Wilkinson, D., Tran, T., Dhahri, W., et al.: Generation of mature compact ventricular cardiomyocytes from human pluripotent stem cells. Nat. Commun. **12**(1), 3155 (2021). https://doi.org/10.1038/s41467-021-23329-z

118. Bao, X., Lian, X., Hacker, T.A., Schmuck, E.G., Qian, T., Bhute, V.J., et al.: Long-term self-renewing human epicardial cells generated from pluripotent stem cells under defined xeno-free conditions. Nature biomedical engineering. **1** (2016). https://doi.org/10.1038/s41551-016-0003

119. Guadix, J.A., Orlova, V.V., Giacomelli, E., Bellin, M., Ribeiro, M.C., Mummery, C.L., et al.: Human pluripotent stem cell differentiation into functional epicardial progenitor cells. Stem cell reports. **9**(6), 1754–1764 (2017). https://doi.org/10.1016/j.stemcr.2017.10.023

120. Iyer, D., Gambardella, L., Bernard, W.G., Serrano, F., Mascetti, V.L., Pedersen, R.A., et al.: Robust derivation of epicardium and its differentiated smooth muscle cell progeny from human pluripotent stem cells. Development. **142**(8), 1528–1541 (2015). https://doi.org/10.1242/dev.119271

121. Zhang, H., Tian, L., Shen, M., Tu, C., Wu, H., Gu, M., et al.: Generation of quiescent cardiac fibroblasts from human induced pluripotent stem cells for in vitro modeling of cardiac fibrosis. Circ. Res. **125**(5), 552–566 (2019). https://doi.org/10.1161/circresaha.119.315491

122. Giacomelli, E., Bellin, M., Sala, L., van Meer, B.J., Tertoolen, L.G., Orlova, V.V., et al.: Three-dimensional cardiac microtissues composed of cardiomyocytes and endothelial cells co-differentiated from human pluripotent stem cells. Development. **144**(6), 1008–1017 (2017). https://doi.org/10.1242/dev.143438

123. Campostrini, G., Meraviglia, V., Giacomelli, E., van Helden, R.W.J., Yiangou, L., Davis, R.P., et al.: Generation, functional analysis and applications of isogenic three-dimensional

self-aggregating cardiac microtissues from human pluripotent stem cells. Nat. Protoc. **16**(4), 2213–2256 (2021). https://doi.org/10.1038/s41596-021-00497-2

124. Noor, N., Shapira, A., Edri R, Gal I, Wertheim L, Dvir T. 3D printing of personalized thick and perfusable cardiac patches and hearts. Advanced Science (Weinheim, Baden-Wurttemberg, Germany). 2019;6(11):1900344. https://doi.org/10.1002/advs.201900344

125. Arai, K., Murata, D., Takao, S., Nakamura, A., Itoh, M., Kitsuka, T., et al.: Drug response analysis for scaffold-free cardiac constructs fabricated using bio-3D printer. Sci. Rep. **10**(1), 8972 (2020). https://doi.org/10.1038/s41598-020-65681-y

126. Xu, P.F., Borges, R.M., Fillatre, J., de Oliveira-Melo, M., Cheng, T., Thisse, B., et al.: Construction of a mammalian embryo model from stem cells organized by a morphogen signalling Centre. Nat. Commun. **12**(1), 3277 (2021). https://doi.org/10.1038/s41467-021-23653-4

127. Hofbauer, P., Jahnel, S.M., Papai, N., Giesshammer, M., Deyett, A., Schmidt, C., et al.: Cardioids reveal self-organizing principles of human cardiogenesis. Cell. (2021). https://doi.org/10.1016/j.cell.2021.04.034

128. Hoang, P., Wang, J., Conklin, B.R., Healy, K.E., Ma, Z.: Generation of spatial-patterned early-developing cardiac organoids using human pluripotent stem cells. Nat. Protoc. **13**(4), 723–737 (2018). https://doi.org/10.1038/nprot.2018.006

129. Drakhlis, L., Biswanath, S., Farr, C.M., Lupanow, V., Teske, J., Ritzenhoff, K., et al.: Human heart-forming organoids recapitulate early heart and foregut development. Nat. Biotechnol. (2021). https://doi.org/10.1038/s41587-021-00815-9

130. Zhang, D., Shadrin, I.Y., Lam, J., Xian, H.Q., Snodgrass, H.R., Bursac, N.: Tissue-engineered cardiac patch for advanced functional maturation of human ESC-derived cardiomyocytes. Biomaterials. **34**(23), 5813–5820 (2013). https://doi.org/10.1016/j.biomaterials.2013.04.026

131. Tsuruyama, S., Matsuura, K., Sakaguchi, K., Shimizu, T.: Pulsatile tubular cardiac tissues fabricated by wrapping human iPS cells-derived cardiomyocyte sheets. Regenerative therapy. **11**, 297–305 (2019). https://doi.org/10.1016/j.reth.2019.09.001

132. Zhao, Y., Rafatian, N., Feric NT, Cox BJ, Aschar-Sobbi R, Wang EY, et al. A platform for generation of chamber-specific cardiac tissues and disease modeling. Cell. 2019;176(4):913–27.e18. https://doi.org/10.1016/j.cell.2018.11.042

133. Goldfracht, I., Protze, S., Shiti, A., Setter, N., Gruber, A., Shaheen, N., et al.: Generating ring-shaped engineered heart tissues from ventricular and atrial human pluripotent stem cell-derived cardiomyocytes. Nat. Commun. **11**(1), 75 (2020). https://doi.org/10.1038/s41467-019-13868-x

134. Miwa, T., Idiris, A., Kumagai, H.: A novel cardiac differentiation method of a large number and uniformly-sized spheroids using microfabricated culture vessels. Regenerative therapy. **15**, 18–26 (2020). https://doi.org/10.1016/j.reth.2020.04.008

135. Breckwoldt, K., Letuffe-Brenière, D., Mannhardt, I., Schulze, T., Ulmer, B., Werner, T., et al.: Differentiation of cardiomyocytes and generation of human engineered heart tissue. Nat. Protoc. **12**(6), 1177–1197 (2017). https://doi.org/10.1038/nprot.2017.033

136. Tiburcy, M., Hudson, J.E., Balfanz, P., Schlick, S., Meyer, T., Chang Liao, M.L., et al.: Defined engineered human myocardium with advanced maturation for applications in heart failure modeling and repair. Circulation. **135**(19), 1832–1847 (2017). https://doi.org/10.1161/circulationaha.116.024145

137. Mills, R.J., Titmarsh, D.M., Koenig, X., Parker, B.L., Ryall, J.G., Quaife-Ryan, G.A., et al.: Functional screening in human cardiac organoids reveals a metabolic mechanism for cardiomyocyte cell cycle arrest. Proc. Natl. Acad. Sci. U. S. A. **114**(40), E8372–E8e81 (2017). https://doi.org/10.1073/pnas.1707316114

138. Thomas, D., Kim, H., Lopez, N., Wu, J.C.: Fabrication of 3D cardiac microtissue arrays using human iPSC-derived cardiomyocytes, cardiac fibroblasts, and endothelial cells. Journal of visualized experiments: JoVE. **169** (2021). https://doi.org/10.3791/61879

139. Kempf, H., Kropp, C., Olmer, R., Martin, U., Zweigerdt, R.: Cardiac differentiation of human pluripotent stem cells in scalable suspension culture. Nat. Protoc. **10**(9), 1345–1361 (2015). https://doi.org/10.1038/nprot.2015.089

140. Tang, F., Barbacioru, C., Wang, Y., Nordman, E., Lee, C., Xu, N., et al.: mRNA-Seq whole-transcriptome analysis of a single cell. Nat. Methods. **6**(5), 377–382 (2009). https://doi.org/10.1038/nmeth.1315

141. Picelli, S., Björklund, Å.K., Faridani, O.R., Sagasser, S., Winberg, G., Sandberg, R.: Smart-seq2 for sensitive full-length transcriptome profiling in single cells. Nat. Methods. **10**(11), 1096–1098 (2013). https://doi.org/10.1038/nmeth.2639

142. Macosko, E.Z., Basu, A., Satija, R., Nemesh, J., Shekhar, K., Goldman, M., et al.: Highly parallel genome-wide expression profiling of individual cells using nanoliter droplets. Cell. **161**(5), 1202–1214 (2015). https://doi.org/10.1016/j.cell.2015.05.002

143. Hashimshony, T., Senderovich, N., Avital, G., Klochendler, A., de Leeuw, Y., Anavy, L., et al.: CEL-Seq2: sensitive highly-multiplexed single-cell RNA-Seq. Genome Biol. **17**, 77 (2016). https://doi.org/10.1186/s13059-016-0938-8

144. Zheng, G.X., Terry, J.M., Belgrader, P., Ryvkin, P., Bent, Z.W., Wilson, R., et al.: Massively parallel digital transcriptional profiling of single cells. Nat. Commun. **8**, 14049 (2017). https://doi.org/10.1038/ncomms14049

145. Guo, C., Kong, W., Kamimoto, K., Rivera-Gonzalez, G.C., Yang, X., Kirita, Y., et al.: CellTag Indexing: genetic barcode-based sample multiplexing for single-cell genomics. Genome Biol. **20**(1), 90 (2019). https://doi.org/10.1186/s13059-019-1699-y

146. Kang, H.M., Subramaniam, M., Targ, S., Nguyen, M., Maliskova, L., McCarthy, E., et al.: Multiplexed droplet single-cell RNA-sequencing using natural genetic variation. Nat. Biotechnol. **36**(1), 89–94 (2018). https://doi.org/10.1038/nbt.4042

147. Stoeckius, M., Zheng, S., Houck-Loomis, B., Hao, S., Yeung, B.Z., Mauck 3rd, W.M., et al.: Cell hashing with barcoded antibodies enables multiplexing and doublet detection for single cell genomics. Genome Biol. **19**(1), 224 (2018). https://doi.org/10.1186/s13059-018-1603-1

148. Li, G., Xu, A., Sim, S., Priest, J.R., Tian, X., Khan, T., et al.: Transcriptomic profiling maps anatomically patterned subpopulations among single embryonic cardiac cells. Dev. Cell. **39**(4), 491–507 (2016). https://doi.org/10.1016/j.devcel.2016.10.014

149. DeLaughter, D.M., Bick, A.G., Wakimoto, H., McKean, D., Gorham, J.M., Kathiriya, I.S., et al.: Single-cell resolution of temporal gene expression during heart development. Dev. Cell. **39**(4), 480–490 (2016). https://doi.org/10.1016/j.devcel.2016.10.001

150. Cui, Y., Zheng, Y., Liu, X., Yan, L., Fan, X., Yong, J., et al.: Single-cell transcriptome analysis maps the developmental track of the human heart. Cell Rep. **26**(7), 1934–50.e5 (2019). https://doi.org/10.1016/j.celrep.2019.01.079

151. Cao, J., Spielmann, M., Qiu, X., Huang, X., Ibrahim, D.M., Hill, A.J., et al.: The single-cell transcriptional landscape of mammalian organogenesis. Nature. **566**(7745), 496–502 (2019). https://doi.org/10.1038/s41586-019-0969-x

152. Asp, M., Giacomello, S., Larsson L, Wu C, Fürth D, Qian X, et al. A spatiotemporal organ-wide gene expression and cell atlas of the developing human heart. Cell. 2019;179(7):1647–60. e19. https://doi.org/10.1016/j.cell.2019.11.025

153. Pijuan-Sala, B., Griffiths, J.A., Guibentif, C., Hiscock, T.W., Jawaid, W., Calero-Nieto, F.J., et al.: A single-cell molecular map of mouse gastrulation and early organogenesis. Nature. **566**(7745), 490–495 (2019). https://doi.org/10.1038/s41586-019-0933-9

154. Lescroart, F., Wang, X., Lin, X., Swedlund, B., Gargouri, S., Sànchez-Dànes, A., et al.: Defining the earliest step of cardiovascular lineage segregation by single-cell RNA-seq. Science. **359**(6380), 1177 (2018). https://doi.org/10.1126/science.aao4174

155. de Soysa, T.Y., Ranade, S.S., Okawa, S., Ravichandran, S., Huang, Y., Salunga, H.T., et al.: Single-cell analysis of cardiogenesis reveals basis for organ-level developmental defects. Nature. **572**(7767), 120–124 (2019). https://doi.org/10.1038/s41586-019-1414-x

156. Hashimshony, T., Senderovich, N., Avital, G., Klochendler, A., de Leeuw, Y., Anavy, L., et al.: CEL-Seq2: sensitive highly-multiplexed single-cell RNA-Seq. Genome Biol. **17**(1), 77 (2016). https://doi.org/10.1186/s13059-016-0938-8

157. Kouno T, Moody J, Kwon AT-J, Shibayama Y, Kato S, Huang Y, et al. C1 CAGE detects transcription start sites and enhancer activity at single-cell resolution. Nature communications. 2019;10(1):360-. https://doi.org/10.1038/s41467-018-08126-5

158. Wang, Y., Waters, J., Leung, M.L., Unruh, A., Roh, W., Shi, X., et al.: Clonal evolution in breast cancer revealed by single nucleus genome sequencing. Nature. **512**(7513), 155–160 (2014). https://doi.org/10.1038/nature13600

159. Zhu, P., Guo, H., Ren, Y., Hou, Y., Dong, J., Li, R., et al.: Single-cell DNA methylome sequencing of human preimplantation embryos. Nat. Genet. **50**(1), 12–19 (2018). https://doi.org/10.1038/s41588-017-0007-6

160. Buenrostro, J.D., Wu, B., Litzenburger, U.M., Ruff, D., Gonzales, M.L., Snyder, M.P., et al.: Single-cell chromatin accessibility reveals principles of regulatory variation. Nature. **523**(7561), 486–490 (2015). https://doi.org/10.1038/nature14590

161. Stoeckius, M., Hafemeister, C., Stephenson, W., Houck-Loomis, B., Chattopadhyay, P.K., Swerdlow, H., et al.: Simultaneous epitope and transcriptome measurement in single cells. Nat. Methods. **14**(9), 865–868 (2017). https://doi.org/10.1038/nmeth.4380

162. Eng, C.L., Lawson, M., Zhu, Q., Dries, R., Koulena, N., Takei, Y., et al.: Transcriptome-scale super-resolved imaging in tissues by RNA seqFISH. Nature. **568**(7751), 235–239 (2019). https://doi.org/10.1038/s41586-019-1049-y

163. Rodriques, S.G., Stickels, R.R., Goeva, A., Martin, C.A., Murray, E., Vanderburg, C.R., et al.: Slide-seq: a scalable technology for measuring genome-wide expression at high spatial resolution. Science. **363**(6434), 1463 (2019). https://doi.org/10.1126/science.aaw1219

164. Stuart, T., Satija, R.: Integrative single-cell analysis. Nat. Rev. Genet. **20**(5), 257–272 (2019). https://doi.org/10.1038/s41576-019-0093-7

165. Lescroart, F., Wang, X., Lin, X., Swedlund, B., Gargouri, S., Sànchez-Dànes, A., et al.: Defining the earliest step of cardiovascular lineage segregation by single-cell RNA-seq. Science. (2018). https://doi.org/10.1126/science.aao4174

166. Tang, F., Barbacioru, C., Wang, Y., Nordman, E., Lee, C., Xu, N., et al.: mRNA-Seq whole-transcriptome analysis of a single cell. Nat. Methods. **6**(5), 377–382 (2009). https://doi.org/10.1038/nmeth.1315

167. Nguyen, Q.H., Lukowski, S.W., Chiu, H.S., Senabouth, A., Bruxner, T.J.C., Christ, A.N., et al.: Single-cell RNA-seq of human induced pluripotent stem cells reveals cellular heterogeneity and cell state transitions between subpopulations. Genome Res. **28**(7), 1053–1066 (2018). https://doi.org/10.1101/gr.223925.117

168. Natarajan, K.N., Teichmann, S.A., Kolodziejczyk, A.A.: Single cell transcriptomics of pluripotent stem cells: reprogramming and differentiation. Curr. Opin. Genet. Dev. **46**, 66–76 (2017). https://doi.org/10.1016/j.gde.2017.06.003

169. Han, X., Chen, H., Huang, D., Chen, H., Fei, L., Cheng, C., et al.: Mapping human pluripotent stem cell differentiation pathways using high throughput single-cell RNA-sequencing. Genome Biol. **19**(1), 47 (2018). https://doi.org/10.1186/s13059-018-1426-0

170. Wu, H., Humphreys, B.D.: Single cell sequencing and kidney organoids generated from pluripotent stem cells. Clin. J. Am. Soc. Nephrol. **15**(4), 550–556 (2020). https://doi.org/10.2215/cjn.07470619

171. Jia, G., Preussner, J., Chen, X., Guenther, S., Yuan, X., Yekelchyk, M., et al.: Single cell RNA-seq and ATAC-seq analysis of cardiac progenitor cell transition states and lineage settlement. Nat. Commun. **9**(1), 4877 (2018). https://doi.org/10.1038/s41467-018-07307-6

172. Combes, A.N., Zappia, L., Er, P.X., Oshlack, A., Little, M.H.: Single-cell analysis reveals congruence between kidney organoids and human fetal kidney. Genome Med. **11**(1), 3 (2019). https://doi.org/10.1186/s13073-019-0615-0

173. Eura, N., Matsui, T.K., Luginbühl, J., Matsubayashi, M., Nanaura, H., Shiota, T., et al.: Brainstem organoids from human pluripotent stem cells. Front. Neurosci. **14**, 538 (2020). https://doi.org/10.3389/fnins.2020.00538

174. Miyoshi, T., Hiratsuka, K., Saiz, E.G., Morizane, R.: Kidney organoids in translational medicine: disease modeling and regenerative medicine. Developmental dynamics: an official

publication of the American Association of Anatomists. **249**(1), 34–45 (2020). https://doi.org/10.1002/dvdy.22

175. Sridhar, A., Hoshino, A., Finkbeiner CR, Chitsazan A, Dai L, Haugan AK, et al. Single-cell transcriptomic comparison of human fetal retina, hPSC-derived retinal organoids, and long-term retinal cultures. Cell Rep. 2020;30(5):1644–59.e4. https://doi.org/10.1016/j.celrep.2020.01.007

176. Hsiao, C.J., Tung, P., Blischak, J.D., Burnett, J.E., Barr, K.A., Dey, K.K., et al.: Characterizing and inferring quantitative cell cycle phase in single-cell RNA-seq data analysis. Genome Res. **30**(4), 611–621 (2020). https://doi.org/10.1101/gr.247759.118

177. Phipson, B., Er, P.X., Combes, A.N., Forbes, T.A., Howden, S.E., Zappia, L., et al.: Evaluation of variability in human kidney organoids. Nat. Methods. **16**(1), 79–87 (2019). https://doi.org/10.1038/s41592-018-0253-2

178. Volpato, V., Smith, J., Sandor, C., Ried, J.S., Baud, A., Handel, A., et al.: Reproducibility of molecular phenotypes after long-term differentiation to human iPSC-derived neurons: a multi-site omics study. Stem cell reports. **11**(4), 897–911 (2018). https://doi.org/10.1016/j.stemcr.2018.08.013

179. Mitchell, J.M., Nemesh, J., Ghosh S, Handsaker RE, Mello CJ, Meyer D, et al. Mapping genetic effects on cellular phenotypes with "cell villages". bioRxiv. 2020:2020.06.29.174383. https://doi.org/10.1101/2020.06.29.174383

180. Cuomo, A.S.E., Seaton, D.D., McCarthy, D.J., Martinez, I., Bonder, M.J., Garcia-Bernardo, J., et al.: Single-cell RNA-sequencing of differentiating iPS cells reveals dynamic genetic effects on gene expression. Nat. Commun. **11**(1), 810 (2020). https://doi.org/10.1038/s41467-020-14457-z

181. Neavin, D., Nguyen, Q., Daniszewski, M.S., Liang, H.H., Chiu, H.S., Wee, Y.K., et al.: Single cell eQTL analysis identifies cell type-specific genetic control of gene expression in fibroblasts and reprogrammed induced pluripotent stem cells. Genome Biol. **22**(1), 76 (2021). https://doi.org/10.1186/s13059-021-02293-3

182. Xiao, Y., Hill, M.C., Zhang M, Martin TJ, Morikawa Y, Wang S, et al. Hippo signaling plays an essential role in cell state transitions during cardiac fibroblast development. Dev Cell. 2018;45(2):153–69.e6. https://doi.org/10.1016/j.devcel.2018.03.019

183. Jia, G., Preussner, J., Chen, X., Guenther, S., Yuan, X., Yekelchyk, M., et al.: Single cell RNA-seq and ATAC-seq analysis of cardiac progenitor cell transition states and lineage settlement. Nat. Commun. **9**(1), 4877 (2018). https://doi.org/10.1038/s41467-018-07307-6

184. Liu, X., Chen, W., Li W, Li Y, Priest JR, Zhou B, et al. Single-cell RNA-seq of the developing cardiac outflow tract reveals convergent development of the vascular smooth muscle cells. Cell Rep. 2019;28(5):1346–61.e4. https://doi.org/10.1016/j.celrep.2019.06.092

185. Cui Y, Zheng Y, Liu X, Yan L, Fan X, Yong J, et al. Single-cell transcriptome analysis maps the developmental track of the human heart. Cell Rep. 2019;26(7):1934–50.e5. https://doi.org/10.1016/j.celrep.2019.01.079

186. Sim, C.B., Phipson, B., Ziemann, M., Rafehi, H., Mills, R.J., Watt, K.I., et al.: Sex-specific control of human heart maturation by the progesterone receptor. Circulation. **143**(16), 1614–1628 (2021). https://doi.org/10.1161/circulationaha.120.051921

187. Suryawanshi, H., Clancy, R., Morozov, P., Halushka, M.K., Buyon, J.P., Tuschl, T.: Cell atlas of the foetal human heart and implications for autoimmune-mediated congenital heart block. Cardiovasc. Res. **116**(8), 1446–1457 (2020). https://doi.org/10.1093/cvr/cvz257

188. Sahara, M., Santoro, F., Sohlmér J, Zhou C, Witman N, Leung CY, et al. Population and single-cell analysis of human cardiogenesis reveals unique LGR5 ventricular progenitors in embryonic outflow tract. Dev Cell. 2019;48(4):475–90.e7. https://doi.org/10.1016/j.devcel.2019.01.005

189. Wang, W., Niu, X., Stuart, T., Jullian, E., Mauck 3rd, W.M., Kelly, R.G., et al.: A single-cell transcriptional roadmap for cardiopharyngeal fate diversification. Nat. Cell Biol. **21**(6), 674–686 (2019). https://doi.org/10.1038/s41556-019-0336-z

190. Churko, J.M., Garg, P., Treutlein, B., Venkatasubramanian, M., Wu, H., Lee, J., et al.: Defining human cardiac transcription factor hierarchies using integrated single-cell heterogeneity analysis. Nat. Commun. **9**(1), 4906 (2018). https://doi.org/10.1038/s41467-018-07333-4

191. Paik, D.T., Tian, L., Lee, J., Sayed, N., Chen, I.Y., Rhee, S., et al.: Large-scale single-cell RNA-Seq reveals molecular signatures of heterogeneous populations of human induced pluripotent stem cell-derived endothelial cells. Circ. Res. **123**(4), 443–450 (2018). https://doi.org/10.1161/circresaha.118.312913

192. Lam, Y.Y., Keung, W., Chan, C.H., Geng, L., Wong, N., Brenière-Letuffe, D., et al.: Single-cell transcriptomics of engineered cardiac tissues from patient-specific induced pluripotent stem cell-derived cardiomyocytes reveals abnormal developmental trajectory and intrinsic contractile defects in hypoplastic right heart syndrome. J. Am. Heart Assoc. **9**(20), e016528 (2020). https://doi.org/10.1161/jaha.120.016528

193. Pezhouman, A., Engel, J.L., Nguyen, N.B., Skelton, R.J.P., Gilmore, W.B., Qiao, R., et al.: Isolation and characterization of hESC-derived heart field-specific cardiomyocytes unravels new insights into their transcriptional and electrophysiological profiles. Cardiovasc. Res. (2021). https://doi.org/10.1093/cvr/cvab102

Recombinant Adeno-Associated Virus for Cardiac Gene Therapy

Cindy Kok, Dhanya Ranvindran, and Eddy Kizana

1 Introduction

The pathogenesis of many congenital and acquired cardiac diseases is closely associated with changes to the genetic and molecular programs in the heart. The discovery of DNA as the biomolecule of genetically inherited diseases and the prospect of long-lasting therapeutic effect to patients was originally introduced over 40 years ago. It is now apparent that gene therapy has broader potential that also includes acquired polygenic diseases, such as ischemic heart disease, arrhythmias, and heart failure (HF). Advances in the understanding of the molecular mechanisms that drive and support these diseases, together with the evolution of increasingly efficient gene transfer technology, has placed some cardiovascular pathophysiology within reach of gene-based therapy as a viable alternative to current treatments. Of equal importance is developing vectors and delivery systems that can efficiently transduce majority of the cardiomyocytes, that can not only offer a long-term expression but also escape the host immune response.

C. Kok · D. Ranvindran
Centre for Heart Research, The Westmead Institute for Medical Research, The University of Sydney, Sydney, NSW, Australia

E. Kizana (✉)
Centre for Heart Research, The Westmead Institute for Medical Research, The University of Sydney, Sydney, NSW, Australia

Department of Cardiology, Westmead Hospital, Westmead, NSW, Australia
e-mail: eddy.kizana@sydney.edu.au

© The Author(s), under exclusive license to Springer Nature Switzerland AG 2022
J. Zhang, V. Serpooshan (eds.), *Advanced Technologies in Cardiovascular Bioengineering*, https://doi.org/10.1007/978-3-030-86140-7_9

2   Gene Therapy Vectors

Gene therapy technology has harnessed the innate ability of viruses to target cells and deliver transgenes for many decades [1, 2]. The predominant use of viral vector systems in preclinical models of gene therapy reflects the increased gene transfer efficiencies achievable with these systems. This efficiency is conferred as a result of including elements of parental virus biology that secure a favorable fate for the transferred gene. Currently, the viral vectors showing greatest efficiency in preclinical and clinical investigations are derived from retroviruses, lentiviruses, adenoviruses, and the adeno-associated viruses (AAV) [3–5]. Of these viral vectors, the recombinant AAV (rAAV) vector family is considered the lead candidate for cardiomyocyte gene transfer, hence the most actively investigated vehicles for clinical gene therapy [3]. Advantages of AAV as a cardiac gene delivery vector include: (a) efficient transduction into cardiomyocytes, (b) eliciting modest cellular immune response, and (c) triggering persistent expression of therapeutic gene of interest, even in the absence of genome integration. As a result of these favorable properties, rAAVs have become a promising therapeutic vehicle for molecular medicine in the future.

3   Wild-Type AAV Biology

Discovered over 50 years ago as a contaminant in an adenovirus-infected cell preparation [6, 7], the AAV is a small non-enveloped parvovirus that requires co-infection with another virus, such as adenovirus or the family of herpes viruses, to successfully replicate within a host [8–11]. The AAV has a short, single-stranded DNA genome around 4.8 kilobases in length, encased within a protective protein shell known as the capsid [12–14] (Fig. 1).

It is composed of 2 genes within separate open reading frames [15]. The *rep* (replication) gene encodes 4 non-structural proteins (Rep 40, Rep 52, Rep 68, and Rep 78) which are crucial for viral replication, transcriptional regulation, genome integration, and packaging [16, 17]. The *cap* (capsid) gene encodes 3 structural proteins, known as viral proteins (VP; VP1, VP2, VP3) which form the capsid that protects the viral genome [12–15]. An additional gene necessary for capsid formation, *aap* (assembly activating protein), encoded on an alternative reading frame has been described [18, 19], which recruits capsid monomers to nucleoli and drives capsid assembly. Recently, the membrane-associated accessory protein *maap* has also been discovered, though its function has not been elucidated [20].

These genes are flanked by 145 nucleotide inverted terminal repeats (ITRs) that are the only cis-elements required for the production of rAAVs [12, 14]. As a result, rAAVs do not encode any viral proteins, an attribute that likely contributes significantly to their limited immunogenicity.

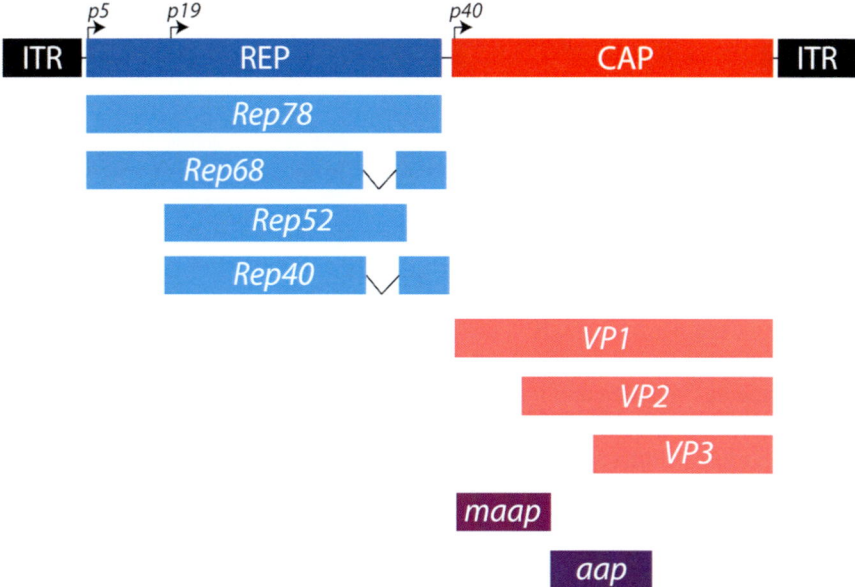

Fig. 1 Schematic of the AAV genome. Viral genes *rep* and *cap* flanked by inverted terminal repeats (ITR). *Rep* encodes the proteins Rep 40, 52, 68 and 78. *Cap* encodes the viral proteins VP1, VP2, and VP3, as well as the membrane accessory protein maap and assembly activating protein aap

4 AAV Capsid Biology

Alongside protecting the viral genome, the capsid is responsible for controlling the efficiency of AAV receptor-mediated cellular entry. Cellular binding is mediated by primary receptor interactions on the target cell surface. Secondary receptor interactions mediate viral internalization and prime the AAV capsid to release its contents. Due to these cellular interactions via surface structures, the capsid confers specific tropism of the virus and the vector derived from it. As the AAV capsid is exclusively encoded by *cap*, small variations in the gene can lead to slight differences in expressed protein receptors on its surface. Therefore, each variation of the capsid is classified as a new AAV variant with its own intrinsically unique transduction and tissue tropism characteristics. To date, hundreds of capsids have been identified or generated in laboratories via bioengineering techniques [21]. However, the first 13 (AAV1–13) are the best characterized [22–24].

5 Cardiac rAAV Tropism in Small Animal Models

The success of preclinical studies in animal models demonstrating high transduction efficiency of rAAV vectors has provided hope for their translation to the clinic. In cardiac research, animal studies are used to explore neurohumoral, hemodynamic, and cellular processes of cardiac diseases, and are extensively used to develop new experimental and clinical approaches for the treatment of a variety of cardiac diseases. However, many considerations should be undertaken in order to model human cardiac disease closely in animals. Such considerations include appropriate choice of animal model, experimental design, and proper analysis of data. Small animal models are still widely used for initial proof-of-concept cardiac gene therapy studies.

Within the rodent cardiac system, several comparative studies have been undertaken to determine the optimal AAV serotype for cardiac tissue. Of the 13 known serotypes, AAV 1, 6, 8, and 9 have demonstrated cardiac tropism in murine models [25]. The current view is that the route of AAV delivery is as important as the choice of serotype. The two primarily used methods of AAV delivery into the heart are intramyocardial injection and intravascular infusion via tail vein. In rodents, AAV1 and AAV6 are more suitable for cardiac gene transfer through intramyocardial, intrapericardial, or intracoronary routes, while AAV8 and AAV9 can achieve more efficient cardiac transduction only via the transvascular route [26]. In support of this, intravenous delivery of 1×10^{11}gc AAV8 and AAV9 in neonatal mice yielded approximately 20- and 200-fold higher transduction levels than AAV1 delivery, respectively [27]. The authors suggested that the high transduction efficiency of AAV9 is most likely attributable to viral capsid, which displays a strong affinity for cardiac tissue. Interestingly, it was found that vector dose significantly affects transduction levels. At a high vector dose (2.5×10^{11}gc), all serotypes (1, 6, 7, 8, and 9) efficiently transduced hearts of mice, as measured by the *LacZ* reporter gene [28]. In contrast, at an intermediate dose (2.5×10^{10} gc), AAV9 was the only serotype that continued to show high transduction levels in the heart [28]. Intracoronary delivery appears more efficient than intravenous delivery in mice, and yields 2.6- to 28-fold higher transgene expression in the heart 3 weeks after AAV injection, depending on AAV serotype [29]. The highest level of cardiac gene expression in rodents was achieved with intracoronary AAV9 delivery (5×10^{11} gc) in comparison to AAV6 and AAV5 [29].

Rodent studies are informative and lay the ground for additional and in-depth investigations in more clinically relevant experimental models. Difficulty in extrapolating results from rodents to humans stems not only from differences in anatomy and physiology, but also because rodent models are limited in their ability to mimic cardiac disease etiology. In contrast, large animals such as pigs and sheep have consistent body vessel anatomy, coronary arterial distribution, and lack of collateral circulation. Taken together, while rodent cardiac studies provide basic insight into cardiac biology and disease, large animal models more closely resemble human cardiac physiology and pathology and thus are more suitable and relevant to study cardiac disease such as myocardial ischemia and post-infarction left ventricle (LV) remodeling.

6   Cardiac rAAV Tropism in Large Animal Models

It was first shown more than 25 years ago that AAV vectors can efficiently transfer reporter genes into myocardial cells via percutaneous intra-arterial infusion in a large animal model [30]. The differences between AAV serotypes 1 and 2 in the mediation of onset of transgene expression was examined in a porcine model, in which the vectors expressing *LacZ* and VEGF, were delivered into the myocardium by intramuscular injection [31]. Both *LacZ* and VEGF expression was detected 3 days after AAV1-mediated gene transfer, but none detected after AAV2-mediated gene transfer [31]. The findings led the authors to conclude that AAV1 mediates earlier onset of transgene expression [31]. In another study, the transduction efficiency of AAV vectors (3.5×10^{10} vg; AAV2 versus AAV6) was compared via infusion in the anterior interventricular vein of a pig [32]. After 28 days, AAV6 vectors demonstrated 48-fold more reporter activity in the target territory compared to AAV2 vectors [32]. Another noteworthy study in a non-human primate model showed that transendocardial delivery of AAV6 was more specific to heart than AAV8 and 9 [33]. Intravenous injection of $2–2.5 \times 10^{14}$ vg/kg of AAV9 in a newborn dog model resulted in widespread gene transfer in skeletal muscle but not in the heart [34]. In contrast, $7.14–9.06 \times 10^{14}$ vg/kg of AAV8 gene transfer resulted in robust expression in both skeletal muscle and the heart [34]. Interestingly, increasing AAV8 dose only moderately improved heart transduction. The authors proposed that the observed changes between cardiac transduction by AAV8 and AAV9 are likely related to the intrinsic property of the viral capsid rather than the vector dose [34, 35].

7   Therapeutic Evaluation of rAAV Vectors in Large Animal Models of Cardiac Disease

The evaluation of AAV vectors in clinically relevant large animal models of HF gained popularity as they can closely mimic the human condition, making them instrumental for the investigation of potential gene therapy strategies. Importantly, these models may also be a pre-requisite for regulatory approval when progressing to human studies.

One of the first AAV-based cardiac gene therapy studies examined AAV2-mediated transduction of VEGF165 in dogs. The study found delivery of rAAV2/VEGF165 (5×10^{12} viral particles) enhanced both angiogenesis and cardiomyocyte viability of the myocardium after permanent coronary occlusion [36]. In another study of myocardial infarction (MI) in pigs, retrograde gene delivery of AAV9-S100A1 (1.5×10^{13} viral particles) prevented cardiac deterioration and reversed ventricular remodeling 14 weeks later [37]. A similar delivery technique was described using AAV6 vector (1×10^{13} vg) encoding βARKct gene 2 weeks post MI. Robust and long-term βARKct expression was found after delivery, leading to significant amelioration of LV hemodynamics and contractile function in pigs with HF [38]. In a different study, the researchers performed intracoronary administration of a novel chimeric vector generated from AAV2 and AAV8 (BNP116), encoding inhibitor-1c, a SERCA2a/phospholamban-complex upstream regulator, in pigs post-MI [39]. Results showed improvement of cardiac function and promising efficacy of this novel vector for cardiac gene transfer [39]. Similar results were demonstrated after intracoronary delivery of rAAV9 carrying I-1c gene post MI in pigs [40].

Numerous studies have documented that self-complementary (sc) AAV vectors have a higher transduction efficiency than conventional single-stranded genome AAV vectors. In a sheep model of acute MI, scAAV6-mediated βARKct gene transfer resulted in preservation of regional and global systolic function, though without arresting ventricular remodeling [41]. The same group demonstrated that delivery of scAAV9/SERCA2a following ovine ischemic cardiomyopathy resulted in superior LV function and attenuated myocyte hypertrophy twelve weeks post MI [42]. The authors also found that HF was significantly less pronounced in sheep transduced with recombinant AAV1/SERCA2a post-MI. Moreover, AAV1/SERCA2a gene delivery restricted the progression of fibrosis and prevented cardiac remodeling [43].

Gene delivery of rAAV9/heme-oxygenase-1 (1×10^{13} viral particles) in a porcine ischemia/reperfusion model demonstrated attenuation of inflammation and improvement of post-ischemic function [44]. An additional study evaluated the impact of small ubiquitin-related modifier 1 (SUMO-1) gene transfer in a swine model of ischemic HF. Antegrade coronary infusion of AAV1/SUMO-1 (5×10^{12} or 1×10^{13} viral particles) demonstrated that improved cardiac function and stabilized LV volumes compared to controls [45].

In a volume overload model in pigs, intracoronary injection of rAAV1/SERCA2a (1×10^{12} viral particles) resulted in positive inotropic effect and improvement of LV remodeling four months post injection [46]. An additional study that used a similar volume overload model found that AAV1/SERCA2a increased eNOS activity and calcium-storage capacity and improved coronary vascular reactivity in the setting of mitral regurgitation [47]. In another study, cardiac delivery of AAV6 encoding SERCA2a (5×10^{12} vg) resulted in significantly better global LV function and lower end-systolic and end-diastolic dimensions [48].

A study in dogs with pacing-induced dilated cardiomyopathy showed that direct myocardial injection of AAV9/VEGF-B_{167} (1×10^{12} gc) delayed the progression towards congestive failure [49]. In a similar study, intracoronary infusion of AAV9/VEGF-B_{167} markedly preserved diastolic and contractile function and attenuated ventricular chamber remodeling, halting the progression of HF [50]. To test the efficiency of AAV2/SERCA2a the effects of different doses of this gene construct on cardiac function was explored after creating pacing-induced HF [51]. Cardiac recirculation delivery of AAV1/SERCA2a elicited a dose-dependent improvement in cardiac performance. In contrast, direct intracoronary infusion did not elicit any effect on ventricular function [51].

In summary, these studies show the significant differences between transduction efficiency as well as the kinetics of transgene expression of AAV serotypes in the different large animal models of HF. Those differences are attributed to inequality in cellular uptake, intracellular trafficking mechanisms of AAV vectors in various species, and structural differences between AAV serotypes at the capsid level. Moreover, AAV transduction greatly depends on the dose and route of delivery.

8 Clinical Trials of rAAV-Based Cardiac Gene Therapy

The first and most extensive clinical trial of gene therapy in patients with HF was launched in the United States in 2007 [52, 53]. It involved the transfer of the SERCA2a cDNA, based on the observation that this sarcoplasmic Ca^{2+} ATPase is downregulated in failing hearts and that restoration of its level improves heart function in both mice and swine [54]. Calcium Up-regulation by Percutaneous Administration of Gene Therapy in Cardiac Disease (CUPID) was a two-part, Phase 1/2 multicenter trial designed to evaluate the safety and the biological effects of gene transfer of the *SERCA2a* cDNA by delivering a recombinant AAV1 (AAV1/ SERCA2a) in patients with advanced HF [52, 53]. Over the past decade, the program has progressed, but failed to demonstrate significant therapeutic efficacy in the most recent phase 2b trial.

Briefly, the first part of the trial (phase 1b) was a dose-escalation open-label study which involved administration of a single intracoronary infusion of AAV1/ SERCA2a to 12 patients [52, 53]. The infusion was spread over multiple coronary arteries in an attempt to provide homogeneous myocardial transduction. Six to twelve-month follow-up of these patients showed an acceptable safety profile [52, 53]. Improvement was detected in several patients, as reflected by symptomatic (five patients), functional (four patients), biomarker (two patients), and LV function/ remodeling (six patients) parameters. Although the phase 1 study involved a small number of patients, the results suggested that AAV1/SERCA2a treatment confers quantitative biological benefit. During this preliminary trial, two patients who failed to improve had pre-existing anti-AAV1 neutralizing antibodies [17, 55] underscoring the importance of this immunologic mechanism in blocking AAV delivery to the cardiomyocyte after intracoronary infusion.

In the phase II trial conducted in 39 patients with advanced HF, patients were randomized to receive intracoronary AAV1/SERCA2a delivery in low, moderate and high doses versus placebo [56, 57]. Treatment outcome was assessed after 12 months of follow-up, where improvement was described according to various parameters including clinical outcomes. The decreased number of CVD events in the high-dose group and a satisfactory safety profile affirmed AAV1/SERCA2a gene therapy as a promising alternative treatment modality in these patients.

Exciting results from early-phase trials paved the way to the initiation of CUPID2, a randomized double-blinded phase II multicenter trial [56]. This trial enrolled 250 patients with stable symptomatic HF, of both ischemic and non-ischemic etiology, with severely depressed ejection fraction, randomized in a 1:1 fashion to receive intracoronary infusion of either placebo or 1×10^{13} AAV1/

SERCA2a viral particles [56]. However, the trial had to be halted early owing to the lack of significant improvement in clinical outcomes. The CUPID2 trial included only a one-time, intracoronary infusion of the rAAV1/SERCA2a. It is likely that the increase in SERCA2a protein was insufficient to provide enough biological effect to provide significant improvements in cardiac function. Assessment of cardiac tissue from treated patients confirmed that lack of benefit was attributed to marginal uptake of the AAV1/SERCA2a vector (< 2%); approximately 1000 times less than that achieved in preclinical models [57, 58]. This hinted at differences in permissiveness of human and porcine cardiomyocytes to the AAV1 capsid, which serves as evidence that the lack of translatability to human might have compromised positive outcomes of the CUPID2 trial.

This discouraging finding also prematurely halted two other related clinical studies using the same vector, the AGENT-HF study [59], investigating the impact of AAV1/SERCA2a on cardiac remodeling parameters in patients with severe HF, and the UK-based SERCA-LVAD study assessing the safety and feasibility of delivering AAV1/SERCA2a to adult patients with a LV assist device inserted for chronic HF. Both studies were prematurely terminated in 2016.

A second variant rAAV-based approach for HF gene therapy is aimed at modulating Ca^{2+} reuptake by overexpression of a constitutively active form of *I-1c*. The ultimate effect of this intervention is to increase the levels of the SERCA2a and restore β-adrenergic stimulation [60]. After experimental success of I-1c gene transfer in large animals, an ongoing Phase I clinical study is assessing the efficacy of intracoronary infusion of BNP116.sc-CMV.I1c (under the name of NAN-101) in 12 patients with congestive HF.

It is apparent that the efficiency of gene delivery to the human heart needs to be improved using advanced vector systems that will lead to safer and more effective clinical trials in gene therapy for cardiac diseases. Although improvements regarding vector delivery, safety and bioengineering have been established, it is reasonable to view current AAV vectors as parent compounds, the development of which should stimulate a rationale and broad generation of cell or tissue-targeted derivatives based on each of the AAV serotypes.

9 Generation of Novel AAV Variants with Increased Cardiotropism

Novel AAV serotypes continue to be identified, predominantly from human liver [61]. This is the process known as natural discovery which has also led to isolation of novel variants with altered tissue tropism despite only slight variations in sequence compared with known AAV serotypes [62, 63]. Interestingly, many natural AAV isolates from human tissues are variants of AAV2 or hybrids of AAV2/3. However, AAV capsids of cardiac origin have not been identified. The majority of recently identified cardiotropic capsids have been obtained by screening highly complex AAV libraries in cardiac or muscle tissues. In many cases, a novel variant superior to the parental AAVs may be identified with increased specificity of delivery to cardiac muscle and enhanced efficacy of post-entry processing. Also, novel variants are attractive as gene therapy candidates due to the potential for reduced risk of pre-existing immunity to the naturally occurring AAV capsids in the human population.

Generation of synthetic AAV capsid variants can be performed using rational design or high throughput strategies, which will be further discussed below. Large scale screening of novel AAVs for variants with desired tropism can be performed by a process known as directed evolution. The stringency and success of candidate capsid selection will be dictated by the selection pressure from the target cells. Therefore, the choice of selection model (cell type, *in vitro* vs *in vivo*, and state of maturation) needs to be made with care to ensure clinical relevance to the targeted patient population and the clinical pathophysiology.

Potential candidates are best tested *in vivo*, as cell culture models do not adequately replicate the architecture of the heart and circulatory system, as well as complex interactions between different organs.

9.1   Liver Detargeting by Ablation of Heparan Binding

A common characteristic of natural AAV variants is the tendency for liver sequestration of vector particles after systemic delivery [64, 65]. Particularly in the case of vectors destined for use in cardiac gene therapy, liver sequestration may be a safety issue as hepatocyte toxicity and inflammation is a known risk following systemic delivery of AAV in high doses [66]. Severe toxicity was observed in piglets and non-human primates treated with 2×10^{14} vector genomes/ kg of an AAV9 variant. More recently in the ASPIRO gene therapy trial aimed at treating X-linked myotubular myopathy, two children tragically passed away after receiving high doses of AAV8 (3×10^{14} vg/kg) [67]. This adverse event is potentially similarly linked to hepatic toxicity. It is therefore desirable to de-target AAVs from the liver to reduce the risk of off-target effects.

Liver sequestration is hypothesized to be attributed to the AAV cellular receptor, membrane-associated heparan sulfate proteoglycan (HSPG), with high affinity being associated with efficient cell entry into murine liver cells [68]. Ablation of heparan sulfate binding is one method which has been shown to reduce liver targeting and thus increase the propensity of modified AAV variants to reach non-hepatic tissues [69].

Asokan et al. used site-directed mutagenesis to create a series of AAV2 variants with mutations in the hexapeptide motif 585-RGNRQA-590, which contains basic amino acid residues implicated in heparan sulfate binding [69] (Fig. 2). This motif forms part of the GH loop within the VP3 subunit of the AAV capsid, which protrudes from the capsid surface and is key to AAV receptor mediated binding to cells. From this panel of modified AAV2 variants, AAV2i8 demonstrated reduced mouse

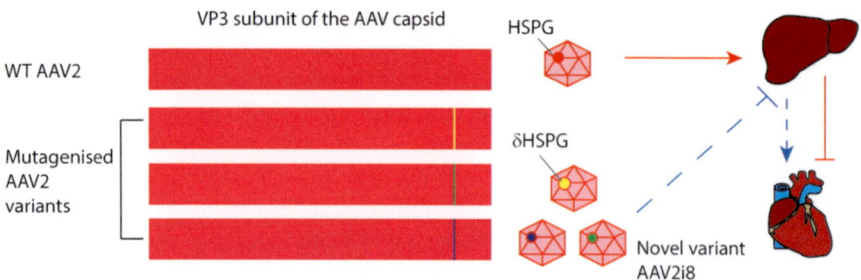

Fig. 2 Generation of liver de-targeted AAV2 variants by ablation of heparan sulfate binding via the heparan sulfate proteoglycan (HSPG). Site directed mutagenesis of the capsid gene to create mutants with single amino changes to the hexapeptide motif 585-RGNRQA-590 within VP3, leading to variants with ablated HSPG binding (δHSPG). Subsequently the liver was detargeted, and more vector was available to reach the cardiac tissue [69]

liver transduction (approximately 40-fold lower than AAV2 and AAV8) and concurrent improvement in transduction of cardiac and skeletal muscle.

A subsequent study which tested the transduction efficiency of AAV2i8 after systemic delivery in rhesus monkeys showed that AAV2i8 was efficient at transducing cardiac tissues. However, this was despite fewer vector copies (vc) detected in cardiac tissue subtypes relative to AAV9. The authors hypothesize that post-entry processing of AAV2i8 may result in increased efficiency of gene expression in cardiac tissues from relatively fewer vc (1.8×10^1 vc of AAV2i8 detected in cardiac tissues versus 1×10^3 vc of AAV9) [70]. Therefore, a relatively lower dose of cardiotropic AAV may be sufficient to transduce cardiac tissue, as a natural consequence of liver de-targeting.

Similarly, a series of AAV9 variants were generated using error prone PCR to induce mutations within the GH loop, predominantly within VP3 [71]. A proportion of vectors generated are non-functional due to deleterious changes to crucial amino acid sequences. Of the remainder which exhibited functional activity, AAV9.45 emerged as a variant which retained the muscle tropism of parental AAV9 but was de-targeted from the liver (3–45-fold less than AAV9). These studies collectively show that rational design can be applied to different parental AAVs to obtain variants which can be further screened for desired patterns of tropism and/or de-targeting.

9.2 Fusion of a Sodium Ion Channel Ligand to the AAV Capsid

Interestingly, one study has explored the utility of fusing a cell targeting ligand to the AAV2 capsid to direct tropism towards the heart. The cardiac sodium channel $Na_v1.5$ is abundantly and specifically expressed on cardiac myocytes. As proof of concept, Finet et al. hypothesized that the sodium channel ligand fused to the AAV capsid would facilitate specific targeting of the modified AAV to the cardiomyocytes [72]. This ligand (modified sea anemone toxin Anthopleurin-B) was inserted into the N terminus of AAV2 VP2 protein, which is another region thought to be permissive to insertion of large peptides (Fig. 3).

This novel variant was then called AAV2-ApB. While infectivity was preserved due to successful binding of AAV2-ApB to the sodium channel in cardiomyocytes, it is interesting to note that when heparan binding is ablated in this variant, inadvertent liver gene transfer was eliminated without impacting heart targeting. This again reaffirms the importance of detargeting from the liver to enhance cardiotropism and safety of novel variants.

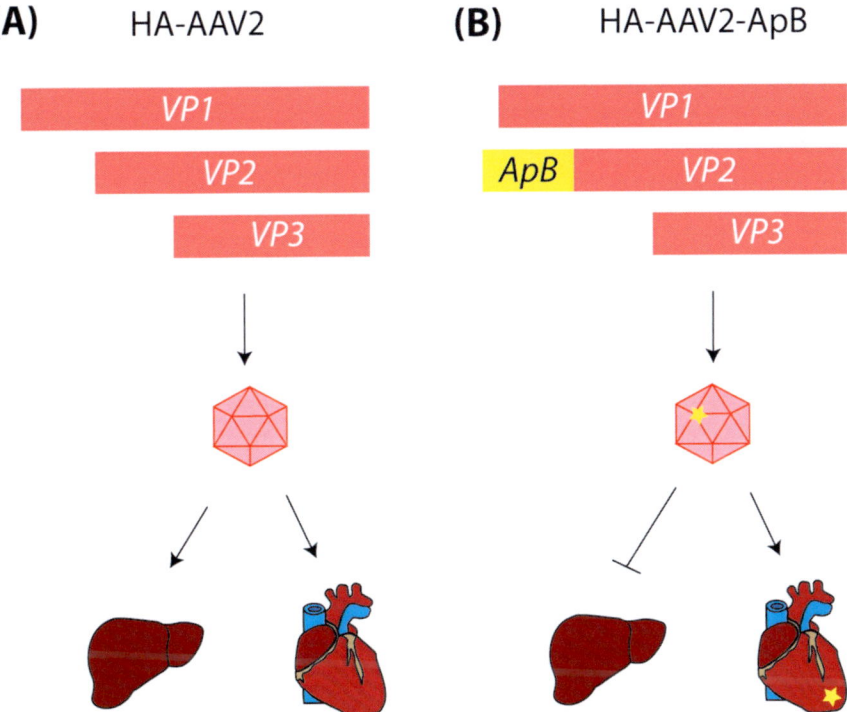

Fig. 3 Heart targeting by fusion of the cardiac sodium channel ligand to the capsid surface. (**a**) A heparan binding ablated AAV2 variant (HA-AAV2) showing residual gene transfer to the rat liver and to the heart. (**b**) Fusion of the sodium ion channel ligand (ApB) to the N terminus of VP2 ablated liver targeting and showed targeted gene transfer to the heart by binding to the cardiac sodium ion channels (yellow star) [72]

9.3 High Throughput Generation of Novel AAV Capsid Variants

One limitation of capsid design with a specific pattern of tropism in mind is the requirement to identify specific cell surface receptors and ligands which correspond to the cellular target. Often this is unknown, and identification of the best receptor may be laborious and subject to bias if pre-selected. To circumvent this challenge, directed evolution has emerged as a strategy for high throughput generation and screening of novel AAV capsids. It is based on the use of complex plasmid pools to generate libraries of AAV variants, which can subsequently be screened for tropism in the cells of interest.

As shown in heparin binding ablated variants (see previous section), regions within VP3 have been identified as being tolerant of mutations. These areas are also able to tolerate insertion of peptides and thus may be amenable to modification of surface epitopes while retaining the ability to be packaged (e.g., amino acid residues 587 and 588 of the AAV2 capsid).

One example is the development of virus display peptide libraries. This technology is based on the insertion of a random 7–9 peptide within the GH loop to create a complex library of unique cap genes [73]. This library plasmid pool is then cloned into the wildtype AAV genome and used to package the AAV peptide display library.

Peptide display libraries have been successfully used to screen for muscle specific variants both *in vitro* and *in vivo* [74]. Ying et al. have used this approach for selection of heart targeted variants by delivering the AAV2 random display peptide library into mice via tail vein injection [75] (Fig. 4).

Hearts were then excised and used to generate organotypic cultures of 300 μm cardiac slices. Subsequent infection of the slices with adenovirus induced replication of AAV variants within the cardiac tissue. The amplified variants were then used to generate a new plasmid library, packaged to create the secondary AAV library and then selection was repeated. The peptides PSVSPRP and VNSTRLP were highly enriched after consecutive rounds of in vivo selection. Functional validation also proved enhanced cardiac targeting compared to parental AAV2, with low levels of transgene expression observed in the lung and pancreas, and significantly reduced liver targeting. However, it is important to note that these variants did not outperform AAV9, as expression of luciferase was 100-fold lower.

Similarly, Yu et al. have subsequently demonstrated that the insertion of a muscle targeting peptide (identified from screening in muscle) after the 587 amino acid residue site not only abolished heparin binding, but also enhanced transduction of the rodent heart (24-fold higher luciferase activity in $AAV_{587}MTP$ than parental AAV2) [76].

Therefore, the use of peptide display libraries is an alternative method which can be used to generate novel cardiotropic variants with ablated heparan binding.

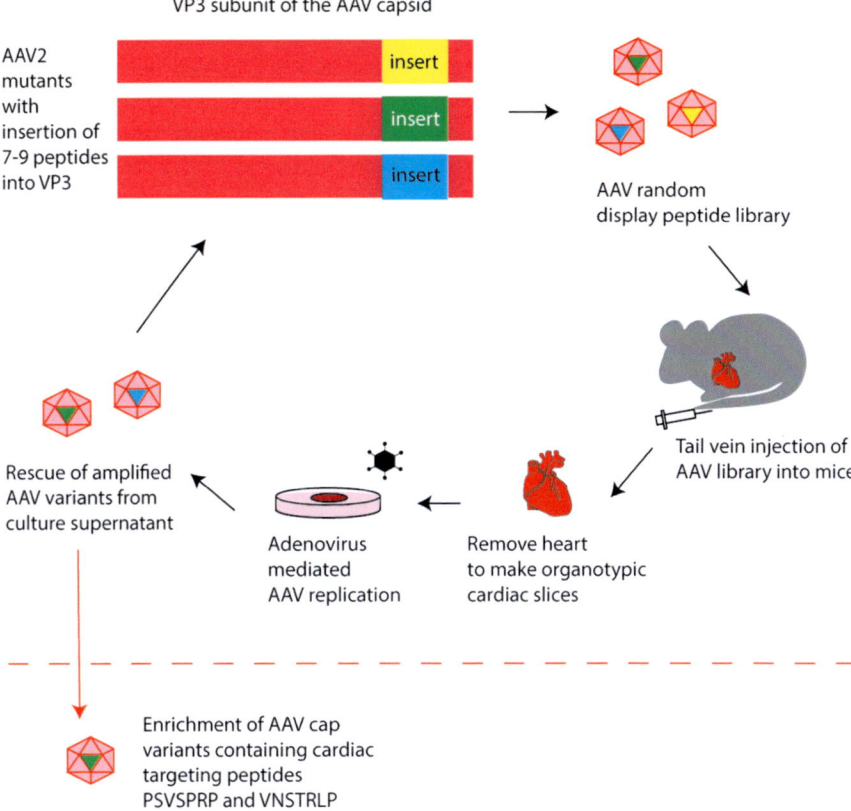

Fig. 4 Generation and screening of an AAV random display peptide library. Insertion of random 7–9 peptide into the GH loop of VP3 to create a library of cap gene variants. Subsequently use of plasmid pool to package AAV, which is then injected into mice. Hearts are then removed and sliced for culture. Cardiac slices are infected with adenovirus to induce replication of cardiotropic variants. Variants are rescued from the culture supernatant and the cap genes rescued by PCR amplification, then subcloned back into the packaging plasmid and used to make a secondary AAV library for further selection [75]

9.4 AAV Capsid Shuffling

The previously described methods both rely on modification or insertion of amino acids into the variable regions of AAV which can specifically tolerate these changes. While effective, this does not account for as yet untested regions of the parental AAV which may potentially allow more diverse AAV variants to be generated when altered.

This limitation can be addressed with the use of capsid shuffling, which relies on the creation of permutations of fragmented parental AAV sequences reconstituted *in vitro* to create an extensively mutagenized library. Increasing the diversity of the mix of parental AAVs may increase the chance of successfully isolating novel AAVs with the desired tropism and ability to evade neutralising antibodies in patients. The pool of parental AAVs may consist of naturally occurring primate isolates as well as variants isolated from non-human sources such as bat or marsupial tissues [77, 78].

Capsid shuffling has been used to select for tropism in a broad range of tissues. Using this approach, Yang et al. generated a shuffled capsid library and obtained a pool of variants which was subsequently selected *in vivo* [79]. The AAV library was injected into mice systemically, with muscle then harvested for PCR based recovery of myotropic AAVs. The recovered capsid sequences were then used to generate a secondary AAV library for further rounds of selection. This method allowed simultaneous selection of muscle tropic variants as well as negative selection for liver targeted AAVs (Fig. 5).

The M41 AAV variant was identified as a result of *in vivo* screening and shown to have an improved heart to liver vector ratio (10:1) compared to AAV9 (1:3) and AAV6 (1:6) after systemic delivery. It was also demonstrated to be resistant to pre-existing neutralizing antibodies in pooled human IgG, which is another attractive characteristic for a novel AAV capsid.

The authors caution that while liver transduction has been reduced in the AAVM41 variant, it can be further improved by increasing stringency during *in vivo* selection. This may be achieved by rescue of mutant AAV sequences from isolated cardiomyocytes, which may be more favorable than isolating from whole tissue which contains a heterogeneous mix of cells. Therefore, careful functional validation in the target cell is required after selection and identification of candidate capsids.

10   Rapid Validation of AAV Variants in Heart Tissue

Improved cell entry is not always correlated with enhanced transduction activity as the capsid structure also influences post entry processing of the AAV variant. Therefore, it is important to select candidates which exhibit efficient gene expression in the target cells. Comparison of performance among vectors *in vivo* is labor intensive as it requires transduction of separate cohorts for each capsid. However, the use of barcoded vectors is an innovative way to competitively test many AAV variants simultaneously in each animal [80]. This ensures that all variants are tested under identical physiological conditions for each biological replicate. It also allows simultaneous testing of variants which were generated using different techniques, such as peptide display, error-prone PCR, capsid shuffling, etc.

To generate the barcoded libraries, each capsid within an individual library of AAV variants is labelled with a short unique sequence of nucleotides (the molecular barcode). The packaged variants are then mixed at equimolar ratio to create the library.

Weinmann *et al* were able to perform large scale simultaneous screening of three different barcoded AAV libraries comprising 183 unique variants [80] (Fig. 6).

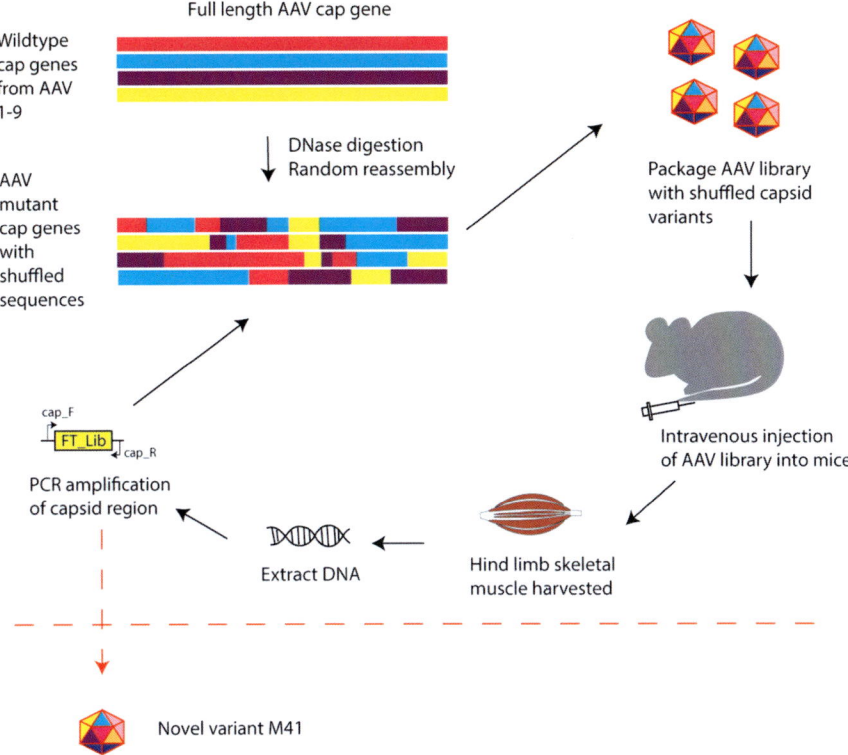

Fig. 5 Capsid shuffling and directed evolution *in vivo*. Parental AAV capsid sequences were digested using DNase I, then randomly reassembled using *pfu* DNA polymerase. The pool of shuffled capsid genes was used to package the AAV library. The AAV library was injected into mice. Hindlimb skeletal muscle was harvested, DNA extracted, and the capsid sequence amplified by PCR. The pool of capsid genes was then subcloned into the packaging plasmid and used to package a secondary AAV library for further rounds of selection [79]

Cohorts of six mice per group were injected with 1–2 x 10^{12} vg of AAV library. After 1–2 weeks, organs were collected to for extraction of DNA and RNA. Next generation sequencing was then utilized for analysis of barcodes detected from each organ of interest. From each nucleic acid fraction, the prevalence of each barcode provides a relative measure of cell entry efficiency (from DNA) and transgene expression efficiency (from RNA).

Interestingly, the authors demonstrated robust selection yielding consistent and superior specificity of the AAVMYO in the entire musculature (skeletal muscle, diaphragm, and heart). This novel AAV9 variant is derived from the peptide display library and contains the P1 (RGDLGLS) peptide insertion, which is hypothesized to have conferred muscle specificity. The authors suggest that the P1 peptide may be transferrable to other parental AAVs to further optimize tissue targeting.

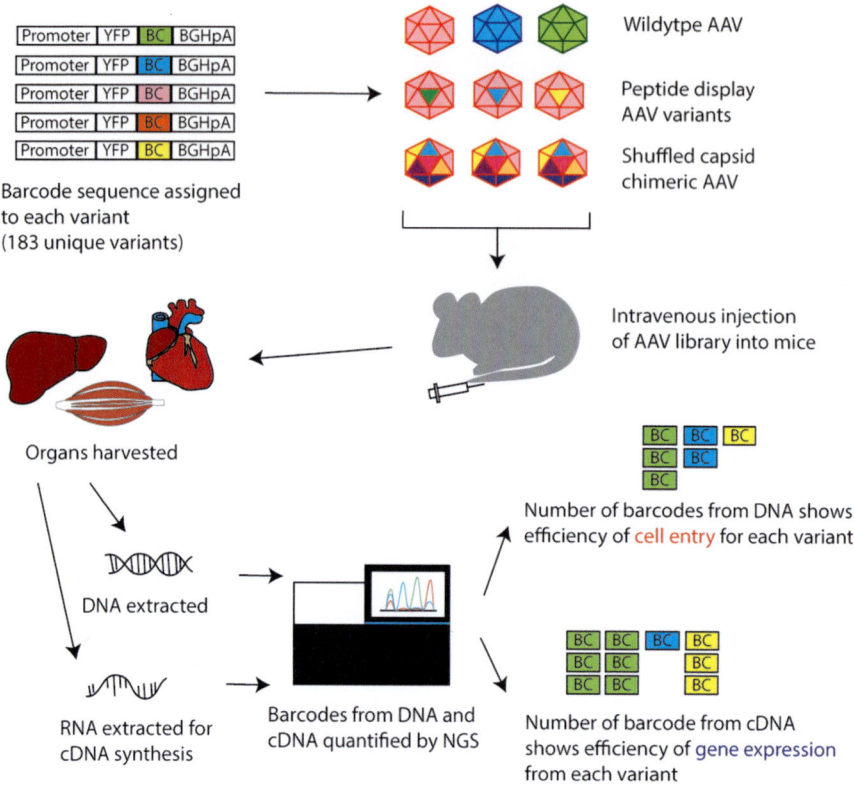

Fig. 6 Large scale screening of AAV variants for novel patterns of tropism *in vivo*. A set of bar-coded libraries were generated from a mix of wildtype, peptide display and shuffled capsid AAV variants. These AAVs were injected into mice, and organs were subsequently harvested. Barcodes were then analyzed by next generation sequencing to identify variants with desired tropism [80]

11 Cardiac Promoters to Further Enhance Target Specificity

It should be emphasized that many of the novel AAVs outlined thus far have been developed as muscle targeting vectors, which have coincidentally shown efficient gene transfer to the heart. In the case of disease models such as heart failure and arrhythmia, it is ideal to restrict gene expression to the heart to avoid off-target effects in non-cardiac muscle tissue.

This can be achieved through the use of cardiac specific promoters, which should only be transcriptionally active in the targeted cell type. Examples of cardiac specific promoters include cardiac myosin light chain, cardiac myosin heavy chain, and cardiac troponin [81–84].

Due to the limited packaging capacity of the AAV genome, it is important to identify promoter elements which are small and easy to package. Depending on the target disease to be treated and the chosen therapeutic transgene, it may also be necessary to control the level of gene expression by choosing promoters with appropriate strength.

12 Towards Clinical Translation

As mentioned previously, the neutral results observed in the CUPID trial have highlighted the importance of efficient gene delivery to the human heart. While the landscape of AAV vectorology is constantly evolving, it should be noted that many of the studies validate the effectiveness of their novel variants in rodent cells and tissues. Therefore, the risk of species-specific differences in transduction efficiency of novel variants is still present. Future work should be aimed towards validation of novel variants using human cells and tissues.

13 Conclusions

The diversity of AAV variants used for pre-clinical cardiac gene therapy studies has rapidly increased over the last ten years. Already there are promising variants which have the potential to precisely target the cells of interest within the heart, and in the process of their isolation, much has been learned about the molecular mechanisms which are likely to influence cardiotropism. This will no doubt lead to more sophisticated rational design of novel AAV variants and selection models to improve the likelihood of successful clinical translation to patients with cardiovascular disease.

References

1. Dunbar, C.E., High, K.A., Joung, J.K., Kohn, D.B., Ozawa, K., Sadelain, M.: Gene therapy comes of age. Science. **359**(6372) (2018). https://doi.org/10.1126/science.aan4672
2. Kotterman, M.A., Chalberg, T.W., Schaffer, D.V.: Viral vectors for gene therapy: translational and clinical outlook. Annu. Rev. Biomed. Eng. **17**, 63–89 (2015). https://doi.org/10.1146/annurev-bioeng-071813-104938
3. Ishikawa, K., Weber, T., Hajjar, R.J.: Human cardiac gene therapy. Circ. Res. **123**(5), 601–613 (2018). https://doi.org/10.1161/CIRCRESAHA.118.311587
4. Lundstrom, K.: Viral vectors in gene therapy. Diseases., 6(2) (2018). https://doi.org/10.3390/diseases6020042
5. Rosenzweig, A.: Vectors for cardiovascular gene therapy. J. Mol. Cell. Cardiol. **35**(7), 731–733 (2003). https://doi.org/10.1016/s0022-2828(03)00144-5
6. Atchison, R.W., Casto, B.C., Hammon, W.M.: Adenovirus-associated defective virus particles. Science. **149**(3685), 754–756 (1965). https://doi.org/10.1126/science.149.3685.754
7. Hoggan, M.D., Blacklow, N.R., Rowe, W.P.: Studies of small DNA viruses found in various adenovirus preparations: physical, biological, and immunological characteristics. Proc. Natl. Acad. Sci. U. S. A. **55**(6), 1467–1474 (1966). https://doi.org/10.1073/pnas.55.6.1467

8. Buller, R.M., Janik, J.E., Sebring, E.D., Rose, J.A.: Herpes simplex virus types 1 and 2 completely help adenovirus-associated virus replication. J. Virol. **40**(1), 241–247 (1981). https://doi.org/10.1128/JVI.40.1.241-247.1981

9. Georg-Fries, B., Biederlack, S., Wolf, J., Zur Hausen, H.: Analysis of proteins, helper dependence, and seroepidemiology of a new human parvovirus. Virology. **134**(1), 64–71 (1984). https://doi.org/10.1016/0042-6822(84)90272-1

10. McPherson, R.A., Rosenthal, L.J., Rose, J.A.: Human cytomegalovirus completely helps adeno-associated virus replication. Virology. **147**(1), 217–222 (1985). https://doi.org/10.1016/0042-6822(85)90243-0

11. Thomson, B.J., Weindler, F.W., Gray, D., Schwaab, V., Heilbronn, R.: Human herpesvirus 6 (HHV-6) is a helper virus for adeno-associated virus type 2 (AAV-2) and the AAV-2 rep gene homologue in HHV-6 can mediate AAV-2 DNA replication and regulate gene expression. Virology. **204**(1), 304–311 (1994). https://doi.org/10.1006/viro.1994.1535

12. Naso, M.F., Tomkowicz, B., Perry 3rd, W.L., Strohl, W.R.: Adeno-associated virus (AAV) as a vector for gene therapy. BioDrugs. **31**(4), 317–334 (2017). https://doi.org/10.1007/s40259-017-0234-5

13. Kotterman, M.A., Schaffer, D.V.: Engineering adeno-associated viruses for clinical gene therapy. Nat. Rev. Genet. **15**(7), 445–451 (2014). https://doi.org/10.1038/nrg3742

14. Samulski, R.J., Muzyczka, N.: AAV-mediated gene therapy for research and therapeutic purposes. Annu Rev Virol. **1**(1), 427–451 (2014). https://doi.org/10.1146/annurev-virology-031413-085355

15. Li, C., Samulski, R.J.: Engineering adeno-associated virus vectors for gene therapy. Nat. Rev. Genet. **21**(4), 255–272 (2020). https://doi.org/10.1038/s41576-019-0205-4

16. Tratschin, J.D., Miller, I.L., Smith, M.G., Carter, B.J.: Adeno-associated virus vector for high-frequency integration, expression, and rescue of genes in mammalian cells. Mol. Cell. Biol. **5**(11), 3251–3260 (1985). https://doi.org/10.1128/mcb.5.11.3251-3260.1985

17. Chaanine, A.H., Kalman, J., Hajjar, R.J.: Cardiac gene therapy. Semin. Thorac. Cardiovasc. Surg. **22**(2), 127–139 (2010). https://doi.org/10.1053/j.semtcvs.2010.09.009

18. Sonntag, F., Schmidt, K., Kleinschmidt, J.A.: A viral assembly factor promotes AAV2 capsid formation in the nucleolus. Proc. Natl. Acad. Sci. U. S. A. **107**(22), 10220–10225 (2010). https://doi.org/10.1073/pnas.1001673107

19. Naumer, M., Sonntag, F., Schmidt, K., Nieto, K., Panke, C., Davey, N.E., et al.: Properties of the adeno-associated virus assembly-activating protein. J. Virol. **86**(23), 13038–13048 (2012). https://doi.org/10.1128/JVI.01675-12

20. Ogden, P.J., Kelsic, E.D., Sinai, S., Church, G.M.: Comprehensive AAV capsid fitness landscape reveals a viral gene and enables machine-guided design. Science (New York, NY). **366**(6469), 1139–1143 (2019). https://doi.org/10.1126/science.aaw2900

21. Kok, C.Y., Alexander, I., Lisowski, L., Kizana, E.: Directed evolution of adeno-associated virus vectors in human cardiomyocytes for cardiac gene therapy. Heart Lung Circ. **27**(11), 1270–1273 (2018). https://doi.org/10.1016/j.hlc.2018.08.014

22. Srivastava, A.: In vivo tissue-tropism of adeno-associated viral vectors. Curr. Opin. Virol. **21**, 75–80 (2016). https://doi.org/10.1016/j.coviro.2016.08.003

23. Asokan, A., Schaffer, D.V., Samulski, R.J.: The AAV vector toolkit: poised at the clinical crossroads. Mol. Ther. **20**(4), 699–708 (2012). https://doi.org/10.1038/mt.2011.287

24. Carter, P.J., Samulski, R.J.: Adeno-associated viral vectors as gene delivery vehicles. Int. J. Mol. Med. **6**(1), 17–27 (2000). https://doi.org/10.3892/ijmm.6.1.17

25. Zincarelli, C., Soltys, S., Rengo, G., Rabinowitz, J.E.: Analysis of AAV serotypes 1-9 mediated gene expression and tropism in mice after systemic injection. Mol. Ther. **16**(6), 1073–1080 (2008). https://doi.org/10.1038/mt.2008.76

26. Asokan, A., Samulski, R.J.: An emerging adeno-associated viral vector pipeline for cardiac gene therapy. Hum. Gene Ther. **24**(11), 906–913 (2013). https://doi.org/10.1089/hum.2013.2515

27. Pacak, C.A., Mah, C.S., Thattaliyath, B.D., Conlon, T.J., Lewis, M.A., Cloutier, D.E., et al.: Recombinant adeno-associated virus serotype 9 leads to preferential cardiac transduction in vivo. Circ. Res. **99**(4), e3–e9 (2006). https://doi.org/10.1161/01.RES.0000237661.18885.f6

28. Bish, L.T., Morine, K., Sleeper, M.M., Sanmiguel, J., Wu, D., Gao, G., et al.: Adeno-associated virus (AAV) serotype 9 provides global cardiac gene transfer superior to AAV1, AAV6, AAV7, and AAV8 in the mouse and rat. Hum. Gene Ther. **19**(12), 1359–1368 (2008). https://doi.org/10.1089/hum.2008.123

29. Fang, H., Lai, N.C., Gao, M.H., Miyanohara, A., Roth, D.M., Tang, T., et al.: Comparison of adeno-associated virus serotypes and delivery methods for cardiac gene transfer. Hum Gene Ther Methods. **23**(4), 234–241 (2012). https://doi.org/10.1089/hgtb.2012.105

30. Kaplitt, M.G., Xiao, X., Samulski, R.J., Li, J., Ojamaa, K., Klein, I.L., et al.: Long-term gene transfer in porcine myocardium after coronary infusion of an adeno-associated virus vector. Ann. Thorac. Surg. **62**(6), 1669–1676 (1996). https://doi.org/10.1016/s0003-4975(96)00946-0

31. Su, H., Yeghiazarians, Y., Lee, A., Huang, Y., Arakawa-Hoyt, J., Ye, J., et al.: AAV serotype 1 mediates more efficient gene transfer to pig myocardium than AAV serotype 2 and plasmid. J. Gene Med. **10**(1), 33–41 (2008). https://doi.org/10.1002/jgm.1129

32. Raake, P.W., Hinkel, R., Muller, S., Delker, S., Kreuzpointner, R., Kupatt, C., et al.: Cardio-specific long-term gene expression in a porcine model after selective pressure-regulated retro-infusion of adeno-associated viral (AAV) vectors. Gene Ther. **15**(1), 12–17 (2008). https://doi.org/10.1038/sj.gt.3303035

33. Gao, G., Bish, L.T., Sleeper, M.M., Mu, X., Sun, L., Lou, Y., et al.: Transendocardial delivery of AAV6 results in highly efficient and global cardiac gene transfer in rhesus macaques. Hum. Gene Ther. **22**(8), 979–984 (2011). https://doi.org/10.1089/hum.2011.042

34. Pan, X., Yue, Y., Zhang, K., Lostal, W., Shin, J.H., Duan, D.: Long-term robust myocardial transduction of the dog heart from a peripheral vein by adeno-associated virus serotype-8. Hum. Gene Ther. **24**(6), 584–594 (2013). https://doi.org/10.1089/hum.2013.044

35. Pan, X., Yue, Y., Zhang, K., Hakim, C.H., Kodippili, K., McDonald, T., et al.: AAV-8 is more efficient than AAV-9 in transducing neonatal dog heart. Hum Gene Ther Methods. **26**(2), 54–61 (2015). https://doi.org/10.1089/hgtb.2014.128

36. Ferrarini, M., Arsic, N., Recchia, F.A., Zentilin, L., Zacchigna, S., Xu, X., et al.: Adeno-associated virus-mediated transduction of VEGF165 improves cardiac tissue viability and functional recovery after permanent coronary occlusion in conscious dogs. Circ. Res. **98**(7), 954–961 (2006). https://doi.org/10.1161/01.RES.0000217342.83731.89

37. Pleger, S.T., Shan, C., Ksienzyk, J., Bekeredjian, R., Boekstegers, P., Hinkel, R., et al.: Cardiac AAV9-S100A1 gene therapy rescues post-ischemic heart failure in a preclinical large animal model. Sci Transl Med. **3**(92), 92ra64 (2011). https://doi.org/10.1126/scitranslmed.3002097

38. Raake, P.W., Schlegel, P., Ksienzyk, J., Reinkober, J., Barthelmes, J., Schinkel, S., et al.: AAV6.betaARKct cardiac gene therapy ameliorates cardiac function and normalizes the catecholaminergic axis in a clinically relevant large animal heart failure model. Eur. Heart J. **34**(19), 1437–1447 (2013). https://doi.org/10.1093/eurheartj/ehr447

39. Ishikawa, K., Fish, K.M., Tilemann, L., Rapti, K., Aguero, J., Santos-Gallego, C.G., et al.: Cardiac I-1c overexpression with reengineered AAV improves cardiac function in swine ischemic heart failure. Mol. Ther. **22**(12), 2038–2045 (2014). https://doi.org/10.1038/mt.2014.127

40. Fish, K.M., Ladage, D., Kawase, Y., Karakikes, I., Jeong, D., Ly, H., et al.: AAV9.I-1c delivered via direct coronary infusion in a porcine model of heart failure improves contractility and mitigates adverse remodeling. Circ. Heart Fail. **6**(2), 310–317 (2013). https://doi.org/10.1161/CIRCHEARTFAILURE.112.971325

41. Swain, J.D., Fargnoli, A.S., Katz, M.G., Tomasulo, C.E., Sumaroka, M., Richardville, K.C., et al.: MCARD-mediated gene transfer of GRK2 inhibitor in ovine model of acute myocardial infarction. J. Cardiovasc. Transl. Res. **6**(2), 253–262 (2013). https://doi.org/10.1007/s12265-012-9418-z

42. Katz, M.G., Fargnoli, A.S., Williams, R.D., Steuerwald, N.M., Isidro, A., Ivanina, A.V., et al.: Safety and efficacy of high-dose adeno-associated virus 9 encoding sarcoplasmic reticulum Ca(2+) adenosine triphosphatase delivered by molecular cardiac surgery with recirculating delivery in ovine ischemic cardiomyopathy. J Thorac Cardiovasc Surg. **148**(3), 1065–1072 (2014) 73e1–2; discussion72–3. https://doi.org/10.1016/j.jtcvs.2014.05.070

43. Katz, M.G., Brandon-Warner, E., Fargnoli, A.S., Williams, R.D., Kendle, A.P., Hajjar, R.J., et al.: Mitigation of myocardial fibrosis by molecular cardiac surgery-mediated gene overexpression. J. Thorac. Cardiovasc. Surg. **151**(4), 1191–1200. e3 (2016). https://doi.org/10.1016/j.jtcvs.2015.11.031

44. Hinkel, R., Lange, P., Petersen, B., Gottlieb, E., Ng, J.K., Finger, S., et al.: Heme oxygenase-1 gene therapy provides cardioprotection via control of post-ischemic inflammation: an experimental study in a pre-clinical pig model. J. Am. Coll. Cardiol. **66**(2), 154–165 (2015). https://doi.org/10.1016/j.jacc.2015.04.064

45. Tilemann, L., Lee, A., Ishikawa, K., Aguero, J., Rapti, K., Santos-Gallego, C., et al.: SUMO-1 gene transfer improves cardiac function in a large-animal model of heart failure. Sci Transl Med. **5**(211), 211ra159 (2013). https://doi.org/10.1126/scitranslmed.3006487

46. Kawase, Y., Ly, H.Q., Prunier, F., Lebeche, D., Shi, Y., Jin, H., et al.: Reversal of cardiac dysfunction after long-term expression of SERCA2a by gene transfer in a pre-clinical model of heart failure. J. Am. Coll. Cardiol. **51**(11), 1112–1119 (2008). https://doi.org/10.1016/j.jacc.2007.12.014

47. Hadri, L., Bobe, R., Kawase, Y., Ladage, D., Ishikawa, K., Atassi, F., et al.: SERCA2a gene transfer enhances eNOS expression and activity in endothelial cells. Mol. Ther. **18**(7), 1284–1292 (2010). https://doi.org/10.1038/mt.2010.77

48. Beeri, R., Chaput, M., Guerrero, J.L., Kawase, Y., Yosefy, C., Abedat, S., et al.: Gene delivery of sarcoplasmic reticulum calcium ATPase inhibits ventricular remodeling in ischemic mitral regurgitation. Circ. Heart Fail. **3**(5), 627–634 (2010). https://doi.org/10.1161/CIRCHEARTFAILURE.109.891184

49. Pepe, M., Mamdani, M., Zentilin, L., Csiszar, A., Qanud, K., Zacchigna, S., et al.: Intramyocardial VEGF-B167 gene delivery delays the progression towards congestive failure in dogs with pacing-induced dilated cardiomyopathy. Circ. Res. **106**(12), 1893–1903 (2010). https://doi.org/10.1161/CIRCRESAHA.110.220855

50. Woitek, F., Zentilin, L., Hoffman, N.E., Powers, J.C., Ottiger, I., Parikh, S., et al.: Intracoronary cytoprotective gene therapy: a study of VEGF-B167 in a pre-clinical animal model of dilated cardiomyopathy. J. Am. Coll. Cardiol. **66**(2), 139–153 (2015). https://doi.org/10.1016/j.jacc.2015.04.071

51. Byrne, M.J., Power, J.M., Preovolos, A., Mariani, J.A., Hajjar, R.J., Kaye, D.M.: Recirculating cardiac delivery of AAV2/1SERCA2a improves myocardial function in an experimental model of heart failure in large animals. Gene Ther. **15**(23), 1550–1557 (2008). https://doi.org/10.1038/gt.2008.120

52. Hajjar, R.J., Zsebo, K., Deckelbaum, L., Thompson, C., Rudy, J., Yaroshinsky, A., et al.: Design of a phase 1/2 trial of intracoronary administration of AAV1/SERCA2a in patients with heart failure. J. Card. Fail. **14**(5), 355–367 (2008). https://doi.org/10.1016/j.cardfail.2008.02.005

53. Jaski, B.E., Jessup, M.L., Mancini, D.M., Cappola, T.P., Pauly, D.F., Greenberg, B., et al.: Calcium upregulation by percutaneous administration of gene therapy in cardiac disease (CUPID trial), a first-in-human phase 1/2 clinical trial. J. Card. Fail. **15**(3), 171–181 (2009). https://doi.org/10.1016/j.cardfail.2009.01.013

54. Kranias, E.G., Hajjar, R.J.: Modulation of cardiac contractility by the phospholamban/SERCA2a regulatome. Circ. Res. **110**(12), 1646–1660 (2012). https://doi.org/10.1161/CIRCRESAHA.111.259754

55. Matkar, P.N., Leong-Poi, H., Singh, K.K.: Cardiac gene therapy: are we there yet? Gene Ther. **23**(8–9), 635–648 (2016). https://doi.org/10.1038/gt.2016.43

56. Greenberg, B., Yaroshinsky, A., Zsebo, K.M., Butler, J., Felker, G.M., Voors, A.A., et al.: Design of a phase 2b trial of intracoronary administration of AAV1/SERCA2a in patients with advanced heart failure: the CUPID 2 trial (calcium up-regulation by percutaneous administration of gene therapy in cardiac disease phase 2b). JACC Heart Fail. **2**(1), 84–92 (2014). https://doi.org/10.1016/j.jchf.2013.09.008

57. Jessup, M., Greenberg, B., Mancini, D., Cappola, T., Pauly, D.F., Jaski, B., et al.: Calcium upregulation by percutaneous administration of gene therapy in cardiac disease (CUPID): a phase 2 trial of intracoronary gene therapy of sarcoplasmic reticulum Ca2+-ATPase in patients

with advanced heart failure. Circulation. **124**(3), 304–313 (2011). https://doi.org/10.1161/CIRCULATIONAHA.111.022889

58. Greenberg, B., Butler, J., Felker, G.M., Ponikowski, P., Voors, A.A., Desai, A.S., et al.: Calcium upregulation by percutaneous administration of gene therapy in patients with cardiac disease (CUPID 2): a randomised, multinational, double-blind, placebo-controlled, phase 2b trial. Lancet. **387**(10024), 1178–1186 (2016). https://doi.org/10.1016/S0140-6736(16)00082-9

59. Hulot, J.S., Salem, J.E., Redheuil, A., Collet, J.P., Varnous, S., Jourdain, P., et al.: Effect of intracoronary administration of AAV1/SERCA2a on ventricular remodelling in patients with advanced systolic heart failure: results from the AGENT-HF randomized phase 2 trial. Eur. J. Heart Fail. **19**(11), 1534–1541 (2017). https://doi.org/10.1002/ejhf.826

60. Pathak, A., del Monte, F., Zhao, W., Schultz, J.E., Lorenz, J.N., Bodi, I., et al.: Enhancement of cardiac function and suppression of heart failure progression by inhibition of protein phosphatase 1. Circ. Res. **96**(7), 756–766 (2005). https://doi.org/10.1161/01.RES.0000161256.85833.fa

61. Cabanes-Creus, M., Hallwirth, C.V., Westhaus, A., Ng, B.H., Liao, S.H.Y., Zhu, E., et al.: Restoring the natural tropism of AAV2 vectors for human liver. Science Translational Medicine. **12**(560), eaba3312 (2020). https://doi.org/10.1126/scitranslmed.aba3312

62. Hsu, H.-L., Brown, A., Loveland, A.B., Lotun, A., Xu, M., Luo, L., et al.: Structural characterization of a novel human adeno-associated virus capsid with neurotropic properties. Nat. Commun. **11**(1), 3279 (2020). https://doi.org/10.1038/s41467-020-17047-1

63. Qin, W., Xu, G., Tai, P.W.L., Wang, C., Luo, L., Li, C., et al.: Large-scale molecular epidemiological analysis of AAV in a cancer patient population. Oncogene. **40**(17), 3060–3071 (2021). https://doi.org/10.1038/s41388-021-01725-5

64. Zincarelli, C., Soltys, S., Rengo, G., Rabinowitz, J.E.: Analysis of AAV serotypes 1–9 mediated gene expression and tropism in mice after systemic injection. Mol. Ther. **16**(6), 1073–1080 (2008). https://doi.org/10.1038/mt.2008.76

65. Wang, L., Wang, H., Bell, P., McCarter, R.J., He, J., Calcedo, R., et al.: Systematic evaluation of AAV vectors for liver directed gene transfer in murine models. Mol. Ther. **18**(1), 118–125 (2010). https://doi.org/10.1038/mt.2009.246

66. Hinderer, C., Katz, N., Buza, E.L., Dyer, C., Goode, T., Bell, P., et al.: Severe toxicity in nonhuman primates and piglets following high-dose intravenous administration of an adeno-associated virus vector expressing human SMN. Hum. Gene Ther. **29**(3), 285–298 (2018). https://doi.org/10.1089/hum.2018.015

67. High-dose AAV gene therapy deaths. Nature biotechnology. 2020;38(8):910. https://doi.org/10.1038/s41587-020-0642-9

68. Cabanes-Creus, M., Westhaus, A., Navarro, R.G., Baltazar, G., Zhu, E., Amaya, A.K., et al.: Attenuation of heparan sulfate proteoglycan binding enhances in vivo transduction of human primary hepatocytes with AAV2. Molecular Therapy - Methods & Clinical Development. **17**, 1139–1154 (2020). https://doi.org/10.1016/j.omtm.2020.05.004

69. Asokan, A., Conway, J.C., Phillips, J.L., Li, C., Hegge, J., Sinnott, R., et al.: Reengineering a receptor footprint of adeno-associated virus enables selective and systemic gene transfer to muscle. Nat. Biotechnol. **28**(1), 79–82 (2010). https://doi.org/10.1038/nbt.1599

70. Tarantal, A.F., Lee, C.C.I., Martinez, M.L., Asokan, A., Samulski, R.J.: Systemic and persistent muscle gene expression in rhesus monkeys with a liver de-targeted adeno-associated virus vector. Hum. Gene Ther. **28**(5), 385–391 (2017). https://doi.org/10.1089/hum.2016.130

71. Pulicherla, N., Shen, S., Yadav, S., Debbink, K., Govindasamy, L., Agbandje-McKenna, M., et al.: Engineering liver-detargeted AAV9 vectors for cardiac and musculoskeletal gene transfer. Mol. Ther. **19**(6), 1070–1078 (2011). https://doi.org/10.1038/mt.2011.22

72. Finet, J.E., Wan, X., Donahue, J.K.: Fusion of anthopleurin-B to AAV2 increases specificity of cardiac gene transfer. Virology. **513**, 43–51 (2018). https://doi.org/10.1016/j.virol.2017.10.006

73. Müller, O.J., Kaul, F., Weitzman, M.D., Pasqualini, R., Arap, W., Kleinschmidt, J.A., et al.: Random peptide libraries displayed on adeno-associated virus to select for targeted gene therapy vectors. Nat. Biotechnol. **21**(9), 1040–1046 (2003). https://doi.org/10.1038/nbt856

74. Samoylova, T.I., Smith, B.F.: Elucidation of muscle-binding peptides by phage display screening. Muscle Nerve. **22**(4), 460–466 (1999). https://doi.org/10.1002/(SICI)1097-4598(199904)22:4<460::AID-MUS6>3.0.CO;2-L

75. Ying, Y., Müller, O.J., Goehringer, C., Leuchs, B., Trepel, M., Katus, H.A., et al.: Heart-targeted adeno-associated viral vectors selected by in vivo biopanning of a random viral display peptide library. Gene Ther. **17**(8), 980–990 (2010). https://doi.org/10.1038/gt.2010.44

76. Yu, C.Y., Yuan, Z., Cao, Z., Wang, B., Qiao, C., Li, J., et al.: A muscle-targeting peptide displayed on AAV2 improves muscle tropism on systemic delivery. Gene Ther. **16**(8), 953–962 (2009). https://doi.org/10.1038/gt.2009.59

77. Smith, R.H., Hallwirth, C.V., Westerman, M., Hetherington, N.A., Tseng, Y.-S., Cecchini, S., et al.: Germline viral "fossils" guide in silico reconstruction of a mid-Cenozoic era marsupial adeno-associated virus. Sci. Rep. **6**(1), 28965 (2016). https://doi.org/10.1038/srep28965

78. Li, Y., Li, J., Liu, Y., Shi, Z., Liu, H., Wei, Y., et al.: Bat adeno-associated viruses as gene therapy vectors with the potential to evade human neutralizing antibodies. Gene Ther. **26**(6), 264–276 (2019). https://doi.org/10.1038/s41434-019-0081-8

79. Yang, L., Jiang, J., Drouin, L.M., Agbandje-McKenna, M., Chen, C., Qiao, C., et al.: A myocardium tropic adeno-associated virus (AAV) evolved by DNA shuffling and in vivo selection. Proc. Natl. Acad. Sci. U. S. A. **106**(10), 3946–3951 (2009). https://doi.org/10.1073/pnas.0813207106

80. Weinmann, J., Weis, S., Sippel, J., Tulalamba, W., Remes, A., El Andari, J., et al.: Identification of a myotropic AAV by massively parallel in vivo evaluation of barcoded capsid variants. Nat. Commun. **11**(1), 5432 (2020). https://doi.org/10.1038/s41467-020-19230-w

81. Phillips, M.I., Tang, Y., Schmidt-Ott, K., Qian, K., Kagiyama, S.: Vigilant vector: heart-specific promoter in an adeno-associated virus vector for cardioprotection. Hypertension. **39**(2), 651–655 (2002). https://doi.org/10.1161/hy0202.103472

82. Aikawa, R., Huggins, G.S., Snyder, R.O.: Cardiomyocyte-specific gene expression following recombinant adeno-associated viral vector transduction. J. Biol. Chem. **277**(21), 18979–18985 (2002). https://doi.org/10.1074/jbc.M201257200

83. Barth, A.S., Kizana, E., Smith, R.R., Terrovitis, J., Dong, P., Leppo, M.K., et al.: Lentiviral vectors bearing the cardiac promoter of the Na+-Ca2+ exchanger report cardiogenic differentiation in stem cells. Mol. Ther. **16**(5), 957–964 (2008). https://doi.org/10.1038/mt.2008.30

84. Prasad, K.M.R., Xu, Y., Yang, Z., Acton, S.T., French, B.A.: Robust cardiomyocyte-specific gene expression following systemic injection of AAV: in vivo gene delivery follows a Poisson distribution. Gene Ther. **18**(1), 43–52 (2011). https://doi.org/10.1038/gt.2010.105

Part IV
Bioengineering Approaches to Cardiovascular Tissue Modeling and Repair

Microfabricated Systems for Cardiovascular Tissue Modeling

Ericka Jayne Knee-Walden, Karl Wagner, Qinghua Wu, Naimeh Rafatian, and Milica Radisic

1 Introduction

Cardiovascular disease (CVD) is the leading cause of death worldwide, including 48.0% of adults (121.5 million in 2016) 20 years of age and older in the United States who suffer from its effects [1, 2]. These statistics may be even greater in developing countries, where the prevalence of CVD is difficult to assess due to limited access to healthcare and challenges around the collection of data and health records [3]. To adequately treat CVD, potentially therapeutic compounds must undergo comprehensive preclinical evaluation and validation before clinical approval [4]. Drugs must show not just that they are safe and effective, but that they demonstrate an improvement over conventional treatments [5]. This process is

E. J. Knee-Walden · Q. Wu
Institute of Biomaterials and Biomedical Engineering, University of Toronto, Toronto, ON, Canada

K. Wagner
Institute of Biomaterials and Biomedical Engineering, University of Toronto, Toronto, ON, Canada

Department of Chemical Engineering and Applied Chemistry, University of Toronto, Toronto, ON, Canada

N. Rafatian
Toronto General Research Institute, University Health Network, Toronto, ON, Canada

M. Radisic (✉)
Institute of Biomaterials and Biomedical Engineering, University of Toronto, Toronto, ON, Canada

Department of Chemical Engineering and Applied Chemistry, University of Toronto, Toronto, ON, Canada

Toronto General Research Institute, University Health Network, Toronto, ON, Canada
e-mail: m.radisic@utoronto.ca

© The Author(s), under exclusive license to Springer Nature Switzerland AG 2022
J. Zhang, V. Serpooshan (eds.), *Advanced Technologies in Cardiovascular Bioengineering*, https://doi.org/10.1007/978-3-030-86140-7_10

expensive, with the median cost per clinical trial coming out at \$3.4 million for phase I, \$8.6 million for phase II, and \$21.3 million for phase III trials [5]. Furthermore, the clinical approval success rate for cardiovascular drugs originating from the pharmaceutical industry is 9% [6]. Most drugs do not make it to market, due mainly to limitations in the physiological and pathophysiological complexity and relevancy that *in vitro* 2D cultures, as well as *in vivo* animal models, possess with respect to native tissue in humans [7]. In fact, the U.S. Environmental Protection Agency (EPA) will stop all studies and funding on mammals by 2035 with financial resources going towards institutions developing non-animal alternatives such as organ-on-a-chip models [8]. Similarly, the Food and Drug Administration (FDA) has proposed efforts to reduce the use of animals for research within its own agency and for research conducted by industry [9].

The need exists for more precise and reliable human tissue models that adequately represent the native 3D microenvironment for pre-clinical drug screening to improve the success rate of clinical trials. Organ-on-a-chip models are microfabricated devices that are used to culture tissues made from cells on a scaffold that delivers structural support [4]. They allow for defined and specific control over the molecular and cellular environment to study the mechanisms of disease progression [10]. In addition, they can be made with the appropriate biomaterials that possess the preferred biological, chemical, and mechanical properties for the tissues [11]. Since cells naturally grow in a 3D environment, the spatial organization of cells, including the extracellular matrix (ECM), greatly contributes to how they interact with one another in their microenvironment [12]. Organ-on-a-chip devices can mimic this microenvironment and provide dynamic flow, vascularization, and electrical stimulation [10, 13]. Given that microfabricated devices can also use patient-specific cells, the role that they could play in the discovery of new and effective drugs, as well as personalized medicine, is promising [14].

This chapter aims to examine the role that materials play in cardiac tissue engineering, the types of microfabrication techniques used, the current microfabricated organ-on-a-chip platforms in cardiac tissue engineering, the types of built-in or functional readouts that exist and why they are important, and how these tissue platforms can be used to help understand the mechanisms of cardiac physiology or pathophysiology.

2 Materials for Microfabricated Systems

For utility in drug screening, it is desirable that cardiac microtissues truly mimic adult heart tissues. To generate such adult-like tissues, cardiac developmental biology is stimulated in culture and the bioreactor environment.

2.1 Cell Types

Heart tissues, *in vivo*, are a mixture of myocytes, endothelial cells, fibroblasts, prevascular cells, macrophages, and neuronal cells [15–17] (Fig. 1). However, to feasibly generate microtissues in a simple laboratory environment, microtissues are commonly only made up of a mixture of myocytes, fibroblasts, and/or endothelial cells (ECs). With the advent of stem cell biology, most microtissues are made from pluripotent stem cell-derived cardiomyocytes (CMs). In commonly used stem cell differentiation protocols (reviewed by [18]), myocytes begin beating at day 8–10 during differentiation. Most tissue engineering platforms to date have used myocytes between days 18–24 post-differentiation. As an exception, Ronaldson et al. applied stem cell-derived myocytes at day 12–14, which were more plastic and were able to generate more mature and adult-like cardiac tissues [19].

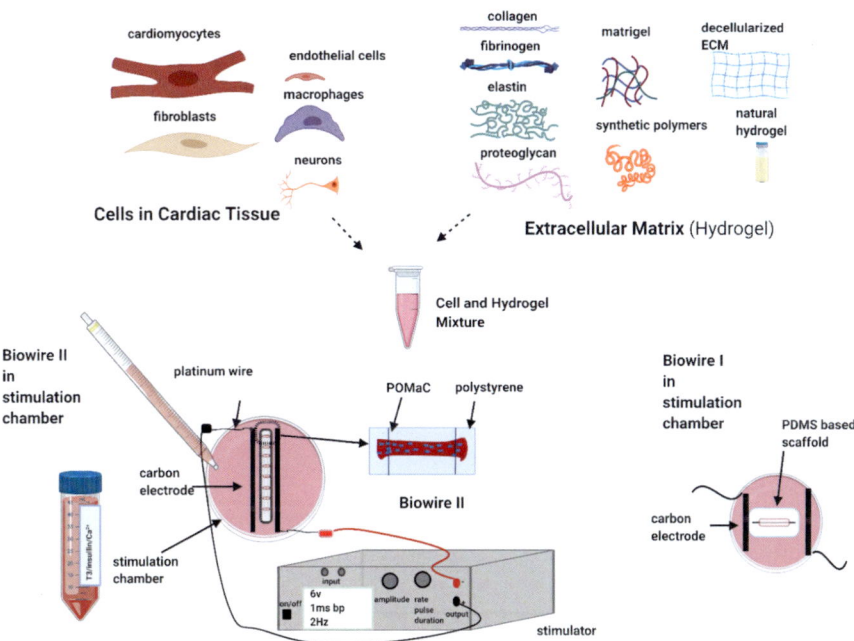

Fig. 1 The heart is a mixture of various types of cells (including myocytes, cardiac fibroblasts, ECs, macrophages, and neurons). Some cell types, mainly fibroblasts or ECs, are mixed with myocytes in at least one of the available natural or synthetic ECM components to engineer cardiac tissue. These cardiac tissues are usually cast in a platform made of PDMS and polystyrene. Here, two of the Radisic Lab cardiac platforms are illustrated schematically. The engineered tissues are kept in defined media containing nutrients or hormones like fatty acids, T3, and insulin with physiological ionic content. Figure created with BioRender.com

Fibroblasts

Fibroblasts are the other essential cells in native cardiac tissues. They generate ECM actively in response to stresses, which are sensed by their cytoskeletal and membrane proteins or via paracrine pathways. Fibroblasts are also crucial in heart conduction; they shield myocytes and allow unidirectional electrical wave propagation through the cardiac tissue, thereby minimizing arrhythmic conditions. Age-matched fibroblasts show better integration into engineered tissues than adult fibroblasts. Different types of fibroblasts (including dermal, cardiac, mesenchymal, and stem cell-derived) have been used in engineering microtissues, but the most consistent results are derived from dermal and cardiac fibroblasts. The presence of 10–30% fibroblasts in engineered tissue makes them more compact, increases their force twitch, and improves their action potential shape [20–23].

Endothelial Cells

Endothelial cells (ECs) are the most abundant cells in the heart. They form the inner layer of capillaries which provide nutrients and oxygen (O_2) to the other cells in the heart. They are also involved in communicating with the ECM and relaying mechanical signals to myocytes. The application of ECs inside engineered cardiac tissues has resulted in the formation of small vessel-like structures [22–24]. The presence of ECs in co-differentiation methods and their existence in engineered heart tissues has enhanced myocyte's maturity and their pharmacological responses. Cardiac-derived ECs are also superior in mimicking the natural cardiac conditions due to the specific developmental origins of the ECs [25, 26].

2.2 Environmental Stimuli

Cardiac microtissues need to recapitulate a desired phenotype, which in most cases is healthy adult-like myocardium. In the case of disease modeling, they must accurately reflect the patient phenotype. As such, they require some essential nutrients and biological compounds in their media. Myocyte maturation is a complex process. In the process of differentiation and maturation, the types of nutrients and biological compounds change over time. Exact events *in vitro* are neither possible to mimic nor are they fully known. However, the inclusion of certain nutrients or biological stimuli are crucial. In the fetal stage, myocytes mainly rely on the anaerobic glycolysis pathway and they largely use glucose and lactate and, to a lesser extent, fatty acids as their source of energy. After birth, the concentration of lactate drops and the concentration of fatty acids rises due to a switch in energy source. These changes have significant impacts on myocyte maturation. Several groups have used defined media with low glucose and high fatty acid content, reporting more adult-like tissues [19, 27]. Other groups (e.g. Radisic Lab) have found commercial StemPro-34 media (Gibco Bioscience), which contains human lipoprotein, to work better than glucose-based media like Roswell Park Memorial Institute 1640 (RPMI) supplemented with B-27 (unpublished results).

The other important compound for myocytes is calcium. The physiological calcium range for cardiac cells is around 1.2 mM. Dulbecco's Modified Eagle Medium (DMEM) based media, unlike RPMI, are within the physiological range [27]. Ascorbic acid or its derivatives are often used in differentiation to promote differentiation efficiency; however, at the tissue level, it has also been used as an antioxidant [28, 29]. Other biological hormones such as triiodothyronine (T3), insulin, or Insulin-Like Growth Factor-1 (IGF-1), have also been applied in media by at least by two groups using defined media for cardiac tissue maturation [27, 30].

2.3 Extracellular Matrix

Cells in cardiac tissue are embedded in a biologically active ECM. ECM provides dynamic support for cells, communicates with them, and helps them remodel their environment (Fig. 1). ECM stiffness and molecular composition are important factors in the engineering of cardiac tissue. ECM guides tissue alignment and directionality. ECM is often introduced into the tissue as a hydrogel at the time of cell seeding. Native tissues have collagen I and collagen III as the main constituents of ECM. While collagen I adds to tensile stress, collagen III relaxes tensile stress. Proteins such as fibrillin, laminin, fibronectin, collagen type IV, and proteoglycans are part of adult cardiac ECM [31]. Both stiffness and the composition of ECM change during development. In adult humans, collagen is more abundant than fibronectin. ECM becomes stiffer in adults. While the early embryonic heart has a stiffness of 1-2 kPa, the adult heart has a stiffness around 20 times higher [32, 33]. A mixture of fibrinogen/thrombin, collagen, and Matrigel are routinely used in cardiac tissue engineering, but the long cross-linking time of the latter two may cause necrotic cores in tissues which hinder their uniformity [30].

Decellularized ECM provides organ-specific ECM composition and structure, with biomimetic mechanical and biochemical cues for cardiac tissue development; however, applying it in small high-throughput microtissues might be a barrier, as it has a 200–300 µM thickness [34, 35]. Natural hydrogel from decellularized myocardium can be used in cardiac tissue engineering. This natural myocardial hydrogel provides the normal cues for development and maturation of region-specific cardiac tissues [36].

Cardiac tissue has a naturally anisotropic structure, which is characterized by fiber alignment, generally in the parallel direction of the applied stress. To facilitate the anisotropy, electrospinning can be used [37]. However, generating anisotropy by this method might be at the expense of a similar pore size for all cellular components of the tissue [38].

Besides natural hydrogel, some synthetic hydrogels can also be used in cardiac tissues, for example poly(ethylene glycol) (PEG)-based hydrogels. PEG, or other polymers like the elastomeric protein elastin, can be conjugated to a conductive material such as graphene oxide (GO) nanoparticles to generate conductive hydrogels. Conductive hydrogels with proper elasticity can improve the structural and functional properties of engineered tissues [39, 40]. Conductive hydrogels using materials like graphene mimic the conductive behaviour of cardiac tissues and thereby promote cardiac function and maturation markers, including structural proteins that work in myocardial force generation (e.g., troponin I, α-actinin) or the electrical conductivity of tissue (such as connexion 43) [41, 42].

2.4   Scaffolds for Microtissues

Microtissues are usually cast into microfabricated platforms. These platforms should provide ideal mechanical supports and unique environments to achieve physiologically relevant tissues. One of the commonly used materials is polydimethylsiloxane (PDMS). PDMS is an elastic material with controllable rigidity through degree of cross-linking. This property makes it suitable for micropillars, stamps, or micropatterning. PDMS is transparent, which helps microtissue imaging on the platform. PDMS is not an ideal platform base however, because it is hydrophobic, which limits its interaction with tissue. The other PDMS shortcoming is its permeability to small molecules, especially those that are hydrophobic. This property does not make PMDS ideal for pharmacological testing since it may carry over compounds or limit the actual available dosage [4, 43]. Diffusion limit and concentration change are factors important for mass transport across a polymer. In microfluidic devices, the concentration of the drug, channel dimensions, fluid velocity, hydrophobic and hydrophilic nature of the drug, and time of the experiment change and affect the mass transport of the drug. Based on these factors, the actual amount of pharmacological compound reaching engineered tissues can be calculated and modeled to make drug testing more accurate [44].

Another material that is used as a scaffold in tissue engineering is polystyrene. Polystyrene is a hard surface with low permeability to gas and hydrophilic or hydrophobic compounds. Polystyrene is commonly used in cell culture plastics and has a rigid structure [45]. Other new materials are also emerging in the field of organ-on-a-chip engineering. These materials are often drug-inert and elastic. Examples of these compounds are butylene–styrene, styrene–ethylene, and polyurethane elastomers [46, 47].

Polyesters such as poly(octamethylene maleate (anhydride) citrate) (POMaC) and poly(octamethylene maleate (anhydride) 1,2,4-butanetricarboxylate) (124 polymer) are other types of flexible, moldable, and biodegradable materials recently used in tissue engineering. The POMaC polymer showed a wide range of Young's moduli between 0.03 and 1.54 MPa, allowing elongation between 48% and 534% strain. The elasticity of both materials is tunable by monomer composition, UV light exposure energy, and porosity of the cured elastomer [48, 49]. Young's modulus in the human heart varies between 10 to 20 kPa. This increases in fibrosis to as high as 130kpa [50]. Therefore, polymers such as POMaC and 124 polymer are ideal for cardiac tissue force and contraction analysis when utilized for scaffolding, as exemplified in the recent work of the Radisic laboratory. Myocyte contractile force is a functional readout in such cardiac microtissues, enabled by bending of POMaC wires and cantilevers. POMaC is an elastic and autofluorescent material that facilitates force tracing in the Biowire™ II and InVADE systems [51, 52]. Moreover, Montgomery et al. used POMaC for building a flexible shape memory scaffold. These patches improved cardiac function following myocardial infarction (MI) [53].

3 Microfabrication Technology for Cardiovascular Tissue Engineering

One primary challenge for generating vascularized cardiac tissues is to develop microfabrication technology that is capable of fabricating polymer-based or cell-laden systems featuring intrinsic vasculature to mimic the structural and biological complexity of native tissue or organs [54, 55]. Recent studies have demonstrated that nano−/micro-topographical cues can affect cell behaviour [56]. It remains a difficult task to engineering complex tissues or organs, such as kidney, heart, and liver, with a vascular supply for the transportation of nutrients, metabolites, and intercellular crosstalk. Two major categories of microfabrication technology (i.e., molding and additive manufacturing) have been carried out to engineer topographical patterns for cardiac engineering. Recent advances in microfabrication techniques, such as soft lithography [57], photolithography, 3D printing [58], and electrospinning [59], have helped further develop the engineering of cardiac tissues and enabled the production of the perfusion culture or co-culture of cardiac tissues in 3D vascular networks.

3.1 Lithography

Lithography has been widely used for generating molds in a wide ranging scale and has the advantages of high efficiency, high resolution, and producing geometry that is compatible with cells [60]. SU-8, a commonly used epoxy-based negative-tone i-line resist, is available in a wide range of viscosities, allowing lithographic fabrication in a variety of thicknesses and geometries [61]. SU-8 master molds fabricated from the micropatterns on the graphic masks applied during soft lithography or photolithography are used for generating hydrogel or polymer-based constructs for engineering cardiac tissues. For example, a microvascular collagen-based network was fabricated by soft lithography and injection molding to investigate the effects of ECs and hydrodynamic stress on the maturation of CMs [62]. The engineered

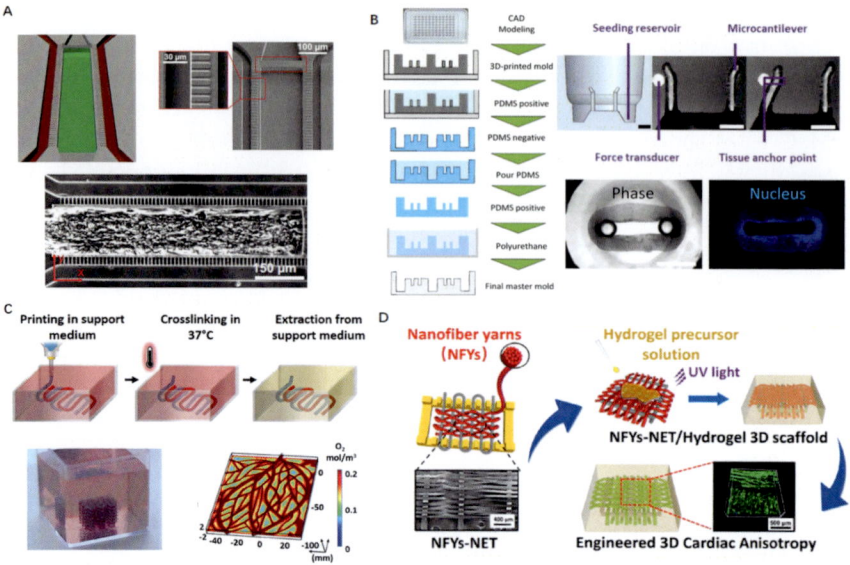

Fig. 2 Microfabrication techniques developed for cardiac tissue engineering. (**a**) Top Image: Microfluidic channels fabricated from soft lithography, showing two nutrient channels (red) and tissue-loading channel (green) with endothelial-like barriers connecting them. Bottom Image: Optical microscopy image of a cardiac tissue generated in the channel. Reproduced with permission from [63]. (**b**) A PDMS plate with microcantilevers was molded by a 3D printed plate and the microcantilevers with tissue anchor points were used to hold cardiac tissue and measure force-displacement. Reproduced with permission from [71]. (C) A 3D printing method developed to perfuse cardiac patches by using a supporting medium. Reproduced with permission from [78]. (D) A 3D hybrid with an interwoven nanofiber yarns network (NFYs-NET) fabricated by a wet-dry electrospinning approach, which was embedded into a hydrogel shell for engineering anisotropic cardiac structures to guide cellular orientation. Reproduced with permission from [85]

hydrogel network could provide temporary mechanical support as naturally derived hydrogel is normally fragile and too soft to support itself. Another interesting study [63] recently demonstrated engineering human induced pluripotent stem cell (hiPSC)-derived cardiac tissues in an ECM-coated microfluidic channel which was fabricated via photolithography (Fig. 2a). This microfabrication technique allowed the formation of parallel microchannels when generating cardiac tissues. To validate the cardiac microphysiological system, pharmacological studies were performed via perfusion of the media channel and drug responses were observed from the aligned cardiac tissues.

In cardiac tissue engineering, a promising strategy for the vascularization of cardiac tissue is co-culture of ECs, CMs, or cardiac fibroblasts to control homotypic and heterotypic cell-cell interactions and allow cellular organization at a microscale level. The Radisic group developed co-culture systems through spatial assembly of honeycomb-like meshes into a loop-hook configuration to form Velcro-like cardiac tissue [64]. Soft lithography was used to create each thin mesh (10–20 μm) made from the biodegradable elastomer POMaC. Due to easy penetration of cell suspension through the mesh structure, an additional EC layer was coated around the cardiac tissue mesh. This co-culture platform was demonstrated to promote cell alignment and guide tissue remodeling.

The same laboratory further demonstrated vascular co-culture by establishing vessel-forming ECs in a 3D scaffold to improve vascularization of engineered cardiac tissues and cell-cell interactions [65]. Direct surgical anastomosis was performed on the perfusable scaffold by connecting it with the femoral vessels of a rat, enabling blood perfusion without clogging or collapsing the scaffold. The soft lithography technique was used to generate specified micro-patterns (10 μm microholes) on the vessel wall for improving permeability and allowing intercellular crosstalk in the tissues. In this study, soft lithography, combined with a 3D stamping technique, produced vascularized cardiac tissues in a scalable manner for building millimeter-scale thick cardiac tissues.

Lithography provides a technique with high resolution that allows for the fabrication of ultrafine structures and vascular 3D networks for generating precise and complex cellular organization. However, this technique contains poorly controlled processes, such as obtaining a good flatness of the resist and removing uncrosslinked resist from narrow channels and edges. It is also difficult to scale up for fabricating high-throughput cardiovascular systems.

3.2   Nonlithography Molding

Molding approaches, including lithography, are commonly used for fabricating micro/nano structures for cardiac tissue engineering. Nonlithography methods such as solvent assisted molding and injection molding are showing their potential for generating vascular grafts. In the context of vascular structure, hydrogels are commonly used for curing in a mold where an annular mold and an internal rod can be used to generate vascular constructs [66]. For example, silk-based vascular grafts were fabricated by solvent-assisted molding combined with electrospinning [67]. The blood compatibility of the grafts was investigated by heparin release after at least one month. Sacrificial molds have been investigated to form vascular networks, through physically or chemically removing sacrificial materials to leave behind networks. Non-toxic solvents are required for this process, otherwise complete evaporation before cell culturing is critical to avoid cytotoxicity for cardiac tissue growth. Water-soluble sacrificial molds were fabricated by 3D printing from sodium alginate [68] and polyvinyl alcohol (PVA) [69, 70], enabling the formation of vascular constructs featuring high cell biocompatibility.

Another recent study [71] demonstrated the use of 3D printed 96-well plates to mold PDMS plates with microcantilevers in each well (Fig. 2b). The microcantilevers could generate hiPSC-derived cardiac microtissues and could be used to measure contractile properties of microtissues.

In another study [72], thick cardiac tissues with prevascular networks were generated from cell sheet stacking. A hydrogel matrix consisting of fibrin and gelatin was coated on a silicone rubber mold, which was used to harvest multilayer cell sheets. A 3D co-culture was formed by generating alternately sandwiched ECs and human skeletal muscle myoblasts.

Injection molding involves the injection of polymeric materials into a predefined mold which contains a precise shape of the final product. The mold needs to be removed once the structures are formed. This technique, used in cardiac tissue engineering, has the advantage of not requiring organic solvents that might be harmful for cell viability and differentiation. Materials such as thermoplastic elastomers, glasses, and thermosetting polymers are commonly applied in this technique. For example, a thermoplastic elastomer of styrene-ethylenebutylene-styrene [73] was fabricated by injection molding and extrusion to generate high-throughput microfluidic devices offering fine features and the elasticity required for potential application in organ-on-a-chip devices. In another study, a multistep injection molding method was used to fabricate fibrin-based heart valves [74]. The fabrication procedure consisted of molding walls and leaflets from different materials and cell phenotypes.

Injection molding provides an excellent opportunity to fabricate vascular networks in a high-throughput manner. However, this technique normally produces a construct with a patterned surface where a thin layer of cardiac tissue can be formed, which makes it difficult to generate real 3D microenvironments for cardiac tissue engineering.

3.3   3D Printing

3D printing technology, integrated with computer-aided design modules, a computer-controlled translation stage, and a three-axis positioning platform, facilitates the fabrication of 3D objects through directly depositing materials in a layer-by-layer fashion. It provides a promising option for engineering vascularized cardiac tissues, due to its capability of customizing predefined geometries featuring fine microstructures and high spatial resolution [75].

Several studies have developed the fabrication of vascularized 3D tissues by 3D printing technology. Miller et al. developed a 3D printing approach to build fully vascularized tissues by using sacrificial carbohydrate materials to generate perfusable channels in engineered tissues [76]. They demonstrated the fabrication of highly spatially organized structures, allowing cell alignment and remodeling in the tissues.

Lind et al. demonstrated a multimaterial 3D printing technique which has an automated design and is able to fabricate cardiac microphysiological devices [77]. This technique offers simple and rapid customization to generate instrumented devices with six materials for higher-throughput and longer-term studies. A recent study [78] demonstrated a simple 3D printing method via a supporting medium followed by hydrogel gelation at 37 °C to fabricate personalized hydrogel structures (Fig. 2c). Two cellular bioinks containing a thermoresponsive collagen hydrogel, hiPSC-CMs and ECs were designed to generate cardiac tissue and blood vessels. This 3D printing strategy allows for the generation of thick, vascularized cardiac patches according to specific patient anatomy, where the architecture can be optimized by mathematically modeling oxygen diffusion.

Bioinks or cell-laden hydrogels can be deposited as droplets or microfilaments through injecting 3D bioprinting or extrusion-based printing techniques to build cell-based blocks for engineering vascular constructs. Kolesky et al. demonstrated 3D bioprinting of human neonatal dermal fibroblast-laden hydrogel inks to engineer tissues with embedded vasculature [79]. 3D vascular networks have been generated by removing 3D printed fugitive ink under mild conditions, allowing the formation of confluent layers of ECs after 48 h of cell perfusion. The same laboratory further developed co-culture of human mesenchymal stem cells (hMSCs), ECs, and human neonatal dermal fibroblasts (hNDFs) to form vascularized human tissues. The thick tissues were printed from multiple bioinks composed of hMSCs and hNDFs using similar bioprinting techniques. The co-culture with ECs was performed by injecting EC suspensions into the microchannels of bioprinted tissues.

The 3D bioprinting technique allows programmable cellular heterogeneity for forming vascular tissues. However, this technique may cause a loss of cell viability because of shear stress during bioprinting. A relatively low viscosity bioink is needed for improving cell viability, but it results in bioprinted vascularized tissues with poor mechanical supports [80]. Thus, it is challenging to 3D print soft and dynamic biological materials with shape fidelity and high resolution for creating complex 3D tissues [81].

A recent study developed a 3D bioprinting method to produce a human cardiac ventricle model [82]. Collagen was directly deposited to form the walls of vascular channels. To promote vascularization, fibronectin and the proangiogenic molecule recombinant vascular endothelial growth factor (VEGF) were incorporated into the collagen bioink. Although 3D bioprinting of a fully functional vascularized heart has not been achieved in research, this technique has presented a strong ability to create components of the human heart as a proof of concept and for partly recapitulating the structural and biological complexity of the native heart.

3.4 Electrospinning

Electrospinning is a promising technique to fabricate nano−/microfibers which requires simple processing and is low in cost. It has been increasingly developed for mimicking the ECM in cardiac tissue engineering. Briefly, a polymeric solution is fed through a needle to form a fiber-based matrix on a collector on the opposite side of the field under a high applied voltage.

Li et al. developed a biocompatible nanofibrous scaffold with randomly distributed nanofibers by electrospinning blends of polyaniline and gelatin for cardiac tissue engineering [83]. In another study, Ruther et al. developed bi-layered patches formed by porous poly(glycerol sebacate) films and electrospun-gelatin membranes [84]. Bi-component patches, with improved hydrophilic properties, layer adhesion, and mechanical strength, were investigated to create cardiac tissues. These cardiac tissue scaffolds were generated from random electrospun matrices unlikely to mimic the native structure of cardiac tissues due to the lack of precise control of fiber arrangement during the deposition process.

An interesting study was presented by Wu et al. who fabricated interwoven aligned nanofiber yarns, consisting of silk fibroin, polycaprolactone, and carbon nanotubes, which were embedded in a hydrogel shell by a wet-dry electrospinning technique (Fig. 2d) [85]. This hybrid scaffold was used to co-culture ECs and CMs to form an endothelialized myocardial tissue construct for improving cardiac tissue function [85].

The alignment of nano−/microfibers can be controlled by the spinning set-up and the type of collector. 3D electrospun scaffolds based on aligned fibers have the potential to mimic anisotropic cardiac structures for controlling cellular orientation and tissue maturation. However, many studies focusing on cardiac engineering with electrospun scaffolds lack cell infiltration in the matrix and result in the relatively poor accessibility of nutrients through the scaffold, though the alignment of the cells on the surface of the matrix can be precisely controlled. Ehler et al. developed a way to engineer primary CMs within electrospun fibers to generate cardiac patches with high cell viability by electrospinning cell-laden fibers [86]. The generation of a cell-laden matrix not only allows the formation of cardiac tissue over a functional depth, but also improves nutrient transfer between the gap junctions in the electrospun fibers.

A recent study developed well-organized ultrafine fibers into 3D scaffolds fabricated by a combination of additive manufacturing and melt electrospinning from a hydroxyl-functionalized polyester [87]. The scaffolds, featuring rectangular patterns, were demonstrated to promote the alignment of cardiac progenitor cells with improved cardiac mechanical relevance [87]. The combined fabrication approaches provide a promising strategy to produce highly organized architectures to regulate cellular organization and activities.

During electrospinning, it may be difficult to precisely control the size and distribution of the fibers during the spinning process, which limits its use in the fabrication of vascular structures for cardiovascular tissue engineering. A growth factor embedded electrospun scaffold was developed by Lakshmanan et al., which loaded dual growth factors VEGF and basic fibroblast growth factor (bFGF) in the nanofibrous matrix. The angiogenic growth factors showed potential to enhance neovascularization in the cardiac tissue as observed by histopathology and immunostaining [88].

4   Types of Microfabricated Systems and Platforms

Many advances have been made in the field of organ-on-a-chip engineering using *in vitro* microfabricated systems. The following section describes the progress made toward cardiac tissue models with the goal of disease modeling, and therapeutic applications such as drug screening and precision medicine.

4.1   Biowire I

One of the challenges with developing an *in vitro*, 3D cardiac tissue model, involves the use of cells that accurately represent adult CMs. Since human CMs are post-mitotic and therefore terminally differentiated cells, they essentially lack the ability to regenerate. This issue has been addressed through the use of human embryonic and induced pluripotent stem cells, hESCs and hiPSCs respectively, the latter of which can be obtained from patient specific donors and differentiated into function-ing CMs. While hESCs and hiPSCs offer a potentially unlimited supply of cells, the CMs generated from these sources are immature and unrepresentative of adult human CMs. Mainly, they display slow conduction velocity, undeveloped action potentials and Ca^{2+} signaling, and a sarcomeric structure that is more representative of fetal tissue [89, 90]. The addition of electrical stimulation can help to mitigate these concerns, as the heart rhythm is naturally maintained by an interconnected network of CMs which communicate through gap junctions and voltage gated ion channels [91].

One method of developing a 3D cardiac tissue model that incorporates electrical stimulation comes from Nunes et al., who cultured CMs derived from hiPSC and hESC directed differentiation, fibroblasts, smooth muscle cells, and ECs around a surgical suture attached at each end to a microfabricated PDMS platform [89]. This *in vitro* model, termed "biowire" or biological wire, showed cell alignment length-wise along the suture, the ability to be electrically stimulated and paced at a specific frequency, and the capability to become mature through the use of a high-frequency regimen after several weeks in culture [89]. The biowire tissues also responded to drug testing by displaying higher Ca^{2+} transient amplitudes when exposed to caf-feine and inhibition of L-type Ca^{2+} channels after the introduction of verapamil or nifedipine [89, 90]. Subsequently, Xiao et al. developed a perfusable biowire using polytetrafluoroethylene (PTFE) tubing as the suture [92] (Fig. 3a). This model dem-onstrated the feasibility of drug testing by running nitric oxide (NO) donor through the wire in culture, thus providing the CMs with biochemical stimulation [92]. The results showed that biowires treated with NO decreased their beating frequency after 24 hours compared to control tissues without treatment, which mimics the vasodilator effect *in vivo* [92].

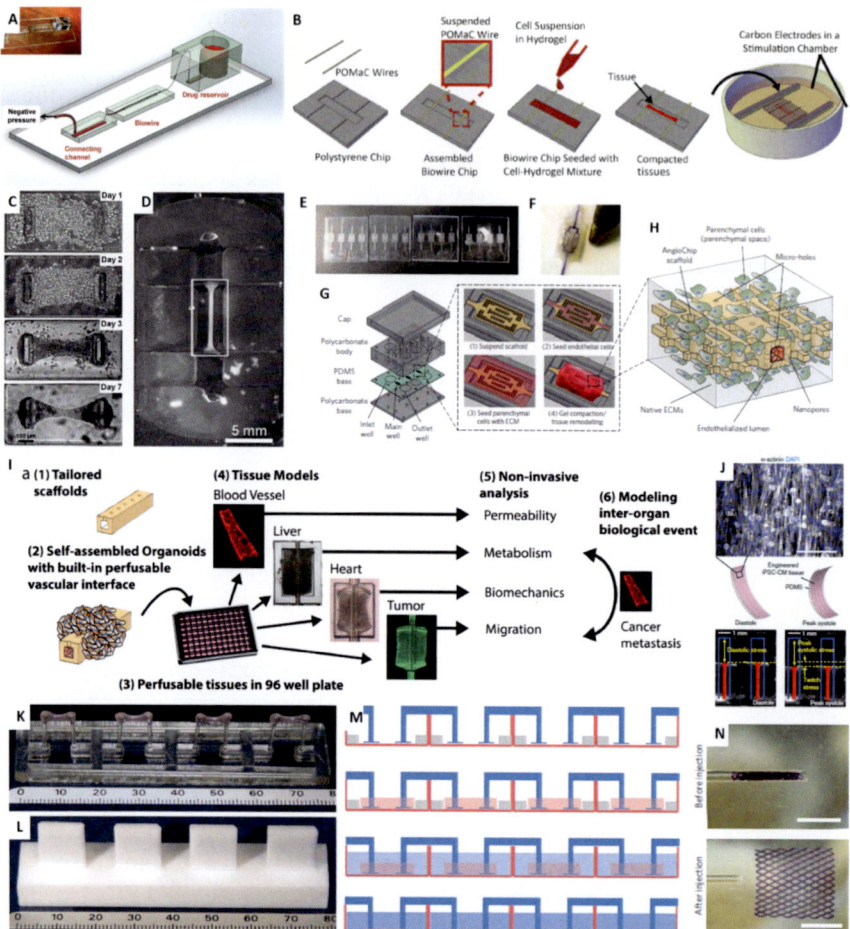

Fig. 3 Representative platforms used in organ-on-a-chip devices. (**a**) Perfusable Biowire I adapted from Xiao et al. [92]. The top corner image illustrates the perfusable system attached to a glass slide. The other image shows the bioreactor set-up with one end containing a drug reservoir channel and the other end with a negative pressure source. Reproduced with permission from [92]. (**b**) Schematic for the Biowire™ II platform adapted from Zhao et al. and reproduced with permission from [52]. (**c**) Neonatal rat cells were mixed with a fibrin and collagen gel and allowed to compact and attach to two cantilevers over 7 days in a post model, adapted from Boudou et al. and reproduced with permission from [91]. (**d**) A PDMS micro-well fitted inside a 6-well plate with two titanium wires where neonatal ventricular rat cells were mixed with a fibrinogen/Matrigel/thrombin hydrogel in the I-Wire platform. Each ridge on either side of the titanium wires provides a place for electrodes during electrical field stimulation. Adapted from Sidorov et al. and reproduced with permission from [100]. (**e**) Glass slides with AngioChip scaffolds. (**f**) An AngioChip containing hepatic tissue and perfused with a color dye. The AngioChip sits next to a ballpoint pen for scale. (**g**) A schematic showing how the AngioChip bioreactor and vascularized tissue are assembled. (**h**) A 3D graphic illustrating part of the inside of an AngioChip tissue. Adapted from Zhang et al. and reproduced with permission from [65]. (**i**) A schematic image of the AngioTube (InVADE) platform adapted from Lai et al. and reproduced with permission from [51]. (**j**) Top image shows-

4.2 Biowire II

The second generation of the biowire platform (Biowire™ II) developed a model where cardiac tissue was suspended between two autofluorescent POMaC polymer wires. The wires were manually placed on a polystyrene chip that was hot embossed from a PDMS master that had been molded from a microfabricated silicon wafer using soft lithography techniques. After the hydrogel mixture (consisting of hiPSC and hESC derived CMs and 10% fibroblasts in a collagen gel matrix) compacted, attached to the wires, and began beating (either spontaneously or after electrical stimulation), a force calculation was made based on the elastic properties of the wires and their displacement [52, 93] (Fig. 3b). The tissues showed CM alignment, expression of sarcomeric proteins, and expected canonical responses to cardiothera-peutic and cardiotoxic agents indicating mature, adult-like properties [94]. This model improves upon the first generation biowire in that the micro-well chambers were made of polystyrene as opposed to PDMS. This makes the Biowire™ II a bet-ter design for drug testing, as PDMS is notorious for absorbing hydrophobic drugs, having the potential to create inconsistency in testing concentrations. The Biowire™ II model was also used with atrial and ventricular CMs and showed gene expression profiles which illustrated strongly enhanced chamber specific identities in direct response to electrical stimulation conditioning [52, 95].

This platform has also been used to test disease models such as cardiac fibrosis, where tissue composition, collagen content, and mechanical properties were con-trolled to directly mimic the condition [96]. The model accurately established char-acteristics of fibrotic myocardium including myofibroblast activation, stiffness, and arrhythmogenesis [96]. Kinase inhibitors have also been tested on the platform as a way of detecting the cardiotoxic effects of cancer targeting agents [97]. The model demonstrated the acute damaging effects on the CM tissues due to the inhibitors compared with cell monolayers [97].

To create higher throughput drug testing using the Biowire™ II platform, a 96-well plate design has been described [95]. A polystyrene sheet molded against a

---

Fig. 3 (continued) tissue from iPSC derived CMs from patients with Barth syndrome stained with α-actinin and DAPI (scale bar is 100 μm). Middle image illustrates the 2D flat PDMS strips plat-form patterned with fibronectin and with myocardial tissue on top in diastole and peak systole. Bottom image shows myocardial tissues in diastole and systole. The difference in the red lines is measured to calculate the stresses. Adapted from Wang et al. and reproduced with permission from [103]. **24-Well Based Platform: (k)** Image of the silicon posts with neonatal rat CM tissues (upside down, scale in mm). **(l)** Image of the Teflon spacer used to make the casting molds (upside down, scale in mm). **(m)** Illustration of tissue generation. Top image shows casting molds which sit in a 24-well plate and use agarose and Teflon spacers. Second image shows the pipetted mixture of the fibrinogen/Matrigel/thrombin hydrogel and CMs in the molds. Third image shows the polymerized tissues attached to the posts at both ends. The tissues are transferred to 24-well plates containing medium. Bottom image illustrates how the tissues compact over a culture period of two to four weeks. Adapted from Hansen et al. and reproduced with permission from [104]. **(n)** Cardiac patch scaffold rolled up before injection and after injection through a 1 mm glass pipette capillary (scale bars are 5 mm). Adapted from Montgomery et al. and reproduced with permission from [53]

PDMS master was hot embossed with carbon electrodes [95]. The polystyrene base was then perfused with POMaC and UV crosslinked [95]. A PDMS mold containing micro-channels was then thermally-bonded to a bottomless 96-well plate [95]. This system illustrated a proof of concept for the scale up of the Biowire™ II platform for drug screening [95].

4.3   Post/Cantilever Model

Another approach to generate cardiac microtissues came from Boudou et al., where neonatal rat cells were cultured in a collagen/fibrin gel around a microfabricated cantilever system which incorporated electrical stimulation [91]. PDMS stamps made from SU-8 masters were used to produce the tissue gauges [91]. Fluorescent microbeads placed on top of the cantilevers were tracked as the microtissues beat under electrical stimulation and based on the spring constant determined from a sensor tip on the cantilever head, the force was calculated [91]. Tissue compaction was observed after 7 days and drug testing was performed with isoproterenol and digoxin [91] (Fig. 3c). The drug testing results showed dose-dependent effects on the tissue contractility and the tissues exhibited aligned sarcomeric structures [91].

An alternative model was developed by Miklas et al. for chronic drug screening that was capable of concurrently applying electrical stimulation and mechanical stimulation in the form of static strain [98]. The added dimension of mechanical strain was supplemented as a way to encourage terminal cell differentiation and an adult-like phenotype, as well as, allowing for different maturation stages or pathological conditions during tissue culture [98]. The method of design involved a PDMS mold obtained from a machined master that contained micro-wells with two built-in post s [98, 99]. The PDMS mold was placed in a bioreactor encompassing a stretch platform and neonatal rat CMs were allowed to compact over 3 days in a hydrogel around the posts [98]. Electrical stimulation was applied and the force of contraction was determined from calculations incorporating the microtissue and post deflection [98, 99]. After 3 days of electrical and mechanical stimulation, the force increased in the microtissues compared to unstimulated controls and improved sarcomeric structure was also observed in comparison to unstimulated controls [98].

This same model was used with hESC derived CMs to test the chronic exposure to isoproterenol, angiotensin II, and endothelin-1, which are therapeutic agents known to have caused nongenetic cardiomyopathy [99]. After 7 days of treatment, the microtissues showed signs of hypertrophy such as increased cell size, a disorganized sarcomere structure, and decreased contractile force [99].

The work of Ronaldson-Bouchard et al. used two PDMS pillars obtained from a machined master to culture hiPSC derived CMs and fibroblasts (75:25 ratio) in a hydrogel [30]. The pillars were enclosed inside a polycarbonate container filled with media for the tissue to culture after attaching to each pillar [30]. Carbon electrodes were aligned along the pillars and the microtissues were electrically stimulated with an increasing frequency of 0.33 Hz per day over 2 weeks after compaction

[30]. Analysis via video of contractile motion and Ca^{2+} transients shows the micro-tissues response to electrical pacing [30]. The results of this model demonstrated that cardiac tissue could be matured to an adult-like phenotype reaching a positive force-frequency relationship [30].

4.4 I-Wire Heart-on-a-Chip

In another organ-on-a-chip device for creating cardiac tissues, neonatal ventricular rat cells were combined in a fibrinogen/Matrigel/thrombin hydrogel and placed into a PDMS mold which consisted of a channel for the 3D tissue and titanium wires on either side in grooves [100] (Fig. 3d). The PDMS mold was created from a master design fabricated on a milling machine. Tensional force was calculated from a poly-ether ether ketone (PEEK) tube (365 μm outer diameter and 120 μm bore) that was attached to a cantilever plate [100].The probe established the same stiffness as the cardiac tissue. As the tissue beat, videos were taken at 200 frames per second and analyzed by the distance the probe tip was displaced to calculate the force. The results showed that after 10–12 days in culture, the sarcomeres elongated and aligned in the direction perpendicular to the two titanium wires. In addition, the length of the tissue was increased as the applied force of the probe increased. When treated with blebbistatin, the Young's modulus decreased significantly, illustrating the effect of the drug on the tissue's elasticity [100].

The same group continued the study of the I-Wire by further investigating bio-mechanical properties of the tissue using a modified Hill model which takes into account the applied tension of the probe to measure resistance and stress inherent in the tissue before, during, and after treatment with blebbistatin and isoproterenol. The results showed a lower passive tension after treatment with blebbistatin and isoproterenol regardless of the transverse load applied [101].

4.5 AngioChip

One issue that needed to be addressed in the 3D *in vitro* organ-on-a-chip platforms was the concern about vascularizing parenchymal tissues to mimic those in the human body [65]. A tissue model that incorporates vasculature can allow for oxygen and nutrients to nourish the surrounding cells [102]. In addition, it more accurately mimics the native physiology of the *in vivo* environment [102]. The AngioChip model from Zhang et al. described a microchannel branched network comprised of a biodegradable scaffold [65] (Fig. 3e and f). The mechanical properties of the scaffold were such that they could support anastomosis and a cardiac tissue that contracted [65]. The scaffold also incorporated micro-pores throughout the vascularized lumen which enabled sprouting to occur [65].

The AngioChip was microfabricated from SU-8 masters using soft lithography [65]. PDMS molds, created from the masters, formed the layers [65]. The layers were placed either on a glass slide (bottom layer) or PDMS (all upper layers), so that POMaC could perfuse through [65] The POMaC was UV-crosslinked, the layers on PDMS were removed, and each layer was aligned and stamped together to form the 3D scaffold on the glass layer [65]. After the finished scaffold was removed from the glass, a bioreactor was manufactured with a polycarbonate base, a PDMS layer on top, a polycarbonate body, which housed the AngioChip scaffold, and a cap on top [65] (Fig. 3g and h). Perfusion of ECs and media through the scaffold occurred between the inlet and outlet wells through pressure-head differences [65]. Once the ECs attached to the lumen, hESC derived CMs were seeded throughout the rest of the scaffold in a hydrogel and allowed to compact. The AngioChip was then surgically implanted in Lewis rats by connecting the femoral vessels. The results of this model demonstrated that cardiac tissue with well-defined vasculature could be used as an *in vivo* implant for surgical anastomosis [65].

4.6   AngioTube

One organ-on-a-chip system attempting to model drug transport through a vascular scaffold is Integrated Vasculature for Assessing Dynamic Events (InVADE) [51]. This platform was built on the previous AngioChip model but required fewer cells and created a greater throughput [51]. The platform provides vascularization through a porous polymer vessel, termed AngioTube, with built in microholes that span the column of a 96-well plate. In this system, a 3D microenvironment is mimicked to trace how a drug interacts with various organs as it would in the human body [51]. In addition, since each organ is connected through a vascular network, cancer cells escaping from a tumour can be tracked as they metastasize by spreading to other organs. In Lai et al., a 96-well plate platform was designed using soft lithography to layer a scaffold with four cantilevers and a vascular channel with micro side holes using PDMS molds to stack the layers [51]. Briefly, during microfabrication the top layer of the PDMS mold was capped to a flat PDMS sheet and the bottom mold to a glass slide [51]. The PDMS mold was then perfused with an elastic polymer, poly(octamethylene maleate (anhydride) 1,2,4-butanetricarboxylate) (124 polymer), and UV-crosslinked [51]. The PDMS sheets were removed and the top and bottom layer of the scaffold were aligned and stamped together using a mask aligner. The scaffold was then incorporated onto a 96-well plate with inlet and outlet channels and designed to hold 10 hepatic and cardiac/tumour scaffolds on one plate [51]. The plate was fabricated using hot embossing with a PDMS mold containing a patterned base and a polystyrene sheet with carbon electrodes incorporated parallel to the scaffold chambers. Polyurethane glue was then used to attach a bottomless 96-well plate to the polystyrene base [51] (Fig. 3i). Human umbilical vein endothelial cells (HUVEC) were used to endothelialize the AngioTube lumen, whereas various parenchymal cells were used to construct different organs in various tissue wells: human hepatocellular carcinoma (HepG2) cells represented a metabolically active liver, GFP expressing MDA-MB-231 cells for a metastatic solid tumour, and hiPSC derived CMs for the free-contracting cardiac muscle [51]. The chemotherapeutic drug Tegafur was tracked through the system to examine the efficacy of the drug after being metabolized in the liver and its effect on the heart. The tumour model was created in series with the liver and after Tegafur was added the system, a greater tumour toxicity was seen in the presence of the liver tissue as opposed to the control model containing the tumour model without the liver tissue. This proof of concept model illustrated the possibility of elucidating the entire cancer invasion-metastasis cascade [51].

4.7 2D Flat PDMS Strips

A platform describing the pathogenesis of Barth syndrome from Wang et al., details a model in which iPSC derived CMs from individuals with the condition were cultured onto a micropatterned fibronectin rectangle with length/width ratios that represented the dimensions of adult CMs [103]. The alignment of sarcomeres was measured after immunostaining with α-actinin and scored using the consistency of the sarcomere spacing with a 2D Fourier spectra. The results illustrated sarcomere disorganization in the Barth syndrome CMs versus the healthy control group. The contractility of the CM tissues was then measured by placing the thin elastomers onto a glass coverslip and allowing them to culture for 5 days before permitting the strips to curl and contract away from the coverslip [103] (Fig. 3j). Movies of the contracting strips allowed the curvature to be used to calculate stresses based on a revised version of Stoney's equation. The strips were also exposed to electrical stimulation and paced from 1 to 5 Hz. The results from this experiment demonstrated that CMs from Barth syndrome patients exhibited lower twitch and peak systolic stress compared to the healthy control CMs, thus validating the phenotype [103].

4.8 24-Well Based Platforms

The work of Hansen, et al., developed fibrin-based mini-engineered heart tissues (FBMEs) using neonatal rat CMs in a fibrinogen/Matrigel/thrombin hydrogel [104]. A silicone scaffold with spaced posts was made from a Sylgard 184 silicone elastomer in a Teflon casting mold [104]. The silicone posts were cultured upside down in a 24-well plate with agarose and Teflon spacers [104] (Fig. 3k and l). The cell suspension and hydrogel mixture were pipetted into the molds and allowed to sit for 2 h to polymerize. The posts with the attached tissues were then placed into culture medium in a 24-well plate and kept for 2–4 weeks [104] (Fig. 3m). Optical analysis was performed by video recording the deflection of the posts and measuring the distance between them during tissue contraction and relaxation [104]. The force was determined based on the elastic modulus of the Sylgard 184 and the geometry of the posts. The histology results showed elongated CMs that were aligned along the direction of force [104]. The force for an average tissue at day 15 was calculated at 0.11 to 0.22mN. The tissues also demonstrated a dose and time dependent decrease in force when the cardiotoxic drug, doxorubicin was delivered [104].The method illustrates a robust protocol for generating 3D cardiac tissue.

The above work continued using hESC derived CMs in Schaaf et al. Embryoid bodies (EBs) were generated and dissociated into a fibrinogen/thrombin hydrogel before being seeded into the silicone post platform [105]. Transcript analysis, immunofluorescence, and electrophysiology were all performed on EBs as well as the cardiac tissues [105]. EBs that were 2–3 weeks old were compared to cardiac tissues that were 5 weeks old. The results showed that cardiac markers such as sarcomeric α-actin, troponin T, and β-myosin heavy chain transcript levels remained the same; however, β-myosin heavy chain expression levels were two to three times greater in the cardiac tissues than the EBs. Immunofluorescence staining showed CMs from EBs were randomly oriented whereas those from the cultured heart tissues demonstrated sarcomeric alignment along the force lines. Mean action potential durations at 80% for tissues were twice as long as those for the EBs. This platform also showed utility in automated drug screening [105].

The same group further developed this platform using rat, mouse, and human CMs in a fibrinogen/Matrigel/thrombin hydrogel under electrical stimulation and perfusion [106]. Tyrode's solution containing fura-2 was incubated with cardiac tissues under continuous perfusion in the 24-well plate [106]. The tissues were then illuminated with ultraviolet light. Ca^{2+} in the tissues was excited at 340 nm and Ca^{2+} free fura-2 was excited at 380 nm [106]. Both emitted at 510 nm. The ratio was calculated (F340-to-F380) from a microscope set-up that also recorded the optical image of the tissue beating. The results showed that the rat cardiac tissues demonstrated the highest force and Ca^{2+} sensitivity, while the hESC derived cardiac tissues had the lowest [106]. hiPSC derived CMs in the form of EBs were used to generate tissues in Mannhardt et al. and also showed good sarcomeric alignment as well as beating frequency changes in response to ryanodine, ivabradine, SEA-0400, TTX, and isoprenaline [107]. This platform thus demonstrates utility as a drug screening method for preclinical testing and disease modeling [107].

4.9   Microfabricated Cardiac Patches

After MI, CMs are lost and replaced with fibroblasts which become myofibroblasts after an immune response is activated, expressing large amounts of α-smooth muscle actin (α-SMA) and collagen. This leads to non-contractile scar tissue that can severely impact the function of the heart by requiring a greater load to pump blood and a steady decline in functional output [108]. Current treatments do not resolve the cardiac issues that stem from MI such as fibrosis, necrosis, and inadequate beating and function. In addition, organ donations are rare, leaving limited options for patients [109]. One treatment, known as cardiomyoplasty, involved the use of muscle cells obtained from other parts of the body, electrically stimulating them, and re-implanting the cells back into the patient's heart in an effort to get it to functionally beat in synchrony again. This method achieved little and was unsuccessful in demonstrating any consistent benefit [109].

A method for treating MI that incorporates microfabrication techniques is the cardiac patch. While technically not an organ-on-a-chip device, the cardiac patch is unique in that its intended purpose involves culturing heart tissue *in vitro* with the overall goal of implanting it *in vivo* after MI. There are several considerations when developing a cardiac patch. For one, will the cells used in the patch be autologous or allogenic? If autologous cells are desired, this requires obtaining cells from the patient after MI then reprogramming them back to iPSCs. Cells would then be differentiated back into CMs and grown on the patch with electrical stimulation to get them beating powerfully enough to contract in synchrony with the native heart once implanted [109]. This option is rather time consuming. An alternative is using cells from a range of donors (akin to blood donation) and having pre-fabricated patches ready upon MI, which brings forth its own complications with biocompatibility, cost, cell sourcing and the logistics of storage and transport. Moreover, tissue engineering methods for cardiac patches need to incorporate the relevant physiology, force generation, and electrical conduction [108]. In addition, the delivery of a cardiac patch to the site of infarction requires that the method of delivery be gentle enough on the patch to ensure cell viability and preferably would be minimally invasive [53].

One method of developing a cardiac patch comes from Liau et al. and involves a hydrogel matrix containing fibrin which was cultured with mouse ESC derived CMs, cardiovascular progenitors (CVPs) which differentiated into CMs, smooth muscle cells, ECs, and neonatal rat ventricular fibroblasts [108]. The matrix was grown for 21 days on a PDMS scaffold with hexagonal posts obtained from soft lithography techniques. The method was also able to incorporate different dimensions, geometry, thickness, and cell alignment which allowed functional anisotropy. The 3D patches showed aligned and electromechanically coupled adult-like CMs which had contractile forces up to 2mN [108].

The work of Zhang et al., also from the Bursac laboratory, continued the work of cardiac tissue patches with human ESC differentiated CMs purified by magnetic-activated cell sorting (MACS) and cultured in a fibrinogen/Matrigel hydrogel [110]. After 2 weeks in culture, the 3D patches exhibited aligned cardiac tissue with matured sarcomeric structures [110]. qRT-PCR expression analysis revealed that OCT4 and NANOG gene expression was present in undifferentiated hESCs, decreased after differentiation to CMs, and vanished in the cardiac patches, while cardiac specific genes such as MLC2v and βMHC increased. Contractile stress of the patches averaged 5.7 +/− 1.0 nN/cell and active and passive tension both increased with applied stretch [110].

The Bursac laboratory then used the cardiac patches developed with neonatal rat hearts in a fibrinogen/thrombin/Matrigel mixture fabricated with PDMS molds generated from a custom PTFE template and porous nylon fabric attached to rectangular posts *in vitro* [111]. They were then implanted into rat epicardium. After 2 weeks in dynamic culture, the cardiac patches had visible contractions. Rats with either chronic MI, acute MI, or no MI were implanted with the cardiac patches. Explants after 4–6 weeks of epicardial implantation showed 11 out of 13 survived and exhibited calcium transients, 18mN of contractile force, and a maximum capture rate (MCR) of 8.7 Hz. Overall, the patch demonstrated a safe and effective therapy that mirrored the contractile function of adolescent myocardium [111].

The microfabricated cardiac patch developed by Montgomery et al. from the Radisic laboratory, was cultured and electrically stimulated *in vitro* before being injected *in vivo* as well [53]. The patch was made by microfabricating various lattice designs using soft lithography and creating a PDMS mold. The best design was chosen based on its ability to be injected and opened successfully from a 1 mm orifice (Fig. 3n). POMaC was then injection molded into the PDMS mold and the elastomeric polymer was UV-crosslinked [53]. The patch mold was attached to a PDMS post holder in a bioreactor containing carbon electrodes on either side for electrical stimulation and cultured for 7 days with human iPSC derived CMs prior to implantation. The patches were rolled up and delivered to a porcine model using a 0.5 cm tool and either fibrin gluing or knotting it to the left ventricle after delivery. The patches were allowed to sit for 7 h before the hearts were harvested. The explanted patches showed successful fixation to the porcine left ventricle and high cell viability compared to non-implanted control patches. This method of cardiac patch fabrication and delivery demonstrated that it is minimally invasive and practical [53].

5 Characterizing Microfabricated Tissues: Incorporating Built-In Readouts into Tissue Platform Design

In order to most effectively utilize microfabricated platforms to model the heart, it is critical to establish relevant parameters and characterization techniques that can produce insight into tissue properties, function, and phenotype. On the surface, built-in readouts provide information on the status of tissues that can be used to assess their functionality, quantify responses to treatments over time, or reveal mechanisms of physiology and pathology. Critically, relevant characterizations performed on microfabricated tissues can also provide the key to establishing connections by extension between *in vitro* observations and the native human heart *in vivo*. Making such extensions is central to realizing the advantages of applying microfabricated tissue platforms over conventional cultures or *in vivo* models. Thus, microfabricated platforms facilitate the measurement of a plethora of detailed outputs that can provide researchers with an accurate picture of how observations in tissues translate to implications for real patients in a clinical setting.

5.1 Advantages of Microfabricated Platforms for In Vitro Tissue Assessments

One of the unique advantages of developing novel microfabricated platforms of human cardiac tissue over animal models and *in vivo* studies of the human heart is the flexibility of such devices to incorporate built-in readouts for tissue assessment. Microfabricated platforms permit a high degree of structural engineering during the design process, making the incorporation of functional and environmental sensors simple. As isolated *ex vivo* systems, these platforms facilitate extensive manipulation and control such that analysis techniques can effectively separate complex processes into individual readouts. The body is a complex environment with many interacting components that can confound analyses that attempt to separate processes into distinct units, and it is difficult to access the cardiac environment in humans or animals to make detailed assessments without highly invasive techniques. Microfabricated platforms also enable many classes of readouts to be generated easily in real-time, a property that can be used to monitor live tissue responses to stimuli or transient features of tissue, permitting enhanced mechanistic insight. Obtaining real-time and repeated measurements on-demand is evidently more complicated *in vivo*.

Increasingly complex microfabricated cardiac platforms afford new opportunities to incorporate different classes of characterizations and built-in sensor systems that cannot be realized in less sophisticated 2D cell cultures. Due to their physiological relevance and flexible engineering control during fabrication, novel readouts can be integrated that are increasingly analogous to clinically significant metrics used to assess *in vivo* heart function. One such example is enabling the measurement of contractile force in microfabricated tissues, which may be useful in predicting human cardiac ejection fraction [112]. Continued advancements in platform fabrication and design have seen equivalent progress in integrating new classes of sensors that exploit a variety of techniques to assess tissue properties, ranging from optics to electrochemistry [4]. Such techniques can assess tissue response to stimuli, functional performance, viability, environmental factors, secretome, and phenotype among other factors. With respect to the assessment of cardiac tissue functionality, contractile force and electrophysiology have been two key areas where the benefits of microfabricated platforms have been realized through the engineering of novel sensory systems. A multitude of assays that are common for assessment of standard *in vitro* cell cultures have also been applied in engineered cardiac tissues; however, the following sections will focus on highlighting those that are most unique and relevant to microfabricated cardiac platforms.

5.2   Tissue Contractile Function

The ability to measure contractile force in microfabricated cardiac tissues has been regarded as one of their most unique and significant advantages in advancing *in vitro* cardiac research [112]. Many animal studies of cardiac disease and therapeutics place significant emphasis on endpoint readouts, such as histological analysis of infarct size, to make general predictions regarding cardiac function. Direct measures of cardiac function are often a primary interest for clinicians, as these may better inform immediate clinical manifestations of disease states [112]. The left ventricle is responsible for pumping oxygenated blood through systemic circulation, and functional changes to parameters such as ejection fraction are often clinically correlated to patient quality of life and prognosis. Echocardiography can be used to assess left ventricular function in animal models with relative accuracy; yet simple 2D and 3D CM culture platforms lack the structural tools required to quantitatively collect information describing cardiac pumping function [113].

Microfabricated platforms have utilized multiple approaches of structural engineering and sensory systems to facilitate the assessment of tissue contractile function. Cantilever-based platforms can measure contractile function by assessing the degree of tissue bending with each synchronous beating action of CMs. Microscope camera recordings or other optical methods can be used to track cantilever deflection, while mechanical equations based on known material parameters can then be used to calculate contractile force [77, 114, 115]. A similar method of optical deflection tracking and mechanical calculations can be performed to assess contractile force of tissues suspended between posts or wires (Fig. 4a and b). Active beating force can be calculated, as can passive tension exerted by tissues related to post-seeding remodeling [42, 52, 105]. Electronic sensors can also be incorporated into engineered platforms to measure contractile force, facilitating improved potential for high-throughput scale-up compared to optical tracking [115]. Imaging methods are generally labor intensive and require specialized equipment, software, and storage of large files whereas electronic measurement can be more compact, efficient, and automated, enhancing potential for assessments in high-throughput screening applications [77, 115]. As one such example, Lind et al. [77] fabricated a cantilever device incorporating an embedded electronic strain sensor within their platform (Fig. 4c). Changes in electrical resistance corresponding to beating function were converted into force readouts that correlated closely with those measured via optical tracking (Fig. 4d) [77]. Other devices have incorporated external probes that can be applied to beating tissues to measure active force via deflection but can also directly apply external mechanical forces to tissues to monitor beating function under loading and observe the characteristic Frank-Starling relationship of cardiac force-tension [100, 101, 115]. The ability to engineer microfabricated devices that can assess cardiac contractile function is one of the unique advances that has contributed to their utility as improved *in vitro* models of the heart [113].

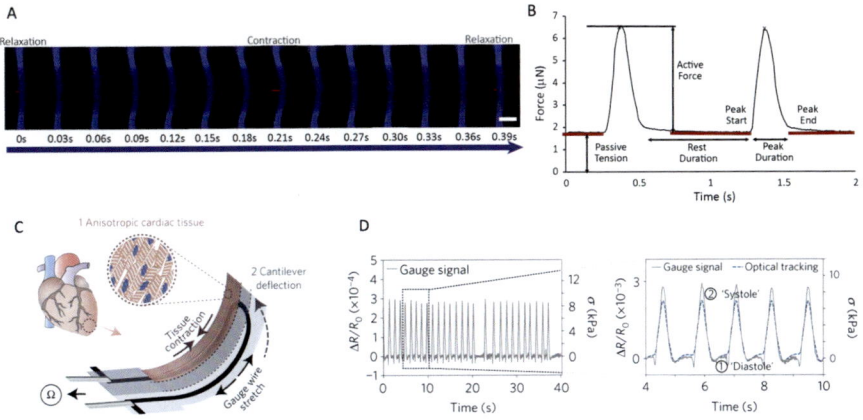

Fig. 4 Multiple methods have been used to assess contractile force in microfabricated cardiac tissue platforms. (**a**) Optical tracking of tissue-induced deflection of polymer wires can be used in mechanical equations to calculate contractile force of beating tissues over time, as depicted in (**b**). Scale bar 200 μm, reproduced with permission from [52]. (**c**) Electronic sensors, such as strain gauges, can be incorporated into microfabricated platforms. Reproduced with permission from [77]. (**d**) Signals can then be converted to contractile force, showing good correlation with values calculated from optical tracking of deflection. Reproduced with permission from [77]

5.3 Electrophysiology

Central to regulating pumping function, electrophysiological characteristics of tissues are also of critical importance in cardiac research. The coordinated flow of ionic currents involved in action potential generation can be characteristic to different species, cell types, degrees of maturity, and physiological states, or impacted by the presence of drugs and diseases [52, 116]. Besides action potential generation, the way that cultured CMs respond functionally to external electrical pacing is also important in assessing behaviours relevant to the human heart *in vivo*, in which synchronous contraction of CMs is driven by the heart's conduction system [116].

Microfabricated platforms possess unique capabilities for performing relevant and meaningful analyses of cardiac electrophysiology *in vitro*, once again balancing increased tissue complexity alongside flexibility and ease of access. In models created using human cells subjected to electrical or mechanical maturation protocols in a 3D environment, tissue electrophysiology has been shown to more accurately reflect *in vivo* reality, as differences in CM maturity and species of origin are known to play significant roles in their electrophysiological behaviour [52, 117].

The shape and properties of action potentials can be directly measured in microfabricated tissues using a standard 'patch-clamping' technique, which has been applied to both atrial and ventricular engineered tissues to replicate well-known *in vivo* differences between these cell types (Fig. 5a and b) [4, 52]. Microelectrode arrays can be incorporated in platform designs to measure 'external field potentials' generated by beating tissues, which can then be correlated with action potential

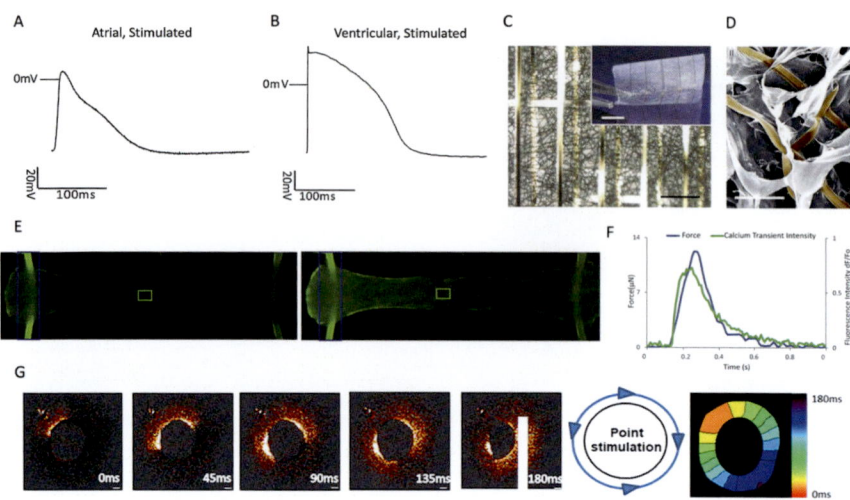

Fig. 5 Engineered cardiac platforms possess unique properties that enable a variety of electrophysiological characterizations of tissue function. Chamber-specific action potentials of electrically matured atrial (**a**) and ventricular (**b**) tissues measured via patch clamping. Reproduced with permission from [52]. (**c**) Optical image of a nanoelectronic scaffold (nanoES), incorporating silicon nanowires into electrospun poly(lactic-co-glycolic acid) fibers, which can be used for real-time monitoring of electrical activity in 3D engineered tissues. Scale bars 200 μm, 5 mm (inset). Reproduced with permission from [118]. (**d**) SEM image of an alginate nanoES. Scale bar 100 μm, reproduced with permission from [118]. (**e**) Ca^{2+} dyes and live imaging can be used to quantify calcium transients in beating tissues, which can then be compared against contractile force curves, as shown in (**f**). Reproduced with permission from [52]. (**g**) Imaging of platforms fabricated using CMs engineered with genetically encoded voltage indicators can be used to visualize cardiac electrophysiology and produce optical maps of tissue activation. Scale bar 0.5 mm, reproduced with permission from [121]

length and used to create activation maps of electrical activity in tissues. Conduction velocity within tissues can also be calculated [105, 117]. In the quest to improve automation and sophistication of built-in electrophysiological readouts, several studies have focused on developing flexible nanoelectronic scaffolds (nanoES) that incorporate 3D wire systems into platforms for simple, effective, and non-invasive measurement of cardiac electrophysiology in thick tissues (Fig. 5c and d) [4, 118, 119].

The incorporation of electronics into microfabricated tissues is useful not only for characterizing their electrical activity, but also for driving tissue stimulation through pulses of applied current. The effects of electrical stimulation on physiological maturation of iPSC-derived CMs have been well documented, and some platforms use external electrical stimulation to drive synchronous contractions to mimic the conduction in the native heart during culture and assessment [52, 89, 120]. Stimulation parameters can be modulated to observe secondary electrophysiological characteristics which are known to change based on tissue phenotype and physiological state. These include excitation threshold (ET), the minimum applied voltage required for synchronous tissue contraction; maximum capture rate (MCR), the maximum driving frequency at which tissue can synchronously beat; force-frequency relationship (FFR), in which increasing driving frequency should cause an increase in contractile force in mature adult tissues; and observation of a characteristic post-rest potentiation (PRP) spike in contractile force [52].

Calcium handling in cardiac tissues is another key characteristic of their electrophysiology, critical to mediating excitation-contraction coupling [116]. Calcium transients can be visualized by incubating tissues with a Ca^{2+} dye followed by fluorescence microscopy to observe the effects of drugs or diseases on CM calcium handling, data which can then be coupled with contractile force curves to holistically assess the contractile process (Fig. 5e and f) [52]. Similar optical techniques have imaged stimulated tissues after applying fluorescent, voltage sensitive dye to create optical maps of cellular activation without the need to incorporate wires or other electronics described earlier [52, 121]. Since dye-mediated optical mapping of tissue electrophysiology is labor intensive and can induce cellular phototoxicity, several recent studies have moved towards genetic engineering to directly encode voltage sensitive fluorescence into CMs before their incorporation into engineered platforms (Fig. 5g) [121, 122].

5.4 Imaging

Microfabricated cardiac platforms have employed numerous other built-in readouts, many of which adapt common *in vitro* assay methods, towards the assessment of tissue structure and function. Traditional optical and fluorescence microscopy techniques can be applied for immunohistochemical analyses of tissue morphology as with 2D cell cultures. Important considerations for imaging microfabricated platforms include optical compatibility of platform materials and tissue thickness, which often necessitates the use of selective plane illumination microscopy, such as confocal, multi-photon, or light sheet imaging, in order to enable live, layer-by-layer, single-cell analyses of thick tissues [123, 124]. Such imaging techniques can also be employed to assess tissue viability when combined with live/dead stains [52]. Other imaging techniques, such as second harmonic generation, have also been employed to assess tissue extracellular matrix content [96, 125].

5.5 Chemical/Environmental Analyses

Chemical and environmental analyses represent another category of readouts that can be assessed by multiple means in microfabricated tissues. Analytes may include tissue metabolites, cytokines, secreted molecules and vesicles, and other media constituents. Oxygen content, glucose concentration, and pH represent examples of chemical and environmental monitoring applied in tissue platforms, utilizing assessment techniques such as imaging with optical probes or through incorporation of electrochemical sensors in platform designs [4, 126]. Electrochemical sensors have also been successfully integrated in platforms for the measurement of trans-epithelial electrical resistance (TEER), a parameter which is often used to quantify the structural integrity of epithelial and EC cultures [127, 128]. Such measurements may prove particularly useful in functional assessments of future vascularized cardiac tissue models. Compounds with known effects on cardiac function, such as caffeine, may also be applied to microfabricated tissues as another measure to assess the extent to which *in vitro* platforms replicate native responses to stimuli *in vivo* [52, 89].

In summary, the ability to incorporate advanced readouts and sensory systems in microfabricated cardiac tissue platforms means that researchers are increasingly able to monitor the function and physiology of live tissues at a high resolution, in real time, and on a single cellular level. Novel modalities for measuring contractile force and electrophysiology have been especially unique to recent microfabricated platforms and offer functional insights that will be critical towards enhancing the clinical relevance and potential applications of such systems. Future work in enhancing built-in readouts for microfabricated cardiac tissues will continue towards improving integration, diversity, resolution, and automation potential of sensory systems in order to advance tissue quantification while supporting higher throughput and lower impact analyses.

6 Applying Microfabricated Systems Towards Challenges in Cardiovascular Research

With significant innovation in materials, stem cell culture, microfabrication, and the emergence of novel platforms incorporating an arsenal of sensors and built-in characterizations, opportunities for applying microfabricated cardiac tissues towards tackling the most significant challenges in cardiovascular research continue to flourish (Fig. 6). With CVD representing the leading cause of death and transplant remaining the only true 'cure' for a failing heart, a significant need exists for improved knowledge in cardiovascular physiology and therapeutic discovery [129, 130]. Firstly, creating new therapies to regenerate heart tissue will require a better understanding of how the healthy heart functions at a cellular level. Due to their ease of accessibility and control as well as improved physiological relevance over conventional cell cultures or animal models, microfabricated cardiac systems possess significant potential for in-depth investigations of cardiac physiology and function. Secondly, the ability to modulate environmental parameters and utilize patient-specific iPSCs in such tissue platforms opens the door to modeling cardiac pathologies to generate a new understanding behind mechanisms of disease initiation and progression. Finally, these devices present exciting implications for drug screening and therapeutic testing with the potential to expedite the translation of novel therapies that may revolutionize quality of life and outcomes for heart patients [115].

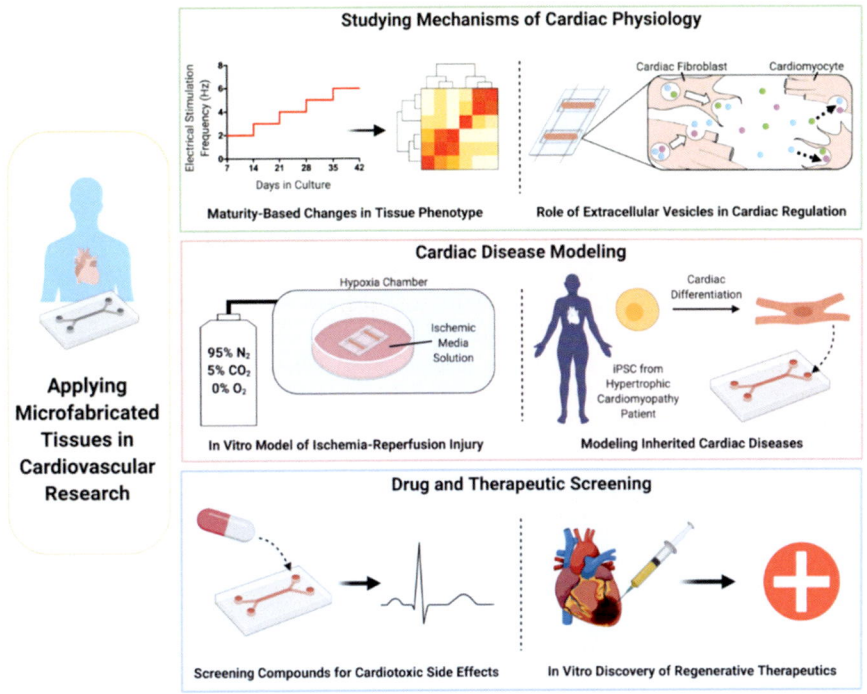

Fig. 6 Examples of current and potential future applications of microfabricated cardiac tissue platforms for improving mechanistic understanding of cardiac physiological processes, modeling cardiac diseases *in vitro*, and improved cardiac drug and therapeutic screening. Figure created with BioRender.com

6.1 Physiological Mechanisms

Human physiology can be very difficult to study at a cellular level *in situ*; this is especially true for investigations of the native human heart. Illuminating cellular mechanisms behind cardiac function is a key first step towards building a broader understanding of cardiac pathologies and developing new therapies to treat heart disease. Not only is the heart difficult to access, but it is also impossible to manipulate or isolate specific parts of the cardiac environment *in vivo*, actions which are necessary for discerning the role of specific components in various physiological scenarios. Effective models are key to overcoming these barriers in order to improve knowledge surrounding human cardiac physiology.

Animal studies and 2D cell cultures currently represent the most commonly utilized platforms in human physiological modelling. Unfortunately, physiological differences between humans and animals can be significant. Organ anatomy, metabolism, and inherent differences in gene expression and regulation have been recognized as some of the major reasons that cellular processes in animals may not accurately reflect those in humans. *In vitro* cultures of human cells can overcome this limitation, however cells grown in 2D environments lack relevant cell-cell and cell-extracellular matrix (ECM) signalling interactions which greatly affect their function and phenotype. Culture materials and media compositions also fail to replicate *in vivo* conditions, therefore contributing to significant differences in genetic expression and regulation compared to cells and tissues *in vivo* [131].

Microfabricated platforms of cardiac tissue aim to overcome the barriers to studying cardiac physiology and the limitations of current models by increasing the complexity and relevance of *in vitro* cardiac cultures. Through the use of biomimetic scaffold materials, multiple cell types, and structural and architectural cues as discussed throughout this chapter, significant progress has been made in accurately replicating cardiac physiology *in vitro*, expanding the potential applications of these platforms in investigating cardiac physiological processes. Recent examples of such applications in literature include the investigation of developmental changes in genetic expression and maturity of atrial and ventricular tissues, and the investigation of the role of extracellular vesicle (EV) signalling in regulating cardiac function, studies which would be difficult or impossible to perform *in vivo* or in simpler models [52, 125, 132].

6.2   Disease Modeling

Ischemic heart disease (IHD) is the most common cause of cardiac injury and death; thus, modeling ischemia-reperfusion injury (IRI) *in vitro* has been a major focus in the creation of "heart attack-on-a-chip" models [130]. Ischemic oxygen deprivation can be recreated in microfabricated tissues via temporary culture in hypoxic incubators. "Ischemic media solutions" may also be applied to tissues during hypoxia to further simulate environmental conditions experienced by the heart during MI *in vivo*, including high levels of potassium, an acidic pH, and the absence of metabolic substrates. Subsequent reperfusion to normoxic conditions has been shown to increase cell death and tissue damage *in vitro*, mirroring known mechanisms of IRI [112, 133]. Beyond building a mechanistic understanding of MI, such models may also present utility in studying subsequent development of chronic heart failure. Since immature iPSC-derived CMs are known to exhibit significant hypoxic resistance compared to adult CMs, microfabricated platforms with high levels of phenotypic maturity may pose the advantage of increased susceptibility to IRI *in vitro* that more closely resembles the mechanisms of damage that occur in adult humans [112]. Besides environmental replication of IRI, vascularized platforms that contain microfluidic channels offer the potential to also replicate the physical process of acute MI. The blockage of coronary arteries can be simulated in such platforms by temporarily cutting off perfusion of nutrient rich media through microchannels, adding a new layer of physiological relevance that can be achieved *in vitro* via microfabricated platforms [112].

Due to its significant role in chronic heart failure post-injury, cardiac fibrosis models have also seen growing interest in recent microfabricated cardiac platform designs. Wang et al. [96] first created a cardiac fibrosis-on-a-chip model by seeding *in vitro* tissues with large populations of cardiac fibroblasts, and Mastikhina et al. [125] subsequently created such platforms with cardiac fibroblasts that were pre-activated in 2D cultures with TGF-β. Scaffold engineering may also be used to induce disease phenotypes, such as through alterations to ECM content or mechanics [134].

Many other myocardial diseases can be recreated in microfabricated tissues thanks to the advent of patient-specific iPSCs. Cardiac cells derived from iPSCs obtained from patients with inherited diseases have been seeded in microfabricated platforms to recreate numerous myocardial disease phenotypes. iPSC-based cardiac platforms have been successfully used to create *in vitro* models of hypertrophic cardiomyopathy [52, 135], dilated cardiomyopathy [136–138], and cardiac arryth-mias [139, 140].

6.3 Drug Screening and the Discovery of Regenerative Therapeutics

Besides their utility in developing an improved understanding of cardiac physiology and pathophysiology, microfabricated tissue platforms have seen a large part of their commercial utility in drug and therapeutic screening. When drugs are developed and investigated for treating a variety of diseases in humans, pre-clinical testing in animal models is standard practice for predicting safety and efficacy before they can advance to the clinical trials stage. It has been estimated that 90% of thera-pies that exhibit efficacy in animals do not provide similar benefits in humans, and two-tiered animal studies can only correctly predict human toxicity of drugs 71% of the time [131]. As a specific example, despite receiving FDA approval, the arthritis drug Vioxx was later withdrawn after being directly linked to 27,000 deaths due to cardiotoxicity [115]. The advent of improved microfabricated tissue models has been a significant development towards the development of the improved drug screening approaches. Microfabricated cardiac tissues may exhibit more realistic physiological responses to drugs, as they are made with human cells and can be structurally and functionally matured in relevant environments. They also enable high-throughput investigations that can generate clinically relevant parameters via online readouts, all contributing to the possibility of improved screening for drug-induced cardiotoxicity. Engineered cardiac platforms have already reached the com-mercialization stage in drug screening applications through start-ups including TARA Biosystems, Myriamed, and EHT-Technologies [115].

Beyond studying cardiotoxicity, *in vitro* platforms that model myocardial disease can also serve as useful conduits for designing and screening novel regenerative cardiac therapeutics that initiate functional recovery in injured and diseased tissues. Due to the lack of existing regenerative therapies to treat heart disease and the cur-rent reliance on transplants [129], investigating mechanisms of cardiac repair in highly relevant microfabricated platforms represents an opportunity to expedite the translation of regenerative cardiac therapies to the clinical stage. The flexibility of microfabricated tissue platforms means that they can be used to screen a variety of different therapeutic strategies, including cell replacement, exogenous EV application, or the use of other modulatory cytokines and compounds. Once again, physiological fidelity, high-throughput capability, and clinically relevant built-in readouts provide additional advantages for therapeutic investigations [115]. A number of the aforementioned myocardial disease models have also been used to test the effects of therapeutic compounds for reversing or preventing tissue damage, and this application of microfabricated cardiac platforms is expected to grow as the push toward clinical translation of regenerative cardiac therapies continues [96, 125, 133].

7   Conclusions

Organ-on-a-chip platforms aim to mimic the native physiological or pathophysiological microenvironment of human tissues *in vivo*. Since 2D cultures and animal models only fulfill some requirements of drug testing in terms of how many actually make it through clinical trials to regulatory approval, it is imperative to develop devices that will accurately represent a drug's safety and efficacy in humans. 3D tissue platforms must be designed in a way that reproduces functional hallmarks of the tissues that one is interested in studying, while defining the parameters and readouts that are most appropriate for determining the model's accuracy. Development of instrumented organ-on-a-chip platforms is particularly promising in terms of delivering information about tissue function in realtime and in a non-invasive manner. Looking forward, differentiation protocols for generating CMs need to produce purer populations to enable creation of more defined tissues. In addition, biomaterials being used in these devices should be non-toxic, free of absorption of drugs and small molecules, and preferably have a tunable Young's modulus. Many advances have been made so far; however, high-throughput platforms and cost-effective scale-up of fabrication should be considered when designing organ-on-a-chip devices that will transform drug discovery.

Acknowledgements Our work is supported by Canadian Institutes of Health Research (CIHR) Operating Grants (MOP-126027, MOP-137107 and MOP-142382), CIHR Foundation Grant (FDN-167274), Natural Sciences and Engineering Research Council of Canada (NSERC) Discovery Grant (RGPIN 326982-10), NSERC-CIHR Collaborative Health Research Grant (CHRP 493737-16). MR is supported by the Canada Research Chair and Killam Fellowship (7025-19-0016).

References

1. Parsa, H., Ronaldson, K., Vunjak-Novakovic, G.: Bioengineering Methods for Myocardial Regeneration, pp. 195–202. Elsevier B.V (2016)
2. Benjamin, E.J., Muntner, P., Alonso, A., Bittencourt, M.S., Callaway, C.W., Carson, A.P., et al.: Heart disease and stroke Statistics-2019 update: a report from the American Heart Association. Circulation. **139**(10), e56–e528 (2019). https://doi.org/10.1161/CIR.0000000000000659
3. Pagidipati, N.J., Gaziano, T.A.: Estimating deaths from cardiovascular disease: a review of global methodologies of mortality measurement. Circulation. **127**(6), 749–756 (2013). https://doi.org/10.1161/CIRCULATIONAHA.112.128413
4. Ahadian, S., Civitarese, R., Bannerman, D., Mohammadi, M.H., Lu, R., Wang, E., et al.: Organ-On-A-Chip Platforms: A Convergence of Advanced Materials, Cells, and Microscale Technologies. Wiley-VCH Verlag (2018)
5. Martin, L., Hutchens, M., Hawkins, C., Radnov, A.: How Much Do Clinical Trials Cost? pp. 381–382. Nature Publishing Group (2017)
6. Dimasi, J.A., Feldman, L., Seckler, A., Wilson, A.: Trends in risks associated with new drug development: success rates for investigational drugs. Clin. Pharmacol. Ther., 272–277 (2010)
7. Moraes, C., Mehta, G., Lesher-Perez, S.C., Takayama, S.: Organs-on-a-Chip: a focus on compartmentalized microdevices. Ann. Biomed. Eng. **40**(6), 1211–1227 (2012). https://doi.org/10.1007/s10439-011-0455-6
8. Grimm, D.U.S.: EPA to Eliminate all Mammal Testing by 2035. (2019)
9. Gottlieb, S.: Statement by FDA Commissioner Scott Gottlieb, M.D., on efforts to reduce animal testing through a study aimed at eliminating the use of dogs in certain trials. (2018)
10. Veldhuizen, J., Migrino, R.Q., Nikkhah, M.: Three-dimensional microengineered models of human cardiac diseases. BioMed Central Ltd. (2019)
11. Khademhosseini, A., Langer, R.: A decade of progress in tissue engineering. Nat. Protoc. **11**(10), 1775–1781 (2016). https://doi.org/10.1038/nprot.2016.123
12. Breslin, S., O'Driscoll, L.: Three-dimensional cell culture: The missing link in drug discovery. Drug Discov. Today, 240–249 (2013)
13. Tissue-engineered disease models. Nat. Publ. Group, 879–880 (2018)
14. Fitzgerald, K.A., Malhotra, M., Curtin, C.M., O'Brien, F.J., O'Driscoll, C.M.: Life in 3D is Never Flat: 3D Models to Optimise Drug Delivery, pp. 39–54. Elsevier (2015)
15. Brack, K.E.: The heart's 'little brain' controlling cardiac function in the rabbit. Exp. Physiol. **100**(4), 348–353 (2015). https://doi.org/10.1113/expphysiol.2014.080168
16. Epelman, S., Liu, P.P., Mann, D.L.: Role of Innate and Adaptive Immune Mechanisms in Cardiac Injury and Repair, pp. 117–129. Nature Publishing Group (2015)
17. Zhou, P., Pu, W.T.: Recounting Cardiac Cellular Composition, pp. 368–370. Lippincott Williams and Wilkins (2016)
18. Leitolis, A., Robert, A.W., Pereira, I.T., Correa, A., Stimamiglio, M.A.: Cardiomyogenesis Modeling Using Pluripotent Stem Cells: The Role of Microenvironmental Signaling. Frontiers Media S.A (2019)
19. Ronaldson-Bouchard, K., Ma, S.P., Yeager, K., Chen, T., Song, L.J., Sirabella, D., et al.: Advanced maturation of human cardiac tissue grown from pluripotent stem cells. Nature. **556**(7700), 239–243 (2018). https://doi.org/10.1038/s41586-018-0016-3
20. Thavandiran, N., Dubois, N., Mikryukov, A., Massé, S., Beca, B., Simmons, C.A., et al.: Design and formulation of functional pluripotent stem cell-derived cardiac microtissues. Proc. Natl. Acad. Sci. U. S. A. **110**(49) (2013). https://doi.org/10.1073/pnas.1311120110
21. Li, Y., Asfour, H., Bursac, N.: Age-dependent functional crosstalk between cardiac fibroblasts and cardiomyocytes in a 3D engineered cardiac tissue. Acta Biomater. **55**, 120–130 (2017). https://doi.org/10.1016/j.actbio.2017.04.027
22. Hookway, T.A., Matthys, O.B., Mendoza-Camacho, F.N., Rains, S., Sepulveda, J.E., Joy, D.A., et al.: Phenotypic variation between stromal cells differentially impacts engineered cardiac tissue function. Tissue Eng. Part A. **25**(9–10), 773–785 (2019). https://doi.org/10.1089/ten.tea.2018.0362

23. Owen, T.J., Harding, S.E.: Multi-cellularity in cardiac tissue engineering, how close are we to native heart tissue? J. Muscle Res. Cell Motil. **40**(2), 151–157 (2019). https://doi.org/10.1007/s10974-019-09528-8

24. Sekine, H., Shimizu, T., Hobo, K., Sekiya, S., Yang, J., Yamato, M., et al.: Endothelial cell coculture within tissue-engineered cardiomyocyte sheets enhances neovascularization and improves cardiac function of ischemic hearts. Circulation. **118**(14 Suppl) (2008). https://doi.org/10.1161/CIRCULATIONAHA.107.757286

25. Ravenscroft Sm PAWAWCMJSJE. Cardiac Non-myocyte Cells Show Enhanced Pharmacological Function Suggestive of Contractile Maturity in Stem Cell Derived Cardiomyocyte Microtissues - PubMed. (2016). p. 99–112

26. Giacomelli, E., Bellin, M., Sala, L., van Meer, B.J., Tertoolen, L.G.J., Orlova, V.V., et al.: Three-dimensional cardiac microtissues composed of cardiomyocytes and endothelial cells co-differentiated from human pluripotent stem cells. Development (Cambridge). **144**(6), 1008–1017 (2017). https://doi.org/10.1242/dev.143438

27. Tiburcy, M., Hudson, J.E., Balfanz, P., Schlick, S., Meyer, T., Liao, M.L.C., et al.: Defined engineered human myocardium with advanced maturation for applications in heart failure modeling and repair. Circulation. **135**(19), 1832–1847 (2017). https://doi.org/10.1161/CIRCULATIONAHA.116.024145

28. Cao, N., Liu, Z., Chen, Z., Wang, J., Chen, T., Zhao, X., et al.: Ascorbic acid enhances the cardiac differentiation of induced pluripotent stem cells through promoting the proliferation of cardiac progenitor cells. Cell Res. **22**(1), 219–236 (2012). https://doi.org/10.1038/cr.2011.195

29. Shiekh, P.A., Singh, A., Kumar, A.: Engineering bioinspired antioxidant materials promoting cardiomyocyte functionality and maturation for tissue engineering application. ACS Appl. Mater. Interfaces. **10**(4), 3260–3273 (2018). https://doi.org/10.1021/acsami.7b14777

30. Ronaldson-Bouchard, K., Yeager, K., Teles, D., Chen, T., Ma, S., Song, L.J., et al.: Engineering of human cardiac muscle electromechanically matured to an adult-like phenotype. Nat. Protoc. **14**(10), 2781–2817 (2019). https://doi.org/10.1038/s41596-019-0189-8

31. Takawale, A., Sakamuri, S.S.V.P., Kassiri, Z.: Extracellular matrix communication and turnover in cardiac physiology and pathology. Compr. Physiol. **5**(2), 687–719 (2015). https://doi.org/10.1002/cphy.c140045

32. Majkut, S., Idema, T., Swift, J., Krieger, C., Liu, A., Discher, D.E.: Heart-specific stiffening in early embryos parallels matrix and myosin expression to optimize beating. Curr. Biol. **23**(23), 2434–2439 (2013). https://doi.org/10.1016/j.cub.2013.10.057

33. Schroer, A., Pardon, G., Castillo, E., Blair, C., Pruitt, B.: Engineering hiPSC Cardiomyocyte In Vitro Model Systems for Functional and Structural Assessment, pp. 3–15. Elsevier Ltd (2019)

34. Kc, P., Hong, Y., Zhang, G.: Cardiac tissue-derived extracellular matrix scaffolds for myocardial repair: advantages and challenges. Regen. Biomater. **6**(4), 185 (2019). https://doi.org/10.1093/RB/RBZ017

35. Blazeski, A., Lowenthal, J., Zhu, R., Ewoldt, J., Boheler, K.R., Tung, L.: Functional properties of engineered heart slices incorporating human induced pluripotent stem cell-derived cardiomyocytes. Stem Cell Rep. **12**(5), 982–995 (2019). https://doi.org/10.1016/j.stemcr.2019.04.002

36. Freytes, D.O., O'Neill, J.D., Duan-Arnold, Y., Wrona, E.A., Vunjak-Novakovic, G.: Natural cardiac extracellular matrix hydrogels for cultivation of human stem cell-derived cardiomyocytes. Methods Mol. Biol. **1181**, 69–81 (2014). https://doi.org/10.1007/978-1-4939-1047-2_7

37. Orlova, Y., Magome, N., Liu, L., Chen, Y., Agladze, K.: Electrospun nanofibers as a tool for architecture control in engineered cardiac tissue. Biomaterials. **32**(24), 5615–5624 (2011). https://doi.org/10.1016/j.biomaterials.2011.04.042

38. Boffito, M., Sartori, S., Ciardelli, G.: Polymeric scaffolds for cardiac tissue engineering: requirements and fabrication technologies. Polym. Int. **63**(1), 2–11 (2014). https://doi.org/10.1002/pi.4608

39. Annabi, N., Shin, S.R., Tamayol, A., Miscuglio, M., Bakooshli, M.A., Assmann, A., et al.: Highly elastic and conductive human-based protein hybrid hydrogels. Adv. Mater. **28**(1), 40–49 (2016). https://doi.org/10.1002/adma.201503255

40. Smith, A.S.T., Yoo, H., Yi, H., Ahn, E.H., Lee, J.H., Shao, G., et al.: Micro-and nano-patterned conductive graphene-PEG hybrid scaffolds for cardiac tissue engineering. Chem. Commun. **53**(53), 7412–7415 (2017). https://doi.org/10.1039/c7cc01988b

41. Hitscherich, P., Aphale, A., Gordan, R., Whitaker, R., Singh, P., Xie, L.-H., et al.: Electroactive graphene composite scaffolds for cardiac tissue engineering. J. Biomed. Mater. Res. A. **106**(11), 2923–2933 (2018). https://doi.org/10.1002/jbm.a.36481

42. Baei, P., Hosseini, M., Baharvand, H., Pahlavan, S.: Electrically conductive materials for in vitro cardiac microtissue engineering. J. Biomed. Mater. Res. A. **108**(5), 1203–1213 (2020). https://doi.org/10.1002/jbm.a.36894

43. Toepke, M.W., Beebe, D.J.: PDMS absorption of small molecules and consequences in microfluidic applications. R. Soc. Chem., 1484–1486 (2006)

44. Shirure, V.S., George, S.C.: Design considerations to minimize the impact of drug absorption in polymer-based organ-on-a-chip platforms. Lab Chip. **17**(4), 681–690 (2017). https://doi.org/10.1039/c6lc01401a

45. Halldorsson, S., Lucumi, E., Gómez-Sjöberg, R., Fleming, R.M.T.: Advantages and Challenges of Microfluidic Cell Culture in Polydimethylsiloxane Devices, pp. 218–231. Elsevier Ltd (2015)

46. Borysiak, M.D., Bielawski, K.S., Sniadecki, N.J., Jenkel, C.F., Vogt, B.D., Posner, J.D.: Simple replica micromolding of biocompatible styrenic elastomers. Lab Chip. **13**(14), 2773–2784 (2013). https://doi.org/10.1039/c3lc50426c

47. Domansky, K., Leslie, D.C., McKinney, J., Fraser, J.P., Sliz, J.D., Hamkins-Indik, T., et al.: Clear castable polyurethane elastomer for fabrication of microfluidic devices. Lab Chip. **13**(19), 3956–3964 (2013). https://doi.org/10.1039/c3lc50558h

48. Tran, R.T., Thevenot, P., Gyawali, D., Chiao, J.C., Tang, L., Yang, J.: Synthesis and characterization of a biodegradable elastomer featuring a dual crosslinking mechanism. Soft Matter. **6**(11), 2449–2461 (2010). https://doi.org/10.1039/c001605e

49. Davenport Huyer, L., Zhang, B., Montgomery, M., Korolj, A., Drecun, S., Conant, G., et al.: A highly elastic and moldable polyester biomaterial for cardiac tissue engineering applications. Front. Bioeng. Biotechnol. **4** (2016). https://doi.org/10.3389/conf.fbioe.2016.01.01951

50. Pandey, P., Hawkes, W., Hu, J., Hone, J., Sheetz, M.: Cardiomyocytes sense matrix rigidity through a combination of muscle and non-muscle myosin contractions. Dev. Cell. **44**, 326–36.e3 (2018). https://doi.org/10.1016/j.devcel.2017.12.024

51. Lai, B.F.L., Huyer, L.D., Lu, R.X.Z., Drecun, S., Radisic, M., Zhang, B.: InVADE: integrated vasculature for assessing dynamic events. Adv. Funct. Mater. **27**(46), 1–11 (2017). https://doi.org/10.1002/adfm.201703524

52. Zhao, Y., Rafatian, N., Feric, N.T., Cox, B.J., Aschar-Sobbi, R., Wang, E.Y., et al.: A platform for generation of chamber-specific cardiac tissues and disease modeling. Cell. **176**(4), 913–27.e18 (2019). https://doi.org/10.1016/j.cell.2018.11.042

53. Montgomery, M., Ahadian, S., Davenport Huyer, L., Lo Rito, M., Civitarese, R.A., Vanderlaan, R.D., et al.: Flexible shape-memory scaffold for minimally invasive delivery of functional tissues. Nat. Mater. **16**(10), 1038–1046 (2017). https://doi.org/10.1038/nmat4956

54. Bannerman, A.D., Ze Lu, R.X., Korolj, A., Kim, L.H., Radisic, M.: The Use of Microfabrication Technology to Address the Challenges of Building Physiologically Relevant Vasculature, pp. 8–16. Elsevier B.V (2018)

55. Arvatz, S., Wertheim, L., Fleischer, S., Shapira, A., Dvir, T.: Channeled ECM-based nanofibrous hydrogel for engineering vascularized cardiac tissues. Nanomaterials. **9**(5) (2019). https://doi.org/10.3390/nano9050689

56. Mengsteab, P.Y., Uto, K., Smith, A.S.T., Frankel, S., Fisher, E., Nawas, Z., et al.: Spatiotemporal control of cardiac anisotropy using dynamic nanotopographic cues. Biomaterials. **86**, 1–10 (2016). https://doi.org/10.1016/j.biomaterials.2016.01.062

57. Fleischer, S., Shapira, A., Feiner, R., Dvir, T.: Modular assembly of thick multifunctional cardiac patches. Proc. Natl. Acad. Sci. U. S. A. **114**(8), 1898–1903 (2017). https://doi.org/10.1073/pnas.1615728114

58. Tsukamoto, Y., Akagi, T., Akashi, M.: Vascularized cardiac tissue construction with orientation by layer-by-layer method and 3D printer. Sci. Rep. **10**(1) (2020). https://doi.org/10.1038/s41598-020-59371-y

59. Augustine, R., Nethi, S.K., Kalarikkal, N., Thomas, S., Patra, C.R.: Electrospun polycaprolactone (PCL) scaffolds embedded with europium hydroxide nanorods (EHNs) with enhanced vascularization and cell proliferation for tissue engineering applications. J. Mater. Chem. B. **5**(24), 4660–4672 (2017). https://doi.org/10.1039/c7tb00518k

60. Kim, J.J., Hou, L., Huang, N.F.: Vascularization of Three-Dimensional Engineered Tissues for Regenerative Medicine Applications, pp. 17–26. Elsevier Ltd (2016)

61. Mukherjee, P., Nebuloni, F., Gao, H., Zhou, J., Papautsky, I.: Rapid prototyping of soft lithography masters for microfluidic devices using dry film photoresist in a non-cleanroom setting. Micromachines. **10**(3) (2019). https://doi.org/10.3390/mi10030192

62. Roberts, M.A., Tran, D., Coulombe, K.L.K., Razumova, M., Regnier, M., Murry, C.E., et al.: Stromal cells in dense collagen promote cardiomyocyte and microvascular patterning in engineered human heart tissue. Tissue Eng. Part A. **22**(7–8), 633–644 (2016). https://doi.org/10.1089/ten.tea.2015.0482

63. Mathur, A., Loskill, P., Shao, K., Huebsch, N., Hong, S.G., Marcus, S.G., et al.: Human iPSC-based cardiac microphysiological system for drug screening applications. Sci. Rep. **5**, 1–7 (2015). https://doi.org/10.1038/srep08883

64. Zhang, B., Montgomery, M., Davenport-H, L., Korolj, A., Radisic, M.: Platform technology for scalable assembly of instantaneously functional mosaic tissues. Sci. Adv. **1**(7) (2015). https://doi.org/10.1126/sciadv.1500423

65. Zhang, B., Lai, B.F.L., Xie, R., Davenport Huyer, L., Montgomery, M., Radisic, M.: Microfabrication of AngioChip, a biodegradable polymer scaffold with microfluidic vasculature. Nat. Protoc. **13**(8), 1793–1813 (2018). https://doi.org/10.1038/s41596-018-0015-8

66. Wang, Z., Mithieux, S.M., Weiss, A.S.: Fabrication Techniques for Vascular and Vascularized Tissue Engineering. Wiley-VCH Verlag (2019)

67. Liu, S., Dong, C., Lu, G., Lu, Q., Li, Z., Kaplan, D.L., et al.: Bilayered vascular grafts based on silk proteins. Acta Biomater. **9**(11), 8991–9003 (2013). https://doi.org/10.1016/j.actbio.2013.06.045

68. Wang, X.Y., Jin, Z.H., Gan, B.W., Lv, S.W., Xie, M., Huang, W.H.: Engineering interconnected 3D vascular networks in hydrogels using molded sodium alginate lattice as the sacrificial template. Lab Chip. **14**(15), 2709–2716 (2014). https://doi.org/10.1039/c4lc00069b

69. Mohanty, S., Larsen, L.B., Trifol, J., Szabo, P., Burri, H.V.R., Canali, C., et al.: Fabrication of scalable and structured tissue engineering scaffolds using water dissolvable sacrificial 3D printed moulds. Mater. Sci. Eng. C. **55**, 569–578 (2015). https://doi.org/10.1016/j.msec.2015.06.002

70. Tocchio, A., Tamplenizza, M., Martello, F., Gerges, I., Rossi, E., Argentiere, S., et al.: Versatile fabrication of vascularizable scaffolds for large tissue engineering in bioreactor. Biomaterials. **45**, 124–131 (2015). https://doi.org/10.1016/j.biomaterials.2014.12.031

71. Thavandiran, N., Hale, C., Blit, P., Sandberg, M.L., McElvain, M.E., Gagliardi, M., et al.: Functional arrays of human pluripotent stem cell-derived cardiac microtissues. Sci. Rep. **10**(1) (2020). https://doi.org/10.1038/s41598-020-62955-3

72. Sasagawa, T., Shimizu, T., Sekiya, S., Haraguchi, Y., Yamato, M., Sawa, Y., et al.: Design of prevascularized three-dimensional cell-dense tissues using a cell sheet stacking manipulation technology. Biomaterials. **31**(7), 1646–1654 (2010). https://doi.org/10.1016/j.biomaterials.2009.11.036

73. Domansky, K., Sliz, J.D., Wen, N., Hinojosa, C., Thompson, G., Fraser, J.P., et al.: SEBS elastomers for fabrication of microfluidic devices with reduced drug absorption by injection

molding and extrusion. Microfluidics Nanofluidics. **21**(6) (2017). https://doi.org/10.1007/s10404-017-1941-4

74. Weber, M., Gonzalez De Torre, I., Moreira, R., Frese, J., Oedekoven, C., Alonso, M., et al.: Multiple-step injection molding for fibrin-based tissue-engineered heart valves. Tissue Eng. Part C: Methods. **21**(8), 832–840 (2015). https://doi.org/10.1089/ten.tec.2014.0396

75. Gao, L., Ma, L., Yin, X.H., Luo, Y.C., Yang, H.Y., Zhang, B.: Nano- and Microfabrication for Engineering Native-Like Muscle Tissues. Wiley (2020)

76. Miller, J.S., Stevens, K.R., Yang, M.T., Baker, B.M., Nguyen, D.H.T., Cohen, D.M., et al.: Rapid casting of patterned vascular networks for perfusable engineered three-dimensional tissues. Nat. Mater. **11**(9), 768–774 (2012). https://doi.org/10.1038/nmat3357

77. Lind, J.U., Busbee, T.A., Valentine, A.D., Pasqualini, F.S., Yuan, H., Yadid, M., et al.: Instrumented cardiac microphysiological devices via multimaterial three-dimensional printing. Nat. Mater. **16**(3), 303–308 (2017). https://doi.org/10.1038/nmat4782

78. Noor, N., Shapira, A., Edri, R., Gal, I., Wertheim, L., Dvir, T.: 3D printing of personalized thick and perfusable cardiac patches and hearts. Adv. Sci. **6**(11) (2019). https://doi.org/10.1002/advs.201900344

79. Kolesky, D.B., Truby, R.L., Gladman, A.S., Busbee, T.A., Homan, K.A., Lewis, J.A.: 3D bioprinting of vascularized, heterogeneous cell-laden tissue constructs. Adv. Mater. **26**(19), 3124–3130 (2014). https://doi.org/10.1002/adma.201305506

80. Hinton, T.J., Jallerat, Q., Palchesko, R.N., Park, J.H., Grodzicki, M.S., Shue, H.J., et al.: Three-dimensional printing of complex biological structures by freeform reversible embedding of suspended hydrogels. Sci. Adv. **1**(9) (2015). https://doi.org/10.1126/sciadv.1500758

81. Wu, Q., Maire, M., Lerouge, S., Therriault, D., Heuzey, M.-C.: 3D printing of microstructured and stretchable chitosan hydrogel for guided cell growth. Adv. Biosyst. **1**(6), 1700058 (2017). https://doi.org/10.1002/adbi.201700058

82. Lee, A., Hudson, A.R., Shiwarski, D.J., Tashman, J.W., Hinton, T.J., Yerneni, S., et al.: 3D bioprinting of collagen to rebuild components of the human heart. Science. **365**, 482–487 (2019)

83. Li, M., Guo, Y., Wei, Y., MacDiarmid, A.G., Lelkes, P.I.: Electrospinning polyaniline-contained gelatin nanofibers for tissue engineering applications. Biomaterials. **27**(13), 2705–2715 (2006). https://doi.org/10.1016/j.biomaterials.2005.11.037

84. Ruther, F., Zimmermann, A., Engel, F.B., Boccaccini, A.R.: Improvement of the layer adhesion of composite cardiac patches. Adv. Eng. Mater. (2019). https://doi.org/10.1002/adem.201900986

85. Wu, Y., Wang, L., Guo, B., Ma, P.X.: Interwoven aligned conductive nanofiber yarn/hydrogel composite scaffolds for engineered 3D cardiac anisotropy. ACS Nano. **11**(6), 5646–5659 (2017). https://doi.org/10.1021/acsnano.7b01062

86. Ehler, E., Jayasinghe, S.N.: Cell electrospinning cardiac patches for tissue engineering the heart. Analyst. **139**(18), 4449–4452 (2014). https://doi.org/10.1039/c4an00766b

87. Castilho, M., Feyen, D., Flandes-Iparraguirre, M., Hochleitner, G., Groll, J., Doevendans, P.A.F., et al.: Melt electrospinning writing of poly-Hydroxymethylglycolide-co-ε-Caprolactone-based scaffolds for cardiac tissue engineering. Adv. Healthc. Mater. **6**(18), 1–9 (2017). https://doi.org/10.1002/adhm.201700311

88. Lakshmanan, R., Kumaraswamy, P., Krishnan, U.M., Sethuraman, S.: Engineering a growth factor embedded nanofiber matrix niche to promote vascularization for functional cardiac regeneration. Biomaterials. **97**, 176–195 (2016). https://doi.org/10.1016/j.biomaterials.2016.02.033

89. Nunes, S.S., Miklas, J.W., Liu, J., Aschar-Sobbi, R., Xiao, Y., Zhang, B., et al.: Biowire: a platform for maturation of human pluripotent stem cell–derived cardiomyocytes. Nat. Methods. **10**(8), 781–787 (2013). https://doi.org/10.1038/nmeth.2524

90. Sun, X., Nunes, S.S.: Biowire Platform for Maturation of Human Pluripotent Stem Cell-Derived Cardiomyocytes, pp. 21–26. Academic Press Inc (2016)

91. Boudou, T., Legant, W.R., Mu, A., Borochin, M.A., Thavandiran, N., Radisic, M., et al.: A microfabricated platform to measure and manipulate the mechanics of engineered cardiac microtissues. Tissue Eng. Part A. **18**(9–10), 910–919 (2012). https://doi.org/10.1089/ten.tea.2011.0341

92. Xiao, Y., Zhang, B., Liu, H., Miklas, J.W., Gagliardi, M., Pahnke, A., et al.: Microfabricated perfusable cardiac biowire: a platform that mimics native cardiac bundle. Lab Chip. **14**(5), 869–882 (2014). https://doi.org/10.1039/c3lc51123e

93. Zhao, Y., Rafatian, N., Wang, E.Y., Feric, N.T., Lai, B.F.L., Knee-Walden, E.J., et al.: Engineering microenvironment for human cardiac tissue assembly in heart-on-a-chip platform. Matrix Biol. **85–86**, 189–204 (2020). https://doi.org/10.1016/j.matbio.2019.04.001

94. Feric, N.T., Pallotta, I., Singh, R., Bogdanowicz, D.R., Gustilo, M.M., Chaudhary, K.W., et al.: Engineered cardiac tissues generated in the biowire II: a platform for human-based drug discovery. Toxicol. Sci. **172**(1), 89–97 (2019). https://doi.org/10.1093/toxsci/kfz168

95. Zhao, Y., Wang, E.Y., Davenport, L.H., Liao, Y., Yeager, K., Vunjak-Novakovic, G., et al.: A multimaterial microphysiological platform enabled by rapid casting of elastic microwires. Adv. Healthcare Mater. **8**(5) (2019). https://doi.org/10.1002/adhm.201801187

96. Wang, E.Y., Rafatian, N., Zhao, Y., Lee, A., Lai, B.F.L., Lu, R.X., et al.: Biowire model of interstitial and focal cardiac fibrosis. ACS Central Sci. **5**, 1146–1158 (2019). https://doi.org/10.1021/acscentsci.9b00052

97. Conant, G., Ahadian, S., Zhao, Y., Radisic, M.: Kinase inhibitor screening using artificial neural networks and engineered cardiac biowires. Sci. Rep. **7**(1) (2017). https://doi.org/10.1038/s41598-017-12048-5

98. Miklas, J.W., Nunes, S.S., Sofla, A., Reis, L.A., Pahnke, A., Xiao, Y., et al.: Bioreactor for modulation of cardiac microtissue phenotype by combined static stretch and electrical stimulation. Biofabrication. **6**(2) (2014). https://doi.org/10.1088/1758-5082/6/2/024113

99. Nunes, S.S., Feric, N., Pahnke, A., Miklas, J.W., Li, M., Coles, J., et al.: Human stem cell-derived cardiac model of chronic drug exposure. ACS Biomater. Sci. Eng. **3**(9), 1911–1921 (2017). https://doi.org/10.1021/acsbiomaterials.5b00496

100. Sidorov VY, Samson PC, Sidorova TN, Davidson JM, Lim CC, Wikswo JP. I-Wire Heart-on-a-Chip I: Three-dimensional cardiac tissue constructs for physiology and pharmacology. Acta Biomater. 2017;48:68–78. doi: https://doi.org/10.1016/j.actbio.2016.11.003

101. Schroer, A.K., Shotwell, M.S., Sidorov, V.Y., Wikswo, J.P., Merryman, W.D.: I-Wire Heart-on-a-Chip II: biomechanical analysis of contractile, three-dimensional cardiomyocyte tissue constructs. Acta Biomater. **48**, 79–87 (2017). https://doi.org/10.1016/j.actbio.2016.11.010

102. Kherani, A.: Revolutionizing drug testing platforms by leveraging angiochip and flight data recorder technologies. STEM Fellowship J. **5**(1), 43–45 (2019). https://doi.org/10.17975/sfj-2019-006

103. Wang, G., McCain, M.L., Yang, L., He, A., Pasqualini, F.S., Agarwal, A., et al.: Modeling the mitochondrial cardiomyopathy of Barth syndrome with induced pluripotent stem cell and heart-on-chip technologies. Nat. Med. **20**(6), 616–623 (2014). https://doi.org/10.1038/nm.3545

104. Hansen, A., Eder, A., Bönstrup, M., Flato, M., Mewe, M., Schaaf, S., et al.: Development of a drug screening platform based on engineered heart tissue. Circ. Res. **107**(1), 35–44 (2010). https://doi.org/10.1161/CIRCRESAHA.109.211458

105. Schaaf, S., Shibamiya, A., Mewe, M., Eder, A., Stöhr, A., Hirt, M.N., et al.: Human engineered heart tissue as a versatile tool in basic research and preclinical toxicology. PLoS ONE. **6**(10) (2011). https://doi.org/10.1371/journal.pone.0026397

106. Stoehr, A., Neuber, C., Baldauf, C., Vollert, I., Friedrich, F.W., Flenner, F., et al.: Automated analysis of contrac-tile force and Ca 2 transients in engineered heart tissue. Am. J. Physiol. Heart Circ. Physiol. **306**, 1353–1363 (2014). https://doi.org/10.1152/ajpheart.00705.2013.-Contraction

107. Mannhardt, I., Breckwoldt, K., Letuffe-Brenière, D., Schaaf, S., Schulz, H., Neuber, C., et al.: Human engineered heart tissue: analysis of contractile force. Stem Cell Rep. **7**(1), 29–42 (2016). https://doi.org/10.1016/j.stemcr.2016.04.011

108. Liau, B., Christoforou, N., Leong, K.W., Bursac, N.: Pluripotent stem cell-derived cardiac tissue patch with advanced structure and function. Biomaterials. **32**(35), 9180–9187 (2011). https://doi.org/10.1016/j.biomaterials.2011.08.050

109. Beans, C.: The race to patch the human heart. Proc. Natl. Acad. Sci. U. S. A. **115**(26), 6518–6520 (2018). https://doi.org/10.1073/pnas.1808317115
110. Zhang, D., Shadrin, I.Y., Lam, J., Xian, H.Q., Snodgrass, H.R., Bursac, N.: Tissue-engineered cardiac patch for advanced functional maturation of human ESC-derived cardiomyocytes. Biomaterials. **34**(23), 5813–5820 (2013). https://doi.org/10.1016/j.biomaterials.2013.04.026
111. Jackman, C.P., Ganapathi, A.M., Asfour, H., Qian, Y., Allen, B.W., Li, Y., et al.: Engineered cardiac tissue patch maintains structural and electrical properties after epicardial implantation. Biomaterials. **159**, 48–58 (2018). https://doi.org/10.1016/j.biomaterials.2018.01.002
112. Chen, T., Vunjak-novakovic, G.: In vitro models of ischemia-reperfusion injury. Regen. Eng. Transl. Med. **4**, 142–153 (2018). https://doi.org/10.1007/s40883-018-0056-0
113. Marwick, T.H.: Ejection fraction pros and cons: JACC state-of-the-art review. J. Am. Coll. Cardiol. **72**(19), 2360–2379 (2018). https://doi.org/10.1016/j.jacc.2018.08.2162
114. Stancescu, M., Molnar, P., McAleer, C.W., McLamb, W., Long, C.J., Oleaga, C., et al.: A phenotypic invitro model for the main determinants of human whole heart function. Biomaterials. **60**, 20–30 (2015). https://doi.org/10.1016/j.biomaterials.2015.04.035
115. Zhang, B., Radisic, M.: Organ-on-A-chip devices advance to market. Lab Chip. **17**(14), 2395–2420 (2017). https://doi.org/10.1039/c6lc01554a
116. Jost, N., Muntean, D.M., Christ, T.: Cardiac arrhythmias: introduction, electrophysiology of the Heart, action potential and membrane currents. In: Jagadeesh, G., Balakumar, P., Maung-U, K. (eds.) Pathophysiology and Pharmacotherapy of Cardiovascular Disease, pp. 977–1002. Springer International Publishing, Cham (2015)
117. Satin, J., Kehat, I., Caspi, O., Huber, I., Arbel, G., Itzhaki, I., et al.: Mechanism of spontaneous excitability in human embryonic stem cell derived cardiomyocytes. J. Physiol. **559**(2), 479–496 (2004). https://doi.org/10.1113/jphysiol.2004.068213
118. Tian, B., Liu, J., Dvir, T., Jin, L., Tsui, J.H., Qing, Q., et al.: Macroporous nanowire nanoelectronic scaffolds for synthetic tissues. Nat. Mater. **11**, 986–994 (2012). https://doi.org/10.1038/nmat3403
119. Feiner, R., Engel, L., Fleischer, S., Malki, M., Gal, I., Shapira, A., et al.: Engineered hybrid cardiac patches with multifunctional electronics for online monitoring and regulation of tissue function. Nat. Mater. **15**(6), 679–685 (2016). https://doi.org/10.1038/nmat4590
120. Radisic, M., Park, H., Shing, H., Consi, T., Schoen, F.J., Langer, R., et al.: Functional assembly of engineered myocardium by electrical stimulation of cardiac myocytes cultured on scaffolds. Proc. Natl. Acad. Sci. U. S. A. **101**(52), 18129–18134 (2004). https://doi.org/10.1073/pnas.0407817101
121. Goldfracht, I., Efraim, Y., Shinnawi, R., Kovalev, E., Huber, I., Gepstein, A., et al.: Engineered heart tissue models from hiPSC-derived cardiomyocytes and cardiac ECM for disease modeling and drug testing applications. Acta Biomater. **92**, 145–159 (2019). https://doi.org/10.1016/j.actbio.2019.05.016
122. Leyton-Mange, J.S., Mills, R.W., Macri, V.S., Jang, M.Y., Butte, F.N., Ellinor, P.T., et al.: Rapid cellular phenotyping of human pluripotent stem cell-derived cardiomyocytes using a genetically encoded fluorescent voltage sensor. Stem Cell Rep. **2**(2), 163–170 (2014). https://doi.org/10.1016/j.stemcr.2014.01.003
123. Huisken, J., Stainier, D.Y.R.: Selective plane illumination microscopy techniques in developmental biology. Development. **136**(12), 1963–1975 (2009). https://doi.org/10.1242/dev.022426
124. Turaga, D., Matthys, O.B., Hookway, T.A., Joy, D.A., Calvert, M., McDevitt, T.C.: Single-cell determination of cardiac microtissue structure and function using light sheet microscopy. Tissue Eng. Part C: Methods. **26**(4) (2020). https://doi.org/10.1089/ten.tec.2020.0020
125. Mastikhina, O., Moon, B.U., Williams, K., Hatkar, R., Gustafson, D., Mourad, O., et al.: Human cardiac fibrosis-on-a-chip model recapitulates disease hallmarks and can serve as a platform for drug testing. Biomaterials. **233**(December 2019), 119741 (2020). https://doi.org/10.1016/j.biomaterials.2019.119741

126. Dmitriev, R.I., Papkovsky, D.B.: Optical probes and techniques for O 2 measurement in live cells and tissue. Cell. Mol. Life Sci. **69**(12), 2025–2039 (2012). https://doi.org/10.1007/s00018-011-0914-0

127. Odijk, M., Van Der Meer, A.D., Levner, D., Kim, H.J., Van Der Helm, M.W., Segerink, L.I., et al.: Measuring direct current trans-epithelial electrical resistance in organ-on-a-chip microsystems. Lab Chip. **15**(3), 745–752 (2015). https://doi.org/10.1039/c4lc01219d

128. Henry, O.Y.F., Villenave, R., Cronce, M.J., Leineweber, W.D., Benz, M.A., Ingber, D.E.: Organs-on-chips with integrated electrodes for trans-epithelial electrical resistance (TEER) measurements of human epithelial barrier function. Lab Chip. **17**(13), 2264–2271 (2017). https://doi.org/10.1039/c7lc00155j

129. Fedak, P.W.M., Verma, S., Weisel, R.D., Li, R.-K.: Cardiac remodeling and failure (I). Cardiovasc. Pathol. **14**(2), 49–60 (2005). https://doi.org/10.1016/j.carpath.2005.01.005

130. Finegold, J.A., Asaria, P., Francis, D.P.: Mortality from ischaemic heart disease by country, region, and age: Statistics from World Health Organisation and United Nations. Int. J. Cardiol. **168**(2), 934–945 (2013). https://doi.org/10.1016/j.ijcard.2012.10.046

131. Fine, B., Vunjak-Novakovic, G.: Shortcomings of animal models and the rise of engineered human cardiac tissue. ACS Biomater. Sci. Eng. **3**(9), 1884–1897 (2017). https://doi.org/10.1021/acsbiomaterials.6b00662

132. Wagner, K.T., Nash, T.R., Liu, B., Vunjak-Novakovic, G., Radisic, M.: Extracellular vesicles in cardiac regeneration: potential applications for tissues-on-a-Chip. Trends Biotechnol., 1–15 (2020). https://doi.org/10.1016/j.tibtech.2020.08.005

133. Chen, T., Vunjak-novakovic, G.: Human tissue-engineered model of myocardial ischemia – reperfusion injury. Tissue Eng. Part A. **00**(00), 1–14 (2018). https://doi.org/10.1089/ten.tea.2018.0212

134. Williams, C., Budina, E., Stoppel, W.L., Sullivan, K.E., Emani, S., Emani, S.M., et al.: Cardiac extracellular matrix-fibrin hybrid scaffolds with tunable properties for cardiovascular tissue engineering. Acta Biomater. **14**, 84–95 (2015). https://doi.org/10.1016/j.actbio.2014.11.035

135. Cashman TJ, Josowitz R, Johnson BV, Gelb BD, Costa KD. Human engineered cardiac tissues created using induced pluripotent stem cells reveal functional characteristics of BRAF-mediated hypertrophic cardiomyopathy. PLoS One 2016;11(1):1–17. doi: https://doi.org/10.1371/journal.pone.0146697

136. Hinson, J.T., Chopra, A., Nafissi, N., Polacheck, W.J., Benson, C.C., Swist, S., et al.: Titin mutations in iPS cells define sarcomere insufficiency as a cause of dilated cardiomyopathy. Science. **349**(6251), 982–986 (2015). https://doi.org/10.1126/science.aaa5458

137. Stillitano, F., Turnbull, I.C., Karakikes, I., Nonnenmacher, M., Backeris, P., Hulot, J.S., et al.: Genomic correction of familial cardiomyopathy in human engineered cardiac tissues. Eur. Heart J. **37**(43), 3282–3284 (2016). https://doi.org/10.1093/eurheartj/ehw307

138. Ma, Z., Huebsch, N., Koo, S., Mandegar, M.A., Siemons, B., Boggess, S., et al.: Contractile deficits in engineered cardiac microtissues as a result of MYBPC3 deficiency and mechanical overload. Nat. Biomed. Eng. **2**(12), 955–967 (2018). https://doi.org/10.1038/s41551-018-0280-4

139. Park, S.J., Zhang, D., Qi, Y., Li, Y., Lee, K.Y., Bezzerides, V.J., et al.: Insights into the pathogenesis of catecholaminergic polymorphic ventricular tachycardia from engineered human heart tissue. Circulation. **140**(5), 390–404 (2019). https://doi.org/10.1161/CIRCULATIONAHA.119.039711

140. Goldfracht I, Protze S, Shiti A, Setter N, Gruber A, Shaheen N, et al. Generating ring-shaped engineered heart tissues from ventricular and atrial human pluripotent stem cell-derived cardiomyocytes. Nat. Commun. 2020;11(1):1–15. doi: https://doi.org/10.1038/s41467-019-13868-x

Bioengineering of Pediatric Cardiovascular Constructs: In Vitro Modeling of Congenital Heart Disease

Holly Bauser-Heaton, Carmen J. Gil, and Vahid Serpooshan

1 Introduction

Congenital heart diseases (CHDs) are rare inherited anatomical and/or electrophysiological abnormalities that affect the whole heart function. To date, a variety of *in silico*, *in vitro*, and *in vivo* modeling platforms have been developed to decipher the pathophysiological mechanisms underlying CHDs [1–3]. Among these, three-dimensional (3D) *in vitro* biomimetic structures can enable precise and controlled regulation of microenvironmental factors, such as flow hemodynamics, tissue composition, and mechanical properties, to study principles that guide our understanding of the development of CHD [3]. Namely, the "no flow no grow" theory states that small perturbations in flow may result in a range of CHDs [4, 5]. Other theories

H. Bauser-Heaton (✉)
Department of Biomedical Engineering, Emory University School of Medicine and Georgia Institute of Technology, Atlanta, GA, USA

Department of Pediatrics, Emory University School of Medicine, Atlanta, GA, USA

Children's Healthcare of Atlanta, Atlanta, GA, USA

Sibley Heart Center at Children's Healthcare of Atlanta, Atlanta, GA, USA
e-mail: BauserH@kidsheart.com

C. J. Gil
Department of Biomedical Engineering, Emory University School of Medicine and Georgia Institute of Technology, Atlanta, GA, USA

V. Serpooshan (✉)
Department of Biomedical Engineering, Emory University School of Medicine and Georgia Institute of Technology, Atlanta, GA, USA

Department of Pediatrics, Emory University School of Medicine, Atlanta, GA, USA

Children's Healthcare of Atlanta, Atlanta, GA, USA
e-mail: vahid.serpooshan@bme.gatech.edu

© The Author(s), under exclusive license to Springer Nature Switzerland AG 2022
J. Zhang, V. Serpooshan (eds.), *Advanced Technologies in Cardiovascular Bioengineering*, https://doi.org/10.1007/978-3-030-86140-7_11

regarding geometric perturbations noted in stenotic valves, arteries, and veins as well as the cellular changes that occur with these alterations can be nimbly studied [5, 6].

Given the challenges in recapitulating the wide variety of complex tissues and environmental factors involved in CHDs, 3D printing and bioprinting have shown great promise to fabricate reproducible and highly tunable models of these birth defects [7–14]. For instance, alterations in stiffness can be explored and can be extrapolated to disease states in which tissues have suffered hypertrophy, atrophy, or have been exposed to altered pressure resulting in thin- walled structures and valves that are thick and stiff [15]. This capacity of alteration of stiffness is vital for the study of CHD and its development as well as future treatment [16]. An additional factor in this population is the age-associated alterations in cellular makeup of tissue. For example, neonates are noted to have thick-walled right ventricles that thin over time [17]. In addition, critical vascular resistance changes during the first few weeks of life resulting in lowered pulmonary arterial pressures and increasing systemic pressures as one ages. Increased elastin in the pulmonary walls is also noted to change into the first decade [18]. As this occurs, the properties of pulmonary arteries and the way they respond to therapies are significantly different as one progresses through each decade of life. All these age-related changes as one matures, alters cardiovascular behavior. Such alteration can be studied by introducing controlled changes in the *in vitro* models, in the cellular and protein composition of the bioink/hydrogel, the scaffold stiffness, and perturbations in flow parameters. Such highly tunable *in vitro* platforms can enable teams to have improved understanding of these important microenvironmental factors even within one disease state. This can only be accomplished in a carefully designed, closely tuned model that can be achieved using recent advancements in the field of 3D bioprinting [19–24].

In addition to the above-mentioned limitations of CHD research, perhaps the most challenging is the growth of our patient population over time resulting in dramatic changes in hemodynamics that result in cellular and tissue dysfunction. Studying a patient longitudinally with these dynamic alterations is impossible in animal models and quite impractical *in vivo*. The use of 3D bioprinting to mimic changes as a patient grows and create potential therapeutic strategies are paramount to our aging and growing population. As 3D printing has progressed, the CHD community has acknowledged its use in several areas. From education to procedural planning in resin printed pathologies to 3D bioprinting cellular layers to understand disease states, this technology serves to improve the outcomes in CHD.

2 Procedural Planning and Education

In vitro modeling of CHD began with simple resin printing initially for pre procedural planning based on ultrasound (US), computed tomography (CT), and magnetic resonance imaging (MRI) data. These models proved promising for procedural planning of surgical and transcatheter procedures [25]. Additionally, 3D printing has been used for surgical simulation allowing complex repairs such as ventricular baffles and transcatheter valve implantation through the use of computer and instrument guided procedure simulation [26]. These 3D printed renderings closely resemble vascular anatomy as in the case of complex pulmonary artery disease, namely major aortopulmonary collaterals [27] amongst other disease types. In addition, baffle creation in double outlet right ventricle from the left ventricle to aorta can be aided by resin-based 3D prints [25]. These synthetic models have proven valuable for stenting procedures and surgical planning for complex heart transplant anastomosis, as well as for educating patients, parents, and students [26, 28–30]. As acceptance increases amongst cardiologists, 3D understanding of CHD has led to improved collective thought processes and continue to gain popularity.

While valuable, the use of these models has limitations in terms of moving from procedural planning to CHD causation study and interruption of pathologic processes. Transition from resin printing to exploration of bioprinting has elevated these models for the investigation of hemodynamic changes, altered flow parameters, mechanical properties, patient and disease specific cellular alterations, and protein changes that directly influence the development of specific disease types of CHD [3, 7, 9, 11, 31].

3 Modeling of the Developing Heart

As the heart develops, most alterations occur quite early in the fetal development within the first 6–8 weeks of gestation. Our understanding of early cardiogenesis relies heavily on the prior works done in animal models, namely chick and mouse embryos [32, 33]. Human genome and genetic alterations that lead to cardiac malformations are also not well understood given prior work primarily done in animal embryogenesis. Currently, there are known human genetic associations, but the disruption in cardiac formation as a result of these genetic associations still remain difficult to elucidate given the early malformation in fetal life [34]. An understanding of fetal cardiac development at the cellular and molecular level, therefore, is paramount to treatment of this disease.

The formation of the heart tube, looping and ultimate development of four chambers can be disrupted early resulting in single ventricle anatomy, transposition of great arteries, heterotaxy, dextrocardia. Additionally, defects in the endocardial cushion may result in atrial septal defects, ventricular septal defects (VSDs), and atrioventricular canal [35]. The challenge of the developing heart is the unique variety of defects that occur and the inability to study at such early time points. Groups have attempted 3D printing early CHD development using tissues as a result of pregnancy loss. Using microCT datasets obtained in post mortem fetuses, a study of feasibility proved ability to print small CHD structures in full 3D datasets [36]. These models printed in rigid resin allowed for reconstruction of tetralogy of Fallot, VSD with hypoplastic aortic arch, borderline or hypoplastic left ventricle and atrioventricular canal at age ranges from 12 + 5 weeks gestation to 21 + 3 weeks. The successful reconstruction of these cases demonstrates the pathologic variety at early gestational age and served as basis for understanding of basic fetal 3D geometric alterations [36]. While the modeling of these early malformations adds to generalizable knowledge and may assist in counseling of future pregnancies for those parents of patients affected, however, the study of microenvironmental alterations in development is not possible in these synthetic platforms.

The use of human induced pluripotent stem cell (iPSC) modeling allows teams to investigate cellular mechanisms of pathophysiologic examples of CHD [37, 38]. *In vitro* modeling of cells derived from human somatic cells of patients with CHD has been utilized for cellular inquiry for mechanisms of disruption in embryogenesis and function [39]. Patients with hypoplastic left heart syndrome (HLHS), for example, exhibit alterations in myosin heavy chin and cardiac muscle alpha isoforms, resulting in dysmorphic sarcomeres and decreased contractility [40]. While utilizing patient cells that are phenotype specific may lend insights into pathology, what is unclear is how closely these iPSCs resemble real life development or even severity of disease.

Other groups have used models of very early heart tube and fetal left ventricle (LV) development with the use of 3D bioprinting (Fig. 1). With biomimetic materials capable of perfusion, various cell types can be seeded, and cellular behavior studied to understand the effects of alterations in flow hemodynamics in normal heart development [7]. Simultaneous study of shear stress within a linear heart tube (embryonic day 22, Fig. 1a–h) and a more advanced fetal stage (week 33, Fig. 1i–p) of LV, identified areas of potential alterations of flow patterns. Bioprinted constructs were perfused at physiological rates, demonstrating adequate levels of endothelial cell viability (Fig. 1o–t). Use of 3D printing allowed the study of microenvironmental factors to understand normal processes and the disruptions that lead to diseases. The basis for this type of *in vitro* study stems from findings of altered biomechanics and shear stress leading to the development of CHD in animal models [41–46]. It is this type of investigative bioprinted modeling that brings together the understanding of properties of flow (further looking at the "no flow no grow" theory), stiffness mechanics, slight modifications of simple geometry, and how each of these lead to cellular disruption. These models can serve as excellent platforms for treatment applications for potential fetal/early intervention to disrupt the development of CHD.

4 Hypoplastic Left Heart Syndrome (HLHS)

The understanding of the developing heart allows for further study of severe heart lesions like HLHS. This CHD is a critical heart defect that is thought to result from small alterations in fetal flow. The "no flow, no grow" theory states that as a structure receives less flow, it does not grow and can result in small percentages of flow change [4, 5]. This theory is the basis for the maldevelopment of any structure downstream from an obstructive lesion. In HLHS the alterations result in the malformation or, in severe forms, atresia of left sided structures, namely the mitral and

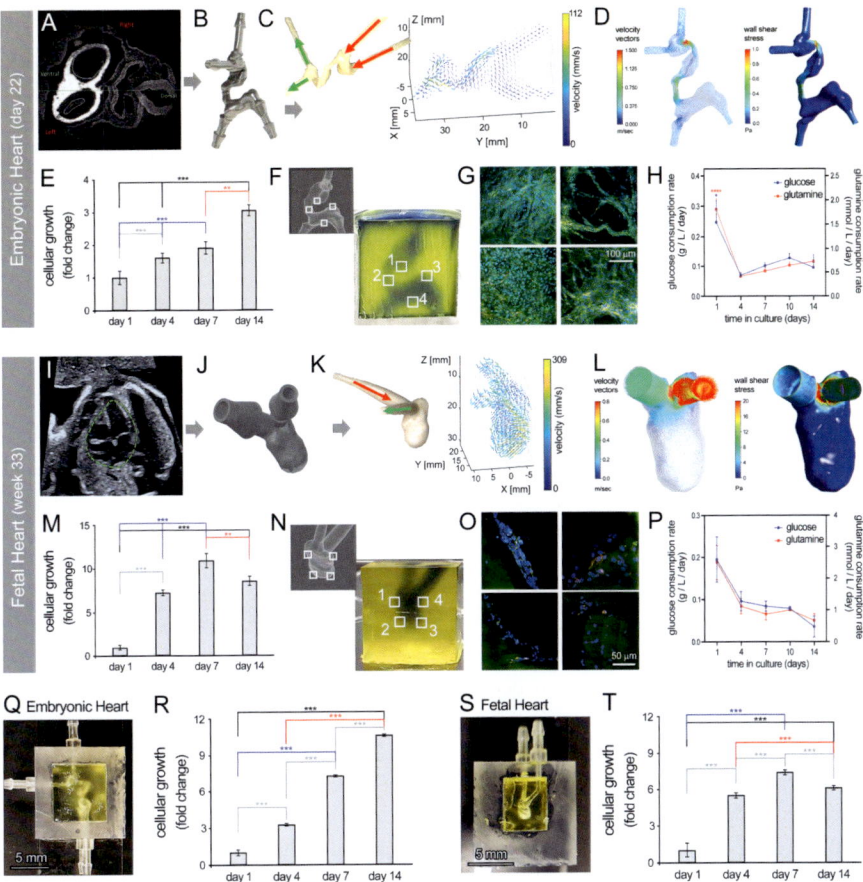

Fig. 1 The use of 3D bioprinting and perfusion bioreactor systems to model various stages of developing human heart. (**a–b**): Human histological data were used to reconstruct a 3D CAD model of the embryonic human heart at Carnegie stage 10 (day 22). (c–d): Flow hemodynamics through the printed model was quantified using ultrasound imaging and particle image velocimetry (PIV) analysis (c), as well as using computational fluid dynamics (CFD) modeling (d) to determine flow velocity and wall shear stress distribution within the 3D geometry. (e–h): Linear heart tubes were bioprinted using gelatin methacrylate (gelMA) hydrogel, seeded with endothelial cells, and studied for luminal space endothelization and metabolomic bioprofiling. (i–j): Fetal echocardiography data of the human heart was used to reconstruct the 3D CAD model of the fetal LV at week 33. (k–l): Flow hemodynamics in fetal LV was analyzed using ultrasound PIV (k) and CFD (l) analyses. (m–p): Endothelial cell growth and metabolic activity in 3D bioprinted fetal LV constructs. (q–t): Bioprinted embryonic (q–r) and fetal heart (s–t) models were endothelialized and, housed within custom-designed perfusion chambers, and perfused at physiological flow rates using for 2-week period. In the confocal images, green: CD31 (endothelial cells), red: connexin-43, and blue: DAPI. Figure reconstructed from [7]

aortic valves. As flow is limited through the left sided heart structures, the aortic arch also becomes markedly hypoplastic and therefore exacerbates the abnormal flow experienced by the fetus. Currently, the true cause of HLHS remain largely unknown [47]. Some genetic preponderance has been shown as have environmental factors [48, 49]. Fetal interventions aimed to disrupt the abnormal flow patterns in some subtypes have been attempted with varied success through the dilation of aortic valves in cases of aortic stenosis. However, the success of these fetal interventions remains difficult to predict [50–52]. At this time, treatment strategies postnatally include a series of three operations; Norwood, Glenn, and Fontan, where pulmonary and systemic blood flow are ultimately balanced. These patients ultimately require transplantation when this palliation fails. The community does not have any additional fetal interventions at this time other than the aforementioned fetal balloon valvuloplasty.

Current medical treatment for HLHS is limited to symptom relief and surgical interventions are palliative [53]. An improved understanding of the development of the disease is therefore required to interrupt the abnormal development and to promote growth of the hypoplastic ventricle. The creation of a 3D printed model for HLHS is complex and requires the same principles outlined above but with additionally complex considerations [14]. Groups have begun to model this disease process starting with a rudimentary left sided structure based on medical imaging via ultrasound and CT [13, 36, 54]. Ruedinger et al. demonstrated a novel approach that creates a patient-specific, 3D printed fetal heart model derived from a fetal echocardiogram, integrates four-dimensional (4D) flow MRI and 3D printing technologies to model and analyze intracardiac flow profiles in normal versus HLHS fetal hearts *in vitro* (Fig. 2) [14]. However, their synthetic models were not able to support dynamic cell culture and analysis. Using stereolithography or embedded bioprinting approaches [55], 3D soft tissue models with good fidelity and adequate biocompatibility can be fabricated to address this issue. For this purpose, STL files can be created from various imaging modalities. Early models of the developing HLHS heart can be derived from fetal ultrasound, requiring complex segmentation based on stacks of 2 dimensional images. The challenge remains to have models of HLHS early in development and often these models as mentioned above, are based on CT following fetal demise. As models improve 3D arrangement of unique forms of HLHS, the study of flow and other microenvironmental factors, such as stiffness and tissue composition, have remained elusive until recent. Alterations in various bioinks based in gelatin [7], allows for altered stiffness that promotes the growth of cells, in this case namely cardiomyocytes. Through the design of perusable modes, the study of the behavior of normal cardiomyocytes and those of patients with HLHS can be compared to identify the role of genetic or patient specific cellular programming versus the current causation of altered flow theory. This allows the community to understand the role of genetic alteration of cellular types as it relates to alterations in perfusion to develop novel therapeutics or development of interventional therapies.

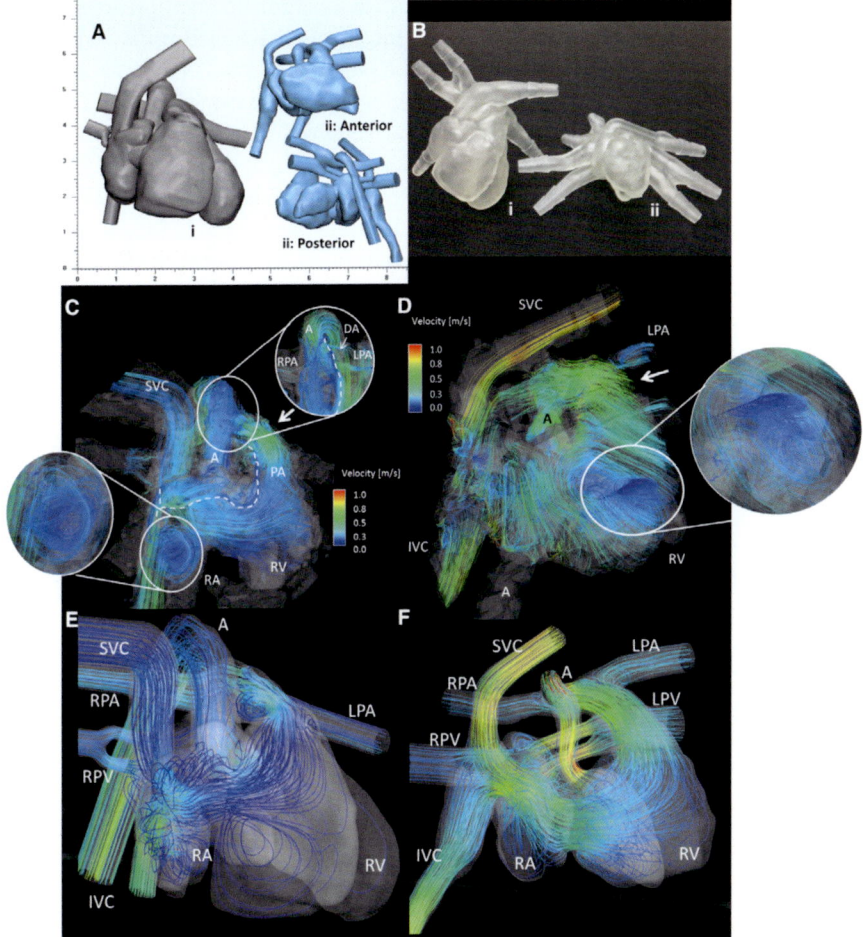

Fig. 2 Using 3D printed models to study hypoplastic left heart syndrome (HLHS). (a): Fetal echocardiogram data was used to generate 3D digital models of (**i**) normal and (**ii**) abnormal fetal heart. The posterior view of the abnormal heart (**ii**) illustrates the HLHS. (**b**): 3D printed models of (**i**) normal and (**ii**) HLHS fetal hearts were printed using synthetic resin. (**c–d**): Constructs were perfused and imaged using 4D MR imaging to assess blood flow velocity streamlines in the normal (**c**) versus HLHS (**d**) fetal heart. Overall, fluid hemodynamics results were consistent with the flow patterns observed from the clinical data. Higher velocities correspond well with the characteristics of HLHS. (**e–f**): Computational fluid dynamic (CFD) modeling of the normal (**e**) and HLHS (**f**) fetal heart models using boundary conditions based on the average flow data from MRI [14]

5 Pulmonary Artery Stenosis

The study of vascular stenosis within the pulmonary vascular tree, both arterial and venous, presents a challenge within the CHD community. The pathology differs for each and every patient and can be related to post-surgical changes, as a modification of flow secondary to shunt physiology, pathologic overgrowth and fibrosis as in the setting of pulmonary hypertension, or idiopathic development of stenosis. In patients with complex congenital malformations of the pulmonary arteries like major aorto-pulmonary collaterals, isolated pulmonary artery of ductal origin, truncus arterio-sus, Williams and Alagilles syndromes, and acquired vascular stenosis, current treatment results in repeated procedures both surgical and transcatheter [56, 57]. The treatments rely on balloon angioplasty, surgical patch angioplasty and stenting, all of which only address the minimal diameter of the vessel and not the underlying pathology leading to repeated stenosis and altered growth [57].

The use of 3D printing and bioprinting in these unique and heterogenous struc-tures enables understanding of procedural planning [58] but also allows a better understanding of the complex nature of these connections. Study of vascular reste-nosis has been embarked upon by the adult congenital groups. Studies have identi-fied areas prone to increased burden of vascular stenosis at areas of bifurcations or bends in vascular courses as a result of alterations in shear stress [59, 60]. While these studies are helpful insights, the nature of coronary stenosis is extremely differ-ent from that of vascular stenosis associated with CHD [61]. Through the use of 3D bioprinting, these complex pulmonary artery stenoses can be explored further with cellularized bioinks[11]. Through the creations of vascular structures with bilayer cellular components, the study of cell-cell interactions as it relates to shear stress can be further achieved. An understanding of endothelial cell signaling to smooth muscle which has been implicated in the endothelial to mesenchymal (EndMT) transition process [62] can be applied to specific patient geometries and flow patterns.

In addition to the work completed in small vascular diameters, such as coronary arteries, the study of pulmonary hypertension (PH), another type of pulmonary arte-rial disease in which cellular layers become hyperplastic, lend further insights and suggestions to alterations in the pulmonary vascular tree. Through the EndMT sig-naling cascades, smooth muscle cells are programmed to overgrow, resulting in luminal obliteration. One group utilized 3D culture to elucidate this mechanism through the exploration of *in vitro* model of pulmonary hypertension [63]. Another utilizes microvessels with layering of smooth muscle cells, endothelial cells, and extracellular matrix to study changes in the distal pulmonary arteries in PH [64].

What is challenging is the need for specific disease models in very rare pulmo-nary artery disease processes, and 3D bioprinting fills that void for the study of pathology in these disease types. Specific geometric abnormalities within the pul-monary vascular tree can therefore be modeled as noted from early works from Tomov et al. where the areas of bifurcation may be noted to have alterations in shear stress that may directly result in cellular abnormalities (Fig. 3) [11]. Other groups have also investigated these geometric considerations with the simulation of patient-specific hemodynamics and examining the effect of PA stenosis in comparison to shunt diameters [65]. These types of investigations are vital to creating novel thera-pies that go beyond surgical treatments to prevent re-stenosis throughout multiple decades.

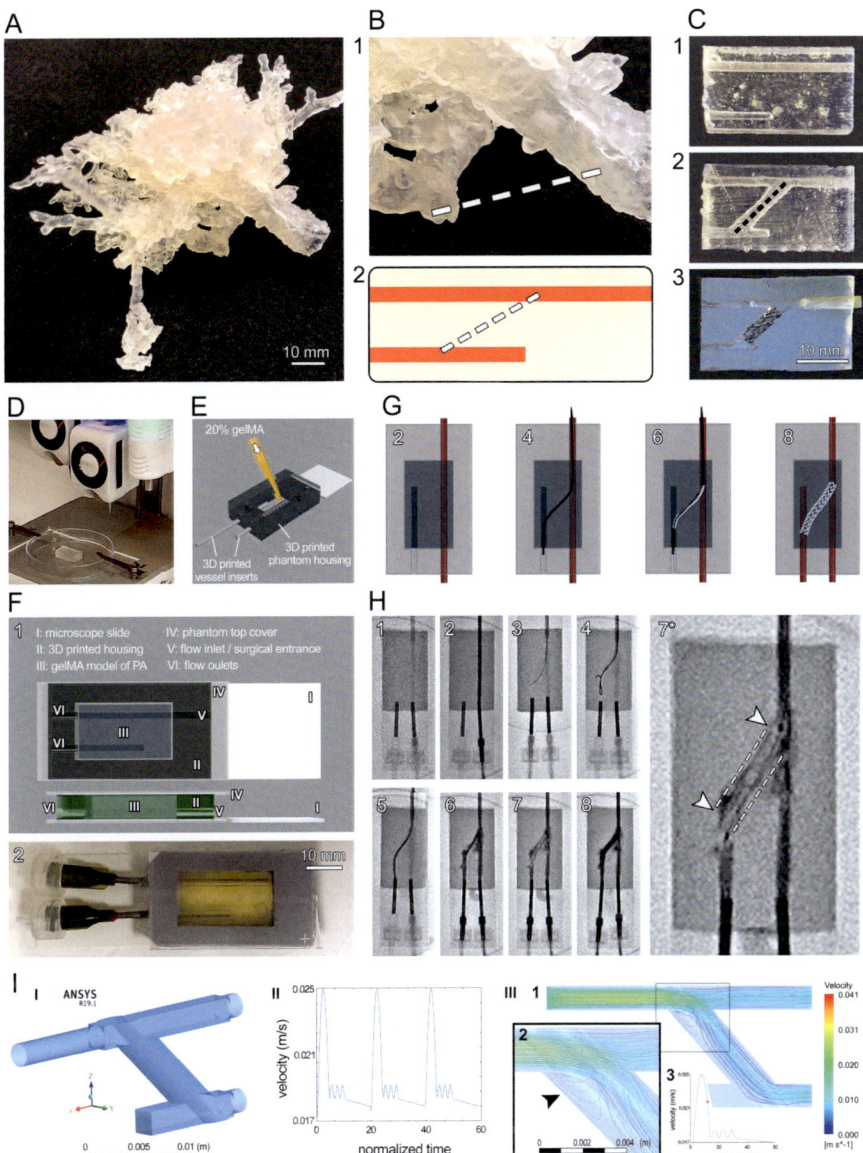

Fig. 3 Patient-specific 3D printed *in vitro* models of pulmonary artery stenosis and its recanalization. (**a**): 3D rotational angiography data of adult patient vasculature with pulmonary artery atresia was used to 3D print a synthetic resin model of the vascular anatomy. (**b**): A simplified phantom was designed by isolating the 3D area that encompassed the occluded artery and a functional patent vessel (1,2). (**c**): Proof-of-concept synthetic models were printed to validate the untreated (1), anastomosed (2), and stented (3) models. (**d–f**): A biological model of simplified pulmonary artery stenosis was subsequently bioprinted using gelatin methacrylate bioink (**d**) and assembled into the 3D printed housing (**e–f**). (**g**): Schematic of the recanalization procedure conducted onto bioprinted samples (**h**): Actual images of catheterization lab procedure to re-establish flow into both vessels. (**i**): Computational fluid dynamics (CFD) modeling of the stented stenotic model. Figure reconstructed from [11]

6   3D Models of Valvular Disease

In addition to the above-mentioned specific disease states, CHD is also complicated by multifactorial valvular disease. While adult valvular disease is often focused on degeneration and calcification, congenital valvular disease is often secondary to malformations resulting in stenosis or regurgitation [66]. Through the lessons taught by the study of adult disease, 3D printing may also prove valuable in congenital valvular lesions. Understanding 3D relationships in valvular disease is critical. The use of 3D printing and bioprinting through patient models allows valvular intervention planning [67, 68]. In addition, modeling and hemodynamic testing may provide clues to mechanisms of obstructive patterns noted in patients following valve replacement. These types of predictions are vital for pediatric patients in which large valves, particularly in the mitral position, can cause left ventricular outflow tract obstruction [69]. Procedural outcomes may be improved and complications decreased with the use of 3D modeling in valvular interventions [70]. Patient-specific modeling is vital for the congenital space as traditional structures and planes are altered by prior surgical intervention or malformation of valves as in atrioventricular canal or truncus arteriosus, amongst others. Having 3D modeling for structural (and congenital) heart disease aids in this understanding and assists in procedural planning [71, 72].

Perhaps one of the seemingly insurmountable challenges in CHD is procedural planning and the accounting for somatic growth. In valvular disease, if unable to be repaired, valve replacement becomes the leading strategy for improvement [73]. Unfortunately, in the congenital patient, options remain limited in patients, leading to short lived palliations. Groups have realized this limitation and 3D bioprinting valves with potential for somatic growth have been a goal. The progress in 3D printing implantable heart valves has led to printing with hydrogels and cellular components required for viability [74, 75]. This understanding also can be applied to CHD valve disease in that cell types and mechanical properties required for functionality remain similar for those of bioprinted variety and native.

7   Pump Failure and Drug Delivery

In addition to disruptions in development, post-surgical and interventional changes, palliative procedures, ventricular failure may result secondary to altered hemodynamics and/or electrical activity, or genetic predisposition. This pump failure can be a result of failure of cardiomyocytes to maintain the ability to contract or can be a result of failure to relax. Most 3D printed therapies have been aimed at post myocardial infarction injury [24, 31, 76–79], which can also be seen in CHD, however these therapies can also be theoretically applied to pump failure in the young heart disease patient. Cardiovascular patches hold promise for regeneration of systolic heart failure in CHD. Groups investigating these myocardial patches have found significant improvement in infarct size after placement of pluripotent stem cell laden patches [80, 81]. The combination of iPSCs introduced into hydrogel bioink allowed for improved regenerative capacity as a vascularized patch [82]. While these patches have not been idealized for CHD use, they hold great promise for small patients with severe pump failure.

In addition to potential restoration of cardiomyocyte function through application of 3D bioprinted patches, bioprinting also may promise delivery of medications in targeted areas [83]. Through the use of 3D bioprinted delivery systems that may be near the surface of the heart, nanoparticles may be designed to release over time, with the goal for maintaining cardiac homeostasis. In addition, 3D bioprinted platforms can be used for drug testing [84], which in the CHD population is needed given small numbers associated with various lesions.

8 Ethical Considerations of 3D Bioprinting and Modeling in Pediatrics

Research in pediatrics is often inherently limited by the population. The use of animal models may not be feasible as the heterogeneity of disease is vast. The pediatric population is also protected as a special group against research that may present harm without generalizable good. Without the appropriate animal models, we therefore are left with few options for testing novel therapeutics. The use of 3D bioprinted models of CHD could allow investigation of altered surgical and transcatheter techniques as well as novel drug therapeutics without putting the pediatric patient at risk.

References

1. Iop, L.: Modeling Cardiac Congenital Diseases: From Mathematic Tools to Human Induced Pluripotent Stem Cells. Conference Papers in Science. (2014); 2014:369246. https://doi.org/10.1155/2014/369246
2. Majumdar, U., Yasuhara, J., Garg, V.: In vivo and in vitro genetic models of congenital heart disease. Cold Spring Harb Perspect Biol. 13(4) (2021). https://doi.org/10.1101/cshperspect.a036764
3. Rufaihah, A.J., Chen, C.K., Yap, C.H., Mattar, C.N.Z.: Mending a broken heart: In vitro, in vivo and in silico models of congenital heart disease. Dis Model Mech. 14(3) (2021). https://doi.org/10.1242/dmm.047522
4. Midgett, M., Thornburg, K., Rugonyi, S.: Blood flow patterns underlie developmental heart defects. Am. J. Physiol. Heart Circ. Physiol. 312(3), H632–HH42 (2017). https://doi.org/10.1152/ajpheart.00641.2016
5. Rugonyi, S.: Genetic and flow anomalies in congenital heart disease. AIMS Genet. 3(3), 157–166 (2016). https://doi.org/10.3934/genet.2016.3.157

6. Woudstra, O.I., Ahuja, S., Bokma, J.P., Bouma, B.J., Mulder, B.J.M., Christoffels, V.M.: Origins and consequences of congenital heart defects affecting the right ventricle. Cardiovasc. Res. **113**(12), 1509–1520 (2017). https://doi.org/10.1093/cvr/cvx155

7. Cetnar, A.D., Tomov, M.L., Ning, L., Jing, B., Theus, A.S., Kumar, A., et al.: Patient-specific 3D bioprinted models of developing human heart. Adv Healthc Mater, e2001169 (2020). https://doi.org/10.1002/adhm.202001169

8. Tomov, M.L., Jing, B., Kumar, A., Do, K., Cetnar, A., Bhamidipati, S.R., et al.: Abstract 405: a personalized, 3D printed <i>in vitro</i> model of vascular anastomosis in single ventricle heart defects. Circ. Res. **127**(Suppl_1), A405-A (2020). https://doi.org/10.1161/res.127.suppl_1.405

9. Serpooshan, V., Tomov, M.L., Kumar, A., Jing, B., Bhamidipati, S.R., Panoskaltsis, N., et al.: Abstract 427: a patient-specific 3D bioprinted platform for <i>in vitro</i> disease modeling and treatment planning in pulmonary vein stenosis. Circ. Res. **127**(Suppl_1), A427-A (2020). https://doi.org/10.1161/res.127.suppl_1.427

10. Jing, B., Tomov, M.L., Wijntjes, A.N., Bhamidipati, S.R., Avazmohammadi, R., Bauser-Heaton, H., et al.: Synthesis of ultrasound-compatible embryonic heart tube phantom using water-soluble 3D printed model for 3D ultrasound flow velocimetry, pp. 1–4. IEEE International Ultrasonics Symposium (IUS)2020 (2020)

11. Tomov, M.L., Cetnar, A., Do, K., Bauser-Heaton, H., Serpooshan, V.: Patient-specific 3-dimensional-bioprinted model for in vitro analysis and treatment planning of pulmonary artery atresia in tetralogy of Fallot and major Aortopulmonary collateral arteries. J. Am. Heart Assoc. **8**(24), e014490 (2019). https://doi.org/10.1161/JAHA.119.014490

12. Yoo, S.-J., Thabit, O., Kim, E.K., Ide, H., Yim, D., Dragulescu, A., et al.: 3D printing in medicine of congenital heart diseases. 3D Print. Med. **2**(1), 3 (2016). https://doi.org/10.1186/s41205-016-0004-x

13. Yoo, S.J., Hussein, N., Peel, B., Coles, J., van Arsdell, G.S., Honjo, O., et al.: 3D Modeling and printing in congenital heart surgery: entering the stage of maturation. Front. Pediatr. **9**, 621672 (2021). https://doi.org/10.3389/fped.2021.621672

14. Ruedinger, K.L., Zhou, H., Trampe, B., Heiser, T., Srinivasan, S., Iruretagoyena, J.I., et al.: Modeling Fetal cardiac anomalies from prenatal echocardiography with 3-dimensional printing and 4-dimensional flow magnetic resonance imaging. Circ. Cardiovasc. Imaging. **11**(9), e007705 (2018). https://doi.org/10.1161/CIRCIMAGING.118.007705

15. Chaturvedi, R.R., Herron, T., Simmons, R., Shore, D., Kumar, P., Sethia, B., et al.: Passive stiffness of myocardium from congenital heart disease and implications for diastole. Circulation. **121**(8), 979–988 (2010). https://doi.org/10.1161/CIRCULATIONAHA.109.850677

16. Hacker, A.L., Reiner, B., Oberhoffer, R., Hager, A., Ewert, P., Muller, J.: Increased arterial stiffness in children with congenital heart disease. Eur. J. Prev. Cardiol. **25**(1), 103–109 (2018). https://doi.org/10.1177/2047487317737174

17. Clark, S.J., Yoxall, C.W., Subhedar, N.V.: Measurement of right ventricular volume in healthy term and preterm neonates. Arch Dis Child Fetal Neonatal Ed. **87**(2), F89–F93.; discussion F-4 (2002). https://doi.org/10.1136/fn.87.2.f89

18. Mecham, R.P.: Elastin in lung development and disease pathogenesis. Matrix Biol. **73**, 6–20 (2018). https://doi.org/10.1016/j.matbio.2018.01.005

19. Elshazly, M.B., Hoosien, M.: Chapter 13 - the future of 3D printing in cardiovascular disease. In: Al'Aref, S.J., Mosadegh, B., Dunham, S., Min, J.K. (eds.) 3D Printing Applications in Cardiovascular Medicine, pp. 243–253. Academic, Boston (2018)

20. El Sabbagh, A., Eleid, M.F., Al-Hijji, M., Anavekar, N.S., Holmes, D.R., Nkomo, V.T., et al.: The various applications of 3D printing in cardiovascular diseases. Curr. Cardiol. Rep. **20**(6), 47 (2018). https://doi.org/10.1007/s11886-018-0992-9

21. Vukicevic, M., Mosadegh, B., Min, J.K., Little, S.H.: Cardiac 3D printing and its future directions. JACC Cardiovasc. Imaging. **10**(2), 171–184 (2017). https://doi.org/10.1016/j.jcmg.2016.12.001

22. Serpooshan, V., Hu, J.B., Chirikian, O., Hu, D.A., Mahmoudi, M., Wu, S.M.: Chapter 8 - 4D printing of actuating cardiac tissue. In: Al'Aref, S.J., Mosadegh, B., Dunham, S., Min, J.K. (eds.) 3D Printing Applications in Cardiovascular Medicine, pp. 153–162. Academic, Boston (2018)

23. Hu, J.B., Tomov, M.L., Buikema, J.W., Chen, C., Mahmoudi, M., Wu, S.M., et al.: Cardiovascular tissue bioprinting: physical and chemical processes. Appl. Phys. Rev. 5(4), 041106 (2018). https://doi.org/10.1063/1.5048807

24. Serpooshan, V., Mahmoudi, M., Hu, D.A., Hu, J.B., Wu, S.M.: Bioengineering cardiac constructs using 3D printing. J. 3D Print. Med. 1(2) (2017)

25. Lau, I.W.W., Sun, Z.: Dimensional accuracy and clinical value of 3D printed models in congenital heart disease: a systematic review and meta-analysis. J Clin Med. 8(9) (2019). https://doi.org/10.3390/jcm8091483

26. Sun, Z.: Clinical applications of patient-specific 3D printed models in cardiovascular disease: current status and future directions. Biomolecules. 10(11) (2020). https://doi.org/10.3390/biom10111577

27. Anwar, S., Rockefeller, T., Raptis, D.A., Woodard, P.K., Eghtesady, P.: 3D printing provides a precise approach in the treatment of tetralogy of Fallot, pulmonary atresia with major Aortopulmonary collateral arteries. Curr. Treat. Options Cardiovasc. Med. 20(1), 5 (2018). https://doi.org/10.1007/s11936-018-0594-2

28. Grab, M., Hopfner, C., Gesenhues, A., Konig, F., Haas, N.A., Hagl, C., et al.: Development and evaluation of 3D-printed cardiovascular phantoms for interventional planning and training. J. Vis. Exp. 167 (2021). https://doi.org/10.3791/62063

29. Anwar, S., Singh, G.K., Miller, J., Sharma, M., Manning, P., Billadello, J.J., et al.: 3D printing is a transformative Technology in Congenital Heart Disease. JACC Basic Transl. Sci. 3(2), 294–312 (2018). https://doi.org/10.1016/j.jacbts.2017.10.003

30. White, S.C., Sedler, J., Jones, T.W., Seckeler, M.: Utility of three-dimensional models in resident education on simple and complex intracardiac congenital heart defects. Congenit. Heart Dis. 13(6), 1045–1049 (2018). https://doi.org/10.1111/chd.12673

31. Tomov, M.L., Gil, C.J., Cetnar, A., Theus, A.S., Lima, B.J., Nish, J.E., et al.: Engineering functional cardiac tissues for regenerative medicine applications. Curr. Cardiol. Rep. 21(9), 105 (2019). https://doi.org/10.1007/s11886-019-1178-9

32. DeLaughter, D.M., Saint-Jean, L., Baldwin, H.S., Barnett, J.V.: What chick and mouse models have taught us about the role of the endocardium in congenital heart disease. Birth Defects Res. A Clin. Mol. Teratol. 91(6), 511–525 (2011). https://doi.org/10.1002/bdra.20809

33. Wittig, J.G., Munsterberg, A.: The early stages of heart development: insights from chicken embryos. J Cardiovasc Dev Dis. 3(2) (2016). https://doi.org/10.3390/jcdd3020012

34. Farr 3rd, G.H., Imani, K., Pouv, D., Maves, L.: Functional testing of a human PBX3 variant in zebrafish reveals a potential modifier role in congenital heart defects. Dis Model Mech. 11(10) (2018). https://doi.org/10.1242/dmm.035972

35. Marx, G.R., Fyler, D.C.: Chapter 38 - endocardial cushion defects. In: Keane, J.F., Lock, J.E., Fyler, D.C. (eds.) Nadas' Pediatric Cardiology, 2nd edn, pp. 663–674. W.B. Saunders, Philadelphia (2006)

36. Sandrini, C., Rossetti, L., Zambelli, V., Zanarotti, R., Bettinazzi, F., Solda, R., et al.: Accuracy of micro-computed tomography in post-mortem evaluation of Fetal congenital heart disease. Comparison between post-mortem micro-CT and conventional autopsy. Front. Pediatr. 7, 92 (2019). https://doi.org/10.3389/fped.2019.00092

37. Yang, C., Xu, Y., Yu, M., Lee, D., Alharti, S., Hellen, N., et al.: Induced pluripotent stem cell modelling of HLHS underlines the contribution of dysfunctional NOTCH signalling to impaired cardiogenesis. Hum. Mol. Genet. 26(16), 3031–3045 (2017). https://doi.org/10.1093/hmg/ddx140

38. Mital, S.: Human pluripotent stem cells to model congenital heart disease. In: Nakanishi, T., Markwald, R.R., Baldwin, H.S., Keller, B.B., Srivastava, D., Yamagishi, H. (eds.) Etiology and

Morphogenesis of Congenital Heart Disease: From Gene Function and Cellular Interaction to Morphology, pp. 321–327, Tokyo (2016)

39. Sharma, A.: Modeling congenital heart disease using pluripotent stem cells. Curr. Cardiol. Rep. **22**(8), 55 (2020). https://doi.org/10.1007/s11886-020-01316-y

40. Kim, M.S., Fleres, B., Lovett, J., Anfinson, M., Samudrala, S.S.K., Kelly, L.J., et al.: Contractility of induced pluripotent stem cell-cardiomyocytes with an MYH6 head domain variant associated with Hypoplastic left heart syndrome. Front. Cell Dev. Biol. **8**, 440 (2020). https://doi.org/10.3389/fcell.2020.00440

41. Hove, J.R., Koster, R.W., Forouhar, A.S., Acevedo-Bolton, G., Fraser, S.E., Gharib, M.: Intracardiac fluid forces are an essential epigenetic factor for embryonic cardiogenesis. Nature. **421**(6919), 172–177 (2003). https://doi.org/10.1038/nature01282

42. Tobita, K., Keller, B.B.: Right and left ventricular wall deformation patterns in normal and left heart hypoplasia chick embryos. Am. J. Physiol. Heart Circ. Physiol. **279**(3), H959–H969 (2000). https://doi.org/10.1152/ajpheart.2000.279.3.H959

43. Groenendijk, B.C., Hierck, B.P., Gittenberger-De Groot, A.C., Poelmann, R.E.: Development-related changes in the expression of shear stress responsive genes KLF-2, ET-1, and NOS-3 in the developing cardiovascular system of chicken embryos. Dev. Dyn. **230**(1), 57–68 (2004). https://doi.org/10.1002/dvdy.20029

44. Groenendijk, B.C., Hierck, B.P., Vrolijk, J., Baiker, M., Pourquie, M.J., Gittenberger-de Groot, A.C., et al.: Changes in shear stress-related gene expression after experimentally altered venous return in the chicken embryo. Circ. Res. **96**(12), 1291–1298 (2005). https://doi.org/10.1161/01.RES.0000171901.40952.0d

45. Groenendijk, B.C., Van der Heiden, K., Hierck, B.P., Poelmann, R.E.: The role of shear stress on ET-1, KLF2, and NOS-3 expression in the developing cardiovascular system of chicken embryos in a venous ligation model. Physiology (Bethesda). **22**, 380–389 (2007). https://doi.org/10.1152/physiol.00023.2007

46. Van der Heiden, K., Groenendijk, B.C., Hierck, B.P., Hogers, B., Koerten, H.K., Mommaas, A.M., et al.: Monocilia on chicken embryonic endocardium in low shear stress areas. Dev. Dyn. **235**(1), 19–28 (2006). https://doi.org/10.1002/dvdy.20557

47. Feinstein, J.A., Benson, D.W., Dubin, A.M., Cohen, M.S., Maxey, D.M., Mahle, W.T., et al.: Hypoplastic left heart syndrome: current considerations and expectations. J. Am. Coll. Cardiol. **59**(1 Suppl), S1–S42 (2012). https://doi.org/10.1016/j.jacc.2011.09.022

48. Gladki, M.M., Skladzien, T., Skalski, J.H.: The impact of environmental factors on the occurrence of congenital heart disease in the form of hypoplastic left heart syndrome. Kardiochir Torakochirurgia Pol. **12**(3), 204–207 (2015). https://doi.org/10.5114/kitp.2015.54455

49. Liu, X., Yagi, H., Saeed, S., Bais, A.S., Gabriel, G.C., Chen, Z., et al.: The complex genetics of hypoplastic left heart syndrome. Nat. Genet. **49**(7), 1152–1159 (2017). https://doi.org/10.1038/ng.3870

50. Laraja, K., Sadhwani, A., Tworetzky, W., Marshall, A.C., Gauvreau, K., Freud, L., et al.: Neurodevelopmental outcome in children after Fetal cardiac intervention for aortic stenosis with evolving Hypoplastic left heart syndrome. J. Pediatr. **184**, 130–6 e4 (2017). https://doi.org/10.1016/j.jpeds.2017.01.034

51. McElhinney, D.B., Tworetzky, W., Lock, J.E.: Current status of fetal cardiac intervention. Circulation. **121**(10), 1256–1263 (2010). https://doi.org/10.1161/CIRCULATIONAHA.109.870246

52. Moon-Grady, A.J., Morris, S.A., Belfort, M., Chmait, R., Dangel, J., Devlieger, R., et al.: International Fetal cardiac intervention registry: a worldwide collaborative description and preliminary outcomes. J. Am. Coll. Cardiol. **66**(4), 388–399 (2015). https://doi.org/10.1016/j.jacc.2015.05.037

53. Ohye, R.G., Schranz, D., D'Udekem, Y.: Current therapy for Hypoplastic left heart syndrome and related single ventricle lesions. Circulation. **134**(17), 1265–1279 (2016). https://doi.org/10.1161/CIRCULATIONAHA.116.022816

54. Huang, J., Shi, H., Chen, Q., Hu, J., Zhang, Y., Song, H., et al.: Three-dimensional printed model fabrication and effectiveness evaluation in Fetuses with congenital heart disease or with a Normal heart. J. Ultrasound Med. **40**(1), 15–28 (2021). https://doi.org/10.1002/jum.15366
55. Ning, L., Mehta, R., Cao, C., Theus, A., Tomov, M., Zhu, N., et al.: Embedded 3D bioprinting of Gelatin Methacryloyl-based constructs with highly Tunable structural Fidelity. ACS Appl. Mater. Interfaces. **12**(40), 44563–44577 (2020). https://doi.org/10.1021/acsami.0c15078
56. Trivedi, K.R., Benson, L.N.: Interventional strategies in the management of peripheral pulmonary artery stenosis. J. Interv. Cardiol. **16**(2), 171–188 (2003). https://doi.org/10.1046/j.1540-8183.2003.08031.x
57. Patel, A.B., Ratnayaka, K., Bergersen, L.: A review: percutaneous pulmonary artery stenosis therapy: state-of-the-art and look to the future. Cardiol. Young. **29**(2), 93–99 (2019). https://doi.org/10.1017/S1047951118001087
58. Witowski, J., Darocha, S., Kownacki, L., Pietrasik, A., Pietura, R., Banaszkiewicz, M., et al.: Augmented reality and three-dimensional printing in percutaneous interventions on pulmonary arteries. Quant Imaging Med Surg. **9**(1), 23–29 (2019). https://doi.org/10.21037/qims.2018.09.08
59. Cunningham, K.S., Gotlieb, A.I.: The role of shear stress in the pathogenesis of atherosclerosis. Lab. Investig. **85**(1), 9–23 (2005). https://doi.org/10.1038/labinvest.3700215
60. Chiu, J.J., Chien, S.: Effects of disturbed flow on vascular endothelium: pathophysiological basis and clinical perspectives. Physiol. Rev. **91**(1), 327–387 (2011). https://doi.org/10.1152/physrev.00047.2009
61. Meot, M., Lefort, B., El Arid, J.M., Soule, N., Lothion-Boulanger, J., Lengelle, F., et al.: Intraoperative stenting of pulmonary artery stenosis in children with congenital heart disease. Ann. Thorac. Surg. **104**(1), 190–196 (2017). https://doi.org/10.1016/j.athoracsur.2016.12.012
62. Piera-Velazquez, S., Jimenez, S.A.: Endothelial to mesenchymal transition: role in physiology and in the pathogenesis of Human diseases. Physiol. Rev. **99**(2), 1281–1324 (2019). https://doi.org/10.1152/physrev.00021.2018
63. Morii, C., Tanaka, H.Y., Izushi, Y., Nakao, N., Yamamoto, M., Matsubara, H., et al.: 3D in vitro model of vascular medial thickening in pulmonary arterial hypertension. Front. Bioeng. Biotechnol. **8**, 482 (2020). https://doi.org/10.3389/fbioe.2020.00482
64. Jin, Q., Bhatta, A., Pagaduan, J.V., Chen, X., West-Foyle, H., Liu, J., et al.: Biomimetic human small muscular pulmonary arteries. Sci Adv. **6**(13), eaaz2598 (2020). https://doi.org/10.1126/sciadv.aaz2598
65. Liu, J., Yuan, H., Zhang, N., Chen, X., Zhou, C., Huang, M., et al.: 3D simulation analysis of central shunt in patient-specific Hemodynamics: effects of varying degree of pulmonary artery stenosis and shunt diameters. Comput. Math. Methods Med. **2020**, 4720908 (2020). https://doi.org/10.1155/2020/4720908
66. Singhi, A.K., Kumar, R.K.: Evaluation of congenital Valvular heart diseases by the Pediatrician: when to follow, when to refer for intervention? Indian J. Pediatr. **82**(11), 1021–1026 (2015). https://doi.org/10.1007/s12098-015-1870-8
67. Ooms, J., Minet, M., Daemen, J., Van Mieghem, N.: Pre-procedural planning of transcatheter mitral valve replacement in mitral stenosis with multi-detector tomography-derived 3D modeling and printing: a case report. Eur. Heart J Case Rep. **4**(3), 1–6 (2020). https://doi.org/10.1093/ehjcr/ytaa098
68. Ooms, J.F., Wang, D.D., Rajani, R., Redwood, S., Little, S.H., Chuang, M.L., et al.: Computed tomography-derived 3D Modeling to guide sizing and planning of Transcatheter mitral valve interventions. JACC Cardiovasc. Imaging. **14** (2021). https://doi.org/10.1016/j.jcmg.2020.12.034
69. Wang, H., Song, H., Yang, Y., Wu, Z., Hu, R., Chen, J., et al.: Morphology display and hemodynamic testing using 3D printing may aid in the prediction of LVOT obstruction after mitral valve replacement. Int. J. Cardiol. **331**, 296–306 (2021). https://doi.org/10.1016/j.ijcard.2021.01.029

70. Garg, R., Zahn, E.M.: Utility of three-dimensional (3D) Modeling for planning structural heart interventions (with an emphasis on Valvular heart disease). Curr. Cardiol. Rep. **22**(10), 125 (2020). https://doi.org/10.1007/s11886-020-01354-6

71. Vukicevic, M., Faza, N.N., Avenatti, E., Durai, P.C., El-Tallawi, K.C., Filippini, S., et al.: Patient-specific 3-dimensional printed Modeling of the tricuspid valve for MitraClip procedure planning. Circ. Cardiovasc. Imaging. **13**(7), e010376 (2020). https://doi.org/10.1161/CIRCIMAGING.119.010376

72. Vukicevic, M., Filippini, S., Little, S.H.: Patient-specific modeling for structural heart intervention: role of 3D printing today and tomorrow (CME). Methodist Debakey Cardiovasc J. **16**(2), 130–137 (2020). https://doi.org/10.14797/mdcj-16-2-130

73. Maganti, K., Rigolin, V.H., Sarano, M.E., Bonow, R.O.: Valvular heart disease: diagnosis and management. Mayo Clin. Proc. **85**(5), 483–500 (2010). https://doi.org/10.4065/mcp.2009.0706

74. Cheung, D.Y., Duan, B., Butcher, J.T.: Current progress in tissue engineering of heart valves: multiscale problems, multiscale solutions. Expert. Opin. Biol. Ther. **15**(8), 1155–1172 (2015). https://doi.org/10.1517/14712598.2015.1051527

75. Kang, L.H., Armstrong, P.A., Lee, L.J., Duan, B., Kang, K.H., Butcher, J.T.: Optimizing photo-encapsulation viability of heart valve cell types in 3D printable composite hydrogels. Ann. Biomed. Eng. **45**(2), 360–377 (2017). https://doi.org/10.1007/s10439-016-1619-1

76. Tomov, M.L., Theus, A., Sarasani, R., Chen, H., Serpooshan, V.: 3D bioprinting of cardiovascular tissue constructs: cardiac bioinks. In: Serpooshan, V., Wu, S.M. (eds.) Cardiovascular Regenerative Medicine: Tissue Engineering and Clinical Applications, pp. 63–77. Springer International Publishing, Cham (2019)

77. Cetnar, A., Tomov, M., Theus, A., Lima, B., Vaidya, A., Serpooshan, V.: 3D bioprinting in clinical cardiovascular medicine. In: Guvendiren, M. (ed.) 3D Bioprinting in Medicine: Technologies, Bioinks, and Applications, pp. 149–162. Springer International Publishing, Cham (2019)

78. Tomov, M.L., Theus, A., Cetnar, A., Bauser-Heaton, H., Serpooshan, V.: Vascularized Multi-Tissue Platform – 3D Bioprinting an Organ Model in Vitro. 2019 AIChE Annual Meeting. Orlando, Florida2019

79. Hu, J.B., Hu, D.A., Buikema, J.W., Chirikian, O., Venkatraman, S., Serpooshan, V., et al.: Bioengineering of vascular myocardial tissue; a 3D bioprinting approach. Tissue Eng. Pt A. **23**, S158–S1S9 (2017)

80. Roche, C.D., Sharma, P., Ashton, A.W., Jackson, C., Xue, M., Gentile, C.: Printability, durability, contractility and vascular network formation in 3D bioprinted cardiac endothelial cells using alginate-Gelatin hydrogels. Front. Bioeng. Biotechnol. **9**, 636257 (2021). https://doi.org/10.3389/fbioe.2021.636257

81. Gao, L., Kupfer, M.E., Jung, J.P., Yang, L., Zhang, P., Da Sie, Y., et al.: Myocardial tissue engineering with cells derived from Human-induced pluripotent stem cells and a native-like, high-resolution, 3-dimensionally printed scaffold. Circ. Res. **120**(8), 1318–1325 (2017). https://doi.org/10.1161/CIRCRESAHA.116.310277

82. Noor, N., Shapira, A., Edri, R., Gal, I., Wertheim, L., Dvir, T.: 3D printing of personalized thick and Perfusable cardiac patches and hearts. Adv Sci (Weinh). **6**(11), 1900344 (2019). https://doi.org/10.1002/advs.201900344

83. Gardin, C., Ferroni, L., Latremouille, C., Chachques, J.C., Mitrecic, D., Zavan, B.: Recent applications of three dimensional printing in cardiovascular medicine. Cells. **9**(3) (2020). https://doi.org/10.3390/cells9030742

84. Peng, W., Datta, P., Ayan, B., Ozbolat, V., Sosnoski, D., Ozbolat, I.T.: 3D bioprinting for drug discovery and development in pharmaceutics. Acta Biomater. **57**, 26–46 (2017). https://doi.org/10.1016/j.actbio.2017.05.025

Biomaterial Interface in Cardiac Cell and Tissue Engineering

Chenyan Wang and Zhen Ma

Abbreviations

2D	Two dimensional
3D	Three dimensional
4D	Four dimensional
ADSCs	Adipose-derived stem cells
ANG-1	Angiopoietin-1
BA	Butyl acrylate
bFGFs	Basic fibroblast growth factors
CAVD	Calcific aortic valve disease
CDCs	Cardiosphere-derived cells
CEST	Chemical exchange saturation transfer
CHP	Collagen hybridizing peptide
CMs	Cardiomyocytes
CNTs	Carbon nanotubes
CPCs	Cardiac progenitor cells
DOX	Doxorubicin
ECM	Extracellular matrix
ECs	Endothelial cells
EHTs	Engineered heart tissues
EPCs	Endothelial progenitor cells

C. Wang · Z. Ma (✉)
Department of Biomedical and Chemical Engineering, Syracuse University,
Syracuse, NY, USA

BioInspired Syracuse: Institute for Material and Living Systems, Syracuse University,
Syracuse, NY, USA
e-mail: zma112@syr.edu

© The Author(s), under exclusive license to Springer Nature
Switzerland AG 2022
J. Zhang, V. Serpooshan (eds.), *Advanced Technologies in Cardiovascular
Bioengineering*, https://doi.org/10.1007/978-3-030-86140-7_12

ESCs	Embryonic stem cells
hESC-CMs	Embryonic stem cell derived cardiomyocytes
hESC-MSCs	Embryonic stem cell derived mesenchymal stem cells
HGF	Hepatocyte growth factor
hiPSC-CMs	Human induced pluripotent stem cell derived cardiomyocytes
hiPSCs	Human induced pluripotent stem cells
IGF-1	Insulin-like growth factor-1
iPSCs	Induced pluripotent stem cells
LVEF	Left ventricular ejection fraction
MCs	Mural cells
MI	Myocardial infarction
MMPs	Matrix metalloproteinases
MRI	Magnetic resonance imaging
MSCs	Mesenchymal stem cells
NPs	Nanoparticles
NRVMs	Neonatal rat ventricular myocytes
PAN	Polyacrylonitrile
PCL	Polycaprolactone
PDMS	Polydimethylsiloxane
PEG	Poly(ethylene glycol)
PEGDA	Poly (ethylene glycol) diacrylate
PF	Polyethylene glycol-fibrinogen
PGS	Poly (glycerol sebacate)
PIPAAm	Poly (N-isopropylacrylamide)
PLGA	Poly (D,L-lactide-co-glycolide)
PTAA	Poly (thiophene-3-acetic acid)
PU	Polyurethane
PVAX	Peroxalate containing vanillyl alcohol
RGD	Arginylglycylaspartic acid
RLP	Resilin-like polypeptide
ROCK	Rho-associated protein kinase
ROS	Reactive oxygen species
SMCs	Smooth muscle cells
SMPs	Shape memory polymers
SWCNTs	Single-wall carbon nanotubes
tBA	*tert*-butyl acrylate
TNF	Tumor necrosis factor
tPA	Tissue plasminogen activator
UPy	Ureido-pyrimid-inone
VEGFs	Vascular endothelial growth factors
YAP	Yes-associated protein
α-CD	α-cyclodextrin

1   Introduction

1.1   *Engineering Cardiac Cell and Tissue Microenvironment*

Cardiovascular diseases are the leading cause of death and continue to have a high prevalence worldwide [1]. The main obstacle of disease treatments is the limited capacity of tissue regeneration [2]. Cardiomyocytes (CMs), which are the major functional component of myocardium, are considered as terminal differentiated cells, thus not able to proliferate to compensate the loss of CMs during disease progression [3]. Instead, the formation of fibrotic tissues at the infarction areas induces higher workload to adjacent cardiomyocytes and interrupts the electrical transmission throughout the heart, leading to arrhythmia [4]. Current clinical interventions focus on the prevention of further pathological cardiac remodeling, especially for patients with end-stage heart diseases [5–7]. Heart transplantation is still the only available way to fully restore the normal heart contractility for severe cardiac dysfunction, but suffers from the limit of appropriate donors [8]. Therefore, there is an urgent need to develop more practical strategies for cardiac disease treatment, relying on the advancement of new technologies of cell and tissue engineering for cardiac repair and regeneration [9].

The field of cardiac cell and tissue engineering has emerged and developed rapidly recent years to establish biomimetic *in vitro* models for comprehensive recognition of heart pathophysiology [10–12]. Comparing to the conventional 2D *in vitro* models, 3D tissue models provide a more physiological relevant microenvironment and may give more realistic evaluation of cardiac behaviors [13, 14]. Moreover, external stimulations are complimentary components for 3D cardiac models to generate CMs with more mature phenotypes. For example, both mechanical and electrical stimulations were found to improve the organization and contractility of immature cardiac cells, which were shown as more aligned sarcomere structure and increased contractile force [11, 15, 16]. For the purposes of disease modeling, previous studies have proved that the alteration of mechanical environments had a significant effect on the manifestation of cardiac dysfunctions using the CMs with disease-causing genetic mutations [17, 18]. For the purposes of drug screening, *in vitro* cardiac models could recapitulate the physiological responses to drug administrations, which will give an opportunity for evaluating the toxicity and efficacy of different drugs to cardiac tissues [19, 20]. One study used an *in vitro* 3D cardiac tissue construct for testing the safety for many antibiotic, antidiabetic and anticancer drugs. The drug-induced cardiotoxicity showed good concordance with clinical observations [21].

Aiming to achieve *in vivo* cardiac tissue repair and regeneration, appropriate environmental signals were needed for designing therapeutic implantable products based on the principles of cardiac cell and tissue engineering [22]. A variety of biomaterial and fabrication techniques have been used to tune the biochemical, mechanical, and electrical properties, in order to modulate the cardiac behaviors, promote the transplanted cell infiltration, activate residual stem cells, and enhance regenerative capabilities. Particularly, stem cells are capable of self-renewal and differentiating into cardiac cells to facilitate the heart regeneration [23]. However, direct cell transplantation has minimal therapeutic outcomes for cardiac diseases, because of poor cell homing, low cell viability at the damaged area, and the limited control of cell fate after transplantation [24]. In cardiac engineering, stem cells can be encapsulated and concentrated in the biocompatible scaffolds, which protect stem cells from complex, dynamic *in vivo* environment and tune their functions for better regenerative outcomes. In one study, injectable hydrogels and patches made of alginate, collagen or chitosan all showed better retention of mesenchymal stem cells (MSCs) than saline controls when they were delivered to the infarct border zone of rat hearts [25].

1.2 Cell-Biomaterial Interface for Cardiac Tissue Engineering

Biomaterials are commonly used to support cell growth and create specific micro-environments to stimulate the structural and functional transformation of cell phenotypes [26]. They are generally categorized as naturally derived and synthetic materials. Natural biomaterials, such as collagen, fibrin and gelatin, promote cell attachment and proliferation [27–29]. However, they are not chemically defined, and difficult to be modified for specific cell-material interface. Synthetic biomaterials, such as polylactic acid, polyglycolic acid and polyurethane, on the other hand, have pre-defined chemical composition and tunable properties, but require further modifications to enable efficient cell attachment [30–32]. The selection of appropriate biomaterials for either *in vitro* or *in vivo* applications, in addition to general biocompatibility, different criteria should be considered. For *in vitro* tissue engineering, biomaterials should allow high tunability on biochemical composition, mechanical properties, conductivity and other relevant functionalities to promote cardiac differentiation and maturation. Besides, disease modeling and drug screening platforms should enable various characterization techniques to evaluate the cardiac phenotypes and functional responses consistently. For *in vivo* cardiac therapy, the main goals are to support the weak and dysfunctional myocardium, inhibit the cell apoptosis, reduce the fibrosis, limit the immunogenicity, and provide a suitable microenvironment for recruiting residual stem cells for tissue regeneration. To achieve these aims, biomaterials need to have sufficient mechanical strength and compliance, degradation ability and active integration with native cardiac tissue.

Integration of biomaterial engineering with other advanced technologies, such as stem cell technology, gene editing and bioreactor designs, offers us great potential to develop more complex systems to meet the need for accurate recapitulation of heart pathophysiology. First, various types of stem cells, including MSCs, embryonic stem cells (ESCs) and induced pluripotent stem cells (iPSCs), showed efficient differentiation into myocardial lineages [33–35]. Human iPSCs are extensively used for cardiac tissue engineering, since it could be reprogrammed from human somatic cells to generate patient-specific iPSCs lines for disease modeling, which lays the foundation of personalized medicine [36]. Meanwhile, autologous human iPSCs is less likely to trigger the immune rejection for cell transplantation studies. Second, CRISPR/Cas9-based genome engineering provides the capability to create specific genetic mutation with isogenic controls with a precise control of the mutation dose [37]. Combination of human iPSCs and CRISPR/Cas9 technologies, diseased cardiac tissue models were developed by growing genome-engineered mutant iPSC-derived cardiomyocytes (iPSC-CMs) on biomaterial scaffolds to investigate the interactive roles of genetic abnormality and environmental factors in disease progression [17, 18]. Last, bioreactors have been used widely to facilitate the nutrient and waste exchange for 3D cardiac tissues under dynamic culture systems, like perfusion or vortex flow [38, 39]. Additionally, bioreactors can integrate mechanical load, cyclic stretch and electrical stimulations, which are important for the construction of functional and mature cardiac tissues [40].

In this chapter, we will discuss the *in vitro* and *in vivo* applications of available types of biomaterials that have been used in cardiac cell and tissue engineering (Fig. 1). Each type of materials has their own unique chemical, mechanical

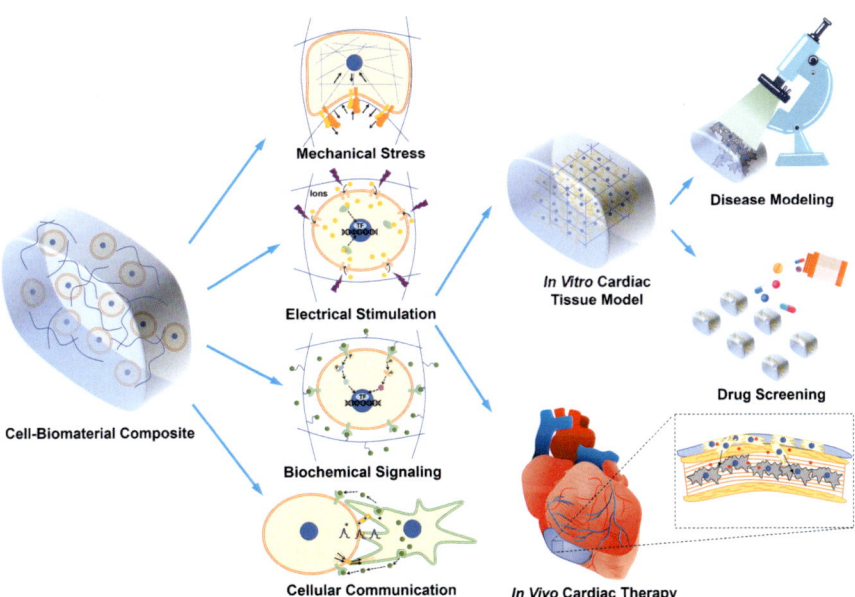

Fig. 1 Applications of biomaterials in cardiac cell and tissue engineering
Myocardial cells modulate their behaviors responding to various environmental stimuli from biomaterials and cellular communications. Engineered biomaterial-cell systems can be used to generate functional cardiac tissue models for disease modeling and drug testing applications. Cell-encapsulated biomaterials could integrate with native damaged cardiac tissues for heart repair and regeneration

properties and geometric structures that have significant impact on different cell types. Bulk hydrogels have outstanding biocompatibility and offer a 3D microenvironment to encapsulated cells. Porous scaffolds allow for cell infiltration and media transport, while fibrous scaffolds promote the biomimetic anisotropic structure of CMs. New stimuli-responsive materials can add temporal control of cell-biomaterial interactions. Furthermore, fabrication techniques are utilized to broaden the properties of biomaterials to guide cell behaviors and create complex biological systems. Accurate speculation of cell and tissue responses to specific microenvironments is essential to understand the biological mechanisms of cardiac disease initiation and progression, and eventually to cure the heart diseases.

2 Bulk Hydrogels

Hydrogels are 3D polymer networks synthesized by physical or chemical crosslinking of hydrophilic polymers [41]. Due to the high hydrophilicity, hydrogels are able to absorb a high volume of water while maintaining their shape, showing similarity with soft tissues. The high-water content of hydrogels makes them favorable for cell adhesion and the porous structure allow for efficient diffusion of nutrients and oxygen [42]. Other outstanding properties, including biocompatibility, mechanical compliance and biodegradability also make hydrogels to be extensively used in tissue engineering applications. Furthermore, the bulk properties of hydrogels could be tuned easily with the control of crosslinking process, physical integration and chemical conjugation [43]. These variations of hydrogel properties can lead to different cell-material crosstalk, which offers an opportunity for directing behaviors of myocardial cells and tissues.

2.1 Hydrogel-Based In Vitro 3D Tissue Constructs

Lots of natural and synthetic hydrogels are utilized as scaffolds for *in vitro* cardiac models and they are compatible with modifications to create more physiological relevant environment. Nowadays, hybrid hydrogels emerged as more advanced culture platform due to their optimized properties for enhancing cardiac maturation and contractility. In this section, we summarize the *in vitro* applications of natural, synthetic hydrogels and discuss advanced properties of hybrid hydrogels (Fig. 2a, b).

Natural Hydrogels Natural hydrogels derived from animal proteins or polysaccharides are the most abundant hydrogels used for *in vitro* cardiac tissue models [42] . They have *in vivo* like chemical compositions and provide fast turnover rate through the active interactions with cell components. However, natural hydrogels usually have poor mechanical performance and uncontrollable biological degradation. These drawbacks limited the application of natural hydrogels for a long-term

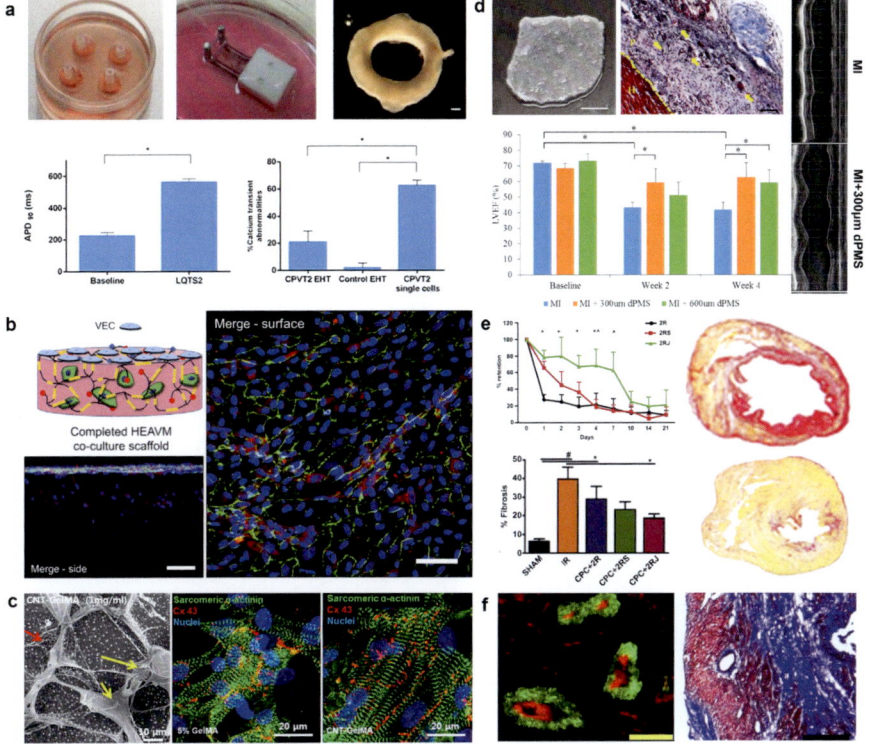

Fig. 2 Hydrogels for *in vitro* and *in vivo* applications
(**a**) Patient-derived EHT were formed based on chitosan-enhanced ECM hydrogels, and recapitulated abnormal action potential and calcium handling (Reprinted from [44] with permission from Elsevier). (**b**) Coculture of valve interstitial cells and valve endothelial cells in RGD-modified PEG hydrogels supports their homeostatic functions (Reprinted from [45] with permission from Elsevier). (**c**) Acellular porcine myocardial slice was used as a cardiac patch to preserve the cardiac functions for MI rats (Reprinted from [46] with permission from American Chemical Society). (**d**) CPCs encapsulated in self-assembly peptide hydrogels showed improved *in vivo* retention and alleviated the pathological fibrosis in MI rats (Reprinted from [47] with permission from Elsevier)

maintenance of contracting cardiac tissues and precise characterization of cell-deposited ECM.

Decellularized ECM, especially cardiac ECM, could support myocardial cells in the most biomimetic way, because they are directly extracted from native cardiac tissues. Das et al., fabricated engineered heart tissues (EHTs) by 3D cell printing using heart tissue derived ECM as bioink [48]. Primary neonatal rat cardiomyocytes in the ECM-based bioink showed more aligned, rod-like arrangement of cardiac regulatory proteins and better Z-disk integrity than cells in the single ECM protein-based scaffolds. Collagen forms the major component of cardiac ECM, and therefore is frequently used to fabricate cardiac tissue constructs. In one study, hESC-CMs and niche cells were mixed within a collagen/Matrigel matrix to fabricate 3D cardiac strips [49]. When uniaxial stretching was applied, hESC-MSCs showed various hallmarks of improved maturation with improvements on active contractile force, passive tension and myofibril alignment. Fibrin gels have been used in developing condensed cardiac tissues, because they have a high cell seeding efficiency and can withstand strong cardiac contractions. In one study, EHTs made by fibrin gel were formed between two elastic posts, which could be reinforced by the metal braces to induce mechanical overload [50]. After the enhancement of mechanical load, CMs exhibited pathological phenotypes, such as reduced contractile force and relaxation velocity, fibrosis and reactivation of fetal genes.

In addition to ECM proteins, naturally derived polysaccharides are also used as scaffolds for cardiac tissue engineering. Alginate derived from seaweed and bacteria shows similarity to ECM, but much easier for processing with complex structures. Agarwal et al., patterned grooves or ridges on alginate substrates using micro-molding technique. Guided by the topographic cues, CMs and smooth muscle cells (SMCs) aligned and elongated along the pattern direction [51]. Resulting cardiac tissues deformed the substrate synchronously and generated contractile stress comparable to healthy myocardium. Chitosan also has better tunable mechanical and biochemical properties than other natural hydrogels. For example, chitosan-enhanced ECM hydrogel was used to generate ring shape cardiac microtissues, shown anisotropic morphology and more mature properties than the CMs in 2D culture. Prolonged action potential and abnormal calcium handling property were recapitulated with patient-specific hiPSC-CMs [44].

Synthetic Hydrogels Synthetic hydrogels emerged in cardiac tissue engineering, because of their great mechanical performance [41]. Most synthetic hydrogels can be functionalized with conjugation of bioactive molecules, such as RGD sequence, which could promote cell adhesion and proliferation [52]. Furthermore, the bulk property can be easily tuned by varying the crosslinking parameters, and the spatial features can be controlled by different microfabrication techniques.

Poly (ethylene glycol) (PEG) is the most widely used synthetic hydrogels with versatile polymer structure and great elasticity. However, their low biodegradability inhibits the breakdown of matrix environment and renewal of ECM. Pueri et al., developed a 3D model of heart valve by co-culturing valve interstitial cells and valve endothelial cells in an RGD-modified PEG hydrogel. Both cell types maintained mature phenotypes, homeostatic functions and produced localized ECM [45]. In another study, cardiac fibroblasts were seeded on the PEG hydrogels with spatially patterned substrate rigidity, which mimicked the mechanical heterogeneity of infarcted myocardium. A directional cellular migration towards the stiff area reflected the persisted progression of fibrosis. After the treatment of ROCK inhibitor, reduction of myofibroblasts was observed, indicating the anti-fibrosis efficacy of this drug [53].

Hybrid Hydrogels Hybrid hydrogels were firstly made by mixing different types of hydrogels to balance the bioactivity and mechanical strength for various tissue engineering applications [54–56]. Later, other types of biomaterials have been incorporated within the hydrogels through copolymerization or physical integration. In this way, hybrid hydrogels can take advantages of different biomaterials to obtain additional functions, such as tunable mechanical stiffness and electrical conductivity.

Hybrid hydrogels have offered great potential to improve the mechanical properties of biological scaffolds for cardiac tissue engineering. A HA/gelatin hybrid gel with tunable stiffness was used to provide different physical signals to the valve interstitial cells. They found that valve interstitial cells differentiated towards myofibroblasts in the hydrogels with valve leaflet mimicked stiffness, but turned into osteogenic cells in the hydrogels with lower stiffness [57]. To improve the elasticity and stretchability of the hydrogels, a resilin-like polypeptide (RLP) hydrogels combined with PEG were fabricated as the hydrogel composites, which exhibited high mechanical resistance and reversible elasticity. Aortic adventitial fibroblasts showed high cell viability, efficient cell growth and spreading, when they were encapsulated in this stretchable hydrogel [58]. To generate the conductive hydrogel, poly (thiophene-3-acetic acid) (PTAA) have been introduced in the gelatin to form a double hydrogel network. ADSCs cultured on this hybrid conductive hydrogel showed the capability of differentiation into cardiac lineages, indicated by upregulated connexin 43 [59]. The efficiency of differentiation was further enhanced by applying external electrical stimulation. In another study, carbon nanotubes (CNTs) were embedded into gelatin-based hydrogels to enhance both mechanical and conductive properties. This conductive scaffold significantly enhanced the CM maturation, by inducing their alignment and spatial organization [60].

2.2 Hydrogel-Based In Vivo Cardiac Therapy

Hydrogels are the most abundant biomaterials used for *in vivo* cardiac therapy, because they have great mechanical compliance to the contracting myocardium, tight adhesion to the damaged tissues, and capability to active resident regenerative cells. In this section, we introduce the therapeutic applications for acellular, cell-laden and molecule-laden hydrogels in forms of preformed cardiac patches or injectable solutions (Fig. 2c, d).

Acellular Hydrogels *In vivo* grafting of hydrogel-based acellular cardiac patches could maintain the integrity of myocardium wall structure and relieve wall stress, which result in preservation of cardiac contraction. Stem cells at damaged area could also contribute to the tissue regeneration by active migration and differentiation. A collagen-based cardiac patch grafted onto infarcted myocardium in adult murine hearts attenuated left ventricular remodeling, diminished fibrosis and facilitated the formation of interconnected blood vessels within infarct [28]. Decellularized ECM from porcine myocardium slice was used for MI treatment by firmly attaching the ECM patch to the host myocardium. A large number of host cells were identified to infiltrate into the implant [46]. Contraction of left ventricle wall and cardiac function parameters were significantly improved in the implanted groups.

In spite of the beneficial effects of cardiac patches for tissue repair, their administration is quite invasive and difficult. Many natural hydrogels with low mechanical strength cannot sustain the suture process. Injectable hydrogels emerged as a solution to overcome this delivery obstacle. They are a unique class of hydrogels that maintain a liquid-like state under ambient conditions, while transfer into solid-like state at physiological conditions [61]. This sol-gel transition makes it possible for injectable hydrogels to be administrated through neighbor vessels as a solution and then solidify to cover the heart defects. In addition to ECM-based hydrogels, alginate has been used as injectable hydrogels due to its less thrombogenesis than other natural materials. Alginate-based hydrogel has become the very first injectable hydrogels in the clinical trials. Sabbah et al., developed an injectable alginate hydrogel for heart failure treatment. After seven weeks, the implant was encapsulated by a thin layer of connective tissue with no evidence of inflammation. Treated dog models showed significantly reduced left ventricle pressure and improved LVEF [62].

Cell-Laden Hydrogels Cell-laden hydrogels has been extensively developed for treating cardiac diseases. CMs and stem cells are two main categories of cell sources for *in vivo* cardiac regeneration. Transplantation of beating CMs could enhance the contraction of damaged myocardium in a direct manner, while stem cells can differentiate into cardiac cells or activate resident progenitor cells through paracrine effects [63]. Other criteria to choose appropriate cell sources should take into consideration of tailored regenerative mechanisms and minimization of the immune responses. Besides, to ensure the normal function of stem cells within hydrogels under *in vivo* condition, cell seeding density and uniformity need to be optimized to have sufficient cell viability and broad intercellular communications.

Both natural and synthetic hydrogels have been used for constructing 3D cell-laden patches and injectable hydrogels. In previous work, alginate-encapsulated MSCs were implanted in the PEG hydrogel patch to ensure the delivery of encapsulated cells to the injured heart [64]. The double-layered hydrogel sustained the presence of stem cells at injured site and prolonged the paracrine secretion of MSCs. As a result, patch treated animal groups exhibited improved vascularization and reduced scar formation. For cell-capsulated injectable hydrogels, stem cells suspension is mixed with an aqueous solution of hydrogel precursors, and then gel *in situ* to form cell-loaded 3D matrices at the injection region. The sol-gel transition should occur without harming encapsulated cells, and the gelation rate should be fast enough to retain transplanted cells at injured sites. In one study, CPCs were encapsulated within self-assembly peptide hydrogels and transplanted to the MI rat model [47]. They found that the activation of CPCs improved acute myocardial retention and hemodynamic function. In addition, CPCs in hydrogels with higher mechanical modulus led to better cardiac performance, which confirmed the importance of suitable mechanical microenvironment for the therapeutic efficacy of stem cells.

Molecule-Laden Hydrogels The bioactive molecules have been found to promote the adhesion, differentiation and migration of resident cells to repair damaged cardiac tissues. Compared to the stem cell therapy, there is less risk of immunogenic rejection and oncogenic response for biomolecule-only delivery [65]. Furthermore, it is easier to control over the protein milieu contents through multiple techniques. In addition to bioactive proteins, genetic modulation also offers a unique opportunity to have a more direct and precise way of modulating cell behaviors. The delivery of specific genes related to cardiac therapeutics into injured myocardium could promote the local healing of damaged myocardial tissues by facilitating vasculature and attenuating fibrosis [66].

Both large and small molecules have been incorporated in hydrogels for targeting release at local myocardium. Vascular endothelial growth factors (VEGFs) are popular growth factors used for promoting *in vivo* angiogenesis. However, single administration of VEGFs is not sufficient to stabilize newly formed blood vessels. Rufaihah et al., used polyethylene glycol-fibrinogen (PF) hydrogel as a carrier for dual controlled release of VEGFs and angiopoietin-1 (ANG-1), which could inhibit the leakage of vasculature [67]. Treated animals showed obvious improvement in the cardiac function, high degree of cardiac muscle preservation and arteriogenesis. To examine the performance of gene therapy for cardiac diseases, one study incorporated VEGF-recombinant DNA plasmid into graphene oxide/gelatin hydrogel. The injection sites showed better cardiac contractions without any obvious inflammatory response [68]. Injectable nanogels and microgels with smaller but more compact polymer network could enhance the encapsulation capability of small molecules and reaction rate. Moreover, the small size of nanogels allows them to penetrate into the cells without many obstructions. In one study, a core-shell colloidal hydrogel was fabricated to achieve dual delivery of small and big molecules [69]. The sequential release of shell encapsulated tissue plasminogen activator (tPA) and core encapsulated Y-27632 showed high efficiency for improving left ventricular ejection fraction and eliminating stress fiber formation.

3 Structural Materials

Comparing to the bulk materials, which mainly create an isotropic microenvironment, the heterogenous features of structural materials could provide anisotropic stimulations to the cardiac cells. Two major categories of structural materials in the cardiac tissue engineering are porous and fibrous biomaterial scaffolds. Porous scaffolds are featured by their solid body, which supports the 3D integrity, and the interior porous structure, which promotes cell adhesion. Fibrous scaffolds could be fabricated in a variety of shapes and have been widely utilized for investigating cell-material interactions. Here we give an introduction for these two types of structural biomaterial scaffolds with a focus on how their tunable architectures affect the cardiac behaviors. Representative examples of structural scaffolds for cardiac cell and tissue engineering are summarized in Fig. 3.

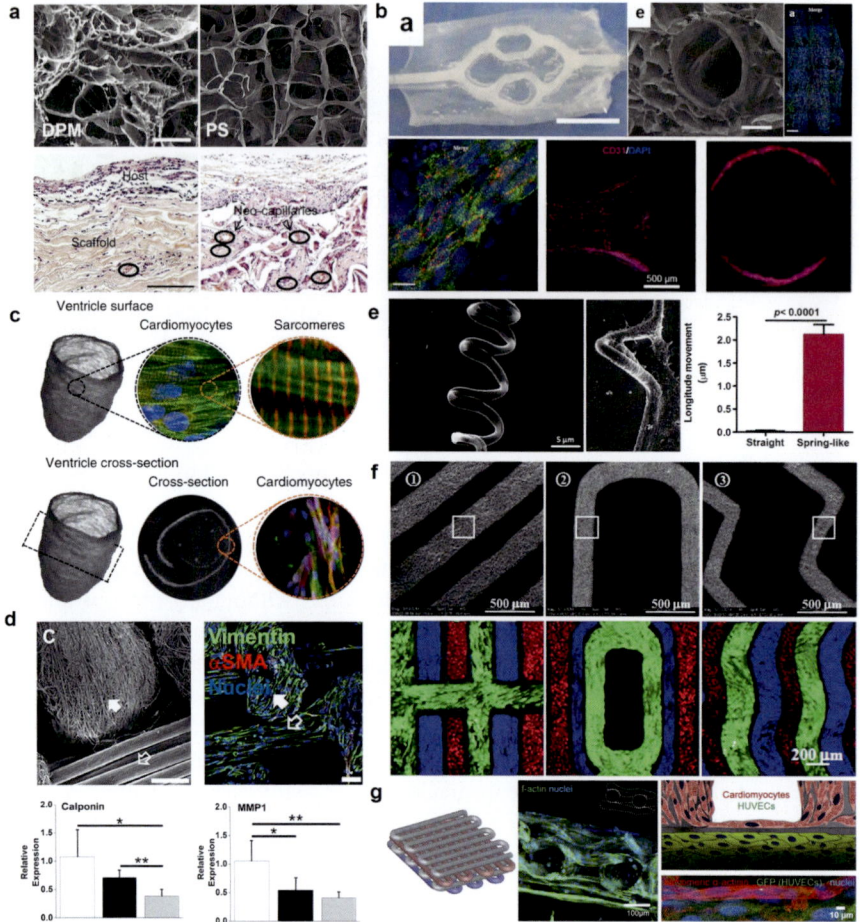

Fig. 3 Porous and fibrous scaffolds used in cardiac tissue engineering
(**a**) The pericardium-derived scaffolds with well-defined architecture and interconnected macropores promoted native angiogenesis compared to decellularized pericardium membranes (Reprinted from [70] with permission from Elsevier). (**b**) ECs and CMs were cultured separately in microchannels or bulk porous chitosan/collagen scaffolds to obtain vascularized cardiac tissues (Reprinted from [71] with permission from IOP Publishing). (**c**) CMs aligned circumferentially in a heart ventricle model guided by the direction of PCL/gelatin nanofibers (Reprinted from [72] with permission from Springer Nature). (**d**) Cardiac tissues were patterned in the strip, oval and wave-like shapes on the fiber mats, and displayed bipolar morphology. (Reprinted from [73] with permission from Royal Society of Chemistry)

3.1 Porous Materials

Porous scaffolds are the earliest forms of 3D scaffolds used in the cardiac tissue engineering, because they allow for more efficient mass transport compared to bulk 3D materials [74]. The porous structures high surface/volume ratio provide sufficient space for cell attachment, migration and proliferation. Pore size, porosity and interconnectivity are the main effective factors for directing the cellular behaviors [75]. Porous structures could be generated with various methods, including gas foaming, freeze drying, particulate leaching, thermally induced phase separation and electrospinning [76]. Although the dimensions of pores and mechanical properties could be tunable by varying the fabrication parameters, it is hard to produce desired pore architectures with high consistency in a time efficient manner. To overcome these limitations, advanced fabrication techniques such as 3D printing, electrohydrodynamic jet printing and melt electrospinning writing are now being applied to fabricate scaffolds with highly controlled, complex porous structures.

Porous Scaffolds for Cell Growth and Alignment Interconnectivity of pores ensures the close intercellular communication throughout the scaffolds. Rajabi-Zeleti et al., developed a pericardium-derived porous scaffold based on decellularization of native pericardium tissues, followed by enzymatic digestion and lyophilization [70]. The highly interconnective macropores enabled CPCs to survive, migrate, proliferate and differentiate at higher rates compared to cells seeded on the decellularized pericardium membranes. Transplanted scaffolds promoted native angiogenesis and cardiac differentiation with limited immunological responses. High porosity and large pore sizes encourage better cell aggregation and penetration due to the improved mass transport, while the mechanical strength is sacrificed due to the great amount of void volume. For example, a gelatin/chitosan hydrogel with increased porosity showed decreased compression modulus but promoted the proliferation and alignment of neonatal rat ventricular myocytes (NRVMs) [77]. In contrast, one study found that a decrease of nanoscale pore size better supported the growth and vascularization of human smooth muscle cells [78]. This phenomenon could be attributed to the improved cell-cell interactions within the smaller pores, leading to an increase of ECM deposition.

Traditional isotropic porous scaffolds have uniform distribution of mechanical stress at the cell-biomaterial interface, leading to cell spreading in random directions. Newly developed 3D scaffolds with anisotropic pores were shown to facilitate *in vivo* like morphological remodeling of CMs. Bursac group successfully engineered elliptical pores within fibrin-based cardiac patches using soft lithography [79]. NVRMs aligned around the pore edges and exhibited greater isometric twitch forces. In another case, accordion-like honeycomb microstructure was fabricated within the poly(glycerol sebacate) (PGS) to produce porous, elastomeric 3D scaffolds with controllable stiffness and anisotropy [80]. NRVMs aligned along the long axis of honeycombs and showed the cross-striations similar to the adult rat ventricular myocardium. As a result, NRVMs exhibited directionally dependent responses to external mechanical loading and electrical stimulation.

Porous Scaffolds for Vascularization To engineer 3D thick cardiac tissues, it is critical to introduce vascular structures to facilitate the exchange of nutrients and waste for resident cells. Porous scaffolds are widely used for vascularization, because their hollow structures allow the invasion and branching of micro-vessels.

To achieve optimal cell-cell interactions, vascular cells and CMs have been cocultured and spatially organized in the physiological-relevant conditions. For *in vivo* transplantation, the efficacy of vascularization is dependent on the creation of a conducive environment to activate the host angiogenesis.

Growth and lining of endothelial cells (ECs) via perfusion through the microchannels have been used extensively to induce vasculature for *in vitro* cardiac tissue constructs. In a previous study, microchannels were fabricated within 3D porous alginate scaffolds using CO_2 laser graving system [81]. Induced by local angiogenetic factors, pre-cultured ECs organized around the microchannels and formed multilayer vessel-like structures. CMs that were added afterward in the bulk porous zone between channels showed elongated morphology and clear sarcomeres. Recently, channel networks produced by sacrificial templates were incorporated in a 3D chitosan/collagen scaffold with oriented micropores fabricated using the "Oriented Thermally Induced Phase Separation" technique [82]. ECs and CMs were cultured separately in the channels and the bulk scaffolds to obtain a cardiac tissue construct mimicking native myocardium anisotropy and vasculature. ECs were able to migrate through the micro-holes on the channel walls to form single or multicellular sprouts into the cardiac tissue.

A bimodal scaffold containing parallel channels and sphere-templated, interconnected porous network was used for engineering transplantable cardiac tissue constructs [83]. CMs grown within the channel networks formed bundle-like large cardiac tissues with support of mass transfer enabled by the surrounding spherical pores. When the acellular scaffolds were implanted in a rat model, the porous structures were filled with small vessels, and the number of perfused vessel lumens was depended on the pore size. Although macropores allowed the infiltration of native micro-vessels, angiogenetic cytokines could further improve the vascularization of the implants. For example, VEGFs were covalently bonded to the porous collagen patches for *in vivo* myocardial repair [84]. Implanted cardiac patches promoted the engraftment of bone marrow cells, which then recruited the endothelial progenitor cells (EPCs) to initiate the angiogenesis process. Blood vessel density was found significantly higher in the VEGF-treated groups than the control groups.

3.2 Fibrous Scaffolds

Native ECM is mainly composed of densely packed fibrillar proteins with nanoscopic topography that could guide the behaviors of different cell types [85]. To enable "bottom up" fabrication of ECM-like scaffolds, micro- and nanofibers are now extensively used in various tissue engineering applications. The fibrous structures support the cell attachment, and the porous intermediate spaces allow cell penetrations. In addition to *in vivo* like nanoscale features, fibrous scaffolds could be made into various shapes on the macroscopic level to generate different cardiac tissues, including myocardium, valves and vessels [86]. Electrospinning is the most prevalent tool used for fabricating fibrous structures due to its low cost and simplicity. Other advanced fabrication techniques, including 3D printing and laser-based polymerization, provide more precise positioning of fibers with high structural complexity [87, 88]. With versatile properties and structures, fibrous scaffolds have been extensively used for studying cell-biomaterial interactions.

Fibrous Scaffolds with Tunable Fiber Morphology Compared to randomly oriented fibers, aligned fibers could direct the unidirectional shaping of the cultured cells, which benefits the differentiation and maturation of CMs from cellular to tissue levels. For example, ESC derived CMs showed more anisotropic organization, improved sarcomere formation on aligned polyurethane (PU) scaffolds compared to the cells cultured on unaligned scaffolds [89]. Another study also proved the aligned PGS/gelatin scaffolds significantly enhanced the elongation of CMs and their synchronous contractions [90]. The aligned fibers could be scaled up to induce physiological-relevant cardiac tissue assembly on the organ level. Parker group reported the establishment of an *in vitro* heart ventricle model composed of pull-spinning polycaprolactone (PCL)/gelatin nanofibers on a rotating ellipsoidal collector [72]. Seeded CMs infiltrated the scaffolds and circumferentially aligned coincident with the nanofiber ultrastructure.

Changing the diameter of individual fibers not only changes the scaffold structural and mechanical properties, but also affect the effective area for cell attachment and growth. For example, electrospun poly(D,L-lactide-co-glycolide) (PLGA)/collagen scaffolds had small fiber diameter, higher hydrophilicity and lower tensile strength than pure PLGA fibers [35], which led to more efficient differentiation of ESCs into CMs. While nanofibers recapitulate the nanoscopic topography of native ECM to promote cell attachment and proliferation, microfibers with large porous structures could increase mechanical integrity and facilitate cell penetration. To take the advantages of both fiber types, an interwoven mesh composed of polyacrylonitrile (PAN) micro and nanofibers was used to replicate the circumferential and radial anisotropy of native aortic valve leaflets [91]. Human aortic valve interstitial cells attached, elongated and aligned in the direction of nano and microfibers, showing fast proliferation and a myofibroblast phenotype.

In addition to the straight uniform fibers, complex fiber structure and morphology have been enabled by different fabrication technologies. For example, coiled fibers could be fabricated by changing the electrospinning rate. In a previous work, a 3D scaffold with spring-like fibers was fabricated to mimic the coiled perimysial fibers of the myocardium [92]. Comparing to the straight fibers that inhibited cardiac contractions due to their rigid surface, the straightening and re-coiling of spring-like fibers gave more compliance to strong contraction forces. Cardiac cells cultivated on the spring-like fibers induced fiber stretching in the contraction direction. Moreover, cardiac tissue exhibited stronger contraction force, higher beating rate and lower excitation threshold. More recently, core-shell fibers were also introduced to enhance the functionality of fibrous scaffolds, in contrast to the uniform characteristics of blended polymers. The core polymers aim to determine the mechanical properties of the fibers, while the shells could determine the biological interfaces, independent to the properties of the cores [93]. The cardiac patch made of core-shell fibers was used to entrap the MSCs for MI treatment [94]. The PGS core was used for imparting suitable mechanical properties, while the fibrinogen shell was used for improving bioactivity. Cardiac contractile functions were improved attributed to active differentiation of MSCs and mechanical support of cardiac patch.

Fibrous Scaffolds with Micropatterned Meshes Geometric control of fibrous meshes adds extra mechanical cues to the cells. In one study, Liu et al., compared the phenotypes of CMs on strip, oval and wave-patterned fiber mats [73]. Compared to cells on the non-patterned aligned fibers, CMs on the patterned scaffolds displayed bipolar morphology and abundant synthesis of sarcomere proteins, which resulted from the better intercellular crosstalk induced by the geometric confinements. Among all different patterned scaffolds, wave-like patterned scaffolds improved cardiac functions to the highest level, because their coiled shape allowed for cardiac beating with minimal energy loss. Moreover, drug responses of CMs showed good correlations with clinical observations on the beating rate and prolonged contractility.

Heterogenous fibrous scaffolds could be obtained by combining fibers with different morphologies to create nonuniform internal structures. For example, hybrid matrices made of fibers with different diameters were used to reproduce the nonuniform mechanical environment, mimicking heart disease progression [95]. Cardiac microtissues under mechanical nonuniformity showed pathological remodeling, which was demonstrated by irregular contractile velocity and inefficient force production. Spatial organization of different cell types could also be achieved by patterning cell-encapsulated fibers using 3D bioprinting. Khademhosseini group successfully developed a spatially patterned endothelialized myocardium based on the 3D bioprinted microfibrous hydrogels, which encapsulated ECs within the printed microfibers and then subsequently deposited with CMs [88]. Homogenously dispersed ECs gradually migrated towards the peripheries of microfibers to be in close contact with CMs. Engineered cardiovascular tissues based on the anisotropic organization of microfibers showed stronger and longer cardiac contraction.

4   Smart Materials

Smart materials, or stimuli-responsive materials, are special polymers that can vary their biochemical or mechanical properties when specific environmental stimuli are present. Common environmental triggers are classified by intrinsic cues (pH, enzymes, redox condition, hydrolysis) that provide by physiological environment and extrinsic cues (temperature, light, ultrasound, magnetic field), which allows spatial and temporal control [96]. Responding to these triggers, smart materials undergo either reversible modulation of physical interactions or permanent cleavage and formation of chemical bonds.

The property switch of smart materials results in a temporal change of microenvironment, which offers a great opportunity for establishing dynamic *in vitro* tissue models with more physiological relevance [97]. Chaudhuri et al., developed a viscoelastic alginate hydrogel that could relax the mechanical stress exerted by cells through matrix reorganization over time [98]. They found MSCs showed enhanced osteogenic differentiation in gels with faster relaxation and formed an interconnected mineralized, collagen-I-rich matrix similar to the bone structure. Smart materials could also serve as efficient carrier vehicles for therapeutic drugs due to their sensitivity to specific *in vivo* conditions. In this way, adverse effects resulting from off-target distribution can be eliminated, and the efficacy of desired drug treatment can be improved. For example, Yang et al., constructed a nanogel exhibiting pH and redox dual responsive property and combined it with doxorubicin (DOX) for targeted cancer therapy [99]. DOX-loaded nanogels accumulated in tumor tissue and inhibited tumor activity significantly with few side effects.

4.1 Smart Materials for In Vitro Dynamic Culture Platforms

Thermo-responsive polymers are the most widely used smart materials for *in vitro* dynamic cardiac culture platforms (Fig. 4a, b) due to the general availability of thermo-sensitive motifs and the easy control of temperature [100]. Triggered by the variation of temperature, these materials could change their surface chemistry, stiffness and topography, which affect the adhesion, morphology and maturation of myocardial cells. In addition to thermo-responsive polymers, magnetic-responsive materials have been generated by incorporating metal particles into elastic substrates.

Change of Surface Chemistry Thermo-responsive hydrogels could undergo a hydrophobic-to-hydrophilic switch due to the formation of hydrogen bond between polymer chains. This switch leads to a disruption of cell-material interactions that is usually utilized for on-demand detachment of cardiac cells without sacrificing intercellular junctions [97]. This technology is recognized as cell sheet engineering, which is a common way of constructing scaffold-free cardiac tissues. Poly (N-isopropylacrylamide) (PIPAAm) is the most developed materials for cell sheet engineering. Shimizu et al., grafted the cell culture surface with PIPAAm, thus neonatal rat cardiomyocytes was able to detach spontaneously from culture surface when temperature was increased [101]. By stacking the cell sheets layer-by-layer, engineered 3D cardiac tissues with controllable thickness were obtained and exhibited spontaneous beating.

In vitro cardiac differentiation always results in heterogenous cell populations, while cell sheet engineering enables the manipulation of cell types within 3D cardiac tissues. Masumoto et al., constructed heterogenous cardiac patches composed of CMs, ECs and mural cells (MCs) collected from PIPAAm grafted surface [105]. Transplanted cardiac patches in MI models showed significant and sustained improvement of systolic function accompanied by neovascularization. However,

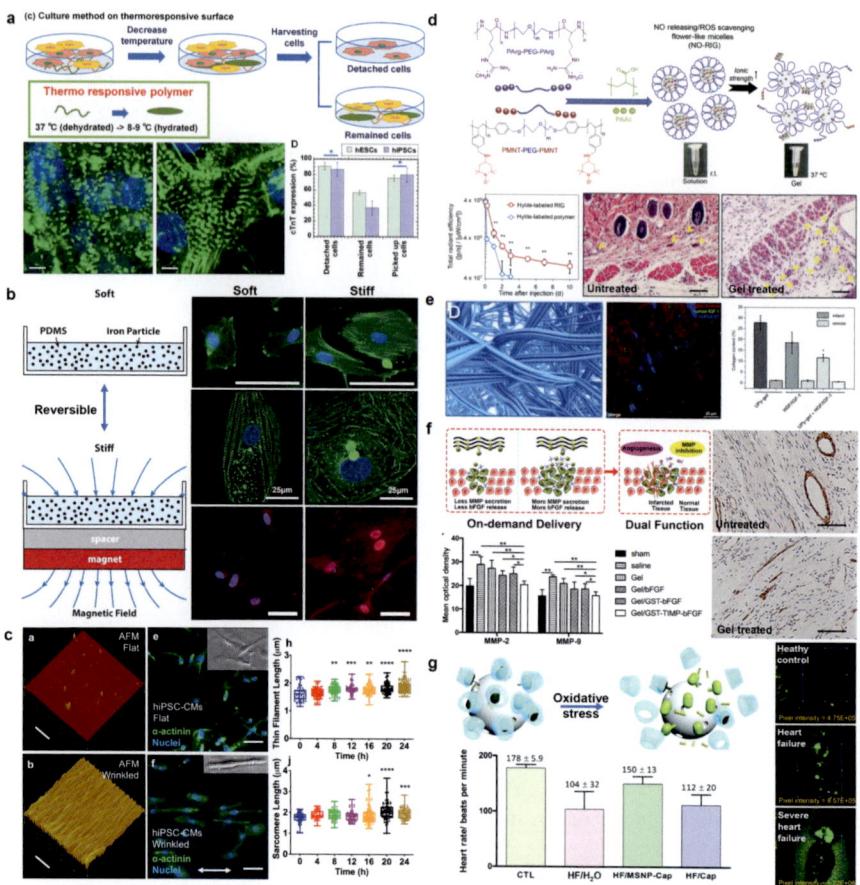

Fig. 4 Applications of smart materials in dynamic culture platforms and drug delivery vehicles (**a**) Stem cell derived CMs demonstrated high purity when detaching from the PIPAAm-coated surface after hydrophilicity transition (Reprinted from [102] with permission from Elsevier). (**b**) hiPSC-CMs showed a sequential remodeling profile for different sarcomere components guided by dynamic topographic cues (Reprinted from [103] with permission from American Chemical Society). (**c**) A temperature-responsive gel with both NO release and ROS scavenging capabilities improved the angiogenesis for MI mice (Reprinted from [104] with permission from Elsevier). (**d**) bFGFs were released after MMP-specific cleavage of their binding sites within the hydrogel, which inhibited the MMP accumulation and promoted angiogenesis in MI rats (Reprinted from [71] with permission from John Wiley and Sons)

without the participation of CMs, cardiac sheets failed to induce neovascularization and subsequent function, which confirmed the major role of CMs in regenerating the injured tissues. ECs and MCs were found to facilitate the sheet function and structural integration. In a recent study, Sung et al., proposed that CMs and non-CMs derived from stem cells have distinct affinity to the PIPAAm polymer [102]. The temperature decrease resulted in specific detachment of CMs, while non-CMs remained on the plate due to their stronger adhesion to PIPAAm. This study contributed a new concept of using PIPAAm as cell sorting platforms.

Change of Surface Stiffness Lots of previous static platforms have proved the fact that myocardial cells are able to adjust their morphology and phenotypes according to different mechanical stiffness. Smart materials with time-dependent modulation of stiffness, could further facilitate our understanding of cardiac mechano-dynamics. The gradual stiffening of hydrogels could be achieved by reinforcing the crosslinking reaction between polymer chains. In 2011, Engler group first investigated the contribution of increased mechanical stiffness to cardiac maturation *in vitro* [106]. The addition of poly (ethylene glycol) diacrylate (PEGDA) to thiolated-HA initiated a slow crosslinking reaction between these two polymers, which induced a gradual increase of elastic modulus. They tuned the molecular weight of PEGDA to achieve a stiffening rate mimicking the microenvironment when CMs were maturing from mesoderm to adult myocardium. Immature CMs grown on stiffened hydrogels showed three-fold increase of mature cardiac specific markers and 60% more maturing muscle fibers than the cells grown on compliant and static polyacrylamide hydrogels.

Most available smart materials with temporal mechanical gradient exhibited unidirectional and irreversible property switch. Recently, efforts have been invested in developing reversible systems to investigate the cardiac remodeling more comprehensively. Corbin et al., introduced a new method of producing substrates with reversible change of stiffness simply by controlling the distance between the magnetic field and a soft polydimethylsiloxane (PDMS) elastomer encapsulating iron particles [107]. They found that the rapid increase of substrate stiffness led to the fast reorganization of actin cytoskeleton, which then induced the transport of YAP from cytoskeleton to the nucleus. Cardiac fibroblasts showed an increase of stress fiber formation and an activated state towards myofibroblast-like phenotype. Besides, CMs showed less organized sarcomeres and lower expression of cardiac markers on stiffer substrates.

Change of Surface Topography Variation of surface topography is another main way of tuning mechanical transductions for myocardial cell. This is largely achieved by utilizing shape memory polymers (SMPs), which could "memorize" a permanent shape, be programmed into a temporary shape and then switch back to permanent shape triggered by external stimuli [108]. For example, making use of micro-molding and the shape memory effect of PCL, Le et al., achieved topographic transition around the physiological temperature [109]. Temporary channel-like patterns fixed on PCL at 28 °C would disappear completely when temperature was increased to 40 °C, leaving a smooth surface. Human MSCs showed great alignment along anisotropic channel patterns at lower temperature while turned into random orientations guided by the removal of nano-topographic cues. In addition to a pattern-to-flat switch, SMP pattern orientations could also be achieved by changing the directions of patterned molds during the program of permanent and temporary shape. Mengsteab et al., found that aligned CM sheets reoriented in both clockwise and counterclockwise directions following the orthogonal switch of nanogrooves on PCL [110]. The alignment of nuclei and focal adhesion complex shared the similar bimodal cell distribution characteristics.

To further pinpoint the time-dependent remodeling profile of CMs during dynamic change of topographic cues, Ma group compared the organization of different sarcomere and cytoskeleton components at various time points during and after the formation of nano-wrinkles on polyelectrolyte coated *tert*-butyl acrylate/butyl acrylate (tBA/BA) SMPs [103]. They found that the structural remodeling of CMs followed a specific order, starting from focal adhesion reassembly, actin filament elongation, to sarcomere extension and final cell alignment. Besides, inhibition of cytoskeletal tension and mechano-sensitivity led to the disruptions of myofibril remodeling responding to dynamic mechanical signals.

4.2 Smart Materials for Drug Delivery Vehicles

The local microenvironment of injured heart tissue is different from that of healthy tissue in various aspects, including temperature, pH, enzymatic activities and redox condition. Therefore, smart materials sensitive to these disease-specific environmental stimuli could serve as targeted delivery vehicles for cardiac therapeutic agents (Fig. 4c, d). Smart hydrogels are the most available types of systems for both stem cell and drug delivery, while more advanced systems, including supramolecular hydrogels and nanoparticles (NPs) are also explored. Different from chemical-crosslinked hydrogels, supramolecular hydrogels are held together by pure noncovalent interactions, which allow for more flexible control of their sol-gel switching and designable drug loading [111, 112]. The high surface/volume ratio of NPs could dramatically increase their drug loading capacity, which makes them attractive drug carriers [113]. Besides, the small diameters of NPs benefit their intravenous injection and systemic transport [114].

Temperature-Responsive Polymers Temperature-responsive hydrogels could transit from a liquid to solid state *in situ* rapidly due to the increased interactions between their hydrophobic chains triggered by elevated temperature. Li et al., incorporated single-wall carbon nanotubes (SWCNTs) to improve the bioactivity of PNIPAAm and facilitate the engraftment of ADSCs [115]. Cultured on this modified hydrogel, ADSCs showed better adhesion, spreading capacity and improved retention in host myocardium. Critical cardiac functions, including fraction shortening and ejection fraction were improved, accompanied by the decrease of scar size and increase of heart wall thickness. In another study, a flower-like micelle made of two functional triblock copolymers was generated to release NO and scavenge reactive oxygen species (ROS) simultaneously [104]. This polymer composite exhibited fast gelation under physiological conditions due to their thermo-sensitive ionic interactions. After injection, liquid triblock copolymers were completely removed after 3 days, while the micelles distributed homogenously within heart tissue and remained for more than 10 days. The stable retention of this gel promoted the sustained release of NO, which would benefit cardiac functions and angiogenesis.

pH-Responsive Polymers The local pH at ischemic myocardium is found to be more acidic than other healthy regions, which make pH-responsive polymers become effective delivery systems. The pH-sensitivity can be attributed to either the protonation of ionizable groups or the degradation of acid-cleavable bonds within smart materials. In one study, to solidify hydrogels locally at infarcted heart, the gelation threshold of a family of smart hydrogels was controlled at around 6.5, while these hydrogels remained liquified in blood with a neutral pH [116]. Cultured cardiosphere-derived cells (CDCs) showed good viability and efficient cardiac lineage differentiation demonstrated by elevated cardiac markers. Bastings et al., fabricated a supramolecular hydrogel using fourfold hydrogen bonding supramolecular ureido-pyrimid-inone (UPy) units coupled via alkyl-urea spacers to PEG chains [111]. Triggered by a subtle change of pH from 8.5 to 7, the crosslinks between fibers broke down, which initiated a fast gelation of hydrogel. Hepatocyte growth factor (HGF) and insulin-like growth factor-1 (IGF-1) embedded within this gel showed a prolonged release for 7 days. After local injection, the collagen content at infarcted regions reduced significantly, and patched-like clusters of viable myocardial tissue were found dispersed between collagen fibers and growth factor loaded hydrogels.

Enzyme-Responsive Polymers Compared to pH and temperature responsive materials, enzymatic-responsive polymers allow for more targeted drug delivery at infarcted heart. These materials usually contain groups that are vulnerable to enzyme digestion and cleavage, and their internal structure would be disrupted upon exposure to specific enzymes. Matrix metalloproteinases (MMPs) that can degrade ECM are the common targets for enzyme-responsive polymers [117]. Fan et al., designed an MMP-responsive hydrogel conjugated with basic fibroblast growth factors (bFGFs) mediated by TIMP peptides [71]. TIMP could specifically bind with abundant MMPs, deactivated them to inhibit further ECM degradation. At the same time, bFGFs would be released due to the cleavage of their binding sites on the hydrogel, which then promoted local angiogenesis. Rat groups treated with this dual functional hydrogel showed obvious increase of ejection fraction and lower collagen deposition compared to controls. Nguyen et al., reported a NP containing peptide sequences for specific recognition of MMPs [118]. Enzymatic cleavage of the peptides resulted in a morphological transition of NPs from discrete micellar structures into network-like scaffolds. Upon this transition, NPs aggregated stably at the infarct and adjacent border zone with minimum accumulation around healthy myocardium. The high specificity of this NP makes it become a promising candidate for drug delivery to treat cardiac diseases.

Redox-Responsive Polymers The high level of ROS in the sites of inflammation and tissue injury indicates that the high oxidative stress could become a disease-specific target [119]. Oxidative-labile linkers, including thioketals, thioethers, arylborinic esters and aryloxylates can be triggered by H_2O_2, superoxide, or hydroxy radicals. For example, NPs fabricated through the single emulsion of peroxalate containing vanillyl alcohol (PVAX) were used for the treatment of DOX-induced cardiomyopathy [120]. Peroxalate has an ester bond that is sensitive to H_2O_2, while vanillyl alcohol possesses antioxidant, anti-inflammatory and anti-apoptotic properties. Injection of PVAX NPs resulted in inhibition in apoptosis within DOX-treated mice and increase of survival rate. Tan et al., attached a fluorescently labeled H_2O_2 probe on the surface of NPs and linked a therapeutic drug of heart failure to the probe by the binding of α-cyclodextrin (α-CD) [121]. When reacted with H_2O_2, the structural change of probes caused the dissociation of α-CD, which then led to probe fluorescence and drug release. Using non-invasive imaging of zebrafish models, they found the fluorescence intensity was proportional to the severity of heart failure. The heartbeat rate and cardiac output were improved after the drug treatment.

5   Conclusion and Future Perspectives

In this book chapter, we summarize the applications of three main categories of popular biomaterials for cardiac cell and tissue engineering, including bulk hydrogels, structural scaffolds, and smart materials. Bulk hydrogels provide a great support for cell culture in either 2D or 3D manner due to their general biocompatibility and softness. At the same time, their biochemical and mechanical properties could be tuned by physical mixing or chemical crosslinking of different components, which result in the flexible control of the microenvironment for myocardial cells. Their efficacy for heart repair, either alone or combined with stem cells and small molecules, have been validated by many MI animal models. However, scaling up cardiac tissue constructs is still challenging, due to the limited mechanical integrity of hydrogels. Structural scaffolds which contain porous or nanofibrous units are featured by their ability to induce cellular morphological remodeling through contact guidance from their heterogenous and anisotropic structures. Nevertheless, the pore walls might limit cell-cell interactions and electronic coupling of CMs. To solve this, conductive materials are now introduced to reinforce the electrical and mechanical properties of porous scaffolds [122]. In the other hand, the compacted nanoscale structure of fibrous scaffolds limits efficient cell infiltration and migration. Recently, multilayer fibrous yarns were embedded within cell-encapsulated bulk hydrogel to provide the mechanical support and promote the cell alignment [91]. Smart materials are newly developed biomaterials with property switch triggered by specific environmental stimuli. They have been developed as dynamic cell culture platforms for real-time observations of cardiac remodeling or smart drug delivery vehicles with significantly improved drug engraftment and specificity for cardiac therapy. A major limitation of smart cell culture substrates is the lack of ability to support the intricate feedback loops that exist *in vivo* between ECM and resident cells. Although smart drug delivery vehicles responding to many different pathological environmental stimuli have been tested for cardiac therapy, the development of more stimuli-responsive functional groups could broaden the utility and improve the specificity of these vehicles to the cellular level.

Current *in vitro* cardiac models are still at the stage of engineering "a piece of cardiac tissue", which usually are lack of the structural and compositional hierarchy to the comparable degree of human heart. This is mainly due to the inevitable trade-off between high fabrication resolution and the need for scaling up models. 3D bioprinting is a versatile tool for scaling up complicated structures using either materials or cells as inks, but most natural materials are not compatible with this technique due to their softness and uncontrollable gelation [123]. The development of printable natural materials would be a critical step towards fabricating off-the-shelf scaffolds for cardiac models. Another challenging issue is the limited specificity and systematic comparison of microenvironments and cell populations amongst various biomaterials and cell resources in the design and generation of cardiac tissue models. Tremendous effort has been focused on conducting cardiac differentiation and manufacturing scaffolds with tissue specificity. Zhao et al., reported the construction of atrial, ventricular and atrioventricular tissues through the spatial patterning of chamber-specific cardiac lineages differentiated from hiPSC lines [124]. Besides, genetic editing technique could be used to create gene variants to engineer mutation-specific cardiac disease models. Through the manipulation of scaffold properties, these models allow for investigating the correlation between genetic mutations and microenvironments during disease progression [125].

On the other hand, biomaterial-based *in vivo* cardiac repair and regeneration is hindered by the poor understanding of material performance and cellular responses under *in vivo* conditions, which results from the complicated and uncontrollable native microenvironment. For example, the endocytosis process is distinct for the cells under *in vitro* and *in vivo* conditions, which makes it hard to predict the *in vivo* uptake of NPs based on *in vitro* screens [126]. Some NPs that are efficient for *in vivo* drug delivery are more likely to be screened out due to their relatively poor *in vitro* performance. Other critical properties of biomaterials, such as degradation rate, toxicity and structural integrity, would also be affected by the transition from *in vitro* to *in vivo* conditions. In addition, reliable tracking tools are now being developed in order to uncover the dynamic *in vivo* profiles of biomaterials and cellular activities. For example, degradation of collagen content is a hallmark of cardiac remodeling, which is commonly assessed by disruptive histology to evaluate the efficacy of cardiac treatment. To track the real-time collagen remodeling, Hwang et al., synthesized a fluorescence tagged collagen hybridizing peptide (CHP), which could specifically hybridize to the degraded collagen chains [127].

Most cardiac tissue models are established following the "top down" strategy, which refers to the culture of myocardial cells on a solid fabricated scaffold. Recently, a "bottom up" strategy has been utilized to scale up cardiac tissue constructs through the modular assembly of living cell/biomaterial building blocks. In this way, we could have versatile macroscopic shapes without losing the control of microscopic structures. Many forms of building blocks, such as single-cell laden microgels, cellularized fibers or cell sheets, have been integrated through thermodynamically driven self-assembly, contact-based direct assembly and force field-mediated remote assembly [128]. The "bottom up" approach is especially necessary for the field of cardiac tissue engineering due to the need of concentrating CMs and reproducing distinct structures of different heart parts. Fleischer et al., reported the construction of a modular cardiac patch consisting of multiple layers with different functions, including accommodating cardiac cells, organizing ECs and releasing bioactive molecules [129]. In another study, cardiac organoids were fabricated by the self-assembly of three different cardiac cells with a ratio corresponding to developing hearts [130]. These organoids could further fuse to form larger tissues while maintaining a high cell viability by pre-forming vascular networks within the organoid units.

Acknowledgement The authors acknowledge the grant support from NIH NICHD (R01HD101130) and NIAMS (R21AR076645), NSF (CBET-1804875 and CBET-1943798) and Syracuse University intramural CUSE Grant, Gerber Grant and BioInspired Institute Seed Grant.

References

1. Benjamin, E.J., Blaha, M.J., Chiuve, S.E., Cushman, M., Das, S.R., Deo, R., et al.: Heart disease and stroke statistics-2017 update a report from the American Heart Association. Circulation. **135**(10), E146–E603 (2017). https://doi.org/10.1161/Cir.0000000000000485
2. Laflamme, M.A., Murry, C.E.: Heart regeneration. Nature. **473**(7347), 326–335 (2011). https://doi.org/10.1038/nature10147
3. Hasan, A., Waters, R., Roula, B., Dana, R., Yara, S., Alexandre, T., et al.: Engineered biomaterials to enhance stem cell-based cardiac tissue engineering and therapy. Macromol. Biosci. **16**(7), 958–977 (2016). https://doi.org/10.1002/mabi.201500396
4. Kazbanov, I.V., ten Tusscher, K.H.W.J., Panfilov, A.V.: Effects of heterogeneous diffuse fibrosis on arrhythmia dynamics and mechanism. Sci Rep-Uk. 2016;6. doi: ARTN 20835; https://doi.org/10.1038/srep20835
5. Ferrario, C.M., Schiffrin, E.L.: Role of mineralocorticoid receptor antagonists in cardiovascular disease. Circ. Res. **116**(1), 206–213 (2015). https://doi.org/10.1161/Circresaha.116.302706
6. Kotecha, D., Holmes, J., Krum, H., Altman, D.G., Manzano, L., Cleland, J.G.F., et al.: Efficacy of beta blockers in patients with heart failure plus atrial fibrillation: an individual-patient data meta-analysis. Lancet. **384**(9961), 2235–2243 (2014). https://doi.org/10.1016/S0140-6736(14)61373-8
7. Sacks, C.A., Jarcho, J.A., Curfman, G.D.: Paradigm shifts in heart-failure therapy – a timeline. New Engl J Med. **371**(11), 989–991 (2014). https://doi.org/10.1056/NEJMp1410241
8. Yacoub, M.: Cardiac donation after circulatory death: a time to reflect. Lancet. **385**(9987), 2554–2556 (2015). https://doi.org/10.1016/S0140-6736(15)60683-3
9. Hirt, M.N., Hansen, A., Eschenhagen, T.: Cardiac tissue engineering state of the art. Circ. Res. **114**(2), 354–367 (2014). https://doi.org/10.1161/Circresaha.114.300522
10. Mathur, A., Ma, Z., Loskill, P., Jeeawoody, S., Healy, K.E.: In vitro cardiac tissue models: current status and future prospects. Adv Drug Deliver Rev. **96**, 203–213 (2016). https://doi.org/10.1016/j.addr.2015.09.011
11. Ronaldson-Bouchard, K., Ma, S.P., Yeager, K., Chen, T., Song, L.J., Sirabella, D., et al.: Advanced maturation of human cardiac tissue grown from pluripotent stem cells. Nature. **556**(7700), 239 (2018). https://doi.org/10.1038/s41586-018-0016-3
12. Wang, G., McCain, M.L., Yang, L.H., He, A.B., Pasqualini, F.S., Agarwal, A., et al.: Modeling the mitochondrial cardiomyopathy of Barth syndrome with induced pluripotent stem cell and heart-on-chip technologies. Nat. Med. **20**(6), 616–623 (2014). https://doi.org/10.1038/nm.3545

13. Marsano, A., Conficconi, C., Lemme, M., Occhetta, P., Gaudiello, E., Votta, E., et al.: Beating heart on a chip: a novel microfluidic platform to generate functional 3D cardiac microtissues. Lab Chip. **16**(3), 599–610 (2016). https://doi.org/10.1039/c5lc01356a

14. Soares, C.P., Midlej, V., de Oliveira, M.E.W., Benchimol, M., Costa, M.L., Mermelstein, C.: 2D and 3D-organized cardiac cells shows differences in cellular morphology, adhesion junctions, presence of myofibrils and protein expression. Plos One. 2012;7(5). doi: ARTN e38147; https://doi.org/10.1371/journal.pone.0038147

15. Nunes, S.S., Miklas, J.W., Liu, J., Aschar-Sobbi, R., Xiao, Y., Zhang, B.Y., et al.: Biowire: a platform for maturation of human pluripotent stem cell-derived cardiomyocytes. Nat. Methods. **10**(8), 781 (2013). https://doi.org/10.1038/Nmeth.2524

16. Ruan, J.L., Tulloch, N.L., Razumova, M.V., Saiget, M., Muskheli, V., Pabon, L., et al.: Mechanical stress conditioning and electrical stimulation promote contractility and force maturation of induced pluripotent stem cell-derived human cardiac tissue. Circulation. **134**(20), 1557 (2016). https://doi.org/10.1161/Circulationaha.114.014998

17. Hinson, J.T., Chopra, A., Nafissi, N., Polacheck, W.J., Benson, C.C., Swist, S., et al.: Titin mutations in iPS cells define sarcomere insufficiency as a cause of dilated cardiomyopathy. Science. **349**(6251), 982–986 (2015). https://doi.org/10.1126/science.aaa5458

18. Ma, Z., Huebsch, N., Koo, S., Mandegar, M.A., Siemons, B., Boggess, S., et al.: Contractile deficits in engineered cardiac microtissues as a result of MYBPC3 deficiency and mechanical overload. Nat Biomed Eng. **2**(12), 955–967 (2018). https://doi.org/10.1038/s41551-018-0280-4

19. Malan, D., Zhang, M., Stallmeyer, B., Muller, J., Fleischmann, B.K., Schulze-Bahr, E., et al.: Human iPS cell model of type 3 long QT syndrome recapitulates drug-based phenotype correction. Basic Res Cardiol. 2016;111(2). doi: ARTN 14; https://doi.org/10.1007/s00395-016-0530-0

20. Shinozawa, T., Nakamura, K., Shoji, M., Morita, M., Kimura, M., Furukawa, H., et al.: Recapitulation of clinical individual susceptibility to drug-induced QT prolongation in healthy subjects using iPSC-derived cardiomyocytes. Stem Cell Rep. **8**(2), 226–234 (2017). https://doi.org/10.1016/j.stemcr.2016.12.014

21. Zhang, Y.S., Yue, K., Aleman, J., Moghaddam, K.M., Bakht, S.M., Yang, J., et al.: 3D bio-printing for tissue and organ fabrication. Ann. Biomed. Eng. **45**(1), 148–163 (2017). https://doi.org/10.1007/s10439-016-1612-8

22. Hashimoto, H., Olson, E.N., Bassel-Duby, R.: Therapeutic approaches for cardiac regeneration and repair. Nat. Rev. Cardiol. **15**(10), 585–600 (2018). https://doi.org/10.1038/s41569-018-0036-6

23. Xin, M., Olson, E.N., Bassel-Duby, R.: Mending broken hearts: cardiac development as a basis for adult heart regeneration and repair. Nat Rev Mol Cell Bio. **14**(8), 529–541 (2013). https://doi.org/10.1038/nrm3619

24. Volarevic, V., Markovic, B.S., Gazdic, M., Volarevic, A., Jovicic, N., Arsenijevic, N., et al.: Ethical and safety issues of stem cell-based therapy. Int. J. Med. Sci. **15**(1), 36–45 (2018). https://doi.org/10.7150/ijms.21666

25. Roche, E.T., Hastings, C.L., Lewin, S.A., Shvartsman, D.E., Brudno, Y., Vasilyev, N.V., et al.: Comparison of biomaterial delivery vehicles for improving acute retention of stem cells in the infarcted heart. Biomaterials. **35**(25), 6850–6858 (2014). https://doi.org/10.1016/j.biomaterials.2014.04.114

26. Lee, E.J., Kasper, F.K., Mikos, A.G.: Biomaterials for tissue engineering. Ann. Biomed. Eng. **42**(2), 323–337 (2014). https://doi.org/10.1007/s10439-013-0859-6

27. McCain, M.L., Agarwal, A., Nesmith, H.W., Nesmith, A.P., Parker, K.K.: Micromolded gelatin hydrogels for extended culture of engineered cardiac tissues. Biomaterials. **35**(21), 5462–5471 (2014). https://doi.org/10.1016/j.biomaterials.2014.03.052

28. Serpooshan, V., Zhao, M.M., Metzler, S.A., Wei, K., Shah, P.B., Wang, A., et al.: The effect of bioengineered acellular collagen patch on cardiac remodeling and ventricular function post

myocardial infarction. Biomaterials. **34**(36), 9048–9055 (2013). https://doi.org/10.1016/j.biomaterials.2013.08.017

29. Williams, C., Budina, E., Stoppel, W.L., Sullivan, K.E., Emani, S., Emani, S.M., et al.: Cardiac extracellular matrix-fibrin hybrid scaffolds with tunable properties for cardiovascular tissue engineering. Acta Biomater. **14**, 84–95 (2015). https://doi.org/10.1016/j.actbio.2014.11.035

30. Aghdam, R.M., Shakhesi, S., Najarian, S., Mohammadi, M.M., Tafti, S.H.A., Mirzadeh, H.: Fabrication of a nanofibrous scaffold for the in vitro culture of cardiac progenitor cells for myocardial regeneration. Int J Polym Mater Po. **63**(5), 229–239 (2014). https://doi.org/10.1080/00914037.2013.800983

31. Chiono, V., Mozetic, P., Boffito, M., Sartori, S., Gioffredi, E., Silvestri, A., et al.: Polyurethane-based scaffolds for myocardial tissue engineering. Interface Focus. 2014;4(1). doi: ARTN 20130045; https://doi.org/10.1098/rsfs.2013.0045

32. Flaig, F., Ragot, H., Simon, A., Revet, G., Kitsara, M., Kitasato, L., et al.: Design of functional electrospun scaffolds based on poly(glycerol sebacate) elastomer and poly(lactic acid) for cardiac tissue engineering. ACS Biomater Sci. Eng. **6**(4), 2388–2400 (2020). https://doi.org/10.1021/acsbiomaterials.0c00243

33. Crowder, S.W., Liang, Y., Rath, R., Park, A.M., Maltais, S., Pintauro, P.N., et al.: Poly(epsilon-caprolactone)-carbon nanotube composite scaffolds for enhanced cardiac differentiation of human mesenchymal stem cells. Nanomedicine-Uk. **8**(11), 1763–1776 (2013). https://doi.org/10.2217/nnm.12.204

34. Karakikes, I., Ameen, M., Termglinchan, V., Wu, J.C.: Human induced pluripotent stem cell-derived cardiomyocytes insights into molecular, cellular, and functional phenotypes. Circ. Res. **117**(1), 80–88 (2015). https://doi.org/10.1161/Circresaha.117.305365

35. Prabhakaran, M.P., Mobarakeh, L.G., Kai, D., Karbalaie, K., Nasr-Esfahani, M.H., Ramakrishna, S.: Differentiation of embryonic stem cells to cardiomyocytes on electrospun nanofibrous substrates. J Biomed Mater Res B. **102**(3), 447–454 (2014). https://doi.org/10.1002/jbm.b.33022

36. Chen, I.Y., Matsa, E., Wu, J.C.: Induced pluripotent stem cells: at the heart of cardiovascular precision medicine. Nat. Rev. Cardiol. **13**(6), 333–349 (2016). https://doi.org/10.1038/nrcardio.2016.36

37. Mandegar, M.A., Huebsch, N., Frolov, E.B., Shin, E., Truong, A., Olvera, M.P., et al.: CRISPR interference efficiently induces specific and reversible gene silencing in human iPSCs. Cell Stem Cell. **18**(4), 541–553 (2016). https://doi.org/10.1016/j.stem.2016.01.022

38. Hatoum, H., Dasi, L.P.: Reduction of pressure gradient and turbulence using vortex generators in prosthetic heart valves. Ann. Biomed. Eng. **47**(1), 85–96 (2019). https://doi.org/10.1007/s10439-018-02128-6

39. Lu, L., Mende, M., Yang, X.G., Korber, H.F., Schnittler, H.J., Weinert, S., et al.: Design and validation of a bioreactor for simulating the cardiac niche: a system incorporating cyclic stretch, electrical stimulation, and constant perfusion. Tissue Eng Pt A. **19**(3–4), 403–414 (2013). https://doi.org/10.1089/ten.tea.2012.0135

40. Paez-Mayorga, J., Hernandez-Vargas, G., Ruiz-Esparza, G.U., Iqbal, H.M.N., Wang, X.C., Zhang, Y.S., et al.: Bioreactors for cardiac tissue engineering. Adv Healthc Mater. 2019;8(7). doi: ARTN 1701504; 10.1002/adhm.201701504

41. Zhu, J.M., Marchant, R.E.: Design properties of hydrogel tissue-engineering scaffolds. Expert Rev Med Devic. **8**(5), 607–626 (2011). https://doi.org/10.1586/Erd.11.27

42. Saludas, L., Pascual-Gil, S., Prosper, F., Garbayo, E., Blanco-Prieto, M.: Hydrogel based approaches for cardiac tissue engineering. Int. J. Pharm. **523**(2), 454–475 (2017). https://doi.org/10.1016/j.ijpharm.2016.10.061

43. Vedadghavami, A., Minooei, F., Mohammadi, M.H., Khetani, S., Kolahchi, A.R., Mashayekhan, S., et al.: Manufacturing of hydrogel biomaterials with controlled mechanical properties for tissue engineering applications. Acta Biomater. **62**, 42–63 (2017). https://doi.org/10.1016/j.actbio.2017.07.028

44. Goldfracht, I., Efraim, Y., Shinnawi, R., Kovalev, E., Huber, I., Gepstein, A., et al.: Engineered heart tissue models from hiPSC-derived cardiomyocytes and cardiac ECM for disease modeling and drug testing applications. Acta Biomater. **92**, 145–159 (2019). https://doi.org/10.1016/j.actbio.2019.05.016

45. Puperi, D.S., Balaoing, L.R., O'Connell, R.W., West, J.L., Grande-Allen, K.J.: 3-dimensional spatially organized PEG-based hydrogels for an aortic valve co-culture model. Biomaterials. **67**, 354–364 (2015). https://doi.org/10.1016/j.biomaterials.2015.07.039

46. Shah, M., Pawan, K.C., Zhang, G.: In vivo assessment of decellularized porcine myocardial slice as an acellular cardiac patch. ACS Appl. Mater. Interfaces. **11**(27), 23893–23900 (2019). https://doi.org/10.1021/acsami.9b06453

47. Boopathy, A.V., Che, P.L., Somasuntharam, I., Fiore, V.F., Cabigas, E.B., Ban, K., et al.: The modulation of cardiac progenitor cell function by hydrogel-dependent Notch1 activation. Biomaterials. **35**(28), 8103–8112 (2014). https://doi.org/10.1016/j.biomaterials.2014.05.082

48. Das, S., Kim, S.W., Choi, Y.J., Lee, S., Lee, S.H., Kong, J.S., et al.: Decellularized extracellular matrix bioinks and the external stimuli to enhance cardiac tissue development in vitro. Acta Biomater. **95**, 188–200 (2019). https://doi.org/10.1016/j.actbio.2019.04.026

49. Zhang, W., Kong, C.W., Tong, M.H., Chooi, W.H., Huang, N., Li, R.A., et al.: Maturation of human embryonic stem cell-derived cardiomyocytes (hESC-CMs) in 3D collagen matrix: effects of niche cell supplementation and mechanical stimulation. Acta Biomater. **49**, 204–217 (2017). https://doi.org/10.1016/j.actbio.2016.11.058

50. Hirt, M.N., Werner, T., Indenbirken, D., Alawi, M., Demin, P., Kunze, A.C., et al.: Deciphering the microRNA signature of pathological cardiac hypertrophy by engineered heart tissue- and sequencing-technology. J. Mol. Cell. Cardiol. **81**, 1–9 (2015). https://doi.org/10.1016/j.yjmcc.2015.01.008

51. Agarwal, A., Farouz, Y., Nesmith, A.P., Deravi, L.F., McCain, M.L., Parker, K.K.: Micropatterning alginate substrates for in vitro cardiovascular muscle on a chip. Adv. Funct. Mater. **23**(30), 3738–3746 (2013)

52. Zhang, X., Xu, B., Puperi, D.S., Yonezawa, A.L., Wu, Y., Tseng, H., et al.: Integrating valve-inspired design features into poly(ethylene glycol) hydrogel scaffolds for heart valve tissue engineering. Acta Biomater. **14**, 11–21 (2015). https://doi.org/10.1016/j.actbio.2014.11.042

53. Zhao, H., Li, X.K., Zhao, S., Zeng, Y., Zhao, L., Ding, H.Y., et al.: Microengineered in vitro model of cardiac fibrosis through modulating myofibroblast mechanotransduction. Biofabrication. 2014;6(4). doi: Artn 045009; https://doi.org/10.1088/1758-5082/6/4/045009

54. Grant, R., Hay, D.C., Callanan, A.: A drug-induced hybrid electrospun poly-capro-lactone: cell-derived extracellular matrix scaffold for liver tissue engineering. Tissue Eng Pt A. **23**(13–14), 650–662 (2017). https://doi.org/10.1089/ten.tea.2016.0419

55. Grover, G.N., Rao, N., Christman, K.L.: Myocardial matrix-polyethylene glycol hybrid hydrogels for tissue engineering. Nanotechnology. 2014;25(1). doi: Artn 014011; https://doi.org/10.1088/0957-4484/25/1/014011

56. Kumar, A., Rao, K.M., Han, S.S.: Synthesis of mechanically stiff and bioactive hybrid hydrogels for bone tissue engineering applications. Chem. Eng. J. **317**, 119–131 (2017). https://doi.org/10.1016/j.cej.2017.02.065

57. Duan, B., Yin, Z.Y., Kang, L.H., Magin, R.L., Butcher, J.T.: Active tissue stiffness modulation controls valve interstitial cell phenotype and osteogenic potential in 3D culture. Acta Biomater. **36**, 42–54 (2016). https://doi.org/10.1016/j.actbio.2016.03.007

58. McGann, C.L., Levenson, E.A., Kiick, K.L.: Resilin-based hybrid hydrogels for cardiovascular tissue engineering. Macromol. Chem. Phys. **214**(2), 203–213 (2013). https://doi.org/10.1002/macp.201200412

59. Yang, B.G., Yao, F.L., Hao, T., Fang, W.C., Ye, L., Zhang, Y.B., et al.: Development of electrically conductive double-network hydrogels via one-step facile strategy for cardiac tissue engineering. Adv. Healthc. Mater. **5**(4), 474–488 (2016). https://doi.org/10.1002/adhm.201500520

60. Shin, S.R., Jung, S.M., Zalabany, M., Kim, K., Zorlutuna, P., Kim, S.B., et al.: Carbon-nanotube-embedded hydrogel sheets for engineering cardiac constructs and bioactuators. ACS Nano. **7**(3), 2369–2380 (2013). https://doi.org/10.1021/nn305559j
61. Radhakrishnan, J., Krishnan, U.M., Sethuraman, S.: Hydrogel based injectable scaffolds for cardiac tissue regeneration. Biotechnol. Adv. **32**(2), 449–461 (2014). https://doi.org/10.1016/j.biotechadv.2013.12.010
62. Sabbah, H.N., Wang, M.J., Gupta, R.C., Rastogi, S., Ilsar, I., Sabbah, M.S., et al.: Augmentation of left ventricular Wall thickness with alginate hydrogel implants improves left ventricular function and prevents progressive remodeling in dogs with chronic heart failure. Jacc-Heart Fail. **1**(3), 252–258 (2013). https://doi.org/10.1016/j.jchf.2013.02.006
63. Waters, R., Alam, P., Pacelli, S., Chakravarti, A.R., Ahmed, R.P.H., Paul, A.: Stem cell-inspired secretome-rich injectable hydrogel to repair injured cardiac tissue. Acta Biomater. **69**, 95–106 (2018). https://doi.org/10.1016/j.actbio.2017.12.025
64. Levit, R.D., Landazuri, N., Phelps, E.A., Brown, M.E., Garcia, A.J., Davis, M.E., et al.: Cellular encapsulation enhances cardiac repair. J Am Heart Assoc. 2013;2(5). doi: ARTN e000367; https://doi.org/10.1161/JAHA.113.000367
65. Tran, C., Damaser, M.S.: Stem cells as drug delivery methods: Application of stem cell secretome for regeneration. Adv Drug Deliver Rev. 2015;82–83:1–11. https://doi.org/10.1016/j.addr.2014.10.007
66. Hasan, A., Khattab, A., Islam, M.A., Abou Hweij, K., Zeitouny, J., Waters, R., et al.: Injectable hydrogels for cardiac tissue repair after myocardial infarction. Adv Sci. 2015;2(11). doi: ARTN 1500122; 10.1002/advs.201500122
67. Rufaihah, A.J., Johari, N.A., Vaibavi, S.R., Plotkin, M., Thien, D.T.D., Kofidis, T., et al.: Dual delivery of VEGF and ANG-1 in ischemic hearts using an injectable hydrogel. Acta Biomater. **48**, 58–67 (2017). https://doi.org/10.1016/j.actbio.2016.10.013
68. Paul, A., Hasan, A., Al Kindi, H., Gaharwar, A.K., Rao, V.T.S., Nikkhah, M., et al.: Injectable graphene oxide/hydrogel-based angiogenic gene delivery system for vasculogenesis and cardiac repair. ACS Nano. **8**(8), 8050–8062 (2014). https://doi.org/10.1021/nn5020787
69. Mihalko, E., Huang, K., Sproul, E., Cheng, K., Brown, A.C.: Targeted treatment of Ischemic and fibrotic complications of myocardial infarction using a dual-delivery microgel therapeutic. ACS Nano. **12**(8), 7826–7837 (2018). https://doi.org/10.1021/acsnano.8b01977
70. Rajabi-Zeleti, S., Jalili-Firoozinezhad, S., Azarnia, M., Khayyatan, F., Vandat, S., Nikeghbalian, S., et al.: The behavior of cardiac progenitor cells on macroporous pericardium-derived scaffolds. Biomaterials. **35**(3), 970–982 (2014). https://doi.org/10.1016/j.biomaterials.2013.10.045
71. Fan CX, Shi JJ, Zhuang Y, Zhang LL, Huang L, Yang W, et al. Myocardial-infarction-responsive smart hydrogels targeting matrix metalloproteinase for on-demand growth factor delivery. Adv Mater. 2019;31(40). doi: ARTN 1902900; https://doi.org/10.1002/adma.201902900
72. MacQueen, L.A., Sheehy, S.P., Chantre, C.O., Zimmerman, J.F., Pasqualini, F.S., Liu, X.J., et al.: A tissue-engineered scale model of the heart ventricle. Nat Biomed Eng. **2**(12), 930–941 (2018). https://doi.org/10.1038/s41551-018-0271-5
73. Liu, W., Zhong, Z., Hu, N., Zhou, Y., Maggio, L., Miri, A., et al.: Coaxial extrusion bioprinting of 3D microfibrous constructs with cell-favorable gelatin methacryloyl microenvironments. Biofabrication. (2017). https://doi.org/10.1088/1758-5090/aa9d44
74. Hollister, S.J.: Porous scaffold design for tissue engineering. Nat. Mater. **4**(7), 518–524 (2005). https://doi.org/10.1038/nmat1421
75. Loh, Q.L., Choong, C.: Three-dimensional scaffolds for tissue engineering applications: role of porosity and pore size. Tissue Eng Part B-Re. **19**(6), 485–502 (2013). https://doi.org/10.1089/ten.teb.2012.0437
76. Ambekar, R.S., Kandasubramanian, B.: Progress in the advancement of porous biopolymer scaffold: tissue engineering application. Ind. Eng. Chem. Res. **58**(16), 6163–6194 (2019). https://doi.org/10.1021/acs.iecr.8b05334

77. Pok, S., Myers, J.D., Madihally, S.V., Jacot, J.G.: A multilayered scaffold of a chitosan and gelatin hydrogel supported by a PCL core for cardiac tissue engineering. Acta Biomater. **9**(3), 5630–5642 (2013). https://doi.org/10.1016/j.actbio.2012.10.032

78. Wang, Y.Y., Hu, J., Jiao, J., Liu, Z.N., Zhou, Z., Zhao, C., et al.: Engineering vascular tissue with functional smooth muscle cells derived from human iPS cells and nanofibrous scaffolds. Biomaterials. **35**(32), 8960–8969 (2014). https://doi.org/10.1016/j.biomaterials.2014.07.011

79. Bian, W.N., Jackman, C.P., Bursac, N.: Controlling the structural and functional anisotropy of engineered cardiac tissues. Biofabrication. 2014;6(2). doi: Artn 024109; 10.1088/1758-5082/6/2/024109

80. Engelmayr, G.C., Cheng, M.Y., Bettinger, C.J., Borenstein, J.T., Langer, R., Freed, L.E.: Accordion-like honeycombs for tissue engineering of cardiac anisotropy. Nat. Mater. **7**(12), 1003–1010 (2008). https://doi.org/10.1038/nmat2316

81. Zieber, L., Or, S., Ruvinov, E., Cohen, S.: Microfabrication of channel arrays promotes vessel-like network formation in cardiac cell construct and vascularization in vivo. Biofabrication. 2014;6(2). doi: Artn 024102; 10.1088/1758-5082/6/2/024102

82. Fang, Y.C., Zhang, T., Zhang, L., Gong, W.F., Sun, W.: Biomimetic design and fabrication of scaffolds integrating oriented micro-pores with branched channel networks for myocardial tissue engineering. Biofabrication. 2019;11(3). doi: ARTN 035004; 10.1088/1758-5090/ab0fd3

83. Madden, L.R., Mortisen, D.J., Sussman, E.M., Dupras, S.K., Fugate, J.A., Cuy, J.L., et al.: Proangiogenic scaffolds as functional templates for cardiac tissue engineering. P Natl Acad Sci USA. **107**(34), 15211–15216 (2010). https://doi.org/10.1073/pnas.1006442107

84. Miyagi, Y., Chiu, L.L.Y., Cimini, M., Weisel, R.D., Radisic, M., Li, R.K.: Biodegradable collagen patch with covalently immobilized VEGF for myocardial repair. Biomaterials. **32**(5), 1280–1290 (2011). https://doi.org/10.1016/j.biomaterials.2010.10.007

85. Jun, I., Han, H.S., Edwards, J.R., Jeon, H.: Electrospun fibrous scaffolds for tissue engineering: viewpoints on architecture and fabrication. Int J Mol Sci. 2018;19(3). Doi: ARTN 745; 10.3390/ijms19030745

86. Capulli, A.K., MacQueen, L.A., Sheehy, S.P., Parker, K.K.: Fibrous scaffolds for building hearts and heart parts. Adv Drug Deliver Rev. **96**, 83–102 (2016). https://doi.org/10.1016/j.addr.2015.11.020

87. Ma, Z., Koo, S., Finnegan, M.A., Loskill, P., Huebsch, N., Marks, N.C., et al.: Three-dimensional filamentous human diseased cardiac tissue model. Biomaterials. **35**(5), 1367–1377 (2014). https://doi.org/10.1016/j.biomaterials.2013.10.052

88. Zhang, Y.S., Arneri, A., Bersini, S., Shin, S.R., Zhu, K., Goli-Malekabadi, Z., et al.: Bioprinting 3D microfibrous scaffolds for engineering endothelialized myocardium and heart-on-a-chip. Biomaterials. **110**, 45–59 (2016). https://doi.org/10.1016/j.biomaterials.2016.09.003

89. Parrag, I.C., Zandstra, P.W., Woodhouse, K.A.: Fiber alignment and coculture with fibroblasts improves the differentiated phenotype of murine embryonic stem cell-derived cardiomyocytes for cardiac tissue engineering. Biotechnol. Bioeng. **109**(3), 813–822 (2012). https://doi.org/10.1002/bit.23353

90. Kharaziha, M., Nikkhah, M., Shin, S.R., Annabi, N., Masoumi, N., Gaharwar, A.K., et al.: PGS:Gelatin nanofibrous scaffolds with tunable mechanical and structural properties for engineering cardiac tissues. Biomaterials. **34**(27), 6355–6366 (2013). https://doi.org/10.1016/j.biomaterials.2013.04.045

91. Wu, S., Duan, B., Qin, X.H., Butcher, J.T.: Living nano-micro fibrous woven fabric/hydrogel composite scaffolds for heart valve engineering. Acta Biomater. **51**, 89–100 (2017). https://doi.org/10.1016/j.actbio.2017.01.051

92. Fleischer, S., Feiner, R., Shapira, A., Ji, J., Sui, X.M., Wagner, H.D., et al.: Spring-like fibers for cardiac tissue engineering. Biomaterials. **34**(34), 8599–8606 (2013). https://doi.org/10.1016/j.biomaterials.2013.07.054

93. Zhao, G.X., Zhang, X.H., Lu, T.J., Xu, F.: Recent advances in electrospun Nanofibrous scaffolds for cardiac tissue engineering. Adv. Funct. Mater. **25**(36), 5726–5738 (2015). https://doi.org/10.1002/adfm.201502142

94. Ravichandran, R., Venugopal, J.R., Mukherjee, S., Sundarrajan, S., Ramakrishna, S.: Elastomeric core/shell nanofibrous cardiac patch as a biomimetic support for infarcted porcine myocardium. Tissue Eng Pt A. **21**(7–8), 1288–1298 (2015). https://doi.org/10.1089/ten.tea.2014.0265

95. Wang, C., Koo, S., Park, M., Vangelatos, Z., Hoang, P., Conklin, B.R., et al.: Maladaptive contractility of 3D human cardiac microtissues to mechanical nonuniformity. Adv. Healthc. Mater. **9**(8), e1901373 (2020)

96. Badeau, B.A., DeForest, C.A.: Programming stimuli-responsive behavior into biomaterials. Annu. Rev. Biomed. Eng. **21**, 241–265 (2019). https://doi.org/10.1146/annurev-bioeng-060418-052324

97. Uto, K., Tsui, J.H., DeForest, C.A., Kim, D.H.: Dynamically tunable cell culture platforms for tissue engineering and mechanobiology. Prog. Polym. Sci. **65**, 53–82 (2017). https://doi.org/10.1016/j.progpolymsci.2016.09.004

98. Chaudhuri, O., Gu, L., Klumpers, D., Darnell, M., Bencherif, S.A., Weaver, J.C., et al.: Hydrogels with tunable stress relaxation regulate stem cell fate and activity. Nat. Mater. **15**(3), 326 (2016). https://doi.org/10.1038/Nmat4489

99. Yang, H., Wang, Q., Huang, S., Xiao, A., Li, F.Y., Gan, L., et al.: Smart pH/redox dual-responsive nanogels for on-demand intracellular anticancer drug release. ACS Appl. Mater. Interfaces. **8**(12), 7729–7738 (2016). https://doi.org/10.1021/acsami.6b01602

100. Sponchioni, M., Palmiero, U.C., Moscatelli, D.: Thermo-responsive polymers: applications of smart materials in drug delivery and tissue engineering. Mat Sci Eng C-Mater. **102**, 589–605 (2019). https://doi.org/10.1016/j.msec.2019.04.069

101. Shimizu, T., Yamoto, M., Kikuchi, A., Okano, T.: Two-dimensional manipulation of cardiac myocyte sheets utilizing temperature-responsive culture dishes augments the pulsatile amplitude. Tissue Eng. **7**(2), 141–151 (2001). https://doi.org/10.1089/107632701300062732

102. Sung, T.C., Su, H.C., Ling, Q.D., Kumar, S.S., Chang, Y., Hsu, S.T., et al.: Efficient differentiation of human pluripotent stem cells into cardiomyocytes on cell sorting thermoresponsive surface. Biomaterials. 2020;253. doi: ARTN 120060; 10.1016/j.biomaterials.2020.120060

103. Sun, S., Shi, H., Moore, S., Wang, C., Ash-Shakoor, A., Mather, P.T., et al.: Progressive myofibril reorganization of human cardiomyocytes on a dynamic nanotopographic substrate. ACS Appl. Mater. Interfaces. **12**(19), 21450–21462 (2020)

104. Vong, L.B., Bui, T.Q., Tomita, T., Sakamoto, H., Hiramatsu, Y., Nagasaki, Y.: Novel angiogenesis therapeutics by redox injectable hydrogel – regulation of local nitric oxide generation for effective cardiovascular therapy. Biomaterials. **167**, 143–152 (2018). https://doi.org/10.1016/j.biomaterials.2018.03.023

105. Masumoto, H., Matsuo, T., Yamamizu, K., Uosaki, H., Narazaki, G., Katayama, S., et al.: Pluripotent stem cell-engineered cell sheets reassembled with defined cardiovascular populations ameliorate reduction in infarct heart function through cardiomyocyte-mediated neovascularization. Stem Cells. **30**(6), 1196–1205 (2012). https://doi.org/10.1002/stem.1089

106. Young, J.L., Engler, A.J.: Hydrogels with time-dependent material properties enhance cardiomyocyte differentiation in vitro. Biomaterials. **32**(4), 1002–1009 (2011). https://doi.org/10.1016/j.biomaterials.2010.10.020

107. Corbin, E.A., Vite, A., Peyster, E.G., Bhoovalam, M., Brandimarto, J., Wang, X., et al.: Tunable and reversible substrate stiffness reveals a dynamic mechanosensitivity of cardiomyocytes. ACS Appl. Mater. Interfaces. **11**(23), 20603–20614 (2019). https://doi.org/10.1021/acsami.9b02446

108. Leng, J.S., Lan, X., Liu, Y.J., Du, S.Y.: Shape-memory polymers and their composites: stimulus methods and applications. Prog. Mater. Sci. **56**(7), 1077–1135 (2011). https://doi.org/10.1016/j.pmatsci.2011.03.001

109. Le, D.M., Kulangara, K., Adler, A.F., Leong, K.W., Ashby, V.S.: Dynamic topographical control of mesenchymal stem cells by culture on responsive poly(epsilon-caprolactone) surfaces. Adv. Mater. **23**(29), 3278 (2011). https://doi.org/10.1002/adma.201100821

110. Mengsteab, P.Y., Uto, K., Smith, A.S.T., Frankel, S., Fisher, E., Nawas, Z., et al.: Spatiotemporal control of cardiac anisotropy using dynamic nanotopographic cues. Biomaterials. **86**, 1–10 (2016). https://doi.org/10.1016/j.biomaterials.2016.01.062

111. Bastings, M.M.C., Koudstaal, S., Kieltyka, R.E., Nakano, Y., Pape, A.C.H., Feyen, D.A.M., et al.: A fast pH-switchable and self-healing supramolecular hydrogel carrier for guided, local catheter injection in the infarcted myocardium. Adv. Healthc. Mater. **3**(1), 70–78 (2014). https://doi.org/10.1002/adhm.201300076

112. Wang, Q., Mynar, J.L., Yoshida, M., Lee, E., Lee, M., Okuro, K., et al.: High-water-content mouldable hydrogels by mixing clay and a dendritic molecular binder. Nature. **463**(7279), 339–343 (2010). https://doi.org/10.1038/nature08693

113. Yang, Y.H., Liu, C.H., Liang, Y.H., Lin, F.H., Wu, K.C.W.: Hollow mesoporous hydroxyapatite nanoparticles (hmHANPs) with enhanced drug loading and pH-responsive release properties for intracellular drug delivery. J. Mater. Chem. B. **1**(19), 2447–2450 (2013). https://doi.org/10.1039/c3tb20365d

114. Kreyling, W.G., Abdelmonem, A.M., Ali, Z., Alves, F., Geiser, M., Haberl, N., et al.: In vivo integrity of polymer-coated gold nanoparticles. Nat. Nanotechnol. **10**(7), 619 (2015). https://doi.org/10.1038/Nnano.2015.111

115. Li, X., Zhou, J., Liu, Z.Q., Chen, J., Lu, S.H., Sun, H.Y., et al.: A PNIPAAm-based thermosensitive hydrogel containing SWCNTs for stem cell transplantation in myocardial repair. Biomaterials. **35**(22), 5679–5688 (2014). https://doi.org/10.1016/j.biomaterials.2014.03.067

116. Li, Z.Q., Fan, Z.B., Xu, Y.Y., Lo, W.S., Wang, X., Niu, H., et al.: pH-sensitive and thermosensitive hydrogels as stem-cell carriers for cardiac therapy. ACS Appl. Mater. Interfaces. **8**(17), 10752–10760 (2016). https://doi.org/10.1021/acsami.6b01374

117. Yabluchanskiy, A., Ma, Y.G., Iyer, R.P., Hall, M.E., Lindsey, M.L.: Matrix metalloproteinase-9: many shades of function in cardiovascular disease. Physiology. **28**(6), 391–403 (2013). https://doi.org/10.1152/physiol.00029.2013

118. Nguyen, M.M., Carlini, A.S., Chien, M.P., Sonnenberg, S., Luo, C.L., Braden, R.L., et al.: Enzyme-responsive nanoparticles for targeted accumulation and prolonged retention in heart tissue after myocardial infarction. Adv. Mater. **27**(37), 5547–5552 (2015). https://doi.org/10.1002/adma.201502003

119. Hao, T., Li, J.J., Yao, F.L., Dong, D.Y., Wang, Y., Yang, B.G., et al.: Injectable fullerenol/alginate hydrogel for suppression of oxidative stress damage in Brown adipose-derived stem cells and cardiac repair. ACS Nano. **11**(6), 5474–5488 (2017). https://doi.org/10.1021/acsnano.7b00221

120. Park, S., Yoon, J., Bae, S., Park, M., Kang, C., Ke, Q.G., et al.: Therapeutic use of H2O2-responsive anti-oxidant polymer nanoparticles for doxorubicin-induced cardiomyopathy. Biomaterials. **35**(22), 5944–5953 (2014). https://doi.org/10.1016/j.biomaterials.2014.03.084

121. Tan, S.Y., Teh, C., Ang, C.Y., Li, M.H., Li, P.Z., Korzh, V., et al.: Responsive mesoporous silica nanoparticles for sensing of hydrogen peroxide and simultaneous treatment toward heart failure. Nanoscale. **9**(6), 2253–2261 (2017). https://doi.org/10.1039/c6nr08869d

122. Lee, J., Manoharan, V., Cheung, L., Lee, S., Cha, B.H., Newman, P., et al.: Nanoparticle-based hybrid scaffolds for deciphering the role of multimodal cues in cardiac tissue engineering. ACS Nano. **13**(11), 12525–12539 (2019). https://doi.org/10.1021/acsnano.9b03050

123. Holzl K, Lin SM, Tytgat L, Van Vlierberghe S, Gu LX, Ovsianikov A. Bioink properties before, during and after 3D bioprinting. Biofabrication. 2016;8(3). doi: Artn 032002; 10.1088/1758-5090/8/3/032002

124. Zhao, Y., Rafatian, N., Feric, N.T., Cox, B.J., Aschar-Sobbi, R., Wang, E.Y., et al.: A platform for generation of chamber-specific cardiac tissues and disease Modeling. Cell. **176**(4), 913 (2019). https://doi.org/10.1016/j.cell.2018.11.042

125. Montag, J., Petersen, B., Flogel, A.K., Becker, E., Lucas-Hahn, A., Cost, G.J., et al.: Successful knock-in of hypertrophic cardiomyopathy-mutation R723G into the MYH7 gene mimics HCM pathology in pigs. Sci Rep-Uk. 2018;8. doi: ARTN 4786; 10.1038/s41598-018-22936-z

126. Paunovska, K., Sago, C.D., Monaco, C.M., Hudson, W.H., Castro, M.G., Rudoltz, T.G., et al.: A direct comparison of in vitro and in vivo nucleic acid delivery mediated by hundreds of nanoparticles reveals a weak correlation. Nano Lett. **18**(3), 2148–2157 (2018). https://doi.org/10.1021/acs.nanolett.8b00432

127. Hwang, J., Huang, Y.F., Burwell, T.J., Peterson, N.C., Connor, J., Weiss, S.J., et al.: In situ imaging of tissue remodeling with collagen hybridizing peptides. ACS Nano. **11**(10), 9825–9835 (2017). https://doi.org/10.1021/acsnano.7b03150

128. Ouyang, L.L., Armstrong, J.P.K., Salmeron-Sanchez, M., Stevens, M.M.: Assembling living building blocks to engineer complex tissues. Adv Funct Mater. 2020;30(26). doi: ARTN 1909009; https://doi.org/10.1002/adfm.201909009

129. Fleischer, S., Shapira, A., Feiner, R., Dvir, T.: Modular assembly of thick multifunctional cardiac patches. P Natl Acad Sci USA. **114**(8), 1898–1903 (2017). https://doi.org/10.1073/pnas.1615728114

130. Richards, D.J., Coyle, R.C., Tan, Y., Jia, J., Wong, K., Toomer, K., et al.: Inspiration from heart development: biomimetic development of functional human cardiac organoids. Biomaterials. **142**, 112–123 (2017). https://doi.org/10.1016/j.biomaterials.2017.07.021

Stem Cell-Based 3D Bioprinting for Cardiovascular Tissue Regeneration

Clara Liu Chung Ming, Eitan Ben-Sefer, and Carmine Gentile

1 Introduction

Cardiovascular disease (CVD) represents the single greatest cause of death in the world, especially in the aging population [1–3]. The increased incidence of CVD has been more recently associated with co-morbidity with other chronic diseases, such as kidney failure and type II diabetes [3–6]. In the last decade, CVD accounted for nearly one third of all deaths worldwide [1, 3, 7]. The global disease burden caused by CVD is estimated to include up to 400 deaths per 100,000 in developed countries and is further driven by an unprecedented growing and aging population, with notable increases in ischaemic heart disease (IHD), stroke and heart failure (HF) [1, 2, 7]. IHD and strokes are caused by a lack of blood supply and oxygen to the heart or brain, respectively, and are the main CVD contributions to the global disease burden accounting for 8–10% of all deaths in Europe [2, 3, 8]. Their treatments have greatly advanced over the past three decades, resulting in improved survival rates [1, 5, 8]. Multiple therapeutic interventions including drugs (such as, cholesterol modifiers and anti-hypertensives) and surgical procedures aiming at repairing or bypassing damaged arteries, have greatly reduced the mortality of CVD

C. Liu Chung Ming · E. Ben-Sefer
School of Biomedical Engineering/FEIT, University of Technology Sydney,
Sydney, NSW, Australia

C. Gentile (✉)
School of Biomedical Engineering/FEIT, University of Technology Sydney,
Sydney, NSW, Australia

Sydney Medical School, The University of Sydney, Sydney, NSW, Australia

Beth Israel Deaconess Medical Center, Harvard Medical School, Boston, MA, USA
e-mail: carmine.gentile@uts.edu.au

© The Author(s), under exclusive license to Springer Nature
Switzerland AG 2022
J. Zhang, V. Serpooshan (eds.), *Advanced Technologies in Cardiovascular
Bioengineering*, https://doi.org/10.1007/978-3-030-86140-7_13

patients and allowed for lifestyle changes to complement these primary interventions [5, 7, 9].

HF is a more complex presentation of CVD, characterized by the heart inability to pump enough blood to meet the body oxygen demand [10]. HF is widely considered to be a chronic phase of cardiac impairment, secondary to other CVD as well as risk factors, including diabetes, obesity, and chronic hypertension [11, 12]. This compromises its contractility and leads to heart failure and death as the current treatments options are limited [13]. Palliative drugs, such as ACE inhibition, beta blockade, and diuretics, or the mechanical assist devices (including ventricular-assist devices, or VAD, pacemakers, defibrillators) only delay the progression of heart failure and do not lead to the regeneration of the heart tissue [14–16]. The gold standard treatment remains a heart transplant, which is available to less than 0.1% of heart failure patients [17]. In the next decade, HF prevalence is estimated to double from 27 million cases worldwide to over 50 million cases, with a 1 in 5 lifetime risk of developing some form of HF [10, 12, 18]. This trend suggests new disease models, treatment options and strategies are vital, as the CVD epidemic continues to grow.

Tissue regeneration in the human body occurs primarily *via* the recruitment of progenitor cells to replace lost cells through differentiation and proliferation [13, 19, 20]. However, adult human heart has limited regenerative capacity as cardiac myocytes lose their ability to proliferate after birth [16]. As the heart does not contain enough stem, precursor, or reserve cells to effectively heal itself after an injury, majority of cardiomyocytes are lost, and the necrotic muscle is replaced with scar tissue [13, 16, 21]. In this context, heart regeneration has been the focus of several studies in the past decades, where cardiovascular researchers developed new approaches to deliver cells to the heart, reactivate the endogenous regenerative capacity through paracrine mechanism or by bio-engineering technologies [14, 22, 23].

In order to overcome the limited survival of cardiac cells following their delivery in a damaged heart, 3D bioprinting technology for cardiac regeneration has gained increasing popularity in the past decade [24]. Cardiovascular tissue engineering (CTE) has emerged to design and manufacture biologically relevant cardiovascular tissues for research applications with the goal of furthering cardiovascular regenerative medicine, and clinical applications to improve CVD patient outcomes and quality of life. A key component of CTE is to accurately recapitulate the human heart microenvironment to promote cell survival, functionality and increase success of potential implantation. Additionally, this process can utilize patient-derived stem cells to improve contractile function and vascularization in bioengineered microtissues with decreased risk of immunological response [24].

2   Molecular, Cellular and Extracellular Approaches to Promote Cardiovascular Regeneration in Humans

This book chapter aims at illustrating a comprehensive overview of the state-of-the-art approaches currently used to promote cardiovascular regeneration (Fig. 1). These include cell-based therapies and cardiac tissue engineering approaches to differentiate and proliferate into functional cardiac myocytes, and through paracrine effects. The cell-free therapy mediates cytoprotection, recruits cells, mediates inflammatory response and prevents fibrotic scar tissue formation. Currently, numerous strategies are investigated thanks to the potential beneficial effects of stem cells on the failing myocardium in the preclinical or clinical settings. This includes bio-engineering methods, to improve the heart microenvironment and promote cell survival, including the use of stem cells and 3D bioprinting technology. Also, the multiple aspects of CTE work in unison to serve this end; this includes areas such as cell types and culture technique, polymers and biopolymers, material scaffolds, vascularization of tissues, hydrogels, bioprinting and bioinks. This book chapter will thoroughly describe all the possible tools for the improved generation of viable and functional heart tissues for *in vitro* and *in vivo* applications. However, several limitations and key unanswered questions prevent their direct application to humans, which will also, be highlighted through this manuscript.

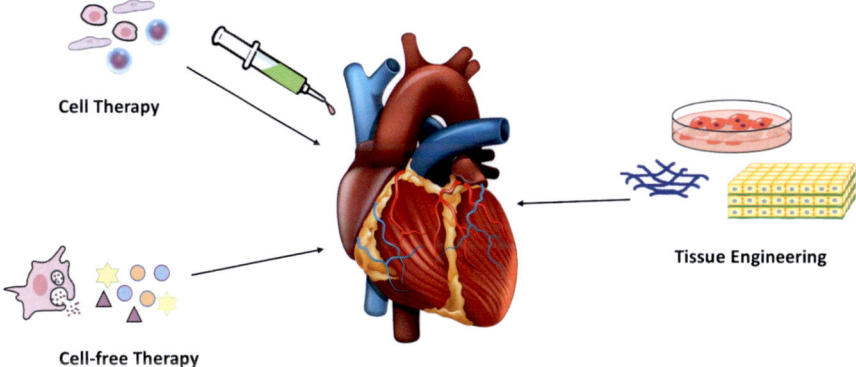

Fig. 1 Approaches for cardiac regeneration to treat a damaged heart
These include: (*1*) Cell therapy by transplantation in the myocardium *via* intracoronary, intramyocardial, intravenous or transendocardial. (*2*) Cell-free therapy secreting paracrine factors, such as, cytokine, growth factors and microRNAs to improve cardiac regeneration. (*3*) Tissue engineering approaches combining cells with biomaterial, such as, cell sheets, scaffolds, spheroids to design cardiac patches or injectable materials for transplantation into the infarcted heart

2.1 Cell-Free Approaches

From the molecular viewpoint, the beneficial effects of cell therapy mainly function through paracrine mechanisms, and they play essential roles during cardiogenesis, cytoprotection, neovascularization and limit inflammatory, profibrotic and apoptosis [14, 17, 25, 26]. Hence, researchers have proposed a novel strategy for heart regeneration and new treatment, that is, cell-free therapy by using paracrine factors [20, 27]. Paracrine factors including growth factors and cytokines are normally released from cells in the myocardium in response to injury [15]. Various types of stem cells, such as, BMCs, CSCs, ESs and MSCs, have shown to mediate cytoprotection via increased expression of fibroblast growth factor (FGF), vascular endothelial growth factor (VEGF), erythropoietin (Epo), and granulocyte-colony stimulating factor (G-CSF) [23, 28]. For example, Lui et al. [29] showed that VEGF-A promotes endothelial specification, engraftment, proliferation and survival of human Isl1+ cardiovascular progenitor cells; hence, suggesting a novel approach for vascular regeneration in the ischemic heart [14, 29]. The main advantage of this cell-free therapy is its safety, that is, avoiding the risks of unlimited cell growth and tumour formation, however, bioactive molecules in the extracellular medium undergo rapid hydrolysis and it is still unclear if they can provide long-term benefit to patients. For example, VEGF lifetime in human blood is less than 30 minutes [30, 31].

Paracrine factors can be secreted in a spatiotemporal manner and enhance regeneration of cardiac myocytes, but the method of delivery still needs to be improved. Various clinical trials have used delivery methods, such as, intracoronary, intravenous and intramyocardial injection of growth factors, but have failed to provide consistent results of significant improvement of myocardial ischemia [14, 32]. For example, in the NORTHERN clinical trial, VEGF gene therapy via intramyocardial injection has failed to improve the perfusion of ischemic myocardium (3 and 6 months) [33]. Another clinical trial based on neuregulin-1 (rhNRG-1) showed no significant difference from the placebo group. However, the same study showed that the short-term administration of rhNRG-1 treatment improved the cardiac function of chronic HF patients by increasing LVEF% and reducing the end-diastolic volume and end-systolic volume [34]. The poor outcomes of growth factor-based approaches could be due to the lack of controlled release, off-target side effects, inappropriate dosage, and the duration of expression [14, 23, 32]. More studies are required to find the ideal delivery methodologies with appropriate dosage and appropriate duration of expression of the growth factors, as prolonged expression might lead to unwanted side effects [14].

Over the past decade, extracellular vesicles (EVs) attracted the interest of studies aiming at regenerating the myocardium as promising tools for the delivery of biologically active molecule to promote new tissue formation [31, 35]. However, more important matters need to be first addressed such as the type and size of vesicles, their content, high cost and time-consuming isolation procedure as well as their potential immunogenicity [23, 31].

2.2 Cell-Based Approaches

Stem cells have been the focus of emerging research to heal or replace damaged cardiac tissues (Table 1) [23, 24]. They are potentially useful in cardiac regeneration as they can self-renew as well as differentiate into multiple types of cells in the body. Various cell types have been studied due to their regenerative potential. These include skeletal myoblasts (SMs), bone marrow- derived cells (BMCs), and mesenchymal stem cells (MSCs) that are known as the first-generation cell types [16, 36]. Despite promising preclinical studies, the transplantation of these cells displayed heterogeneous clinical outcomes, which could be due to differences in study design, including, cell preparation, delivery route, dose, and follow-up methods [23, 37]. Due to the inconsistencies of the first-generation cell types, the field has tried to use other cell types known as the second-generation cell types (Fig. 2). These include cardiac stem/progenitor cells (CSCs/CPCs), pluripotent stem cells (embryonic stem cells (ESCs) and induced pluripotent stem cells (iPSCs) [23]. The

Table 1 Cell therapy types to treat heart diseases

	Cell types	Type of delivery	Advantages	Disadvantages	References
First-generation	Skeletal myoblasts (SMs)	Intramyocardial injection	– Positive data and excellent engraftment	– Higher rate of arrhythmias	[40, 41]
	Bone marrow-derived cells (BMMCs)	Delivered either systemically or intramyocardially	– Safe, abundant, and easy to isolate	– Limited differentiation	[42, 43]
	Mesenchymal stem cells (MSCs)	LV injection	– Multipotent – Reduce any risk of immune rejection	– Little differentiation to cardiomyocytes – Minimal cell engraftment	[44–46]
Second generation	Cardiac stem cells (CSCs)	LV injection	– Multipotent, self-renewing – Reduce myocardial damage	– Limited amount – Minimal cell engraftment	[47, 48]
	Embryonic stem cells (ESCs)	Intramyocardial injection	– Pluripotent, self-renewing Improve cardiac function	– Ethical concerns – Increase risk of ventricular arrhythmias	[21, 49, 50]
	Human induced pluripotent stem cells (hiPSCs)	Intramyocardial injection	– Pluripotent, self-renewing – Same benefits as ESCs	– Less mature than adult cardiomyocytes	[51, 52]

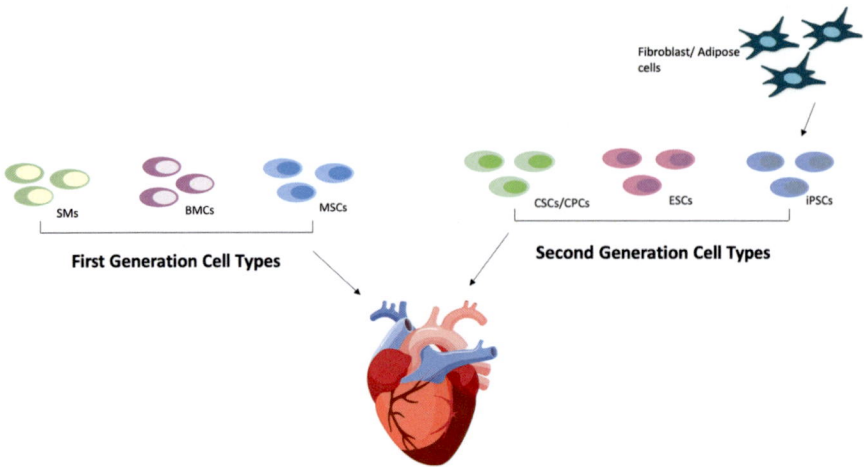

Fig. 2 Evolution of cell therapy for cardiac regeneration
First-generation cell types, such as, SMs, BMCs and MSCs, showed feasibility and safety outcomes, but limited efficacy in clinical setting. Second-generation cell type approaches consist of
CSCs/CPCs, ESCs and iPSCs that demonstrated more efficient and improve therapeutic effects on
the failing heart

second-generation cell types have shown to be more efficient and better therapeutic
effects compared to the first generation in terms of improving engraftment and cardiac function and reducing angiogenesis and scar size. However, the second generation cell types have some limitations [38, 39]. This section provides an overview of
the first-generation and second-generation cell types.

Skeletal Myoblasts (SMs)

The first type of cell used for cardiac regeneration as cellular therapy was skeletal
myoblasts delivered by intramyocardial injection during coronary artery bypass
grafting (CABG) surgery [28, 41]. The attributes of SMs are that they are resistant
to ischemia, highly abundant, no potential drawbacks of undifferentiated stem cells
and no immune rejection from the patients due to their autologous origin [16]. The
clinical trials in both MI and HF by Menasché [41] showed evidence of positive
safety data and excellent engraftment but found concerns of being pro-arrhythmic
and discovered a lack of electrophysiology coupling due to N-cadherin and con-
nexin-43 downregulation. Consequently, the investigations using SMs have ceased
due to evidence that skeletal myoblast might increase the risk of ventricular arrhythmias [23, 28, 40].

Bone Marrow-Derived Cells (BMCs)

Previously, it was reported that BMCs derived from adult bone marrow could give rise to cardiomyocytes, vascular endothelium and smooth muscle both *in vitro* and *in vivo* [16, 43, 53]. Bone marrow mononuclear cells have mostly been used for the investigations as the cell source is safe, abundant, and easy to isolate [17, 23]. The advantages of using BMCs are to avoid ethical and clinical issues. Furthermore, there were prominent results on the preclinical trials using large animal studies, the delivery of BMCs either systemically or intramyocardially, have demonstrated an increase of cardiomyogenesis and a significant improvement in heart function [25, 43]. Some of the clinical trials showed significant improvement of cardiac function in MI and HF patients, however, others did not find significant beneficial effects of the cell therapy [54–56]. The possible reasons could be due to low cell engraftment and limited differentiation potential; for example, a study by Kajstura *et al.* [42] showed that the BMC derived-cardiomyocytes (BMC-CMs) promote regeneration of the infarcted myocardium, however the developed cardiomyocytes are not fully mature, are poorly coupled and could not be oriented to endogenous myocytes [42, 43].

Mesenchymal Stem Cells (MSCs)

MSCs, also known as bone marrow stromal cells, are adult cells. They are precursors of non-hematopoietic tissues, such as, muscles, tendons, bones, fibroblast, and adipose tissues. MSCs are multipotent, have a high expansion rate and have immunomodulatory properties [17, 23, 43]. Human MSCs are easily isolated from patients, hence, reducing any risk of immune rejection from the patient during transplantation [16]. They have proven to be promising for cardiac repair in numerous pre-clinical trials including reversing post-MI remodelling and restored tissue perfusion leading to a significant reduction in infarct size and increased the left ventricle ejection fraction (LVEF) [44–46]. However, this cell source has raised safety concern in the clinical trials as most MSCs studies showed that the cells die within a week or two post-transplantation and little differentiation to cardiomyocytes [17, 23, 45]. Hence, further investigations of the MSCs in clinical trials need to be done with a larger number of patients to fully understand the paracrine mechanisms of MSCs on humans [23].

Cardiac Stem Cells (CSCs)

Cardiac stem cells (CSCs) and cardiac progenitor cells (CPCs) were first identified in 2003, located in the myocardium upon expression of c-kit and presenting the ability to undergo cardiomyogenic differentiation [57]. They are multipotent, self-renewing and have the capacity of forming myocytes, vascular cells and smooth muscle cells [28, 57]. However, it is likely that the number and function of CSCs and CPCs are impaired and limited with increasing age and due to multiple factors, including CVD, environmental changes and senescent changes within the cells [43]. According to Beltrami *et al.* [57], injection of c-kit+ CSCs/CPCs into ischemic heart/post-MI in a rat model reduced the extent of myocardial damage. Following few weeks, more than 50% of the cardiomyocytes and vascular cells that are normally present in the heart were recovered [57]. Nevertheless, the concern regarding c-kit+ CSCs/CPCs is that they have a predilection to differentiate more towards vascular cells rather than cardiomyocytes [28]. Hence, more studies need to be performed to identify the signals required to drive differentiation to specific types of cells [58]. Despite that they are autologous and tissue-specific to cardiovascular lineage, the isolation and expansion of CSCs/CPCs are available in very limited amount [16, 25]. Another type of CSCs used are cardiac-derived cells (CDCs), they are derived from myocardial biopsies and cardiac explants in culture to form cardiospheres, they have proven to have positive outcomes in infarcted heart [17, 23]. For example, CDCs were transplanted *via* intracoronary delivery in MI pig model, and the results showed positive safety data and significant cardioprotection with reduced microvascular obstruction, infarct size and attenuates adverse acute remodelling [59]. The initial clinical trials with CSCs/CPCs (CDCs) showed promising results and demonstrated signs of efficacy. However, other studies showed that these cells, like BMCs, have minimal long-term engraftment and cardiac differentiation [47, 60, 61]. Further studies need to be executed to fully elucidate their paracrine mechanisms of cardiac repair and with a larger cohort [23].

Embryonic Stem Cells (ESCs)

ESCs are derived from blastocysts taken from day 4 to 6 before the implantation into the uterine wall. They are pluripotent, that is, they are capable to form all cell types in the body [21, 28]. ESCs provide a renewable source of cardiomyocytes for basic research and pharmacological testing [26]. Mouse ESCs were first derived in 1988 and human ESCs were derived in 1998. However, research using ESCs have been controversial since the derivation of ESCs destroys the embryo, raising ethical concerns [21, 49, 50]. Mouse and human ESCs can be easily differentiated into cardiomyocytes and human-derived cardiomyocytes have shown to improve the function of infarcted rodent heart and non-human primates' hearts [21]. However, their clinical use is limited, ESC-derived cardiomyocytes could lead to immunogenic and teratogenic side effects and increase the risk of ventricular arrhythmias [25]. Despite that, ESCs can direct the specific cellular differentiation pathway for different cell types, there is lack of understanding of the molecular and genetic signals that regulate cell proliferation and differentiation [43]. Therefore, more studies are required to avoid contamination of undifferentiated ESC as the inherent risk of residual undifferentiated stem cells could induce teratoma formation. Proper quality control measures must be executed to minimize the risk of the formation of tumour [62].

Induced Pluripotent Stem Cells (iPSCs)

iPSCs can be obtained from differentiated cells including cardiac fibroblast and adult somatic cells, they are generated by reprogramming protocols that use four genes, Oct3/4, Sox2, c-Myc, and Kfl4 [63–65]. IPSCs are an alternative to ESCs as they display similar characteristics while avoiding the ethical tensions. They are pluripotent, can differentiate into cardiac lineage and self-renewing but in culture only [23, 65]. Derived iPSCs from human somatic cells have been one of the most remarkable discoveries in cardiovascular research as they can be derived from patients with complex genetic defects and create disease models [66]. Human iPSCs (hiPSCs) can be an autologous cell source for cardiac repair and increase cardiac function as they can differentiate into functional cardiomyocytes [67, 68]. For example, Ye *et al.* (2014) demonstrated hiPSC- derived cardiomyocytes (hiPSC-CMs) co-cultured with endothelial cells, and smooth muscles transplanted intramyocardially in a MI porcine model improved LV function, myocardial metabolism, arteriole density, and reduced infarction size and cell death without inducing ventricular arrhythmias [69]. Nevertheless, hiPSC-CMs are less mature than adult cardiomyocytes, based on the ultrastructure, electrophysiological and metabolically characteristics which can be improved in long-term cultures; however, the ideal level of maturation still need to be found [51, 52]. The limitation of the autologous hiPSCs approach is the financial feasibility as the process of obtaining patient-specific somatic cells, reprogramming them to iPSCs then differentiating into cardiomyocytes as well as doing expensive quality control experiments, could take over four months and doing this for each patient is cost-prohibitive [17]. Further studies need to be done to address any rejection or teratoma formation, and iPSC method requires the use of viral vectors before advancement into clinics [16, 23].

2.3 2D vs 3D Cultures

Cardiac cells are typically cultured either as 2D monolayers or more recently, 3D cell cultures, such as spheroids, engineered heart tissues and cell sheets. Both cell culturing techniques have advantages and disadvantages, though previous downsides of 3D culturing are easing in recent years.

Monolayer cell cultures are cheap with access to high throughput assays but often display biological activity that deviates from the *in vivo* response [70, 71]. This is largely due to access of transformed cell lines, genetically modified cells that lower semblance to *in vivo* counterparts but allow unrestricted proliferation. For this reason, transformed cells are often used as the foundation for drug discovery and cardiotoxicity studies [72]. Primary cardiac cells derived from animals or humans are difficult to isolate and therefore not a reliable source for CTE, though the experiments conducted with such cells may be considered necessary as proof-of-concept studies [73]. While useful for looking at certain biochemical, genetic and functional mechanisms, 2D cultures ultimately lack compete physiology such as cell-cell and cell-ECM interactions. This is especially significant when recapitulating the human heart microenvironment. As factors such as electrical conductivity and mechanical contraction are unique to the heart and serve specialized functions, generating and carrying the cardiac action potential and acting as a pump for the cardiovascular system, respectively.

3D cell cultures and technologies are the response to this drawback of 2D cardiac models and are rapidly gaining recognition for their potential to model heart tissues and diseases [19, 24, 70, 74, 75]. EHTs have been used through numerous studies as a therapeutic tool as they have been demonstrated to improve cardiac function following myocardial injury. Another EHT-application is for disease modelling using the patient-specific hiPSC-SM to evaluate mutations, drug screening and individual risk of a patient such as drug-induced side effects [76–79]. Also, compared to 2D cultures, EHTs can be a promising model to study cardiac function and contractility, and are accessible to perform all types of evaluations as a cardiac muscle tissue in the heart. This includes contraction kinetics, rhythm and rate, genetic and protein analyses, and histological analyses of semithin, paraffin or ultrathin sections [79, 80]. While there are limitations in 2D cell systems in term of viability, proliferation, differentiation and function of cardiomyocytes; the advantages of using EHTs are that they are easy to execute and provide a great quality of research outcome as well as allow long-term experimentation, repeated measurements under steady and controlled conditions [76, 79].

Cells can be cultured in scaffolds, scaffold-free or matrices environment aiming to mimic the ECM aspects of the heart. For example, biomaterial scaffolds, such as, collagen and fibrin; provide a 3D environment for cells to attach, interact with each other and conduct electrical signals [81, 82]. Cardiac myocytes cultured in 3D often employ a biomaterial such as a hydrogel or biocompatible polymer to mimic the ECM, providing a 3D architecture for cell to interact in all spatial dimensions, both with other cells and their environment. This allows for fine-tuning of the microenvironment by modifying properties such as elasticity, stiffness, conductivity and porosity [72]. These are core aspects of CTE as the utility of 3D culturing and bioengineering to simulate blood flow, observe contractile forces and relaxation velocity in cardiac myocytes with variable mechanical and electrical cues that are the tools necessary to create a complex and accurate microenvironment [75, 83–86]. With the increase in controllable parameters there is also an increase in complexity. Lack of standardised protocols compared to 2D culturing means experimental design is more demanding and without high-throughput testing.

Cell spheroids can be generated in several ways, both with and without a scaffold to support the development of the spheroid depending on the method chosen such as hanging-drop or low surface adhesion plates [87, 88]. Spheroids display phenotypes that are conditional or absent in 2D cultures such as contractile activity in cardiac myocytes and T-tubule formation with endothelial cells. Engineered heart tissues were first introduced by Eschenhagen et al. [89] and are still utilized today in several variants. These tissue constructs contain cells seeded onto a biomaterial scaffold (see below) and is subjected to mechanical forces, aligning cells along the force lines [82]. Cell sheets utilize temperature sensitive surfaces to culture monolayers of cells that can be detached as a sheet of cells and continuously stacked over each other, resulting in a thick sheet of cells [90]. Cell sheets are seeing increasing work in vascularized tissue studies, which will be covered in greater detail below.

2.4 Biomaterials for Cardiovascular Tissue Engineering: Polymers, Scaffolds & Hydrogels

Advancements in cell therapies have allowed the direct introduction of cells to the damaged heart to ascertain any therapeutic benefits of long-term regenerative effects. However, these studies have demonstrated a consistent deficit in exogenous cell survival following transplantation [91–93]. As biomaterials have demonstrated a utility in increasing cell retention, survival and proliferation, various natural and synthetic polymers are now being explored to enhance current therapies and innovate novel approaches as scaffolds for recapitulating human heart physiology.

Typically, *in vitro* tissue engineering methods involve the use of porous scaffolds to either a. provide structural support for a diseased tissue or b. transfer cells to the damaged tissue [84, 94, 95]. Determining a biomaterial for CTE is largely dependent on the specific goal of the research or clinical outcome; if a construct is to regenerate a tissue over a longer period of time, a highly porous structure is necessary that allows nutrients to properly exchange between cells and their environment while promoting tissue vascularization [96]. Each biomaterial currently used in tissue engineering has its own profile of benefits and drawbacks inherent in the properties of the material when applied to cardiac cells and tissues.

Natural Biomaterials

Naturally-derived biomaterials are desirable options when attempting to recapitulate human physiology as they can be used to produce biomimetic organ scaffolds [97, 98]. Well-characterised biomaterials, such as, alginate, chitosan, gelatin and decellularized extracellular matrix (dECM) are commonly used in CTE due to their strong profile of benefits with relatively small drawbacks. Alginate, generally sourced from brown algae, can generate highly porous structures in the form of a hydrogel that allow high cell seeding numbers at physiologically relevant cell densities with little-to-no immunological response [99]. Chitosan is a polysaccharide derived from the crustacean exoskeleton polymer chitin, the latter can be used to generate a number of scaffolds and hydrogels when combined with other polymers [100, 101]. Liu *et al.* [102] reported chitosan hydrogels may improve the myocardial environment following MI via reactive oxygen species regulation and by recruiting chemokines. Gelatin is produced from denaturing the ECM protein collagen, resulting in a bioactive protein capable of enhancing cell-scaffold interactions [103]. For this reason, gelatin and collagen-based scaffolds are commonly utilized in the production of cardiac patches that have seen success in multiple animal studies of regenerating damaged cardiac tissue following MI [77, 104]. Silk fibroin (SF) has been also explored as a natural source to 3D bioprint cardiac cells, but SF's intrinsic properties prevent its use by itself [105]. A more recent method of producing favourable scaffolds for cardiac cells is isolating the ECM from tissue without damaging, while removing any cells normally found within [106, 107]. This can be performed by enzymatically treating human cadaver hearts to recover human cardiac ECM and re-seed with patient-specific cells, without fear of major immunological response [108]. Natural biomaterials generally perform well in CTE. However, they are less mechanically stable, possess variable biodegradation rates and are difficult to tweak when possible.

Synthetic Biomaterials

Synthetic biomaterials generally offer favourable and diverse properties at the cost of less semblance to *in vivo* tissue. Plastics such as polylactic acid (PLA) and polyglycolic acid (PGA) have already been widely used in surgical therapies as products such as sutures and stents. Synthetic biomaterials are primarily used as structural scaffolds for cell seeding specific components such as the heart valves engineered on PLA, PGA and polycaprolactone (PCL) scaffolds [109]. As synthetic compounds, tweaking the properties are possible by chemically altering structure and combining multiple polymers together for novel desirable features. Recently, conductive and elastic properties have been added to polymers with promising results [110–113]. Spearman *et al.* [114] reported a polymer blend of conductive polymer polypyrrole (PPy) and PCL, was assessed for cardiac cell sheet development and yielded electrical resistance resembling human heart tissue. Similarly, Davenport Huyer *et al.* [115] suggested the potential application of a novel polymer, named 124 polymer, that yielded elastic properties mimicking adult heart myocardium and supported rat cardiac cell attachment, as a scaffold comparable to PLA scaffolds. Furthermore, the 124 polymer elastic properties can be modified before polymerisation and before UV cross-linking. The advancement in synthetic biomaterials is the foundation for innovating novel semi-synthetic scaffolds that incorporate the tissue modelling of natural biomaterials with the versatility of synthetics.

Hybrid Biomaterials

No individual biomaterial, natural or synthetic, can faithfully recapitulate the human heart microenvironment [116]. Hybrid biomaterials, that is, natural biomaterials that have either been biochemically altered or combined with a synthetic biomaterial, can perform a vast range of functions in CTE. Though this does not provide a single hybrid biomaterial that perfectly mimics the cardiac microenvironment, it does provide the foundation for future work. Park *et al.* [117] developed a hybrid scaffold with improved capacity to seed and attach cells using the synthetic biomaterials poly-lactic-co-glycolic-acid (PLGA) and poly(DL-lactide-co-caprolactone) in conjunction with collagen coating to incorporate binding factors found in natural ECM, and found the hybrid scaffold promoted cardiac tissue contractile and metabolic performance when compared to either biomaterial alone. Since Park *et al.* [117] study, approaches with similar methodologies have reported promising results, such as studies by Sapir *et al.* [118] and Rai *et al.* [119] utilising alginate and poly-glycerol-sebacate (PGS), respectively, bound with additional cell adhesion binding domains found naturally on collagen, fibronectin and laminin. These types of hybrid scaffold are vital to the development of functional cardiac patches for clinical use due to their ability to maintain cell viability at high populations while promoting cardiac gene expression and metabolic activity [116].

The physical characteristics of scaffolds can also be enhanced utilising composites of natural and synthetic biomaterials to tweak mechanical properties and closely resemble *in vivo* myocardium [116]. Engineered scaffolds require mechanical features to support cell viability, proliferation and function of contractile tissue that will not weather when introduced to the native myocardium which is constantly beating. Hybrid biomaterials address this challenge by incorporating biomaterials with stiffness and elasticity akin to the *in vivo* myocardium supplemented with natural ECM proteins. Kai *et al.* [120] utilized electrospinning, a method to produce fibres at nanoscale using electrical force, to mix gelatin and PCL resulting in improved cardiac myocyte attachment and alignment compared to electrospun PCL fibres alone. Kharaziha *et al.* [103] electrospun gelatin and PGS at varying ratios and chemically cross-linked the resulting hybrid gelatin matrix demonstrated elasticity like native myocardium and improved cardiomyocyte contraction. Similarly, PGS has been used as a core for scaffold with gelatin, fibrinogen or collagen shell for cell adhesion, all resulting in improved elastic properties and increased expression of cardiomyocyte contractile proteins troponin-T and α-actinin [121, 122]. Furthermore, cardiac dECM hydrogels are desirable but lack properties such as appropriate stiffness and degradation rates. In response to this, Lee *et al.* [123] developed a gelatin hydrogel with tuneable stiffness and degradation via varying degrees of vinyl sulfone polymerization and reported stiffness at 9 kPa resulted in improved cardiomyocyte network formation and contractile velocity with enhanced α-actinin and connexin-43 expression.

The development of natural, synthetic and hybrid scaffolds have advanced CTE by demonstrating novel methods to both produce and combine biomaterials that can better mimic *in vivo* tissue properties and recapitulate aspects of the heart microenvironment. The next limitation of optimising the performance of engineered cardiac tissues is providing a supporting vascular network for oxygen and nutrients transport.

2.5   The Vascularization Problem

During *in vitro* cell culture, nutrients can be easily supplied to cells, though once an engineered tissue is implanted this becomes severely limited without vasculature in close proximities to cells. CTE builds upon research already aimed at restoring function to ischemic tissue by innovating and testing methods of inducing vasculogenesis, formation of new blood vessels *de novo*, and angiogenesis, formation of new blood vessels from existing vasculature, primarily focusing on the latter. While biomaterial scaffolds have advanced greatly, allowing for clinically relevant tissues to be produced, vascularization remains a great challenge in the field of tissue engineering. An adult human heart houses approximately 10% of the total capillaries in the body, resulting in a densely vascularized structure [124]. Therefore, an engineered cardiac tissue requires not only a degree of vascularization to survive in the host, but a highly dense vascular network capable of meeting the metabolic demands of *in vivo* cardiac tissue.

Current strategies for promoting vascularization in engineered tissue primarily consist of either: tissue grafts that are progressively vascularized by the host, or pre-vascularizing engineered tissue constructs [125, 126]. There is a large body of evidence that demonstrates a host's vascular system will slowly extend into a non-vascular construct following transplantation and this is largely dependent on the presence of healthy and functional vessels at the implantation site [127–129]. Therefore, a damaged myocardium is an unideal region for relying on endogenous vascularization alone due to poor pre-existing vasculature. The addition of vascular endothelial cells has previously demonstrated self-assembly into tubular structures without external stimuli, and are capable of integrating with host vasculature after transplantation [130]. Though this is promising, endogenous vascularization both with and without endothelial cells takes days, while cell death without oxygen occurs in minutes. Omentum tissue has been explored as another biomaterial to support vascular myocardial regeneration [131]. A systematic review by Wang *et al.* [131] concluded that bioengineered cardiac tissue with omentum support improved cell retention and induced angiogenesis of transplanted tissues. Pro-angiogenic growth factors, notably VEGF, has seen utility in both 2D and 3D cultures to stimulate vascular formation, though the vessels are immature and risk leakage of plasma resulting in further complications [126, 132, 133]. Additionally, VEGF is known to be the cause of vascular tumours in multiple tissue types. While these strategies can increase the rate of endogenous vascularization, other strategies have explored prolonging cell survival in unfavourable conditions [134]. Cocktails of enzymes and factors such as caspase inhibitors and insulin-like growth factor-1, have seen success in rats by giving cardiomyocytes protection from a range of cell death mechanisms [135]. It is important to note that while these appear promising for CTE, this strategy is yet to be clinically verified in large scale trials.

Pre-vascularization may be equally as important as endogenous vascularization for engineered cardiac tissues [24]. The premise with pre-vascularization is similar to endogenous vascularization, to reduce the amount of time transplanted tissues are without vasculature and subject to harsh conditions such as hypoxic environments. Multiple biomaterial scaffolds mentioned previously, have demonstrated formation of vascular networks when seeded with endothelial cells and some have incorporated dECM to produce complex vascularization that fused with endogenous vasculature following transplantation [107, 136]. Other cell vascularization methods include spheroids and cell sheets. Spheroids are a round 3D clusters of cells that can be pre-vascularized when cultivated in specific growth methods. Cardiac spheroids can be pre-vascularized by culturing cardiac endothelial cells, fibroblasts and myocytes. Vascularized cardiac spheroids have previously been used to study the heart microenvironment and have been proposed as a method to improve transplanted tissue survival [137, 138]. Cell sheets are a scaffold free, monolayer culture technique and utilise temperature sensitive surfaces to allow cells to detach while maintaining cell-cell connections and ECM [139]. Culturing cells in this manner allows for multiple layers of cell sheets to be produced and harvested for tissue implantation. Multiple studies have established cardiac cell sheets as promising for therapeutic applications as they can be pre-vascularized with co-cultured endothelial cells and cardiac myocytes that anastomose to host tissue after transplantation [90, 140, 141]. More recently, newer techniques to pre-vascularize cardiac tissue have been explored. Song et al. [142] utilized cell sheets with reprogrammed fibroblasts via known cardiac cell-fate factors (Gata4, Mef2c, Tbx5, Hand2, Myocd) on nanoporous PLGA membranes with layers of endothelial cells between cell sheets. This approach proved effective in implantation on the epicardium and preventing adverse cardiac remodelling in rat hearts post-myocardial infarction. Another method of vascularization is the combination of 3D printing technology with cells and biomaterials called bioprinting.

2.6   3D Bioprinting of Heart Tissues

Like all technologies, different bioprinting methods have unique advantages and disadvantages [143]. Extrusion based bioprinting dispenses material continuously in a pre-defined shape using either pneumatic or mechanical forces. This technique is the most common, and accordingly the least expensive. Additionally, extrusion bioprinting allows rapid printing times and can print clinically relevant high cell densities such as those required for cardiac tissue implants and patches. However, material choice is somewhat limited as highly viscous bioinks are required and therefore, high printing pressure can disrupt cell viability. Kolesky et al. [144] used pneumatic bioprinting to construct preformed vascular networks followed by lining with human umbilical vein endothelial cells (HUVECs) resulting in thick vasculature that were maintained for 6 weeks. Inkjet bioprinting involves the release of fluid droplets at precise locations via thermal or acoustic forces. This method yields high print resolutions as low as 20 μm, is compatible with a large range of bioinks and results in high cell viability. Because there are low pressures on the bioink, delicate cell types can be used, but this comes at the cost of lower structural integrity and therefore only lower cell densities can be printed. Xu *et al.* [145] used primary feline and H1 cardiomyocytes on a controlled porosity alginate hydrogel with viable cell populations, indicating inkjet bioprinting may be useful in engineering patient specific designed cardiac tissues. Stereolithography is unlike the previous methods as instead of placement of material, the construct is hardened via light from a vessel of fluid containing photoactive polymers. This method is rapid and removes physical stress on the cells and bioinks resulting in moderately high resolutions as low as 50 μm while maintaining high cell viability. The range of materials accessible are extremely limited as photoactive polymers are a requirement and are further restricted by an ultraviolet wavelength to activate the photopolymer that does not harm the cells [109]. Stereolithography bioprinting has been used to generate patient specific models to assist in surgical planning and to create vascularized tissue constructs from photoactive polymers with modifiable elasticity and tensile strength [146, 147].

Bioprinted cardiac tissues have been generated to mimic several features of the cardiac microenvironment for both *in vitro* and *in vivo* applications (Fig. 3) [24, 148]. These applications include modelling for diseases, drug screening and potential transplantation to replace or support the regeneration of damaged myocardium. The vascularization problem previously discussed, has seen progress with the advent of bioprinting by generating vascular networks through several methods including vascular structures via simultaneous bioprinting of cells and biomaterials;

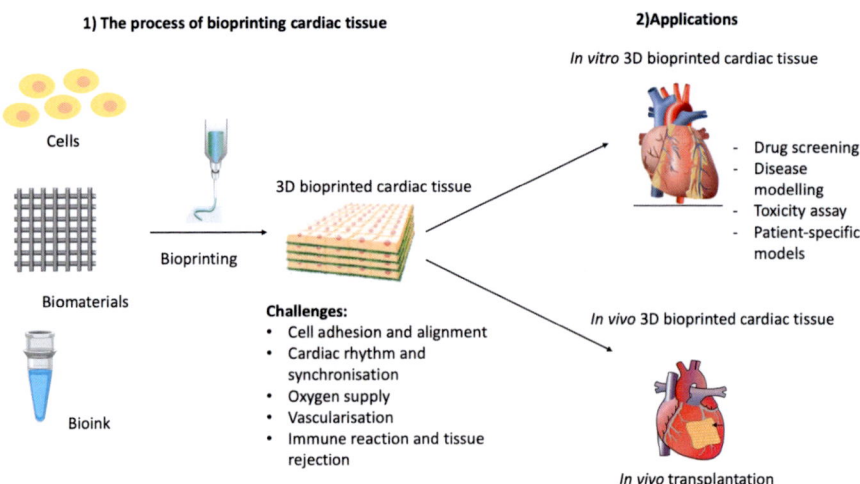

Fig. 3 Applications of 3D bioprinted cardiac tissues
1) The process of bioprinting cardiac tissue, the challenges of EHTs are cell adhesion and alignment, the cardiac tissue contractility and vascularization, oxygen supply, and can cause tissue rejection and immune reaction during transplantation. 2) The applications of 3D bioprinted cardiac tissue for *in vitro* testing are drug screening, disease modelling, toxicity assays and patient-specific models. For *in vivo* testing, 3D bioprinted cardiac tissue are transplanted on ischemic *in vivo* models to promote cardiac regeneration

addition of angiogenic factors in bioprinted constructs; and bioprinting of channel-based constructs for pre-fabricated vascular networks [148]. Previously generated spheroids have been bioprinted, used as building blocks and subsequently fused into vascular constructs with a range of cell types including human smooth muscle cells, human dermal fibroblasts and more importantly cardiac fibroblasts, endothelial cells (EC) and hiPSC-CM [24, 144, 149, 150]. Ink-jet bioprinters can be utilized to deposit biomaterial scaffolds and EC simultaneously to form microvasculature scaffolds allowing EC proliferation into tubular structures with clinically relevant cell viabilities and stable structural integrity of vasculature [151, 152]. Angiogenic growth factors have been explored in bioprinted constructs with some success. HUVECs cultured in VEGF before bioprinting with iPSC-CM were reported to integrate with host vasculature when transplanted in mice [153]. Furthermore, VEGF slowly released into scaffolds (demonstrated with both Matrigel and alginate) promotes angiogenic vessel formation and CD31 expression, which similarly have seen promising results in mice transplants [154]. Clinically relevant constructs require vasculature that supports flow throughout the entire structure. Bioprinting uniquely offers this feature of manufacturing complex and organized networks to promote nutrient, waste and oxygen transport [155]. This can be accomplished by bioprinting a hydrogel containing a removable internal structure, yielding hollow networks that can be populated with ECs to mimic *in vivo* vasculature [156, 157]. Additionally, direct bioprinting of perfusable constructs is possible using multiple print-heads containing an outer cross-linkable hydrogel (e.g. Gel-MA) and an inner head with the appropriate cross-linking solution [148, 155].

Contractility of bioprinted cardiac tissues remains a challenge to overcome before wide-use of clinically relevant constructs are readily available [24]. Cardiac spheroids from iPSC-CM already have been demonstrated to spontaneously contract, but are limited by immature phenotypes [138, 158, 159]. Improving the environment of iPSC-CM demonstrates an increase in cell contractility and contributes to overall cardiac tissue development [149]. Strategies to improve bioprinted cardiac environments include addition of conductive polymers for electrical propagation support and elastic polymers for mechanical support unique to the myocardium [115, 146, 160, 161]. Though these biomaterials have been used in CTE previously, optimising such biomaterials for bioprinting is possible and slowly progressing but still a work-in-progress [24, 162].

In Vitro Testing of 3D Bioprinted Cardiac Tissues

Cardiac bioprinting provides an alternative approach to regenerate infarcted heart by integrating cardiac cells and 3D biomaterials/ biomaterial-free. The application of the cardiac bioprinting has shown promises as an option to create functional cardiac tissue to regenerate or replaced damaged tissue in the myocardium [149, 163]. The bioengineered cardiac tissue can mimic the structural, physiological, and functional features of native myocardium but also, can be used for disease modelling of myocardial infarction, ischemia-reperfusion injury and heart failure [151, 153].

This approach allows a 3D structure of the complex arrangement of cells and ECM, supporting the cells and enhance their reorganization into functional cardiac tissues [163, 164].The engineered cardiac patch has to undergo *in vitro* maturation and testing before being transplanted onto the defected heart. For example, in this research, the *in vitro* testing done on the 3D cardiac patch were maturation testing and vascularization testing [165]. Jang *et al.* [165] developed a 3D pre-vascularized myocardium patch using cardiac progenitors and MSCs with the decellularized extracellular matrix (dECM) bioink, they showed that dECM enhanced structural maturation of cells and promote vascular formation as well as the functionality of cells for tissue regeneration. Also, other *in vitro* tests are done on the 3D cardiac patch for quality control and to look at the cardiac patch structural, mechanical, and electrical properties. The tests performed are flow cytometry, immunohistochemistry, immunofluorescence, cell viability assays and optical-electrical mapping [150, 153, 165, 166].

3D bioprinted cardiac tissue approach could give the possibility to have patient-specific tissue models that could be used to test therapeutic schemes, help in clinical diagnosis and treatment of diseases through replacement of the injured tissues [74]. Considerable limitations in currently available *in vitro* and *in vivo* models of myocardial infarction are related to their limited ability to recapitulate the complex pathophysiology typical of the human heart tissue and their inability to directly translate from the bench to the bedside [167]. Furthermore, a study demonstrated that 3D cardiac tissues derived from hiPSC-CMs coated with extracellular matrix (fibronectin and gelatin nanofilm) have the potential to be used as a drug screening system for drug discovery and cardiotoxicity assay [168]. However, 3D tissue culturing requires not only the biological factors but also, the mechanical and electrical simulations as they are required to accustom the engineered cardiac tissues to their new functions. Such functions consist of contraction, reception, delivery of blood and electrical signalling [169].

Bioreactors or microfluidic "organ-on-chips" devices provide precise control by mimicking the various mechanical and chemical factors *in vivo* and, recapitulate the cellular microenvironment of the heart and monitor the critical parameters such as pH, nutrient supply and oxygen level [169–171]. Bioreactors and microfluidic devices promote and maintain cellular morphology and cell-specific functionality of the 3D biofabricated tissue but also, prevent shear stress and control the flow rates [172, 173]. These devices allow the bioprinted tissue to develop a perusable microvascular network which is fundamental for treating ischemic diseases [174]. For example, Zhang *et al.* [175] encapsulated endothelial cells with microfibrous hydrogel bioink which was then seeded with either hiPCS-cardiomyocytes or neonatal rat cardiomyocytes to get spontaneous and synchronous contractions. The 3D engineered endothelialized myocardial tissues were then added to a microfluidic perfusion bioreactor to evaluate cardiovascular toxicity [175]. Hence, they demonstrated that the organ-on-a-chip model could be used as drug screening and could act as a 3D organ model system to improve treatment efficacy, but more research needs to be done to find the ideal device for the 3D bio fabricated cardiac tissue.

The limitation of this approach is that many biofabricated cardiac tissues are either not conductive for perfusion or do not translate properly from bioprinter to bioreactors/microfluidic systems. Therefore, the development of cardiac engineered tissues that are nurtured in those devices, are not yet suitable for animal models or clinical use [176]. For example, the perfusion bioreactor design is able to generate and cultured biofabricated/bioprinted vascular networks from human mesenchymal stem cells (hMSCs) cocultured with fibroblast, however, transferring those vascular networks to an animal or patient lead to the destruction of those delicate networks [176, 177]. The goal is for the bioprinted cardiac tissues to be cultured and transferred from the setting (bioreactor/microfluidic systems) to be used for clinical purposes.

In Vivo Testing of 3D Bioprinted Cardiac Tissues

The recent advancements of 3D bioprinted cardiac tissues have showed great promises for cardiac regeneration in ischemic *in vivo* models after the engraftment of bioprinted cardiac patches in small animals such as mice and rats [24, 150–153, 178]. In particular, *in vivo* models of permanent left anterior descending (LAD) ligation seem suitable for the preclinical testing of 3D bioprinted patches for cardiac regeneration [179]. Research of cardiovascular tissue regeneration using bioprinting mainly focuses on the vasculature, myocardium and heart valves [151]. To fully investigate whether the cardiac patches improve cardiac function in ischemic models, the *in vivo* tests done are echocardiography, hemodynamics, histological analysis on the infarct size and fibrosis, cell viability, flow cytometry, gene expression and protein analyses [150, 152, 153, 165].

Jang *et al.* [165] developed a 3D-vascularized stem cell patch with cardiac progenitor and MSCs using heart tissue-derived dECM (hdECM) bioink to mimic tissue-specific ECM composition. The cardiac patches were implanted subcutaneously in mouse model. The results showed that the developed 3D stem cell patch promoted significant vascularization and tissue matrix formation in the infarcted heart model *in vivo* (Fig. 4a). The patch also shown to improve cardiac functions, reduced cardiac hypertrophy and fibrosis and increased migration from patch to

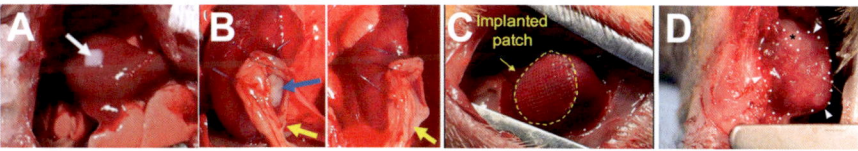

Fig. 4 Types of 3D bioprinted cardiac patch for *in vivo* implantation
(**a**) 3D bioprinted pre-vascularized stem cell sheet patch using hdECM bioink with hCPCs and human MSCs on the mouse heart [165]. (**b**) 3D bioprinted biomaterial-free cardiac patch made from spheroids of co-cultured hiPSC-CM:FB:EC onto the rat heart [150]. (**c**) 3D bioprinted biomaterial-free cardiac tissue followed by omentum patch. Cardiac patch made from spheroids of co-cultured hiPSC-CM:FB:EC (blue arrow) and omentum patch (yellow arrow) implanted in the rat model [152]. (**d**) 3D bioprinted hydrogel-based cardiac patch was transplanted on the mouse heart (hydrogel made of alginate 4% and gelatine 8% in media) [178]

infarct area [165]. In another study, Maiullari *et al.* [153] integrated the use of 3D bioprinting (PEG-Fibrinogen) with iPSC-derived cardiomyocytes and HUVECs, and the 3D bioprinted cardiac patch was subcutaneously implanted in mice models. They demonstrated that the bioprinted pre-vascularized stem cell patch can effectively develop vasculature in the transplanted tissues and support the host's vasculature. Also, it can provide a cardiac niche-like microenvironment which leads to beneficial results for cardiac repair [153]. However, more pre-clinical studies using additional bioinks and hydrogels are needed to gain a better understanding of their potential to effectively treat myocardial infarction before translating them into human clinical trials [169]. Furthermore, the use of biomaterials faces challenges such as adverse host responses, that is, inflammatory response, immunogenicity, but also fibrous tissue formation, biomaterial degradation and toxicity of the biomaterial products that could affect the long term function of the 3D bioprinted cardiac tissue construct [150, 180].

Consequently, other pre-clinical studies are using 3D bioprinted biomaterial-free cardiac tissue on *in vivo* model and are yielding satisfactory outcomes. For example, Ong et al. [150] bioprinted tissue spheroids composed of hiPSC-CMs, fibroblast and endothelial cells into myocardial patches (Fig. 4b). These cardiac patches were implanted onto rat models showing vascularization and engraftment into native rat myocardium, suggesting the therapeutic regenerative potential of this 3D biomaterial-free method [150, 169]. But the limitation of this biomaterial free cardiac patch is that the heart rate of the rats were different from the cardiac patches. Hence, more studies need to be executed and extended to understand the effect of the cardiac patch in rat myocardium and the long-term regenerative potential in the damaged heart before being clinically studied on humans. Wang *et al.* [149] demonstrated that the omentum can be used to support the engraftment of cardiac patches to overcome those challenges. Yeung *et al.* [152] implanted the 3D bioprinted biomaterial-free cardiac patch followed by the omentum patch into rat myocardial infarction model resulting in an increased in blood vessels, smaller scar area and improvement of the cardiac function (via echocardiography study) (Fig. 4c). Concluding that the 3D bioprinted biomaterial–free cardiac patches have the potential to improve regeneration in cardiac tissue, promote angiogenesis and reduce scar tissue formation in the infarcted area followed by the omentum patch [152]. But more studies need to be executed and extended to understand the effect of the cardiac patch in rat myocardium and the long-term regenerative potential in the damaged heart before being clinically studied on humans. Also, further studies are required to understand the omentum mechanism to improve functional cardiac benefit which is crucial for clinical translation [131]. Despite the improvement and new avenues of 3D bioprinted cardiac tissue, limitations remain in *in vivo* models such as, poor vascularization and not synchronous contractile activity [24].

3 Discussion

In the past decades, cardiovascular regeneration has been the focus to improve cardiac function and prevent heart failure in ischemic adult heart. New approaches such as reactivating the endogenous regenerative capacity through paracrine mechanisms and deliver cells to the heart are undergoing pre-clinical studies and clinical studies. Cell-free therapies have given researchers a potential strategy for regenerating tissue endogenously, yet further studies are required. Notably, most studies require further validation, safety confirmation for the method of delivery, appropriate dosage and ideal duration of expression of the paracrine factors as prolonged expression might lead to unwanted side effects. On the other hand, approaches utilising cells and biomaterials have seen mixed results. Investigation of some cellular therapies involving SMBs have ceased due to potential complications while other stem-cell based therapies that is, MSCs, BMCs, CSCs, and iPSCs are all still being considered for their clinical potential. However, the current limitation of cell-based therapies is their limited amount and little differentiations to cardiomyocytes. More studies need to be done to elucidate their paracrine mechanisms for cardiac repair and with a larger cohort. HiPSCs in particular, have propelled cell-based therapy and CTE forward as an abundant source of previously lacking cardiomyocytes. Furthermore, HiPSCs can be differentiated into functional cardiomyocytes and be utilized to study patient-specific disease model but is cost-prohibitive. Furthermore, iPSCs are less mature than adult cardiomyocytes and the optimum level of maturation still need to be found. The major concerns of cell therapies are the patient's immunological response, teratoma formation caused by pluripotent stem cells and arrhythmias. These concerns have been somewhat addressed via numerous strategies such as patient-derived iPSC and biomaterials to support regeneration of myocardium that have limited-to-no immune response. However, with the cell-based and cell-free approaches, more studies need to be executed to understand the long-term regenerative potential and their mechanism effects to improve cardiac function and improve cardiac myocytes proliferation.

CTE has allowed the use of each component, that is, cell types, biomaterials, bioinks and bioprinting as well as growth factors to recapitulate the microenvironment of the human heart. Leading the ability to design and manufacture cardiovascular tissues that could improve CVD patient outcome and quality of life. The tools and concepts currently studied and utilized in CTE are beginning to transition from scientific research to clinical applications [97, 109, 164, 181]. This is undoubtedly due to the multi-disciplinary actions of biology and engineering that have yielded several avenues to explore methods of cardiac regeneration using 3D printing. Bioprinting has emerged as a technology that stands out, largely due to the incorporation of aspects from across the field of CTE and the possibility of producing functional cardiac tissues with complex architecture and reproducibility. Currently there are challenges to be answered before the promise of fully functional bioprinted cardiac tissues can be realized. Tailor-made bioinks that uniquely suit the cardiac environment are still some time away from recapitulating cardiac physiology in terms of electrical signalling and physical properties specific to the cardiac niche. Furthermore, engineered heart tissues currently in development need validation from large scale *in vivo* studies to determine the best method of properly integrating with host vasculature after transplantation. At present, both pre-vascularization and activation of endogenous vascularization mechanisms have been promising. Contractility, both synchronicity and magnitude, of bioprinted tissues is another hurdle that must be overcome before therapeutic interventions like bioprinted cardiac patches can be readily available.

4   Conclusions

The future of cardiac engineering requires a more integrated approach to recapitulate the niche environment of both the physiological myocardium and the pathophysiological. Unique and complex bioinks that are tailored to providing these biochemical and structural intricacies and highly specialized biofabrication technologies such as advanced 3D culturing platforms and bioprinters capable of ultrahigh resolutions while maintaining high cell viability and populations are significant milestones for translating research to clinical practice and supportive therapy to truly regenerative.

The 3D bioprinted cardiac tissue could replace *in vitro* 2D cultures model system and could be used as disease modelling, drug and toxicity testing. Furthermore, bioprinted cardiac patch have been studied in small animals and have shown that they have therapeutic regenerative potential. However, the limitations of grafting 3D cardiac patches on small animals are poor vascularization and not synchronous contractile activity. Omentum could be used to overcome those challenges as well as support the engraftment of cardiac patches. However, further studies need to be done before bioprinted cardiac tissues can be tested clinically on humans.

Acknowledgments CG was supported by a University of Sydney Kick-Start Grant, CDIP Grant, Cardiothoracic Surgery Research Grant, UTS Seed Funding and Catholic Archdiocese of Sydney Grant for Adult Stem Cell Research. A particular thank to Dr. Christopher D. Roche (The University of Sydney/UTS) for the critical review provided.

References

1. Thomas, H., Diamond, J., Vieco, A., Chaudhuri, S., Shinnar, E., Cromer, S., et al.: Global atlas of cardiovascular disease. Global Heart. **13**(3) (2018)
2. Joseph, P., Leong, D., McKee, M., Anand, S.S., Schwalm, J.-D., Teo, K., et al.: Reducing the global burden of cardiovascular disease, part 1: the epidemiology and risk factors. Circ. Res. **121**(6), 677–694 (2017)
3. Wang, H., Naghavi, M., Allen, C., Barber, R.M., Bhutta, Z.A., Carter, A., et al.: Global, regional, and national life expectancy, all-cause mortality, and cause-specific mortality for

249 causes of death, 1980–2015: a systematic analysis for the Global Burden of Disease Study 2015. Lancet. **388**(10053), 1459–1544 (2016)

4. Stevens, L.A., Viswanathan, G., Weiner, D.E.: Chronic kidney disease and end-stage renal disease in the elderly population: current prevalence, future projections, and clinical significance. Adv. Chronic Kidney Dis. **17**(4), 293–301 (2010)

5. Heidenreich, P.A., Trogdon, J.G., Khavjou, O.A., Butler, J., Dracup, K., Ezekowitz, M.D., et al.: Forecasting the future of cardiovascular disease in the United States: a policy statement from the American Heart Association. Circulation. **123**(8), 933–944 (2011)

6. Manuel, D.G., Tuna, M., Hennessy, D., Bennett, C., Okhmatovskaia, A., Finès, P., et al.: Projections of preventable risks for cardiovascular disease in Canada to 2021: a microsimulation modelling approach. CMAJ Open. **2**(2), E94 (2014)

7. Roth, G.A., Johnson, C., Abajobir, A., Abd-Allah, F., Abera, S.F., Abyu, G., et al.: Global, regional, and national burden of cardiovascular diseases for 10 causes, 1990 to 2015. J. Am. Coll. Cardiol. **70**(1), 1–25 (2017)

8. Palomeras Soler, E., Casado, R.V.: Epidemiology and risk factors of cerebral ischemia and ischemic heart diseases: similarities and differences. Curr. Cardiol. Rev. **6**(3), 138–149 (2010)

9. Deaton, C., Froelicher, E.S., Wu, L.H., Ho, C., Shishani, K., Jaarsma, T.: The global burden of cardiovascular disease. European Journal of Cardiovascular Nursing. **10**(2_suppl), S5–S13 (2011)

10. Ferreira, J.P., Kraus, S., Mitchell, S., Perel, P., Piñeiro, D., Chioncel, O., et al.: World heart federation roadmap for heart failure. Glob. Heart. **14**(3), 197 (2019)

11. Ambrosy, A.P., Fonarow, G.C., Butler, J., Chioncel, O., Greene, S.J., Vaduganathan, M., et al.: The global health and economic burden of hospitalizations for heart failure: lessons learned from hospitalized heart failure registries. J. Am. Coll. Cardiol. **63**(12), 1123–1133 (2014)

12. Ziaeian, B., Fonarow, G.C.: Epidemiology and aetiology of heart failure. Nat. Rev. Cardiol. **13**(6), 368–378 (2016)

13. Go, A.S., Mozaffarian, D., Roger, V.L., Benjamin, E.J., Berry, J.D., Borden, W.B., et al.: Heart disease and stroke statistics – 2013 update: a report from the American Heart Association. Circulation. **127**(1), e6–e245 (2013). https://doi.org/10.1161/CIR.0b013e31828124ad

14. Lui, K.O., Zangi, L., Chien, K.R.: Cardiovascular regenerative therapeutics via synthetic paracrine factor modified mRNA. Stem Cell Res. **13**(3), 693–704 (2014)

15. Burchfield, J.S., Dimmeler, S.: Role of paracrine factors in stem and progenitor cell mediated cardiac repair and tissue fibrosis. Fibrogenesis Tissue Repair. **1**(1), 4 (2008). https://doi.org/10.1186/1755-1536-1-4

16. Li, S.C., Wang, L., Jiang, H., Acevedo, J., Chang, A.C., Loudon, W.G.: Stem cell engineering for treatment of heart diseases: potentials and challenges. Cell Biol. Int. **33**(3), 255–267 (2009)

17. Gerbin, K.A., Murry, C.E.: The winding road to regenerating the human heart. Cardiovasc. Pathol. **24**(3), 133–140 (2015). https://doi.org/10.1016/j.carpath.2015.02.004

18. Bui, A.L., Horwich, T.B., Fonarow, G.C.: Epidemiology and risk profile of heart failure. Nat. Rev. Cardiol. **8**(1), 30 (2011)

19. Günter, J., Wolint, P., Bopp, A., Steiger, J., Cambria, E., Hoerstrup, S.P., et al.: Microtissues in cardiovascular medicine: regenerative potential based on a 3D microenvironment. Stem Cells Int. **2016**, 9098523 (2016). https://doi.org/10.1155/2016/9098523

20. Witman, N., Zhou, C., Beverborg, N.G., Sahara, M., Chien, K.R.: Cardiac progenitors and paracrine mediators in cardiogenesis and heart regeneration. Seminars in Cell & Developmental Biology: Elsevier, 29–51 (2020)

21. Lee, R.T., Walsh, K.: The future of cardiovascular regenerative medicine. Circulation. **133**(25), 2618–2625 (2016)

22. Lee, J.S., Romero, R., Han, Y.M., Kim, H.C., Kim, C.J., Hong, J.S., et al.: Placenta-on-A-chip: a novel platform to study the biology of the human placenta. J. Matern. Fetal Neonatal Med. **29**(7), 1046–1054 (2016). https://doi.org/10.3109/14767058.2015.1038518

23. Cambria, E., Pasqualini, F.S., Wolint, P., Günter, J., Steiger, J., Bopp, A., et al.: Translational cardiac stem cell therapy: advancing from first-generation to next-generation cell types. npj Regenerative Medicine. **2**(1), 17 (2017). https://doi.org/10.1038/s41536-017-0024-1

24. Roche, C.D., Brereton, R.J.L., Ashton, A.W., Jackson, C., Gentile, C.: Current challenges in three-dimensional bioprinting heart tissues for cardiac surgery. Eur. J. Cardiothorac. Surg. (2020). https://doi.org/10.1093/ejcts/ezaa093

25. Bollini, S., Smart, N., Riley, P.R.: Resident cardiac progenitor cells: at the heart of regeneration. J. Mol. Cell. Cardiol. **50**(2), 296–303 (2011)

26. He, J.Q., Ma, Y., Lee, Y., Thomson, J.A., Kamp, T.J.: Human embryonic stem cells develop into multiple types of cardiac myocytes: action potential characterization. Circ. Res. **93**(1), 32–39 (2003). https://doi.org/10.1161/01.Res.0000080317.92718.99

27. Noseda, M., Peterkin, T., Simões, F.C., Patient, R., Schneider, M.D.: Cardiopoietic factors: extracellular signals for cardiac lineage commitment. Circ. Res. **108**(1), 129–152 (2011)

28. Hansen, T., Saleh, S., Figtree, G.A., Gentile, C.: The role of redox signalling in cardiovascular regeneration. In: Chakraborti, S., Dhalla, N.S., Ganguly, N.K., Dikshit, M. (eds.) Oxidative stress in heart diseases, pp. 19–37. Springer Singapore, Singapore (2019)

29. Lui, K.O., Zangi, L., Silva, E.A., Bu, L., Sahara, M., Li, R.A., et al.: Driving vascular endothelial cell fate of human multipotent Isl1+ heart progenitors with VEGF modified mRNA. Cell Res. **23**(10), 1172–1186 (2013). https://doi.org/10.1038/cr.2013.112

30. Eppler, S.M., Combs, D.L., Henry, T.D., Lopez, J.J., Ellis, S.G., Yi, J.H., et al.: A target-mediated model to describe the pharmacokinetics and hemodynamic effects of recombinant human vascular endothelial growth factor in humans. Clin. Pharmacol. Ther. **72**(1), 20–32 (2002). https://doi.org/10.1067/mcp.2002.126179

31. Gomzikova, M.O., Rizvanov, A.A.: Current trends in regenerative medicine: from cell to cell-free therapy. BioNanoScience. **7**(1), 240–245 (2017). https://doi.org/10.1007/s12668-016-0348-0

32. Henry, T.D., Annex, B.H., McKendall, G.R., Azrin, M.A., Lopez, J.J., Giordano, F.J., et al.: The VIVA trial: vascular endothelial growth factor in ischemia for vascular angiogenesis. Circulation. **107**(10), 1359–1365 (2003)

33. Stewart, D.J., Kutryk, M.J., Fitchett, D., Freeman, M., Camack, N., Su, Y., et al.: VEGF gene therapy fails to improve perfusion of ischemic myocardium in patients with advanced coronary disease: results of the NORTHERN trial. Mol. Ther. **17**(6), 1109–1115 (2009)

34. Gao, R., Zhang, J., Cheng, L., Wu, X., Dong, W., Yang, X., et al.: A phase II, randomized, double-blind, Multicenter, based on standard therapy, placebo-controlled study of the efficacy and safety of recombinant human neuregulin-1 in patients with chronic heart failure. J. Am. Coll. Cardiol. **55**(18), 1907–1914 (2010). https://doi.org/10.1016/j.jacc.2009.12.044

35. Raik, S., Kumar, A., Bhattacharyya, S.: Insights into cell-free therapeutic approach: role of stem cell "soup-ernatant". Biotechnol. Appl. Biochem. **65**(2), 104–118 (2018)

36. Segers, V.F., Lee, R.T.: Stem-cell therapy for cardiac disease. Nature. **451**(7181), 937–942 (2008). https://doi.org/10.1038/nature06800

37. Behfar, A., Crespo-Diaz, R., Terzic, A., Gersh, B.J.: Cell therapy for cardiac repair – lessons from clinical trials. Nat. Rev. Cardiol. **11**(4), 232–246 (2014). https://doi.org/10.1038/nrcardio.2014.9

38. Rossini, A., Frati, C., Lagrasta, C., Graiani, G., Scopece, A., Cavalli, S., et al.: Human cardiac and bone marrow stromal cells exhibit distinctive properties related to their origin. Cardiovasc. Res. **89**(3), 650–660 (2011). https://doi.org/10.1093/cvr/cvq290

39. Citro, L., Naidu, S., Hassan, F., Kuppusamy, M.L., Kuppusamy, P., Angelos, M.G., et al.: Comparison of human induced pluripotent stem-cell derived cardiomyocytes with human mesenchymal stem cells following acute myocardial infarction. PLoS One. **9**(12), e116281 (2014). https://doi.org/10.1371/journal.pone.0116281

40. Gavira, J.J., Herreros, J., Perez, A., Garcia-Velloso, M.J., Barba, J., Martin-Herrero, F., et al.: Autologous skeletal myoblast transplantation in patients with nonacute myocardial infarction: 1-year follow-up. J. Thorac. Cardiovasc. Surg. **131**(4), 799–804 (2006). https://doi.org/10.1016/j.jtcvs.2005.11.030

41. Menasché, P., Hagège, A.A., Scorsin, M., Pouzet, B., Desnos, M., Duboc, D., et al.: Myoblast transplantation for heart failure. Lancet. **357**(9252), 279–280 (2001). https://doi.org/10.1016/s0140-6736(00)03617-5

42. Kajstura, J., Rota, M., Whang, B., Cascapera, S., Hosoda, T., Bearzi, C., et al.: Bone marrow cells differentiate in cardiac cell lineages after infarction independently of cell fusion. Circ. Res. **96**(1), 127–137 (2005)
43. Ballard, V.L., Edelberg, J.M.: Stem cells and the regeneration of the aging cardiovascular system. Circ. Res. **100**(8), 1116–1127 (2007)
44. Mazo, M., Hernández, S., Gavira, J.J., Abizanda, G., Araña, M., López-Martínez, T., et al.: Treatment of reperfused ischemia with adipose-derived stem cells in a preclinical swine model of myocardial infarction. Cell Transplant. **21**(12), 2723–2733 (2012). https://doi.org/1 0.3727/096368912x638847
45. Schuleri, K.H., Feigenbaum, G.S., Centola, M., Weiss, E.S., Zimmet, J.M., Turney, J., et al.: Autologous mesenchymal stem cells produce reverse remodelling in chronic ischaemic cardiomyopathy. Eur. Heart J. **30**(22), 2722–2732 (2009). https://doi.org/10.1093/ eurheartj/ehp265
46. Valina, C., Pinkernell, K., Song, Y.H., Bai, X., Sadat, S., Campeau, R.J., et al.: Intracoronary administration of autologous adipose tissue-derived stem cells improves left ventricular function, perfusion, and remodelling after acute myocardial infarction. Eur. Heart J. **28**(21), 2667–2677 (2007). https://doi.org/10.1093/eurheartj/ehm426
47. Bolli, R., Chugh, A.R., D'Amario, D., Loughran, J.H., Stoddard, M.F., Ikram, S., et al.: Cardiac stem cells in patients with ischaemic cardiomyopathy (SCIPIO): initial results of a randomised phase 1 trial. Lancet. **378**(9806), 1847–1857 (2011). https://doi.org/10.1016/ s0140-6736(11)61590-0
48. Doppler, S.A., Deutsch, M.-A., Lange, R., Krane, M.: Cardiac regeneration: current therapies-future concepts. J. Thorac. Dis. **5**(5), 683–697 (2013). https://doi.org/10.3978/j. issn.2072-1439.2013.08.71
49. Evans, M.J., Kaufman, M.H.: Establishment in culture of pluripotential cells from mouse embryos. Nature. **292**(5819), 154–156 (1981). https://doi.org/10.1038/292154a0
50. Thomson, J.A., Itskovitz-Eldor, J., Shapiro, S.S., Waknitz, M.A., Swiergiel, J.J., Marshall, V.S., et al.: Embryonic stem cell lines derived from human blastocysts. Science. **282**(5391), 1145–1147 (1998). https://doi.org/10.1126/science.282.5391.1145
51. Jonsson, M.K., Vos, M.A., Mirams, G.R., Duker, G., Sartipy, P., de Boer, T.P., et al.: Application of human stem cell-derived cardiomyocytes in safety pharmacology requires caution beyond hERG. J. Mol. Cell. Cardiol. **52**(5), 998–1008 (2012). https://doi.org/10.1016/j. yjmcc.2012.02.002
52. Kamakura, T., Makiyama, T., Sasaki, K., Yoshida, Y., Wuriyanghai, Y., Chen, J., et al.: Ultrastructural maturation of human-induced pluripotent stem cell-derived cardiomyocytes in a long-term culture. Circ. J. **77**(5), 1307–1314 (2013). https://doi.org/10.1253/circj. cj-12-0987
53. Orlic, D., Kajstura, J., Chimenti, S., Jakoniuk, I., Anderson, S.M., Li, B., et al.: Bone marrow cells regenerate infarcted myocardium. Nature. **410**(6829), 701–705 (2001). https://doi. org/10.1038/35070587
54. Nasseri, B.A., Ebell, W., Dandel, M., Kukucka, M., Gebker, R., Doltra, A., et al.: Autologous CD133+ bone marrow cells and bypass grafting for regeneration of ischaemic myocardium: the Cardio133 trial. Eur. Heart J. **35**(19), 1263–1274 (2014). https://doi.org/10.1093/ eurheartj/ehu007
55. Schächinger, V., Assmus, B., Britten, M.B., Honold, J., Lehmann, R., Teupe, C., et al.: Transplantation of progenitor cells and regeneration enhancement in acute myocardial infarction: final one-year results of the TOPCARE-AMI trial. J. Am. Coll. Cardiol. **44**(8), 1690–1699 (2004). https://doi.org/10.1016/j.jacc.2004.08.014
56. Tendera, M., Wojakowski, W., Ruzyłło, W., Chojnowska, L., Kepka, C., Tracz, W., et al.: Intracoronary infusion of bone marrow-derived selected CD34+CXCR4+ cells and non-selected mononuclear cells in patients with acute STEMI and reduced left ventricular ejection fraction: results of randomized, multicentre myocardial regeneration by intracoronary infusion of selected population of stem cells in acute myocardial infarction (REGENT) trial. Eur. Heart J. **30**(11), 1313–1321 (2009). https://doi.org/10.1093/eurheartj/ehp073

57. Beltrami, A.P., Barlucchi, L., Torella, D., Baker, M., Limana, F., Chimenti, S., et al.: Adult cardiac stem cells are multipotent and support myocardial regeneration. Cell. **114**(6), 763–776 (2003)

58. van Vliet, P., Roccio, M., Smits, A.M., van Oorschot, A.A., Metz, C.H., van Veen, T.A., et al.: Progenitor cells isolated from the human heart: a potential cell source for regenerative therapy. Neth Heart J. **16**(5), 163–169 (2008). https://doi.org/10.1007/bf03086138

59. Kanazawa, H., Tseliou, E., Malliaras, K., Yee, K., Dawkins, J.F., De Couto, G., et al.: Cellular postconditioning: allogeneic cardiosphere-derived cells reduce infarct size and attenuate microvascular obstruction when administered after reperfusion in pigs with acute myocardial infarction. Circ. Heart Fail. **8**(2), 322–332 (2015). https://doi.org/10.1161/circheartfailure.114.001484

60. Makkar, R.R., Smith, R.R., Cheng, K., Malliaras, K., Thomson, L.E., Berman, D., et al.: Intracoronary cardiosphere-derived cells for heart regeneration after myocardial infarction (CADUCEUS): a prospective, randomised phase 1 trial. Lancet. **379**(9819), 895–904 (2012). https://doi.org/10.1016/s0140-6736(12)60195-0

61. Malliaras, K., Makkar, R.R., Smith, R.R., Cheng, K., Wu, E., Bonow, R.O., et al.: Intracoronary cardiosphere-derived cells after myocardial infarction: evidence of therapeutic regeneration in the final 1-year results of the CADUCEUS trial (CArdiosphere-derived aUtologous stem CElls to reverse ventricUlar dySfunction). J. Am. Coll. Cardiol. **63**(2), 110–122 (2014). https://doi.org/10.1016/j.jacc.2013.08.724

62. Nussbaum, J., Minami, E., Laflamme, M.A., Virag, J.A., Ware, C.B., Masino, A., et al.: Transplantation of undifferentiated murine embryonic stem cells in the heart: teratoma formation and immune response. FASEB J. **21**(7), 1345–1357 (2007). https://doi.org/10.1096/fj.06-6769com

63. Qian, L., Huang, Y., Spencer, C.I., Foley, A., Vedantham, V., Liu, L., et al.: In vivo reprogramming of murine cardiac fibroblasts into induced cardiomyocytes. Nature. **485**(7400), 593–598 (2012)

64. Takahashi, K., Tanabe, K., Ohnuki, M., Narita, M., Ichisaka, T., Tomoda, K., et al.: Induction of pluripotent stem cells from adult human fibroblasts by defined factors. Cell. **131**(5), 861–872 (2007). https://doi.org/10.1016/j.cell.2007.11.019

65. Takahashi, K., Yamanaka, S.: Induction of pluripotent stem cells from mouse embryonic and adult fibroblast cultures by defined factors. Cell. **126**(4), 663–676 (2006). https://doi.org/10.1016/j.cell.2006.07.024

66. Braam, S.R., Passier, R., Mummery, C.L.: Cardiomyocytes from human pluripotent stem cells in regenerative medicine and drug discovery. Trends Pharmacol. Sci. **30**(10), 536–545 (2009)

67. Yan, B., Singla, D.K.: Transplanted induced pluripotent stem cells mitigate oxidative stress and improve cardiac function through the Akt cell survival pathway in diabetic cardiomyopathy. Mol. Pharm. **10**(9), 3425–3432 (2013). https://doi.org/10.1021/mp400258d

68. Zhang, J., Wilson, G.F., Soerens, A.G., Koonce, C.H., Yu, J., Palecek, S.P., et al.: Functional cardiomyocytes derived from human induced pluripotent stem cells. Circ. Res. **104**(4), e30–e41 (2009). https://doi.org/10.1161/circresaha.108.192237

69. Ye, L., Chang, Y.H., Xiong, Q., Zhang, P., Zhang, L., Somasundaram, P., et al.: Cardiac repair in a porcine model of acute myocardial infarction with human induced pluripotent stem cell-derived cardiovascular cells. Cell Stem Cell. **15**(6), 750–761 (2014). https://doi.org/10.1016/j.stem.2014.11.009

70. Duval, K., Grover, H., Han, L.-H., Mou, Y., Pegoraro, A.F., Fredberg, J., et al.: Modeling physiological events in 2D vs. 3D cell culture. Physiology. **32**(4), 266–277 (2017)

71. Mathur, A., Loskill, P., Shao, K., Huebsch, N., Hong, S., Marcus, S.G., et al.: Human iPSC-based cardiac microphysiological system for drug screening applications. Sci. Rep. **5**, 8883 (2015)

72. Novakovic, G.V., Eschenhagen, T., Mummery, C.: Myocardial tissue engineering: in vitro models. Cold Spring Harb. Perspect. Med. **4**(3), a014076 (2014)

73. Zimmermann, W.-H., Eschenhagen, T.: Cardiac tissue engineering for replacement therapy. Heart Fail. Rev. **8**(3), 259–269 (2003)
74. Jang, J.: 3D bioprinting and in vitro cardiovascular tissue modeling. Bioengineering. **4**(3), 71 (2017)
75. Fitzgerald, K.A., Malhotra, M., Curtin, C.M., O'Brien, F.J., O'Driscoll, C.M.: Life in 3D is never flat: 3D models to optimise drug delivery. J. Control. Release. **215**, 39–54 (2015)
76. Katare, R.G., Ando, M., Kakinuma, Y., Sato, T.: Engineered heart tissue: a novel tool to study the ischemic changes of the heart in vitro. PloS One. **5**(2), e9275-e (2010). https://doi.org/10.1371/journal.pone.0009275
77. Zimmermann, W.-H., Melnychenko, I., Wasmeier, G., Didié, M., Naito, H., Nixdorff, U., et al.: Engineered heart tissue grafts improve systolic and diastolic function in infarcted rat hearts. Nat. Med. **12**(4), 452–458 (2006)
78. Naito, H., Melnychenko, I., Didié, M., Schneiderbanger, K., Schubert, P., Rosenkranz, S., et al.: Optimizing engineered heart tissue for therapeutic applications as surrogate heart muscle. Circulation. **114**(1 Suppl), I72–I78 (2006). https://doi.org/10.1161/circulationaha.105.001560
79. Eder, A., Vollert, I., Hansen, A., Eschenhagen, T.: Human engineered heart tissue as a model system for drug testing. Adv. Drug Deliv. Rev. **96**, 214–224 (2016). https://doi.org/10.1016/j.addr.2015.05.010
80. Hirt, M.N., Werner, T., Indenbirken, D., Alawi, M., Demin, P., Kunze, A.-C., et al.: Deciphering the microRNA signature of pathological cardiac hypertrophy by engineered heart tissue- and sequencing-technology. J. Mol. Cell. Cardiol. **81**, 1–9 (2015). https://doi.org/10.1016/j.yjmcc.2015.01.008
81. Vunjak Novakovic, G., Eschenhagen, T., Mummery, C.: Myocardial tissue engineering: in vitro models. Cold Spring Harb. Perspect. Med. **4**(3), a014076 (2014). https://doi.org/10.1101/cshperspect.a014076
82. Eschenhagen, T., Mummery, C.: Myocardial tissue engineering: in vitro models. Cold Spring Harbor Perspectives in Medicine. **4**(3) (2014)
83. Hirt, M.N., Sörensen, N.A., Bartholdt, L.M., Boeddinghaus, J., Schaaf, S., Eder, A., et al.: Increased afterload induces pathological cardiac hypertrophy: a new in vitro model. Basic Res. Cardiol. **107**(6), 307 (2012)
84. Bouten, C., Dankers, P., Driessen-Mol, A., Pedron, S., Brizard, A., Baaijens, F.: Substrates for cardiovascular tissue engineering. Adv. Drug Deliv. Rev. **63**(4–5), 221–241 (2011)
85. Wang, Y., Hill, J.A.: Electrophysiological remodeling in heart failure. J. Mol. Cell. Cardiol. **48**(4), 619–632 (2010)
86. Ryan, A.J., Brougham, C.M., Garciarena, C.D., Kerrigan, S.W., O'Brien, F.J.: Towards 3D in vitro models for the study of cardiovascular tissues and disease. Drug Discov. Today. **21**(9), 1437–1445 (2016)
87. Sharma, P., Gentile, C.: Cardiac spheroids as in vitro bioengineered heart tissues to study human heart pathophysiology. J. Vis. Exp. (2021). https://doi.org/10.3791/61962
88. Zuppinger, C.: 3D culture for cardiac cells. Biochimica et Biophysica Acta (BBA)-Molecular. Cell Res. **1863**(7), 1873–1881 (2016)
89. Eschenhagen, T., Fink, C., Remmers, U., Scholz, H., Wattchow, J., Weil, J., et al.: Three-dimensional reconstitution of embryonic cardiomyocytes in a collagen matrix: a new heart muscle model system. FASEB J. **11**(8), 683–694 (1997)
90. Sakaguchi, K., Shimizu, T., Horaguchi, S., Sekine, H., Yamato, M., Umezu, M., et al.: In vitro engineering of vascularized tissue surrogates. Sci. Rep. **3**, 1316 (2013)
91. Menasche, P.: Cardiac cell therapy: lessons from clinical trials. J. Mol. Cell. Cardiol. **50**(2), 258–265 (2011)
92. Oh, H., Ito, H., Sano, S.: Challenges to success in heart failure: cardiac cell therapies in patients with heart diseases. J. Cardiol. **68**(5), 361–367 (2016)

93. Yanamandala, M., Zhu, W., Garry, D.J., Kamp, T.J., Hare, J.M., Jun, H.-W., et al.: Overcoming the roadblocks to cardiac cell therapy using tissue engineering. J. Am. Coll. Cardiol. **70**(6), 766–775 (2017)
94. Giraud, M.-N., Armbruster, C., Carrel, T., Tevaearai, H.T.: Current state of the art in myocardial tissue engineering. Tissue Eng. **13**(8), 1825–1836 (2007)
95. Jawad, H., Ali, N., Lyon, A., Chen, Q., Harding, S., Boccaccini, A.: Myocardial tissue engineering: a review. J. Tissue Eng. Regen. Med. **1**(5), 327–342 (2007)
96. Chen, Q., Harding, S., Ali, N., Jawad, H., Boccaccini, A.: Cardiac tissue engineering. Tissue engineering using ceramics and polymers, pp. 335–356. Elsevier (2007)
97. Wang, H., Roche, C.D., Gentile, C.: Omentum support for cardiac regeneration in ischaemic cardiomyopathy models: a systematic scoping review. Eur. J. Cardiothorac. Surg. **58**(6), 1118–1129 (2020). https://doi.org/10.1093/ejcts/ezaa205
98. Huyer, L.D., Montgomery, M., Zhao, Y., Xiao, Y., Conant, G., Korolj, A., et al.: Biomaterial based cardiac tissue engineering and its applications. Biomed. Mater. **10**(3), 034004 (2015)
99. Dar, A., Shachar, M., Leor, J., Cohen, S.: Optimization of cardiac cell seeding and distribution in 3D porous alginate scaffolds. Biotechnol. Bioeng. **80**(3), 305–312 (2002)
100. Ahmadi, F., Oveisi, Z., Samani, S.M., Amoozgar, Z.: Chitosan based hydrogels: characteristics and pharmaceutical applications. Res. Pharma. Sci. **10**(1), 1 (2015)
101. Lam, M.T., Wu, J.C.: Biomaterial applications in cardiovascular tissue repair and regeneration. Expert. Rev. Cardiovasc. Ther. **10**(8), 1039–1049 (2012)
102. Liu, Z., Wang, H., Wang, Y., Lin, Q., Yao, A., Cao, F., et al.: The influence of chitosan hydrogel on stem cell engraftment, survival and homing in the ischemic myocardial microenvironment. Biomaterials. **33**(11), 3093–3106 (2012)
103. Kharaziha, M., Nikkhah, M., Shin, S.-R., Annabi, N., Masoumi, N., Gaharwar, A.K., et al.: PGS: Gelatin nanofibrous scaffolds with tunable mechanical and structural properties for engineering cardiac tissues. Biomaterials. **34**(27), 6355–6366 (2013)
104. Serpooshan, V., Zhao, M., Metzler, S.A., Wei, K., Shah, P.B., Wang, A., et al.: The effect of bioengineered acellular collagen patch on cardiac remodeling and ventricular function post myocardial infarction. Biomaterials. **34**(36), 9048–9055 (2013)
105. Vettori, L., Sharma, S., Rnjak-Kovacina, J., Gentile, C.: 3D bioprinting of cardiovascular tissues for in vivo and in vitro applications using hybrid hydrogels containing silk fibroin: state of the art and challenges. Curr Tissue Microenviron Rep. **1**, 261–276 (2020). https://doi.org/10.1007/s43152-020-00026-5
106. Gilbert, T.W.: Strategies for tissue and organ decellularization. J. Cell. Biochem. **113**(7), 2217–2222 (2012)
107. Ott, H.C., Matthiesen, T.S., Goh, S.-K., Black, L.D., Kren, S.M., Netoff, T.I., et al.: Perfusion-decellularized matrix: using nature's platform to engineer a bioartificial heart. Nat. Med. **14**(2), 213–221 (2008)
108. Guyette, J.P., Gilpin, S.E., Charest, J.M., Tapias, L.F., Ren, X., Ott, H.C.: Perfusion decellularization of whole organs. Nat. Protoc. **9**(6), 1451–1468 (2014)
109. Theus, A.S., Tomov, M.L., Cetnar, A., Lima, B., Nish, J., McCoy, K., et al.: Biomaterial approaches for cardiovascular tissue engineering. Emergent Materials., 1–15 (2019)
110. Balint, R., Cassidy, N.J., Cartmell, S.H.: Conductive polymers: towards a smart biomaterial for tissue engineering. Acta Biomater. **10**(6), 2341–2353 (2014)
111. Sales, V.L., Engelmayr Jr., G.C., Johnson Jr., J.A., Gao, J., Wang, Y., Sacks, M.S., et al.: Protein precoating of elastomeric tissue-engineering scaffolds increased cellularity, enhanced extracellular matrix protein production, and differentially regulated the phenotypes of circulating endothelial progenitor cells. Circulation. **116**(11_supplement), I-55–I-63 (2007)
112. Stella, J.A., Liao, J., Hong, Y., Merryman, W.D., Wagner, W.R., Sacks, M.S.: Tissue-to-cellular level deformation coupling in cell micro-integrated elastomeric scaffolds. Biomaterials. **29**(22), 3228–3236 (2008)
113. Zhao, F., Shi, Y., Pan, L., Yu, G.: Multifunctional nanostructured conductive polymer gels: synthesis, properties, and applications. Acc. Chem. Res. **50**(7), 1734–1743 (2017)

114. Spearman, B.S., Hodge, A.J., Porter, J.L., Hardy, J.G., Davis, Z.D., Xu, T., et al.: Conductive interpenetrating networks of polypyrrole and polycaprolactone encourage electrophysiological development of cardiac cells. Acta Biomater. **28**, 109–120 (2015)
115. Davenport Huyer, L., Zhang, B., Korolj, A., Montgomery, M., Drecun, S., Conant, G., et al.: Highly elastic and moldable polyester biomaterial for cardiac tissue engineering applications. ACS Biomater Sci. Eng. **2**(5), 780–788 (2016)
116. Shapira, A., Feiner, R., Dvir, T.: Composite biomaterial scaffolds for cardiac tissue engineering. Int. Mater. Rev. **61**(1), 1–19 (2016)
117. Park, H., Radisic, M., Lim, J.O., Chang, B.H., Vunjak-Novakovic, G.: A novel composite scaffold for cardiac tissue engineering. In Vitro Cellular & Developmental Biology-Animal. **41**(7), 188–196 (2005)
118. Sapir, Y., Kryukov, O., Cohen, S.: Integration of multiple cell-matrix interactions into alginate scaffolds for promoting cardiac tissue regeneration. Biomaterials. **32**(7), 1838–1847 (2011)
119. Rai, R., Tallawi, M., Barbani, N., Frati, C., Madeddu, D., Cavalli, S., et al.: Biomimetic poly (glycerol sebacate)(PGS) membranes for cardiac patch application. Mater. Sci. Eng. C. **33**(7), 3677–3687 (2013)
120. Kai, D., Prabhakaran, M.P., Jin, G., Ramakrishna, S.: Guided orientation of cardiomyocytes on electrospun aligned nanofibers for cardiac tissue engineering. J. Biomed. Mater. Res. B Appl. Biomater. **98**(2), 379–386 (2011)
121. Ravichandran, R., Venugopal, J.R., Sundarrajan, S., Mukherjee, S., Ramakrishna, S.: Poly (glycerol sebacate)/gelatin core/shell fibrous structure for regeneration of myocardial infarction. Tissue Eng. Part A. **17**(9–10), 1363–1373 (2011)
122. Ravichandran, R., Venugopal, J.R., Sundarrajan, S., Mukherjee, S., Sridhar, R., Ramakrishna, S.: Expression of cardiac proteins in neonatal cardiomyocytes on PGS/fibrinogen core/shell substrate for cardiac tissue engineering. Int. J. Cardiol. **167**(4), 1461–1468 (2013)
123. Lee, S., Serpooshan, V., Tong, X., Venkatraman, S., Lee, M., Lee, J., et al.: Contractile force generation by 3D hiPSC-derived cardiac tissues is enhanced by rapid establishment of cellular interconnection in matrix with muscle-mimicking stiffness. Biomaterials. **131**, 111–120 (2017)
124. Stoker, M.E., Gerdes, A.M., May, J.F.: Regional differences in capillary density and myocyte size in the normal human heart. Anat. Rec. **202**(2), 187–191 (1982)
125. Roche, C., Sharma, P., Ashton, A., Jackson, C., Xue, M., Gentile, C.: Printability, durability, contractility and vascular network formation in 3D bioprinted cardiac endothelial cells using alginate-gelatin hydrogels. Front. Bioeng. Biotechnol. (2021). https://doi.org/10.3389/fbioe.2021.636257
126. Esser, T.U., Roshanbinfar, K., Engel, F.B.: Promoting vascularization for tissue engineering constructs: current strategies focusing on HIF-regulating scaffolds. Expert. Opin. Biol. Ther. **19**(2), 105–118 (2019)
127. Riegler, J., Tiburcy, M., Ebert, A., Tzatzalos, E., Raaz, U., Abilez, O.J., et al.: Human engineered heart muscles engraft and survive long term in a rodent myocardial infarction model. Circ. Res. **117**(8), 720–730 (2015)
128. Tiburcy, M., Hudson, J.E., Balfanz, P., Schlick, S., Meyer, T., Chang Liao, M.-L., et al.: Defined engineered human myocardium with advanced maturation for applications in heart failure modeling and repair. Circulation. **135**(19), 1832–1847 (2017)
129. Zhang, M., Methot, D., Poppa, V., Fujio, Y., Walsh, K., Murry, C.E.: Cardiomyocyte grafting for cardiac repair: graft cell death and anti-death strategies. J. Mol. Cell. Cardiol. **33**(5), 907–921 (2001)
130. Stoehr, A., Hirt, M.N., Hansen, A., Seiffert, M., Conradi, L., Uebeler, J., et al.: Spontaneous formation of extensive vessel-like structures in murine engineered heart tissue. Tissue Eng. Part A. **22**(3–4), 326–335 (2016)
131. Wang, H., Roche, C.D., Gentile, C.: Omentum support for cardiac regeneration in ischaemic cardiomyopathy models: a systematic scoping review. Eur. J. Cardiothorac. Surg. (2020). https://doi.org/10.1093/ejcts/ezaa205

132. Fagiani, E., Christofori, G.: Angiopoietins in angiogenesis. Cancer Lett. **328**(1), 18–26 (2013)
133. Visconti, R.P., Kasyanov, V., Gentile, C., Zhang, J., Markwald, R.R., Mironov, V.: Towards organ printing: engineering an intra-organ branched vascular tree. Expert. Opin. Biol. Ther. **10**(3), 409–420 (2010)
134. Don, C.W., Murry, C.E.: Improving survival and efficacy of pluripotent stem cell–derived cardiac grafts. J. Cell. Mol. Med. **17**(11), 1355–1362 (2013)
135. Laflamme, M.A., Chen, K.Y., Naumova, A.V., Muskheli, V., Fugate, J.A., Dupras, S.K., et al.: Cardiomyocytes derived from human embryonic stem cells in pro-survival factors enhance function of infarcted rat hearts. Nat. Biotechnol. **25**(9), 1015–1024 (2007)
136. Yang, X., Pabon, L., Murry, C.E.: Engineering adolescence: maturation of human pluripotent stem cell–derived cardiomyocytes. Circ. Res. **114**(3), 511–523 (2014)
137. Caspi, O., Lesman, A., Basevitch, Y., Gepstein, A., Arbel, G., Habib, I.H.M., et al.: Tissue engineering of vascularized cardiac muscle from human embryonic stem cells. Circ. Res. **100**(2), 263–272 (2007)
138. Polonchuk, L., Chabria, M., Badi, L., Hoflack, J.-C., Figtree, G., Davies, M.J., et al.: Cardiac spheroids as promising in vitro models to study the human heart microenvironment. Sci. Rep. **7**(1), 1–12 (2017)
139. Masuda, S., Shimizu, T.: Three-dimensional cardiac tissue fabrication based on cell sheet technology. Adv. Drug Deliv. Rev. **96**, 103–109 (2016)
140. Sekine, H., Shimizu, T., Hobo, K., Sekiya, S., Yang, J., Yamato, M., et al.: Endothelial cell coculture within tissue-engineered cardiomyocyte sheets enhances neovascularization and improves cardiac function of ischemic hearts. Circulation. **118**(14_suppl_1), S145–SS52 (2008)
141. Sekiya, S., Shimizu, T., Yamato, M., Kikuchi, A., Okano, T.: Bioengineered cardiac cell sheet grafts have intrinsic angiogenic potential. Biochem. Biophys. Res. Commun. **341**(2), 573–582 (2006)
142. Song, S.Y., Kim, H., Yoo, J., Kwon, S.P., Park, B.W., Kim, J.-J., et al.: Prevascularized, multiple-layered cell sheets of direct cardiac reprogrammed cells for cardiac repair. Biomater. Sci. **8**(16), 4508–4520 (2020)
143. Serpooshan, V., Mahmoudi, M., Hu, D.A., Hu, J.B., Wu, S.M.: Bioengineering cardiac constructs using 3D printing. J 3D Print Med. **1**(2), 123–139 (2017)
144. Kolesky, D.B., Homan, K.A., Skylar-Scott, M.A., Lewis, J.A.: Three-dimensional bioprinting of thick vascularized tissues. Proc. Natl. Acad. Sci. **113**(12), 3179–3184 (2016)
145. Xu, T., Baicu, C., Aho, M., Zile, M., Boland, T.: Fabrication and characterization of bioengineered cardiac pseudo tissues. Biofabrication. **1**(3), 035001 (2009)
146. Baudis, S., Pulka, T., Steyrer, B., Wilhelm, H., Weigel, G., Bergmeister, H., et al.: 3D-printing of urethane-based photoelastomers for vascular tissue regeneration. MRS Online Proc Libr Arch. **1239** (2010)
147. Dankowski, R., Baszko, A., Sutherland, M., Firek, L., Kałmucki, P., Wróblewska, K., et al.: 3D heart model printing for preparation of percutaneous structural interventions: description of the technology and case report. Kardiologia Polska (Polish Heart Journal). **72**(6), 546–551 (2014)
148. Duan, B.: State-of-the-art review of 3D bioprinting for cardiovascular tissue engineering. Ann. Biomed. Eng. **45**(1), 195–209 (2017)
149. Wang, Z., Lee, S.J., Cheng, H.-J., Yoo, J.J., Atala, A.: 3D bioprinted functional and contractile cardiac tissue constructs. Acta Biomater. **70**, 48–56 (2018). https://doi.org/10.1016/j.actbio.2018.02.007
150. Ong, C.S., Fukunishi, T., Zhang, H., Huang, C.Y., Nashed, A., Blazeski, A., et al.: Biomaterial-free three-dimensional bioprinting of cardiac tissue using human induced pluripotent stem cell derived cardiomyocytes. Sci. Rep. **7**(1), 4566 (2017). https://doi.org/10.1038/s41598-017-05018-4

151. Cui, H., Miao, S., Esworthy, T., Zhou, X., Lee, S.-J., Liu, C., et al.: 3D bioprinting for cardiovascular regeneration and pharmacology. Adv. Drug Deliv. Rev. **132**, 252–269 (2018). https://doi.org/10.1016/j.addr.2018.07.014

152. Yeung, E., Fukunishi, T., Bai, Y., Bedja, D., Pitaktong, I., Mattson, G., et al.: Cardiac regeneration using human-induced pluripotent stem cell-derived biomaterial-free 3D-bioprinted cardiac patch in vivo. J. Tissue Eng. Regen. Med. **13**(11), 2031–2039 (2019)

153. Maiullari, F., Costantini, M., Milan, M., Pace, V., Chirivì, M., Maiullari, S., et al.: A multicellular 3D bioprinting approach for vascularized heart tissue engineering based on HUVECs and iPSC-derived cardiomyocytes. Sci. Rep. **8**(1), 13532 (2018). https://doi.org/10.1038/s41598-018-31848-x

154. Kuss, M., Duan, B.: 3D bioprinting for cardiovascular tissue engineering. Rapid prototyping in cardiac disease, pp. 167–182. Springer (2017)

155. Jia, W., Gungor-Ozkerim, P.S., Zhang, Y.S., Yue, K., Zhu, K., Liu, W., et al.: Direct 3D bioprinting of perfusable vascular constructs using a blend bioink. Biomaterials. **106**, 58–68 (2016)

156. Lee, V.K., Kim, D.Y., Ngo, H., Lee, Y., Seo, L., Yoo, S.-S., et al.: Creating perfused functional vascular channels using 3D bio-printing technology. Biomaterials. **35**(28), 8092–8102 (2014)

157. Bertassoni, L.E., Cecconi, M., Manoharan, V., Nikkhah, M., Hjortnaes, J., Cristino, A.L., et al.: Hydrogel bioprinted microchannel networks for vascularization of tissue engineering constructs. Lab Chip. **14**(13), 2202–2211 (2014)

158. Birket, M.J., Mummery, C.L.: Pluripotent stem cell derived cardiovascular progenitors–a developmental perspective. Dev. Biol. **400**(2), 169–179 (2015)

159. Gentile, C.: Filling the gaps between the in vivo and in vitro microenvironment: engineering of spheroids for stem cell technology. Curr. Stem Cell Res. Ther. **11**(8), 652–665 (2016)

160. Jiang, L., Gentile, C., Lauto, A., Cui, C., Song, Y., Romeo, T., et al.: Versatile fabrication approach of conductive hydrogels via copolymerization with vinyl monomers. ACS Appl. Mater. Interfaces. **9**(50), 44124–44133 (2017)

161. Mawad, D., Mansfield, C., Lauto, A., Perbellini, F., Nelson, G.W., Tonkin, J., et al.: A conducting polymer with enhanced electronic stability applied in cardiac models. Sci. Adv. **2**(11), e1601007 (2016)

162. Mehrotra, S., Moses, J.C., Bandyopadhyay, A., Mandal, B.B.: 3D printing/bioprinting based tailoring of in vitro tissue models: recent advances and challenges. ACS Applied Bio Materials. **2**(4), 1385–1405 (2019)

163. Noor, N., Shapira, A., Edri, R., Gal, I., Wertheim, L., Dvir, T.: 3D printing of personalized thick and perfusable cardiac patches and hearts. Advanced Science. **6**(11), 1900344 (2019)

164. Birla, R.K., Williams, S.K.: 3D bioprinting and its potential impact on cardiac failure treatment: an industry perspective. APL bioengineering. **4**(1), 010903 (2020). https://doi.org/10.1063/1.5128371

165. Jang, J., Park, H.-J., Kim, S.-W., Kim, H., Park, J.Y., Na, S.J., et al.: 3D printed complex tissue construct using stem cell-laden decellularized extracellular matrix bioinks for cardiac repair. Biomaterials. **112**, 264–274 (2017)

166. Tijore, A., Irvine, S.A., Sarig, U., Mhaisalkar, P., Baisane, V., Venkatraman, S.: Contact guidance for cardiac tissue engineering using 3D bioprinted gelatin patterned hydrogel. Biofabrication. **10**(2), 025003 (2018)

167. Sharma, P., Wang, X., Liu Chung Ming, C., Vettori, L., Figtree, G., Boyle, A., et al.: Considerations for the bioengineering of advanced cardiac in vitro models of myocardial infarction. Small. (2021). https://doi.org/10.1002/smll.202003765

168. Takeda, M., Miyagawa, S., Fukushima, S., Saito, A., Ito, E., Harada, A., et al.: Development of in vitro drug-induced cardiotoxicity assay by using three-dimensional cardiac tissues derived from human induced pluripotent stem cells. Tissue Eng. Part C Methods. **24**(1), 56–67 (2018)

169. Alonzo, M., AnilKumar, S., Roman, B., Tasnim, N., Joddar, B.: 3D bioprinting of cardiac tissue and cardiac stem cell therapy. Transl. Res. **211**, 64–83 (2019)

170. Jin, G., Yang, G.H., Kim, G.: Tissue engineering bioreactor systems for applying physical and electrical stimulations to cells. J. Biomed. Mater. Res. B Appl. Biomater. **103**(4), 935–948 (2015)
171. Zhang, Y.S., Yue, K., Aleman, J., Mollazadeh-Moghaddam, K., Bakht, S.M., Yang, J., et al.: 3D bioprinting for tissue and organ fabrication. Ann. Biomed. Eng. **45**(1), 148–163 (2017)
172. Paez-Mayorga, J., Hernández-Vargas, G., Ruiz-Esparza, G.U., Iqbal, H.M., Wang, X., Zhang, Y.S., et al.: Bioreactors for cardiac tissue engineering. Adv. Healthc. Mater. **8**(7), 1701504 (2019)
173. Visone, R., Talò, G., Lopa, S., Rasponi, M., Moretti, M.: Enhancing all-in-one bioreactors by combining interstitial perfusion, electrical stimulation, on-line monitoring and testing within a single chamber for cardiac constructs. Sci. Rep. **8**(1), 16944 (2018). https://doi.org/10.1038/s41598-018-35019-w
174. Qasim, M., Haq, F., Kang, M.-H., Kim, J.-H.: 3D printing approaches for cardiac tissue engineering and role of immune modulation in tissue regeneration. Int. J. Nanomedicine. **14**, 1311–1333 (2019). https://doi.org/10.2147/IJN.S189587
175. Zhang, Y.S., Arneri, A., Bersini, S., Shin, S.-R., Zhu, K., Goli-Malekabadi, Z., et al.: Bioprinting 3D microfibrous scaffolds for engineering endothelialized myocardium and heart-on-a-chip. Biomaterials. **110**, 45–59 (2016)
176. Smith, L.J., Li, P., Holland, M.R., Ekser, B.: FABRICA: a bioreactor platform for printing, perfusing, observing, & stimulating 3D tissues. Sci. Rep. **8**(1), 7561 (2018). https://doi.org/10.1038/s41598-018-25663-7
177. Ball, O., Nguyen, B.B., Placone, J.K., Fisher, J.P.: 3D printed vascular networks enhance viability in high-volume perfusion bioreactor. Ann. Biomed. Eng. **44**(12), 3435–3445 (2016). https://doi.org/10.1007/s10439-016-1662-y
178. Roche, C.D., Gentile, C.: Transplantation of a 3D bioprinted patch in a murine model of myocardial infarction. Journal of visualized experiments: JoVE. **163** (2020)
179. Roche, C.D., Gentile, C.: Transplantation of a 3D bioprinted patch in a murine model of myocardial infarction. J. Vis. Exp. **163** (2020). https://doi.org/10.3791/61675
180. Norotte, C., Marga, F.S., Niklason, L.E., Forgacs, G.: Scaffold-free vascular tissue engineering using bioprinting. Biomaterials. **30**(30), 5910–5917 (2009). https://doi.org/10.1016/j.biomaterials.2009.06.034
181. Roche, C.D., Brereton, R.J., Ashton, A.W., Jackson, C., Gentile, C.: Current challenges in three-dimensional bioprinting heart tissues for cardiac surgery. Eur. J. Cardio-Thoracic Surg. (2020)

Creating and Validating New Tools to Evaluate the Electrical Integration and Function of hPSC-Derived Cardiac Grafts In Vivo

Wahiba Dhahri, Fanny Wulkan, and Michael A. Laflamme

1 Introduction

Over the past two decades, we and other investigators in the field have reported efficient methods for the guided cardiac differentiation of human pluripotent stem cells (hPSCs) [1–3].The resultant hPSC-derived cardiomyocytes (hPSC-CMs) have been delivered to injured hearts both as injectable single-cell suspensions [1, 4–9] and implantable myocardial tissue "patches" [10–13], and both approaches have been shown to mediate the partial remuscularization of the infarct scar and beneficial effects on left ventricular (LV) contractile function in preclinical models of myocardial infarction (MI). While the in vitro electrophysiological (EP) phenotype of hPSC-CMs has been the subject of intense study [14–18], in vivo experiments have largely focused on histological and contractile endpoints, so comparatively little is known about the electrical function of hPSC-CMs following transplantation or how these parameters might change with increasing duration in vivo. In this chapter, we will review the various approaches that have been applied to specifically assess the host-graft electromechanical integration and EP phenotype of hPSC-CM graft tissue.

An important early goal for the field was to resolve whether hPSC-CM grafts were even capable of electromechanical integration and synchronous activation

W. Dhahri · F. Wulkan
McEwen Stem Cell Institute, University Health Network, Toronto, ON, Canada

M. A. Laflamme (✉)
McEwen Stem Cell Institute, University Health Network, Toronto, ON, Canada

Peter Munk Cardiac Centre, University Health Network, Toronto, ON, Canada

Department of Laboratory Medicine & Pathobiology, University of Toronto, Toronto, ON, Canada
e-mail: Michael.Laflamme@uhnresearch.ca

© The Author(s), under exclusive license to Springer Nature
Switzerland AG 2022
J. Zhang, V. Serpooshan (eds.), *Advanced Technologies in Cardiovascular Bioengineering*, https://doi.org/10.1007/978-3-030-86140-7_14

313

with host myocardium during systole, as the presence or absence of such coupling had major implications on the potential mechanistic basis for the beneficial effects of cardiomyocyte transplantation. Many adult stem cell types have no capacity for myogenic differentiation and yet some have been reported to indirectly enhance LV contractile function via passive mechanical buttressing, immunomodulation or pro-angiogenic signaling [19]. A regenerative therapy based on hPSC-CMs has potential to *directly* enhance systolic function by contributing new force-generating units, but this obviously requires that the graft cardiomyocytes couple electrically and contract synchronously with host muscle. As described further below, our group has used graft-autonomous reporters of graft activation to unambiguously demonstrate that hPSC-CM grafts are indeed capable of 1:1 coupling with host myocardium, but we have also found that their integration is typically far from perfect (with many recipient hearts showing little or no coupled graft). Given this situation, we predict that future efforts to further enhance the functionality of hPSC-CM graft tissue will depend on the availability of tools to quantitatively assess the presence and extent of host-graft electromechanical integration.

Beyond the simple question as to whether hPSC-CM grafts can couple to host myocardium, it has also become increasingly clear that the field will also require improved tools to probe the EP phenotype of hPSC-CM graft tissue more comprehensively and acquire fundamental parameters like graft conduction velocity (CV), action potential (AP) duration and restitution properties. Underscoring this need, the transplantation of hPSC-CMs in porcine and non-human primate MI models has been found to result in transient bouts of ventricular tachycardia (VT) [6, 8, 9, 20], and the risk of graft-related arrhythmias has emerged as arguably the most important barrier to translation. While some mechanisms for this phenomenon have been suggested based on first principles [21–23] it is particularly challenging to determine their relative contribution and design appropriate mitigation strategies without an improved understanding of graft EP phenotype and how it is affected by duration in vivo and host factors. Moreover, going forward, the field will need reliable, standardized methods to compare the in vivo EP function of different hPSC-CM products (e.g. progeny of different hPSC lines, hPSC-CMs at different stages of maturation, or single cell suspensions versus cardiac patches).

With these considerations in mind, we describe here the various techniques that our group and other investigators in the field have employed to investigate the electromechanical integration, EP phenotype and pro-arrhythmic behavior of hPSC-CM graft tissue. In general, these efforts can be divided into three broad approaches: (1) direct electrical recordings, typically via epicardial electrode pairs or arrays placed on engrafted hearts; (2) optical mapping of engrafted hearts using fluorescent signals from either small-molecules voltage-sensitive dyes or genetically encoded calcium- or voltage-sensitive reporters; or (3) catheter-based electroanatomic mapping, a standard clinical technique that is only practical in large-animal models (see Fig. 1 for overview of these techniques and the insights that can be gleaned from each). As outlined further below, each of these approaches has its own advantages and disadvantages, and they have all meaningfully contributed to our current understanding of hPSC-CM graft EP function.

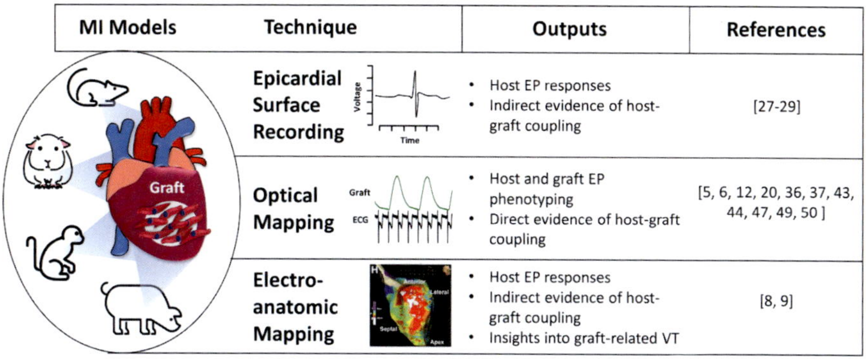

MI Models	Technique	Outputs	References
	Epicardial Surface Recording Voltage / Time	• Host EP responses • Indirect evidence of host-graft coupling	[27-29]
	Optical Mapping Graft / ECG	• Host and graft EP phenotyping • Direct evidence of host-graft coupling	[5, 6, 12, 20, 36, 37, 43, 44, 47, 49, 50]
	Electro-anatomic Mapping	• Host EP responses • Indirect evidence of host-graft coupling • Insights into graft-related VT	[8, 9]

Fig. 1 Methods to assess hPSC-CM graft electromechanical integration and electrophysiological function in animal model of myocardial infarction. Multiple techniques have been used in the field to investigate graft electrical function and host responses in preclinical models including epicardial voltage recordings via electrodes or multi-electrode arrays, optical mapping of $[Ca^{2+}]_i$- or voltage-sensitive fluorescent reporters, and catheter-based endocardial electroanatomic mapping

2 Evaluation of hPSC-CM Graft EP Function Via Direct Epicardial Surface Recordings

Arguably, the most straightforward method to assess the EP function of hPSC-CM grafts is via direct recording of electrical activity from the epicardial surface of the engrafted heart using an open-chest or ex vivo Langendorff heart preparation. Unipolar or bipolar electrograms can be acquired using either electrode pairs or custom-built electrode arrays suitable for multisite contact mapping [24–26]. In early work, Zimmermann and co-workers showed the utility of this approach in an experiment designed to establish the electromechanical integration of engineered heart tissues (EHTs) formed using primary neonatal rat cardiomyocytes [27]. In this study, the authors implanted EHTs onto the epicardial surface of infarcted rats then later mapped recipient hearts ex vivo using a 256-channel electrode array. They reported that infarcted but sham-operated rats exhibited conduction delays and QRS amplitude attenuation that were partially rescued in EHT recipients, thereby providing at least indirect evidence of some degree of host-graft electromechanical integration. Similar methods have been employed in a small number of studies to probe the electrical activity of hearts with epicardial implants formed using pluripotent stem cell derivatives [28, 29]. For example, Higuchi and co-workers implanted cell sheets populated with either fibroblasts or mouse induced pluripotent stem cell (iPSC)-derived cardiomyocytes onto infarcted rat hearts, then performed multisite electrode mapping at time-points up to 7-days post-transplantation [28]. While their evidence for electromechanical integration of the cardiomyocyte sheets was limited, they described reduced activation recovery intervals and a trend toward increased premature ventricular contractions (PVCs) in the cardiomyocyte sheet recipients versus fibroblast sheet or sham controls. Their finding of ectopic excitations over the cardiomyocyte sheet area is concerning for graft-related tachyarrhythmias, but it was not reported whether these impulses were propagated into host myocardium or whether the cardiomyocyte recipients showed spontaneous electrocardiographic (ECG) abnormalities.

While conceptually straightforward, this approach of evaluating graft EP function by directly measuring epicardial electrical activity has significant limitations that have led many investigators to pursue alternative strategies. First, even high-density electrode array systems often lack sufficient spatial and temporal resolution to elucidate the mechanism of cardiac arrhythmias, particularly in smaller rodent hearts [30]. Second, estimates of activation and repolarization times based on contact mapping can sometimes be challenging under relevant experimental conditions including ischemia [31, 32]. Finally, and most importantly, there is no way to reliably discriminate between graft- and host-derived electrical signals by this method. This essentially limits its relevance to experiments involving EHTs implanted onto the epicardial surface of the heart, and even here one needs carefully blinded comparisons with sham-operated controls to attribute any observed electrical behavior to the presence of graft tissue. Such recordings should not be considered reliable for the detection of intramyocardial hPSC-CM grafts formed via direct injection, as any graft-derived signals would be expected to be dwarfed by that of host subepicardial muscle.

3 Evaluation of hPSC-CM Graft EP Function Via Optical Mapping

The inherent limitations of direct electrode mapping led to the development of optical mapping techniques for the study of myocardial activation and repolarization, and this technology has been more recently applied for the specific purpose of investigating the EP function and pro-arrhythmic behavior of hPSC-CM-engrafted hearts. The fundamental principles of optical mapping have been the subject of multiple excellent reviews [33]. In brief, optical mapping typically involves the labeling of an excised heart with a potentiometric dye that changes its spectral properties in response to a change in transmembrane voltage. The epicardial surface of the dye-labeled heart is then imaged with an appropriately high-speed optical detection system (e.g. a charge-coupled device (CCD) camera or photodiode array system). While optical mapping presents its own set of technical challenges (including artifacts from contractile motion or the need to pharmacologically induce mechanical arrest, poor depth of field, photobleaching, and phototoxicity), it allows one to track ventricular activation and repolarization times with a spatiotemporal resolution that is generally well beyond that achievable by electrode arrays. Optical mapping has also been successfully combined with other physiological readouts (e.g. fluorescent indicators of intracellular calcium $[Ca^{2+}]_i$ handling) [34], and it has proven particularly powerful for elucidating mechanisms of arrhythmogenesis [35].

Gepstein and co-workers were the first to apply optical mapping with potentiometric dyes to specifically investigate the electromechanical integration of hPSC-CM graft tissue [36]. In this study, the authors labeled hPSC-CMs with the green fluorescent lipophilic dye 3,3′-dioctadecycloxacarbocyanin (DiO) and directly injected these cells into LV wall of naïve (uninfarcted) rat hearts. They later performed optical mapping of the engrafted hearts using an ex vivo ex vivo Langendorff preparation and the spectrally distinct potentiometric dye PGH-I and identified hPSC-CM graft tissue based on static DiO fluorescence. They observed normal propagation across the epicardially visible DiO+ graft region in four of six recipients and only minimal CV slowing in two of six recipients. While their study was a landmark in that it was the first to apply optical mapping for this purpose, the authors' approach suffers from the same potential limitation as surface electrograms. Because the PGH-I dye could label both graft and host tissue, there was no way to be certain that PGH-I-derived optical action potentials (oAPs) were arising from hPSC-CM graft as opposed to host muscle located either deep or superficial to the graft. Indeed, in work described in more detail below, our group later confirmed that factitious optical signals are quite possible under comparable experimental conditions [5, 37]. One can probably overcome this limitation and reliably discriminate graft- and host-derived signals by using an intravital imaging modality that permits single-cell resolution (e.g. two-photon microscopy [38, 39]), but this alternative is technically challenging and obviously comes at the expense of breadth of field.

3.1 Optical Mapping Using Genetically Encoded Calcium-Sensitive Fluorescent Reporters

Because of the inherent difficulty in discriminating graft- and host-derived optical signals from engrafted hearts stained with small-molecule potentiometric dyes, we and others in the field have instead turned to transgenic hPSC-CMs expressing genetically encoded calcium- or voltage-sensitive fluorescent proteins. In this case, because the fluorescent protein is not expressed by the recipient myocardium, it serves as a dynamic, *graft-autonomous* reporter of graft activation. By far the most frequently used reporter for this purpose has been the calcium-sensitive fluorescent protein GCaMP. GCaMP is a fusion protein that combines a circularly permuted form of enhanced green fluorescent protein (GFP) and moieties from the calcium-binding protein calmodulin and myosin light chain kinase, and multiple GCaMP variants are now available that have been engineered for improved brightness, kinetics, signal-to-noise ratio, or subcellular compartmentalization [40, 41]. In presence of calcium, the calmodulin moiety of the GCaMP protein undergoes a conformational change, resulting in the C- and N-terminal ends of GFP being brought into proximity with each other and an increase in green fluorescence. One next-generation GCaMP variant, GCaMP6, exhibits as much as a 20-fold increase in fluorescence with a change in $[Ca^{2+}]_i$ from 0 to ~1 µM [42].

Roell and co-workers demonstrated how GCaMP fluorescence could be used to assess the electromechanical integration of intramyocardial grafts in a landmark study involving the transplantation of primary cardiomyocytes [43]. These authors injected infarcted mouse hearts with embryonic cardiomyocytes that were either isolated from a GCaMP-expressing transgenic mouse or transduced with a lentivirus expressing GCaMP, then engrafted hearts were imaged using an open-chest preparation. GCaMP-derived $[Ca^{2+}]_i$ signals arising from the graft muscle were found to be entrained to the host sinus rhythm, but they often activated with either a 1:2 or 1:4 block (i.e. one GCaMP fluorescence transient for every second or fourth host QRS complex). This result directly proved the possibility of host-graft coupling, although it also indicated that the integration of the graft muscle was often imperfect.

Inspired by the latter study, our group has used an analogous approach to quantitatively examine the electromechanical integration of hPSC-CM grafts in uninjured and cryoinjured guinea pig hearts [5, 37, 44]. For this purpose, we used zinc finger nuclease- or CRISPR/Cas9-mediated gene-editing to create transgenic hPSCs that stably express GCaMP (via insertion of a GCaMP expression cassette into the AAVS1 "safe harbor" locus [45, 46]). As described in a series of recent reports [5, 37, 44, 47], we have transplanted GCaMP+ hPSC-CMs from this transgenic line into both naïve (uninjured) guinea pig hearts as well as hearts at either 10- or 28-days following cardiac cryoinjury. Because of its extensive collaterals, the guinea pig heart is not amenable to MI via coronary ligation as with other rodent models [48]. In most of these experiments, recipient hearts were imaged using a custom dual-channel, high-speed imaging system (Fig. 2A), which enabled the simultaneous acquisition of GCaMP fluorescence transients and a second, spectrally distinct fluorescent reporter. By correlating the timing of graft-derived $[Ca^{2+}]_i$ signals (GCaMP fluorescence) with host-derived activation signals (either using the host ECG or oAPs from host myocardium labeled with a spectrally distinct potentiometric dye), we can determine whether and what fraction of hPSC-CM graft tissue is 1:1 coupled with host myocardium in any given recipient heart. Figure 2B shows representative graft-derived $[Ca^{2+}]_i$ and simultaneously acquired ECG traces from an imaging experiment with GCaMP+ hPSC-CM graft tissue in an uninjured heart (during this particular experiment, the second emission channel was not utilized).

This work in the guinea pig model has taught us several important lessons about activation and propagation in hPSC-CM graft tissue. First, we have consistently observed 1:1 host-graft coupling in the case of GCaMP+ hPSC-CMs transplanted into uninjured hearts, and propagation in graft tissue appears very rapid and synchronous in this context [5]. On the other hand, host-graft electromechanical integration has proven less consistent in the case of GCaMP+ hPSC-CMs implanted in injured hearts. While there has been some variation that appears to be at least partially attributed to hPSC line-to-line differences [37], overall only about one-third of engrafted hearts in a guinea pig subacute cryoinjury model (i.e. cell transplantation at 10-days post-injury) have shown at least some graft regions with 1:1 host-graft coupling [37]. Host-graft integration appears to be even worse following transplantation in chronic cryoinjury model (with cell transplantation at 28-days post-injury).

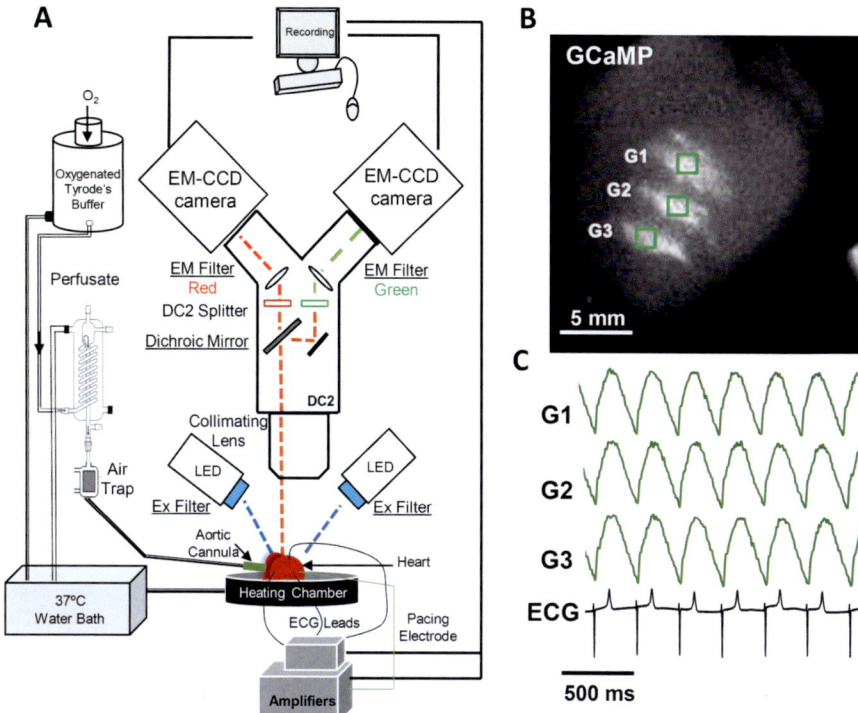

Fig. 2 Optical mapping of graft-derived GCaMP $[Ca^{2+}]_i$ signals from engrafted hearts. A. Schematic of the high-speed dual-channel imaging system used in laboratory to acquire GCaMP+ fluorescent transients and a spectrally distinct second fluorescent reporter. As depicted, this system is configured for simultaneous imaging of GCaMP and di-2-ANEPEQ fluorescent signals, but filter sets can be readily exchanged to accommodate other fluorophores. In brief, the engrafted heart is maintained ex vivo on a Langendorff apparatus and excited with LED spotlights bandpass-filtered to 450–490 nm. Emitted light from the heart is focused and split into separate "green" and "red" channels using a dichroic system with a 565 nm longpass dichroic mirror. The "green" channel (i.e. GCaMP signal) is further bandpass-filtered to 500–530 nm before detection by a high-speed EM-CCD camera, while the "red" channel (i.e. potentiometric dye signal) signal is longpass filtered at >700 nm before detection by a second EM-CCD camera. B. Representative image of an uninjured guinea pig heart with GCaMP+ hPSC-CM graft on the "green" channel. Three graft regions of interest (ROIs) are labeled G1-G3. C. GCaMP-derived fluorescent $[Ca^{2+}]_i$ signals arising from each of the preceding ROIs (green traces) along with the simultaneously acquired host ECG (black trace). Note that GCaMP fluorescent traces occurred in 1:1 synchrony with QRS complex from the host ECG, thereby proving entrainment of hPSC-CM graft and host myocardium

Here, somewhat less than 20% of engrafted hearts have shown any regions of 1:1 coupled graft [44]. Taken collectively, these data suggest that intervening scar tissue or other factors present in the injured heart act as a major barrier to hPSC-CM integration, an outcome that is perhaps unsurprising. A second important outcome from this work is that, as noted above, conventional potentiometric dyes like RH237 and di-4-ANEPPS are simply not reliable for the evaluation of graft EP function in this experimental preparation. We concluded this after staining engrafted hearts with these dyes and observing frequent instances in which the graft-autonomous GCaMP reporter *proved* that a given graft was completely uncoupled from the host (with GCaMP transients occurring independently of the host ECG), and yet potentiometric dye-derived oAPs from the same region of interest appeared to occur in 1:1 synchrony with the host ECG. We subsequently went on to show that this phenomenon resulted from oAPs from host myocardium located deep to the graft that shined through the scar to the epicardial surface, an outcome that underscores the need for a genetically encoded reporter to discriminate between graft and host signals [37].

The GCaMP reporter has been used more recently to evaluate the electromechanical integration of hPSC-CM grafts in other animal models or following other delivery methods. For example, working independently, both our group and the Shiba lab found that GCaMP+ cardiomyocytes show fairly reliable 1:1 host-graft coupling after transplantation in non-human primate MI model (in our study using hPSC-CMs [9], and in the Shiba study using allogenic monkey iPSC-CMs [20]). This finding of seemingly better host-graft integration following transplantation into monkey versus guinea pig hearts may reflect better coupling given the primate's slower heart rate, geometric or biological differences between scars formed by ischemia-reperfusion versus cryoinjury, or simply our inability to comparably scan the entirety of the graft tissue in the much larger primate heart. In another interesting study, Weinberger and co-workers used hPSC-CMs transduced with a lentiviral vector expressing GCaMP to probe the integration of EHTs implanted onto the epicardial surface of cryoinjured guinea pig hearts [12]. Perhaps paradoxically, while the amount of surviving cardiac grafts per input cardiomyocyte achieved by this delivery method appeared substantially greater than that reported historically with hPSC-CMs injected intramyocardially in cell suspensions, electromechanical integration of the EHTs appeared quite limited (with only 2 of 7 hearts with GCaMP+ EHTs showing synchronous activation with host myocardium). Similar results were obtained by Jackman and colleagues, who created engineered cardiac patches with impressive structural and functional maturation and yet failed to detect any coupling between GGaMP+ patches and host myocardium after epicardial implantation in uninjured rat hearts [49]. Finally, in a head-to-head study comparing outcomes with GCaMP+ hPSC-CMs delivered via epicardial patch versus transepicardial injection, Gerbin and colleagues also found that patches showed limited or no host-graft coupling [50]. Taken collectively, these data suggest that the epicardium may act as a meaningful barrier to electromechanical integration, so other interventions (e.g., disruption of the epicardium to facilitate host-graft contact or exogenous pacing of the patch) may be necessary to achieve synchronous activation of cardiac patches and host muscle.

3.2 Optical Mapping of GCaMP and a Simultaneously Imaged Water-Soluble Voltage Dye

As noted above, during dual-camera optical mapping of guinea pig hearts implanted with GCaMP+ hPSC-CMs and labeled with conventional potentiometric dyes like RH237, our group has routinely observed discordance between GCaMP fluorescence transients and dye-derived oAPs acquired in the same location. We subsequently showed that ostensible graft-derived oAPs from these sites actually arose from subendocardial *host* muscle located deep to the graft; in fact, hPSC-CM graft tissue was not appreciably stained at all by the dye [5, 37]. We hypothesized that the failure of conventional potentiometric dyes to label the graft was because these dyes, which are highly lipophilic, and loaded by perfusion, and likely fully partition into host muscle before reaching the poorly perfused graft. Consistent with this hypothesis, we found that much better outcomes were possible by instead labeling engrafted hearts with less commonly used water-soluble potentiometric dyes like di-2-ANEPEQ [51]. Indeed, by perfusing engrafted hearts with di-2-ANEPEQ, we were able to reliably stain both host and hPSC-CM graft myocardium and, for the first time, acquire unambiguously *graft*-derived oAPs that, as predicted, occurred synchronously with simultaneously acquired GCaMP $[Ca^{2+}]_i$ signals. Moreover, because labeling with water-soluble di-2-ANEPEQ was reversible with different wash-in and wash-out kinetics between graft and host tissue, we were able to transiently isolate graft-derived oAPs with an even greater degree of confidence. Using these optical mapping methods, we measured previously unattainable EP parameters in hPSC-CM graft tissue (e.g. graft CV and oAP durations), compared the EP phenotype of grafts formed from different parental hPSC lines and in different injury contexts, and directly confirmed that graft ectopy is a potential arrhythmia mechanism [37].

3.3 Optical Mapping Using Genetically Encoded Voltage-Sensitive Fluorescent Reporters

While the preceding techniques suffice to reliably distinguish hPSC-CM graft and host oAPs, they are admittedly somewhat cumbersome, involving tedious wash-out experiments and the simultaneous imaging of di-2-ANEPEQ and a graft-autonomous reporter of graft activation (e.g., GCaMP). With this in mind, our group has recently pursued a more direct strategy to probe hPSC-CM graft EP function based on the transplantation of transgenic cardiomyocytes expressing genetically encoded voltage-sensitive fluorescent proteins (VSPs). Multiple VSP designs have been explored including fluorescent biosensors based on microbial rhodopsins as well as fusion proteins incorporating the transmembrane voltage-sensing domain of the ascidian *Ciona intestinalis* voltage-sensing phosphatase (Ci-VSD) [52, 53]. These reporters have been engineered further through an iterative process of mutagenesis and selection, and now a wide range of spectral VSP variants with improved brightness, dynamic range, and/or kinetics are now available. VSPs have proven to be an invaluable tool for neuroscientists interested in probing electrical responses within the central nervous system (CNS) [54–56], but they have seen comparatively little application in the cardiovascular system [57, 58]. Perhaps contributing to this poor uptake, VSPs have a few technical limitations, some of which are especially problematic for intravital imaging in the heart. For example, many VSPs are relatively dim and/or exhibit small changes in fluorescence ($\Delta F/F$) with depolarization compared to calcium-sensitive fluorescent proteins (e.g., GCaMP). These factors can contribute to a low signal-to-noise ratio, particularly in myocardium which is significantly more light scattering than CNS tissues. Most VSPs also exhibit slow kinetics in response to voltage changes relative to small-molecule potentiometric dyes, although this is arguably more problematic for neuronal cell types that generally have much shorter AP durations than ventricular cardiomyocytes.

On the other hand, this approach of optical mapping cells engineered to express VSPs has several important advantages for probing hPSC-CM EP function. In this case, the genetically encoded voltage indicator directly and exclusively reports oAPs from the graft, while host-derived oAPs can be simultaneously acquired by labeling the heart with a spectrally distinct potentiometric dye like RH237 that we have previously shown uniquely labels host myocardium. To test the feasibility of this approach, our group used CRISPR/Cas9-based gene-editing to create a series of transgenic hPSC lines that each express one of four different VSPs: ArcLight, a green fluorescent Ci-VSD-based reporter [59], ASAP1, a green fluorescent reporter based on the *Gallus gallus* voltage-sensitive phosphatase [60], and ArchD95H [61] and QuasAr2 [62], two red-shifted archaerhodpsin-based indicators. Figure 3 depicts the targeting strategy that was used to create these various VSP reporter lines as well as representative photomicrographs and oAPs recorded ASAP1+ hPSC-CM in vitro. In brief, while we have successfully detected oAPs from cardiomyocytes expressing all four VSP+ lines in vitro, the ASAP1 reporter has exhibited the best balance between brightness, spectral separation, kinetics, and signal-to-noise ratio in our hands.

ASAP1 ("Accelerated Sensor of Action Potential 1"), a relatively bright VSP based on the fusion of circularly permuted GFP and a voltage-sensing domain derived from the chicken *Gallus gallus* voltage-sensitive phosphatase [60], has relatively high voltage sensitivity (~15% $\Delta F/F$ per 100 mV depolarization) and acceptable on-off kinetics (<5 ms) in cardiomyocytes. Moreover, in ongoing experiments, we have found no evidence that expression of ASAP1 or any of the other tested VSPs exerts significant off-target effects on hPSC-CM cellular function (data not shown).

During optical mapping experiments with VSP-expressing hPSC-CMs transplanted into both naïve and injured guinea pig hearts, we have been able to successfully image graft-derived oAPs using the same dual-camera system previously employed to image GCaMP+ graft tissue. Although we could detect fluorescent signals from all four VSP lines in vivo, the ASAP1 reporter again worked best in our hands for this application, and we have employed this VSP in all our recent

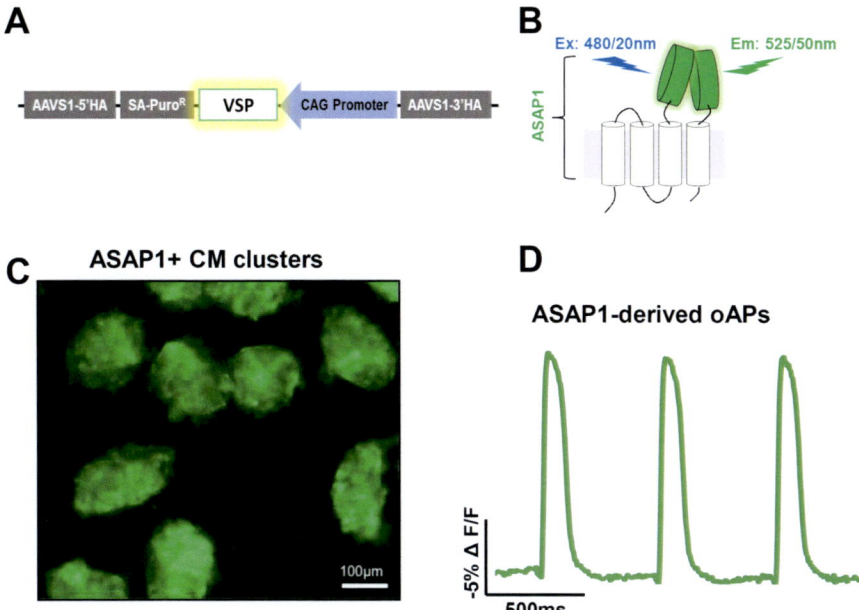

Fig. 3 Generation and in vitro imaging of VSP-expressing hPSC-CMs. A. To generate transgenic hPSC lines stably expressing the tested VSPs, wildtype hPSCs were co-transfected with the depicted donor vector as well as plasmids encoding for site-specific nucleases targeting the AAVS1 safe harbor locus [5, 45, 46]. The donor vector is comprised of 5' and 3' AAVS1 homology arms flanking a transgene in which the strong constitutive CAG promoter drives expression of the VSP as well as puromycin resistance. In successfully targeted cells, this construct is inserted in the AAVS1 locus by homologous recombination. B. ASAP1 is a fusion construct in which circularly permuted GFP is inserted into an extracellular loop of the *Gallus gallus* voltage-sensing domain. Peak excitation is at 488 nm; emission is at 525/50 nm. C. Representative photomicrograph of green fluorescent ASAP1+ hPSC-CM clusters. D. Representative optical action potentials acquired from ASAP1+ hPSC-CMs during spontaneous beating

experiments. (ArcLight+ grafts also had acceptable brightness and signal-to-noise ratio but had AP waveforms that were distorted by ArcLight's slow temporal kinetics [59, 63]). Figure 4 depicts our typical workflow and representative traces from an optical mapping experiment with an ASAP1+ hPSC-CM engrafted heart. During these experiments, we first record graft-derived oAPs (green fluorescent ASAP1 signals) in the absence of any exogenous fluorophores, then pause and re-image the heart after staining with the red-shifted potentiometric dye RH237, which labels only host muscle. Host electrical activity is monitored using RH237-derived oAPs, surface electrograms and intracellular voltage recordings via sharp electrodes implanted into host muscle has become a routine experimental preparation in our laboratory, and we have used this approach to quantitatively assess the presence and extent of host-graft electromechanical coupling and acquire key host and graft EP parameters (e.g. CV, AP duration, restitution properties) in a much more streamlined and reproducible fashion than other methods. In unpublished work, we have

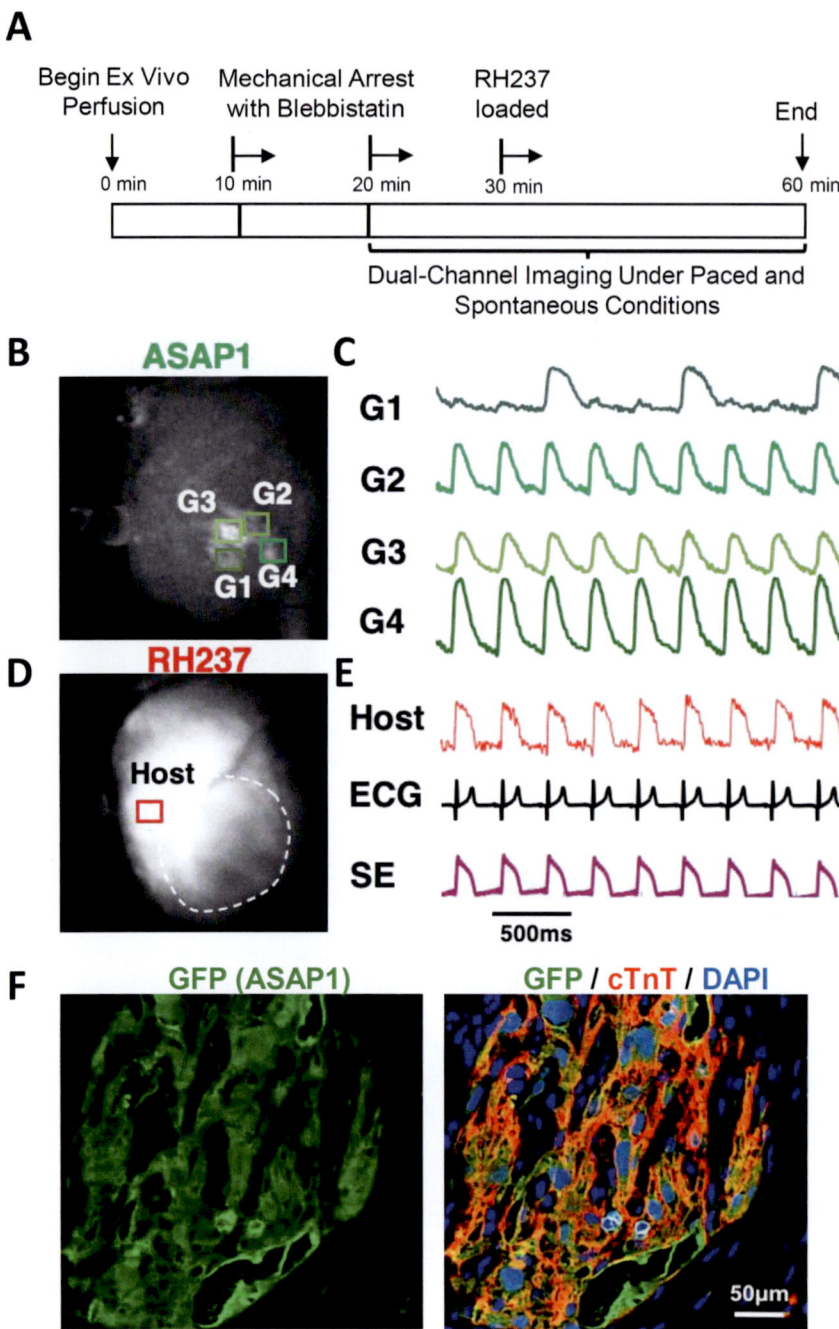

Fig. 4 Optical mapping data of engrafted hearts with ASAP1⁺ hPSC-CMs. A. Overview of the experimental protocol used for ex vivo imaging of guinea pig hearts with ASAP1+ hPSC-CM graft

also used these techniques to quantitatively compare hPSC-CM populations transplanted at varying stages of maturation and to investigate mechanisms of graft-related arrhythmias.

4 Evaluation of hPSC-CM Graft EP Function in Large-Animal Models

In recent years, hPSC-CMs have advanced to transplantation studies in larger, more translationally relevant preclinical models including infarcted non-human primates [6, 64] and swine [9, 11]. While this work in large-animal MI models has yielded encouraging data that hPSC-CM transplantation can mediate beneficial effects on LV contractile function [8, 11], it has also revealed an elevated risk of graft-related arrhythmias in hPSC-CM recipients with multiple reports describing frequent VT during the initial weeks following intra-myocardial hPSC-CM delivery [6, 8, 9, 20]. Of note, this phenomenon of graft-related arrhythmias was not predicted by earlier experiments in smaller, faster-rated rodent models, underscoring the importance of studying the electrical integration and EP function of hPSC-CM grafts in a more clinically relevant species whenever possible. Indeed, while our laboratory has recently observed potentially pro-arrhythmic behavior during optical mapping of engrafted guinea pig hearts [37], we cannot directly correlate these findings with the VT risk in large animals because guinea pig recipients do not exhibit the same spontaneous tachyarrhythmias [5].

In the future, we hope to overcome these gaps in understanding by applying the genetically encoded VSPs and optical mapping tools described in the preceding section to hPSC-CMs in large-animal models. While we expect this approach will prove feasible, it will be technically challenging for several reasons: large-animal models obviously have higher costs and lower throughput, it is significantly more

---

Fig. 4 (continued) before and after labeling with the potentiometric dye RH237. In brief, hearts are excised, mounted on a modified Langendorff apparatus and allowed to stabilize electrically. After mechanical arrest with blebbistatin to prevent motion artifacts, ASAP1-derived oAPs are acquired from the graft tissue at baseline, i.e., prior to loading with any exogenous dye. Next, the heart is loaded with RH237, which allows the simultaneous acquisition of graft (ASAP1) and host (RH237) oAPs. This ex vivo preparation is typically stable for >60 minutes, and electrical activity is imaged under both spontaneous and paced conditions. B-E. Data from a representative imaging experiment. In this case, a cryoinjured guinea pig heart was transplanted with ASAP1+ hPSC-CMs, and optical mapping was performed at 14 days post-transplantation. B. Image of the engrafted heart acquired on the ASAP1 (green) channel with scar tissue (white line) and four graft regions of interest (ROIs) labeled G1-G4. C. Graft-derived oAPs (ASAP1 fluorescence transients) from these four ROIs. D. Corresponding still image on RH237 (red) channel marked with a single host ROI (red box). E. Simultaneously acquired host electrical signals including host-derived oAPs from the host ROI (red trace), host ECG (black trace) and sharp electrode (SE) recordings (purple trace). F. Representative confocal images showing ASAP1+ hESC-CM graft tissue in injured guinea pig heart immunostained for GFP (ASAP1, green), cardiac troponin T (cTnT, red) and nuclei (DAPI, blue)

difficult to maintain larger hearts ex vivo [65, 66], and the greater tissue distances and light scattering properties of myocardium will attenuate graft-derived optical signals from larger hearts (although we have successfully overcome this in the past with GCaMP+ grafts [6]). We predict that optical mapping with VSP reporters will prove readily amenable for the evaluation of epicardially-implanted EHTs, where there should be minimal signal attenuation by intervening host tissue. On the other hand, intramural optical mapping via fiber-optrodes [67] may ultimately prove necessary in the case of VSP+ intramyocardial grafts located deep to scar and subepicardial host myocardium.

While working to develop these novel optical approaches, our group and others in the field have turned to electroanatomic mapping (EAM) systems as a more immediately deployable technology in large animals. During EAM, a mapping catheter or multi-electrode array is guided percutaneously into the LV (or cardiac chamber of interest) and used to record electrical activation in relation to anatomic location. This allows one to create three-dimensional maps of endocardial activation that can then be used to target ablations or similar interventions. While EAM suffers from some of the same disadvantages as those described for epicardial surface electrograms and mapping systems described above in Sect. 2 (principally, no ability to distinguish graft and host-derived electrical signals), it has proven an invaluable tool for probing arrhythmia mechanisms clinically and has very recently provided useful insights into the mechanisms of graft-related VT in hPSC-CM recipients.

For example, our group recently reported the successful EAM of infarcted pigs with hPSC-CM grafts using a high-density, grid-patterned multi-electrode array [9, 68]. In this experiment, we mapped hPSC-CM or vehicle recipients at 10-days post-transplantation, a time-point that corresponded to the peak incidence of arrhythmias in cardiomyocyte recipients. As expected, infarcted vehicle controls in this study were all in normal sinus rhythm and showed a normal pattern of endocardial activation. They were also resistant to induced VT. By contrast, all hPSC-CM recipients were spontaneous monomorphic VT during the procedure and exhibited an abnormal pattern of endocardial activation. While we had hypothesized that VT in hPSC-CM recipients was driven by reentry, endocardial activation maps from the cell-engrafted hearts showed no evidence of a reentrant circuit and instead had a focal activation pattern with earliest activation at the LV apex (located <1 mm from histologically confirmed graft tissue) (Fig. 5). These outcomes from the endocardial mapping studies were supported by epicardial mapping and clinical pacing maneuvers, all of which implicated a focal automatic mechanism rather than reentry.

Working independently, Liu and co-workers performed catheter-based endocardial EAM on infarcted macaques implanted with hPSC-CMs and also found VT originated from a point-source with no evidence of reentry [8]. It should be emphasized that neither of these studies can exclude the possibility of micro-reentry (i.e. reentrant phenomenon below the spatial resolution of the mapping system); but, taken collectively, they suggest a focal mechanism driven by graft ectopy is more likely than the conventional re-entrant mechanism that predominates in post-MI VT [69]. We predict that the greater spatiotemporal resolution afforded by the previously described optical mapping approaches will provide a more definitive answer

Activation Maps from hPSC-CM Recipient

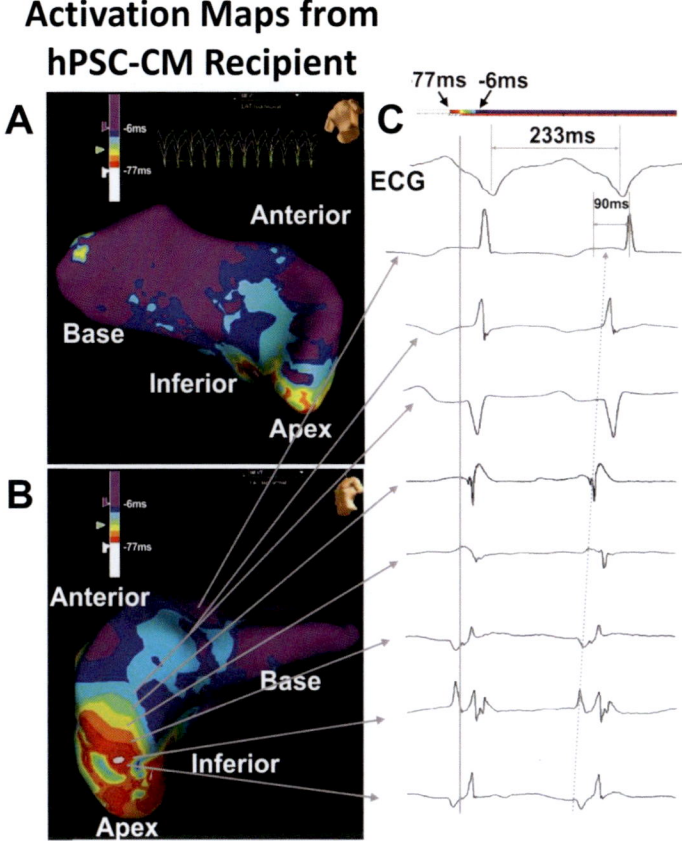

Fig. 5 Electroanatomic mapping of infarcted porcine hearts with hPSC-CM graft. For this experiment, infarcted pigs were transplanted with either hPSC-CMs or vehicle then were subjected to catheter-based endocardial electroanatomic mapping (EAM) at 10-days post-transplantation. While local activation time (LAT) maps from vehicle recipients showed a normal pattern of endocardial activation (data not shown), corresponding maps from hPSC-CM recipients consistently had an abnormal pattern. A-B. Septal (A) and anterolateral apical (B) views of LV endocardial LAT maps acquired from a representative hPSC-CM recipient during VT showing a focal activation pattern and no evidence of macro-reentry. The white and red areas in the map represent sites of earliest activation, while later-activating sites are depicted in a gradient from orange to yellow to green to blue. C. Bipolar electrogram recordings from the corresponding sites indicated in the anterolateral apical view LAT map (panel B). (All images are taken from Romagnuolo et al. [9])

to this issue. These outcomes will have major ramifications on the safety of hPSC-CM transplantation, what types of pharmacological interventions that might prevent graft-related arrhythmias, and whether the latter are a cell-autonomous phenomenon that might be overcome by the implantation of a more electrical quiescent cell population (as implied by graft ectopy).

5 Conclusions

In summary, several complementary approaches have been described for assessing both the electromechanical integration and EP phenotype of hPSC-CM grafts in preclinical MI models. While all of these approaches have their advantages and disadvantages, in our opinion, only systems involving a graft-autonomous reporter of graft activation (i.e. genetically-encoded $[Ca^{2+}]_i$ or voltage biosensor) should be considered fully reliable for demonstrating 1:1 host-graft coupling. Surface electrograms cannot distinguish graft from host, and optical readouts based on conventional fluorescent potentiometric dyes can give rise to factitious graft-derived signals [37]. Our laboratory has very recently validated optical mapping methods based on graft-expressed fluorescent reporters, and we have used these tools to evaluate host-graft coupling and acquire key EP parameters (i.e. graft CV and AP characteristics) from hPSC-CMs implanted in rodent models. At present, the best available data from more relevant large-animal models is that hPSC-CM graft electromechanical integration in these species is at least as good or better than in rodents models [6, 20]. Perhaps consequently, hPSC-CMs have also shown an elevated risk of graft-related VT in multiple large-animal models [6, 8, 9, 20]. We look forward to future studies that will integrate more powerful optical methods with standard EAM readouts and anticipate that these will inform the development of safer and more efficacious cell therapies.

Acknowledgements This work was supported by the McEwen Stem Cell Institute, the Peter Munk Cardiac Centre and the University of Toronto's Medicine by Design/Canada First Research Excellence Fund initiative. Dr. Dhahri was supported by a Mitacs Elevate postdoctoral fellowship. The authors would also like to thank Dr. Rasha Al-attar for constructive comments on the manuscript.

Disclosures MAL is a scientific founder and consultant for BlueRock Therapeutics.

References

1. Laflamme, M.A., Chen, K.Y., Naumova, A.V., Muskheli, V., Fugate, J.A., Dupras, S.K., et al.: Cardiomyocytes derived from human embryonic stem cells in pro-survival factors enhance function of infarcted rat hearts. Nat. Biotechnol. **25**(9), 1015–1024 (2007). https://doi.org/10.1038/nbt1327

2. Kattman, S.J., Witty, A.D., Gagliardi, M., Dubois, N.C., Niapour, M., Hotta, A., et al.: Stage-specific optimization of activin/nodal and BMP signaling promotes cardiac differentiation of mouse and human pluripotent stem cell lines. Cell Stem Cell. **8**(2), 228–240 (2011). https://doi.org/10.1016/j.stem.2010.12.008

3. Zhu, W.-Z., Van Biber, B., Laflamme, M.A.: Methods for the derivation and use of cardiomyocytes from human pluripotent stem cells. Methods Mol. Biol. **767**, 419–431 (2011). https://doi.org/10.1007/978-1-61779-201-4_31

4. Caspi, O., Huber, I., Kehat, I., Habib, M., Arbel, G., Gepstein, A., et al.: Transplantation of human embryonic stem cell-derived cardiomyocytes improves myocardial performance in

infarcted rat hearts. J. Am. Coll. Cardiol. **50**(19), 1884–1893 (2007). https://doi.org/10.1016/j.jacc.2007.07.054

5. Shiba, Y., Fernandes, S., Zhu, W.-Z., Filice, D., Muskheli, V., Kim, J., et al.: Human ES-cell-derived cardiomyocytes electrically couple and suppress arrhythmias in injured hearts. Nature. **489**(7415), 322–325 (2012). https://doi.org/10.1038/nature11317

6. Chong, J.J.H., Yang, X., Don, C.W., Minami, E., Liu, Y.-W., Weyers, J.J., et al.: Human embryonic-stem-cell-derived cardiomyocytes regenerate non-human primate hearts. Nature. **510**(7504), 273–277 (2014). https://doi.org/10.1038/nature13233

7. Funakoshi, S., Miki, K., Takaki, T., Okubo, C., Hatani, T., Chonabayashi, K., et al.: Enhanced engraftment, proliferation, and therapeutic potential in heart using optimized human iPSC-derived cardiomyocytes. Sci. Rep. **6**, 19111 (2016). https://doi.org/10.1038/srep19111

8. Liu, Y.-W., Chen, B., Yang, X., Fugate, J.A., Kalucki, F.A., Futakuchi-Tsuchida, A., et al.: Human ESC-derived cardiomyocytes restore function in infarcted hearts of non-human primates. Nat. Biotechnol. **36**(7), 597–605 (2018). https://doi.org/10.1038/nbt.4162

9. Romagnuolo, R., Masoudpour, H., Porta-Sánchez, A., Qiang, B., Barry, J., Laskary, A., et al.: Human embryonic stem cell-derived cardiomyocytes regenerate the infarcted pig heart but induce ventricular Tachyarrhythmias. Stem Cell Rep. **12**(5), 967–981 (2019). https://doi.org/10.1016/j.stemcr.2019.04.005

10. Shadrin, I.Y., Allen, B.W., Qian, Y., Jackman, C.P., Carlson, A.L., Juhas, M.E., et al.: Cardiopatch platform enables maturation and scale-up of human pluripotent stem cell-derived engineered heart tissues. Nat. Commun. **8** (2017). https://doi.org/10.1038/s41467-017-01946-x

11. Gao, L., Gregorich, Z.R., Zhu, W., Mattapally, S., Oduk, Y., Lou, X., et al.: Large cardiac muscle patches engineered from human induced-pluripotent stem cell-derived cardiac cells improve recovery from myocardial infarction in swine. Circulation. **137**(16), 1712–1730 (2018). https://doi.org/10.1161/CIRCULATIONAHA.117.030785

12. Weinberger, F., Breckwoldt, K., Pecha, S., Kelly, A., Geertz, B., Starbatty, J., et al.: Cardiac repair in guinea pigs with human engineered heart tissue from induced pluripotent stem cells. Sci. Transl. Med. **8**(363), 363ra148 (2016). https://doi.org/10.1126/scitranslmed.aaf8781

13. Tiburcy, M., Hudson, J.E., Balfanz, P., Schlick, S., Meyer, T., Chang Liao, M.-L., et al.: Defined engineered human myocardium with advanced maturation for applications in heart failure modeling and repair. Circulation. **135**(19), 1832–1847 (2017). https://doi.org/10.1161/CIRCULATIONAHA.116.024145

14. Binah, O., Dolnikov, K., Sadan, O., Shilkrut, M., Zeevi-Levin, N., Amit, M., et al.: Functional and developmental properties of human embryonic stem cells-derived cardiomyocytes. J. Electrocardiol. **40**(6 Suppl), S192–S196 (2007). https://doi.org/10.1016/j.jelectrocard.2007.05.035

15. Dolnikov, K., Shilkrut, M., Zeevi-Levin, N., Danon, A., Gerecht-Nir, S., Itskovitz-Eldor, J., et al.: Functional properties of human embryonic stem cell-derived cardiomyocytes. Ann. N. Y. Acad. Sci. **1047**, 66–75 (2005). https://doi.org/10.1196/annals.1341.006

16. Dolnikov, K., Shilkrut, M., Zeevi-Levin, N., Gerecht-Nir, S., Amit, M., Danon, A., et al.: Functional properties of human embryonic stem cell-derived cardiomyocytes: intracellular Ca2+ handling and the role of sarcoplasmic reticulum in the contraction. Stem Cells. **24**(2), 236–245 (2006). https://doi.org/10.1634/stemcells.2005-0036

17. Lundy, S.D., Zhu, W.-Z., Regnier, M., Laflamme, M.A.: Structural and functional maturation of cardiomyocytes derived from human pluripotent stem cells. Stem Cells Dev. **22**(14), 1991–2002 (2013). https://doi.org/10.1089/scd.2012.0490

18. Sartiani, L., Bettiol, E., Stillitano, F., Mugelli, A., Cerbai, E., Jaconi, M.E.: Developmental changes in cardiomyocytes differentiated from human embryonic stem cells: a molecular and electrophysiological approach. Stem Cells. **25**(5), 1136–1144 (2007). https://doi.org/10.1634/stemcells.2006-0466

19. Laflamme, M.A., Murry, C.E.: Heart regeneration. Nature. **473**(7347), 326–335 (2011). https://doi.org/10.1038/nature10147

20. Shiba, Y., Gomibuchi, T., Seto, T., Wada, Y., Ichimura, H., Tanaka, Y., et al.: Allogeneic transplantation of iPS cell-derived cardiomyocytes regenerates primate hearts. Nature. **538**(7625), 388–391 (2016). https://doi.org/10.1038/nature19815

21. Chen, H.-S.V., Kim, C., Mercola, M.: Electrophysiological challenges of cell-based myocardial repair. Circulation. **120**(24), 2496–2508 (2009). https://doi.org/10.1161/CIRCULATIONAHA.107.751412

22. Paci, M., Penttinen, K., Pekkanen-Mattila, M., Koivumäki, J.T.: Arrhythmia mechanisms in human induced pluripotent stem cell-derived cardiomyocytes. J. Cardiovasc. Pharmacol. **77**(3), 300–316 (2020). https://doi.org/10.1097/FJC.0000000000000972

23. Smith, R.R., Barile, L., Messina, E., Marbán, E.: Stem cells in the heart: what's the buzz all about? Part 2: arrhythmic risks and clinical studies. Heart Rhythm. **5**(6), 880–887 (2008). https://doi.org/10.1016/j.hrthm.2008.02.011

24. Burgess, M.J., Lux, R.L., Wyatt, R.F., Abildskov, J.A.: The relation of localized myocardial warming to changes in cardiac surface electrograms in dogs. Circ. Res. **43**(6), 899–907 (1978). https://doi.org/10.1161/01.res.43.6.899

25. Gallagher, J.J., Ticzon, A.R., Wallace, A.G., Kasell, J.: Activation studies following experimental hemiblock in the dog. Circ. Res. **35**(5), 752–763 (1974). https://doi.org/10.1161/01.res.35.5.752

26. Ideker, R.E., Smith, W.M., Wolf, P.D.: Cardiac mapping at Duke medical center. Am. J. Cardiol. **63**(Suppl), 17F–30F (1989)

27. Zimmermann, W.-H., Melnychenko, I., Wasmeier, G., Didié, M., Naito, H., Nixdorff, U., et al.: Engineered heart tissue grafts improve systolic and diastolic function in infarcted rat hearts. Nat. Med. **12**(4), 452–458 (2006). https://doi.org/10.1038/nm1394

28. Higuchi, T., Miyagawa, S., Pearson, J.T., Fukushima, S., Saito, A., Tsuchimochi, H., et al.: Functional and electrical integration of induced pluripotent stem cell-derived cardiomyocytes in a myocardial infarction rat heart. Cell Transplant. **24**(12), 2479–2489 (2015). https://doi.org/10.3727/096368914X685799

29. Lü, S., Li, Y., Gao, S., Liu, S., Wang, H., He, W., et al.: Engineered heart tissue graft derived from somatic cell nuclear transferred embryonic stem cells improve myocardial performance in infarcted rat heart. J. Cell. Mol. Med. **14**(12), 2771–2779 (2010). https://doi.org/10.1111/j.1582-4934.2010.01112.x

30. Attin, M., Clusin, W.T.: Basic concepts of optical mapping techniques in cardiac electrophysiology. Biol. Res. Nurs. **11**(2), 195–207 (2009). https://doi.org/10.1177/1099800409338516

31. Rosenbaum, D., Jalife, J. (eds.): Optical Mapping of Cardiac Excitation and Arrhythmias, 1st edn. Wiley-Blackwell, Armonk (2001)

32. Haws, C.W., Lux, R.L.: Correlation between in vivo transmembrane action potential durations and activation-recovery intervals from electrograms. Effects of interventions that alter repolarization time. Circulation. **81**(1), 281–288 (1990). https://doi.org/10.1161/01.cir.81.1.281

33. Efimov, I.R., Nikolski, V.P., Salama, G.: Optical imaging of the heart. Circ. Res. **95**(1), 21–33 (2004). https://doi.org/10.1161/01.RES.0000130529.18016.35

34. Choi, B.R., Salama, G.: Simultaneous maps of optical action potentials and calcium transients in Guinea-pig hearts: mechanisms underlying concordant alternans. J. Physiol. **529**(Pt 1), 171–188 (2000). https://doi.org/10.1111/j.1469-7793.2000.00171.x

35. Davidenko, J.M., Kent, P.F., Chialvo, D.R., Michaels, D.C., Jalife, J.: Sustained vortex-like waves in normal isolated ventricular muscle. Proc. Natl. Acad. Sci. U. S. A. **87**(22), 8785–8789 (1990). https://doi.org/10.1073/pnas.87.22.8785

36. Gepstein, L., Ding, C., Rahmutula, D., Rehemedula, D., Wilson, E.E., Yankelson, L., et al.: In vivo assessment of the electrophysiological integration and arrhythmogenic risk of myocardial cell transplantation strategies. Stem Cells. **28**(12), 2151–2161 (2010). https://doi.org/10.1002/stem.545

37. Filice, D., Dhahri, W., Solan, J.L., Lampe, P.D., Steele, E., Milani, N., et al.: Optical mapping of human embryonic stem cell-derived cardiomyocyte graft electrical activity in injured hearts. Stem Cell Res Ther. **11**(1), 417 (2020). https://doi.org/10.1186/s13287-020-01919-w

38. Rubart, M., Pasumarthi, K.B.S., Nakajima, H., Soonpaa, M.H., Nakajima, H.O., Field, L.J.: Physiological coupling of donor and host cardiomyocytes after cellular transplantation. Circ. Res. **92**(11), 1217–1224 (2003). https://doi.org/10.1161/01.RES.0000075089.39335.8C
39. Yang, T., Rubart, M., Soonpaa, M.H., Didié, M., Christalla, P., Zimmermann, W.-H., et al.: Cardiac engraftment of genetically-selected parthenogenetic stem cell-derived cardiomyocytes. PLoS One. **10**(6), e0131511 (2015). https://doi.org/10.1371/journal.pone.0131511
40. Nagai, T., Sawano, A., Park, E.S., Miyawaki, A.: Circularly permuted green fluorescent proteins engineered to sense Ca2+. Proc. Natl. Acad. Sci. U. S. A. **98**(6), 3197–3202 (2001). https://doi.org/10.1073/pnas.051636098
41. Tallini, Y.N., Ohkura, M., Choi, B.-R., Ji, G., Imoto, K., Doran, R., et al.: Imaging cellular signals in the heart in vivo: cardiac expression of the high-signal Ca2+ indicator GCaMP2. Proc. Natl. Acad. Sci. U. S. A. **103**(12), 4753–4758 (2006). https://doi.org/10.1073/pnas.0509378103
42. Chen, T.-W., Wardill, T.J., Sun, Y., Pulver, S.R., Renninger, S.L., Baohan, A., et al.: Ultrasensitive fluorescent proteins for imaging neuronal activity. Nature. **499**(7458), 295–300 (2013). https://doi.org/10.1038/nature12354
43. Roell, W., Lewalter, T., Sasse, P., Tallini, Y.N., Choi, B.-R., Breitbach, M., et al.: Engraftment of connexin 43-expressing cells prevents post-infarct arrhythmia. Nature. **450**(7171), 819–824 (2007). https://doi.org/10.1038/nature06321
44. Shiba, Y., Filice, D., Fernandes, S., Minami, E., Dupras, S.K., Biber, B.V., et al.: Electrical integration of human embryonic stem cell-derived cardiomyocytes in a Guinea pig chronic infarct model. J. Cardiovasc. Pharmacol. Ther. **19**(4), 368–381 (2014). https://doi.org/10.1177/1074248413520344
45. Hockemeyer, D., Soldner, F., Beard, C., Gao, Q., Mitalipova, M., DeKelver, R.C., et al.: Efficient targeting of expressed and silent genes in human ESCs and iPSCs using zinc-finger nucleases. Nat. Biotechnol. **27**(9), 851–857 (2009). https://doi.org/10.1038/nbt.1562
46. Hockemeyer, D., Wang, H., Kiani, S., Lai, C.S., Gao, Q., Cassady, J.P., et al.: Genetic engineering of human pluripotent cells using TALE nucleases. Nat. Biotechnol. **29**(8), 731–734 (2011). https://doi.org/10.1038/nbt.1927
47. Zhu, W.-Z., Filice, D., Palpant, N.J., Laflamme, M.A.: Methods for assessing the electromechanical integration of human pluripotent stem cell-derived cardiomyocyte grafts. Methods Mol. Biol. **1181**, 229–247 (2014). https://doi.org/10.1007/978-1-4939-1047-2_20
48. Maxwell, M.P., Hearse, D.J., Yellon, D.M.: Species variation in the coronary collateral circulation during regional myocardial ischaemia: a critical determinant of the rate of evolution and extent of myocardial infarction. Cardiovasc. Res. **21**(10), 737–746 (1987). https://doi.org/10.1093/cvr/21.10.737
49. Jackman, C.P., Ganapathi, A.M., Asfour, H., Qian, Y., Allen, B.W., Li, Y., et al.: Engineered cardiac tissue patch maintains structural and electrical properties after epicardial implantation. Biomaterials. **159**, 48–58 (2018). https://doi.org/10.1016/j.biomaterials.2018.01.002
50. Gerbin, K.A., Yang, X., Murry, C.E., Coulombe, K.L.K.: Enhanced electrical integration of engineered human myocardium via Intramyocardial versus Epicardial delivery in infarcted rat hearts. PLoS One. **10**(7), e0131446 (2015). https://doi.org/10.1371/journal.pone.0131446
51. Antić, S., Zecević, D.: Optical signals from neurons with internally applied voltage-sensitive dyes. J. Neurosci. **15**(2), 1392–1405 (1995)
52. Bando, Y., Sakamoto, M., Kim, S., Ayzenshtat, I., Yuste, R.: Comparative evaluation of genetically encoded voltage indicators. Cell Rep. **26**(3), 802-13.e4 (2019). https://doi.org/10.1016/j.celrep.2018.12.088
53. Xu, Y., Zou, P., Cohen, A.E.: Voltage imaging with genetically encoded indicators. Curr. Opin. Chem. Biol. **39**, 1–10 (2017). https://doi.org/10.1016/j.cbpa.2017.04.005
54. Chen, Z., Truong, T.M., Ai, H.-W.: Illuminating Brain Activities with Fluorescent Protein-Based Biosensors. Chemosensors (Basel). **5**(4) (2017). https://doi.org/10.3390/chemosensors5040032

55. Knöpfel, T., Song, C.: Optical voltage imaging in neurons: moving from technology development to practical tool. Nat. Rev. Neurosci. **20**(12), 719–727 (2019). https://doi.org/10.1038/s41583-019-0231-4

56. Milosevic, M.M., Jang, J., McKimm, E.J., Zhu, M.H., Antic, S.D.: In vitro testing of voltage indicators: Archon1, ArcLightD, ASAP1, ASAP2s, ASAP3b, Bongwoori-Pos6, BeRST1, FlicR1, and Chi-VSFP-Butterfly. eNeuro. **7**(5) (2020). https://doi.org/10.1523/ENEURO.0060-20.2020

57. Chang Liao, M.-L., de Boer, T.P., Mutoh, H., Raad, N., Richter, C., Wagner, E., et al.: Sensing cardiac electrical activity with a cardiac myocyte – targeted Optogenetic voltage indicator. Circ. Res. **117**(5), 401–412 (2015). https://doi.org/10.1161/CIRCRESAHA.117.306143

58. Leyton-Mange, J.S., Mills, R.W., Macri, V.S., Jang, M.Y., Butte, F.N., Ellinor, P.T., et al.: Rapid cellular phenotyping of human pluripotent stem cell-derived cardiomyocytes using a genetically encoded fluorescent voltage sensor. Stem Cell Rep. **2**(2), 163–170 (2014). https://doi.org/10.1016/j.stemcr.2014.01.003

59. Jin, L., Han, Z., Platisa, J., Wooltorton, J.R.A., Cohen, L.B., Pieribone, V.A.: Single action potentials and subthreshold electrical events imaged in neurons with a fluorescent protein voltage probe. Neuron. **75**(5), 779–785 (2012). https://doi.org/10.1016/j.neuron.2012.06.040

60. St-Pierre, F., Marshall, J.D., Yang, Y., Gong, Y., Schnitzer, M.J., Lin, M.Z.: High-fidelity optical reporting of neuronal electrical activity with an ultrafast fluorescent voltage sensor. Nat. Neurosci. **17**(6), 884–889 (2014). https://doi.org/10.1038/nn.3709

61. Park, J., Werley, C.A., Venkatachalam, V., Kralj, J.M., Dib-Hajj, S.D., Waxman, S.G., Cohen, A.E., et al.: Screening fluorescent voltage indicators with spontaneously spiking HEK cells. PloS one. **8**(12) (2013). https://doi.org/10.1371/journal.pone.0085221

62. Hochbaum, D.R., Zhao, Y., Farhi, S.L., Klapoetke, N., Werley, C.A., Kapoor, V., et al.: All-optical electrophysiology in mammalian neurons using engineered microbial rhodopsins. Nat. Methods. **11**(8), 825–833 (2014). https://doi.org/10.1038/nmeth.3000

63. Han, Z., Jin, L., Platisa, J., Cohen, L.B., Baker, B.J., Pieribone, V.A.: Fluorescent protein voltage probes derived from ArcLight that respond to membrane voltage changes with fast kinetics. PLoS One. **8**(11), e81295 (2013). https://doi.org/10.1371/journal.pone.0081295

64. Blin, G., Nury, D., Stefanovic, S., Neri, T., Guillevic, O., Brinon, B., et al.: A purified population of multipotent cardiovascular progenitors derived from primate pluripotent stem cells engrafts in postmyocardial infarcted nonhuman primates. J. Clin. Invest. **120**(4), 1125–1139 (2010). https://doi.org/10.1172/JCI40120

65. Brook, J., Kim, M.-Y., Koutsoftidis, S., Pitcher, D., Agha-Jaffar, D., Sufi, A., et al.: Development of a pro-arrhythmic ex vivo intact human and porcine model: cardiac electrophysiological changes associated with cellular uncoupling. Pflugers Arch. Eur. J. Physiol. **472**(10), 1435–1446 (2020). https://doi.org/10.1007/s00424-020-02446-6

66. Schechter, M.A., Southerland, K.W., Feger, B.J., Linder, D., Ali, A.A., Njoroge, L., et al.: An isolated working heart system for large animal models. J. Visual. Exp. JoVE. **88** (2014). https://doi.org/10.3791/51671

67. Kong, W., Ideker, R.E., Fast, V.G.: Intramural optical mapping of V(m) and ca(i)2+ during long-duration ventricular fibrillation in canine hearts. Am. J. Physiol. Heart Circ. Physiol. **302**(6), H1294–H1305 (2012). https://doi.org/10.1152/ajpheart.00426.2011

68. Deno, D.C., Bhaskaran, A., Morgan, D.J., Goksu, F., Batman, K., Olson, G.K., et al.: High-resolution, live, directional mapping. Heart Rhythm. **17**(9), 1621–1628 (2020). https://doi.org/10.1016/j.hrthm.2020.04.039

69. Benito, B., Josephson, M.E.: Ventricular tachycardia in coronary artery disease. Revista Espanola De Cardiologia (English Ed). **65**(10), 939–955 (2012). https://doi.org/10.1016/j.recesp.2012.03.027

Part V
Clinical Perspectives

Understanding the Molecular Interface of Cardiovascular Diseases and COVID-19: A Data Science Approach

Dibakar Sigdel, Dylan Steinecke, Ding Wang, David Liem, Maya Gupta, Alex Zhang, Wei Wang, and Peipei Ping

1 Introduction

In order to extract insights from the ever-growing number of biomedical reports, efficient computing strategies and new data science algorithms are necessary. Several algorithms and applications have been applied in text mining [1–3] and machine learning [4, 5] including literature-based knowledge discovery [6, 7], molecular mechanism-based disease classification [8] and drug repurposing [8–10]. Our main motivation in this project is twofold: one is to dissect relationships among molecular phenotypes, diseases, and symptoms; and another is to study the interface between COVID-19 and CVDs (Fig. 1). Here we demonstrate an example of biomedical text mining application, where we seek to illustrate the utility of text mining approach in biomedical application; by elucidating the molecular mechanisms

D. Sigdel · D. Liem
Department of Physiology, University of California, Los Angeles, CA, USA

NHLBI Integrated Cardiovascular Data Science Training Program (iDISCOVER), University of California, Los Angeles, CA, USA

D. Steinecke · M. Gupta · A. Zhang
NHLBI Integrated Cardiovascular Data Science Training Program (iDISCOVER), University of California, Los Angeles, CA, USA

D. Wang
Department of Physiology, University of California, Los Angeles, CA, USA

W. Wang
Department of Computer Science, University of California, Los Angeles, CA, USA

Department of Computational Medicine, University of California, Los Angeles, CA, USA

Scalable Analytics Institute (ScAi), University of California, Los Angeles, CA, USA

Bioinformatics Interdepartmental Graduate Program, University of California, Los Angeles, CA, USA

© The Author(s), under exclusive license to Springer Nature Switzerland AG 2022
J. Zhang, V. Serpooshan (eds.), *Advanced Technologies in Cardiovascular Bioengineering*, https://doi.org/10.1007/978-3-030-86140-7_15

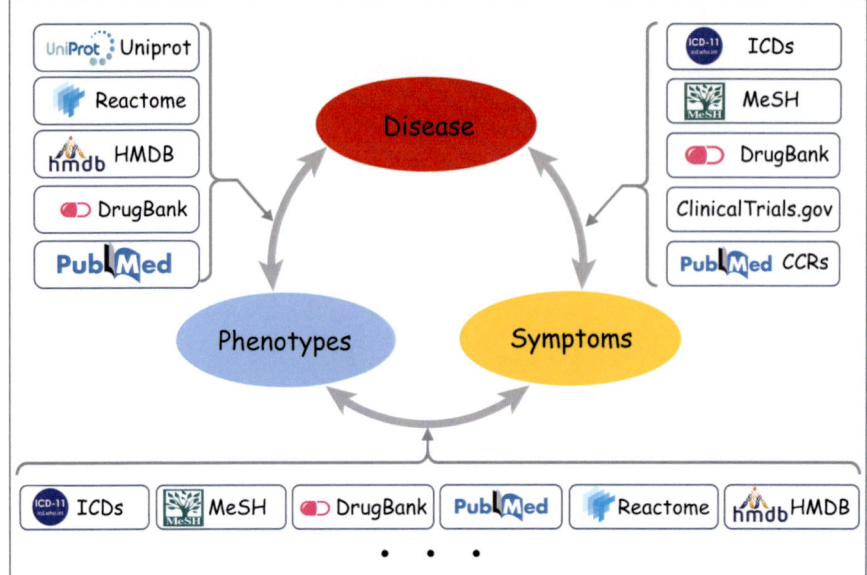

Fig. 1 Characterizing relationships among molecular phenotypes, diseases, and symptoms.
Understanding disease pathology through molecular phenotyping is a rapidly growing approach in precision medicine; this complements the classic clinical perspective diagnosis of disease through signs and symptoms. Bridging molecular phenotypes to disease signs and symptoms remains a challenging task. Overcoming this challenge is a central motivation for our project using methods of biomedical informatics. Intuitively, these relationships can be represented in a graph structure informed by multiple databases. Some of these databases are listed in the boxes in the figure above

at the intersection of COVID-19 and CVD, we aim to provide insights into how pre-existing CVD conditions may exacerbate COVID-19 outcomes [11–14]. The different components involved in data preparation, text mining, and analysis are depicted in the workflow (Fig. 2).

We assembled our data sources from several public domains, including biomedical publications from PubMed, the National Library of Medicine's (NLM) Medical

P. Ping (✉)
Department of Physiology, University of California, Los Angeles, CA, USA

NHLBI Integrated Cardiovascular Data Science Training Program (iDISCOVER), University of California, Los Angeles, CA, USA

Scalable Analytics Institute (ScAi), University of California, Los Angeles, CA, USA

Bioinformatics Interdepartmental Graduate Program, University of California, Los Angeles, CA, USA

Biomedical Informatics IDP, UCLA, Los Angeles, USA

Department of Medicine/Cardiology, University of California, Los Angeles, CA, USA
e-mail: pping38@g.ucla.edu

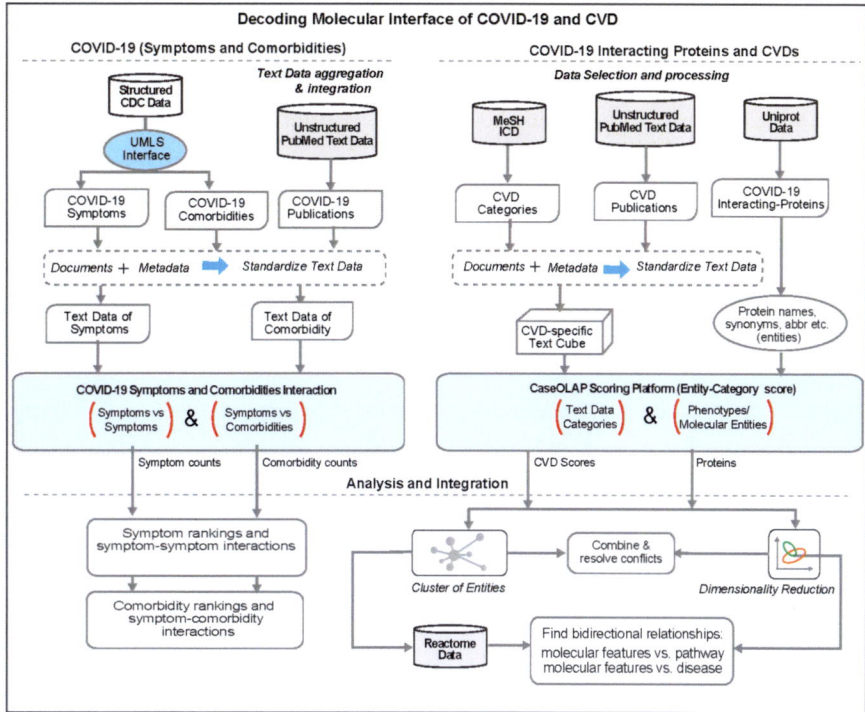

Fig. 2 Workflow on decoding the molecular interface of COVID-19 with CVDs. The data science platform for integrating both unstructured and structured data through entity-category association and machine learning approaches to explore the molecular interface of COVID-19 and CVDs. We built our data corpus from multiple data sources. The assembled datasets are normalized, transformed, and classified into the categories under domain knowledge. Molecular entities of COVID-19 bearing CaseOLAP scores across eight CVD categories are ranked. This information offered an opportunity to construct a score-based machine learning pipeline to harness the highlighted molecular pathogenesis as well as hidden and unappreciated relationships

Subject Headings (MeSH), International Classification of Disease (ICD) codes, and Reactome (see **Materials and Methods**). We show how COVID-19 symptoms interact with each other (Table 1, Fig. 3) and disease comorbidities (Table 2, Figs. 4 and 5). We explore one of the significant comorbidities, CVD, and analyze the COVID-19-interacting proteins (Table 4) as entities and 8 CVDs (Table 3) as categories through entity-category association [15].

By design, most of our current text documents have been prepared for human readers, not for machines; thus, most of them are not machine readable. They are inconveniently formatted (e.g., text paragraphs, pdf formats) for computational tasks. Many methods, tools, and databases have been developed to overcome the challenge of data organization and conducting machine learning [15–19]. Here we obtain structured and unstructured data (see **Materials and Methods**); index documents; search COVID-19 symptoms (Fig. 3) and comorbidities (Figs. 4 and 5) in

Table 1 COVID-19 symptoms obtained from CDC website

Symptoms	Occurrence	MeSH	MeSH ID	ICD-11	ICD-11
Fever	80–90%	Fever	D005334	Fever of other or unknown origin	MG26
Chills	10–15%	Chills	D023341	Chills	MG21
Cough	70–80%	Cough	D003371	Cough	MD12
Fatigue	40–50%	Fatigue	D005221	Fatigue	MG22
Shortness of breath	20–30%	Dyspnea	D004417	Dyspnea	MD11.5
Sore throat	10–15%	Pharyngitis	D010612	Pain in throat	MD36.0
Rhinorrhea	5–10%	Rhinorrhea	D000086722	Other specified upperrespiratory tract disorders	CA0Y
Diarrhea	4–14%	Diarrhea	D003967	Diarrhoea	ME05.1
Vomiting	5–10%	Vomiting	D014839	Vomiting	MD90.1
Nausea	5–10%	Nausea	D009325	Nausea	MD90.0
Headache	10–15%	Headache	D006261	Other specified symptoms or signs involving nervous system	MB6Y
Myalgia	10–15%	Myalgia	D063806	Myalgia	FB56.2
Ageusia	5%	Ageusia	D000370	Dysgeusia	MB41.2
Confusion	<10%	Confusion	D003221	Disorientation	MB21.4
Chest pain	<10%	Chest pain	D002637	Pain in throat or chest	MD30
Anosmia	<10%	Olfaction Disorders	D000857	Anosmia	MB41.0
Cyanosis	1–5%	Cyanosis	D003490	Cyanosis	ME64.1
Nasal congestion	5–10%	Rhinorrhea	D000086722	Nasal congestion	MD11.9
Hypersomnia	5–10%	Disorders of Excessive Somnolence	D006970	Hypersomnia due to a medical condition	7A23

Both MeSH tree terms and ICD codes are assembled along with extra information in percentage occurrences

publications (See **Result and Discussion**); identify COVID-19-interacting proteins in CVDs in all of PubMed's CVD-related publications; and subsequently apply text mining analyses with ML approaches to explore underlying molecular mechanisms (Figs. 6 and 7).

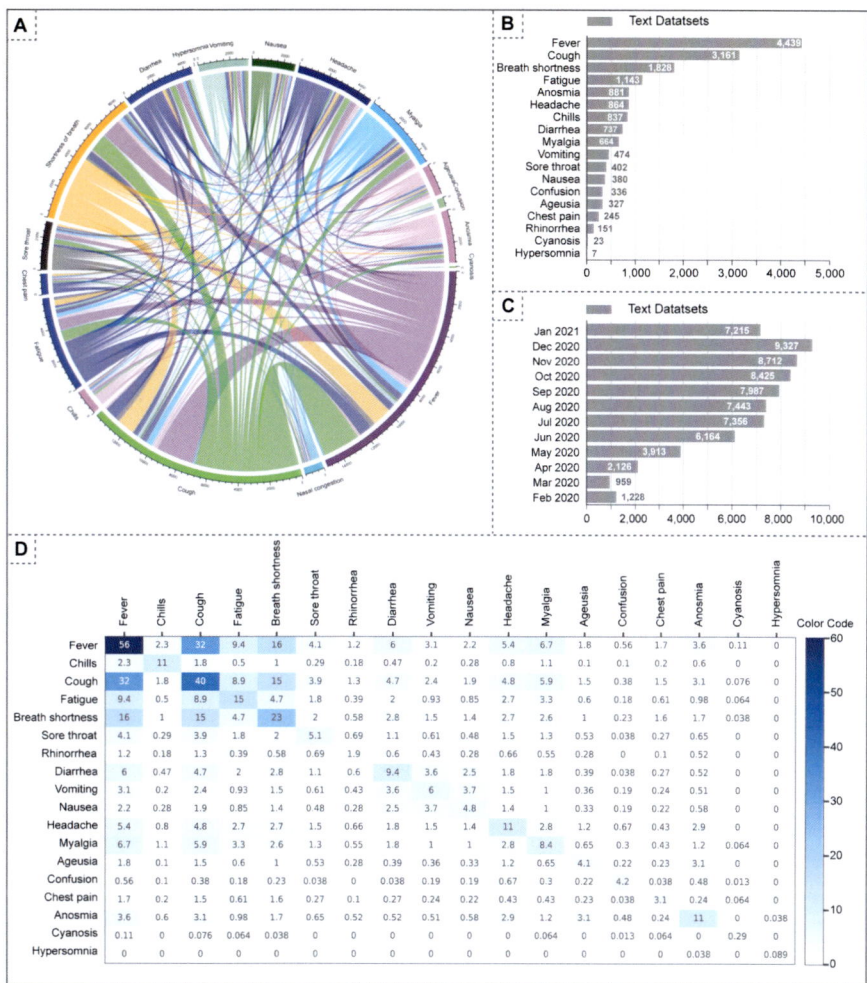

Fig. 3 Our data corpora contain COVID-19 symptoms as defined by CDC. The CDC defined COVID-19 symptoms were applied to screen 80,732 publications (PubMed); we found a total of 7,872 reports containing COVID-19 and their associated symptoms. (**a**) A circular chord diagram displays the COVID-19 symptoms as reported in a publication. The perimeter corresponds to the number of publications referring to this symptom (not perfectly to scale due to double-counting). The sizes of the chords connecting symptoms correspond to publications where both symptoms co-existed; the two symptoms are on either ends of the chord. (**b**) A bar chart displays the total number of documents where the symptom was reported (allows 'double-counting': for a publication to be counted in multiple bars if a publication reported multiple symptoms). "Fever" and "cough" are the two most frequently reported symptoms in our data corpora. (**c**) A bar chart visualizing the total number of COVID-19 publications per month analyzed in our study (Feb 2020–Jan 2021). (**d**) A heatmap displays the information in 3A. The percentage of publications containing the symptoms is displayed in each heatmap cell. As with 3A, there is double counting in each cell. The heatmap is symmetric along the diagonal (top left to bottom right)

Table 2 COVID-19 Comorbidities obtained from CDC website

Major comorbidities	Minor comorbidities
Pregnancy	Asthma
Transplantation	Hypertension
Smoking	Thalassemia
Cardiovascular Disease	Diabetes Mellitus, Type 1
Obesity	Cerebrovascular Disorders
Pulmonary Disease, Chronic Obstructive (COPD)	Liver Diseases
Neoplasms	HIV
Diabetes Mellitus, Type 2 (T2DM)	Immunosuppressive Agents
Kidney Diseases	Bone Marrow Transplantation
Down Syndrome	Cystic Fibrosis
	Pulmonary Fibrosis
	Overweight
	Dementia
	Acquired Immunodeficiency Syndrome (AIDS)

Both major and minor COVID-19 comorbidities informed by the CDC are listed on the table

2   Materials and Methods

We provide a systematic overview of the unstructured data sources (e.g., PubMed publications), the structured data sources which partially structured the unstructured data (e.g., MeSH trees, ICD codes, CDC), and the structured biomedical entity datasets (e.g., COVID-19-interacting proteins, pathways). A diagrammatic presentation of the computation and analysis pipeline is detailed in Fig. 2. This workflow demonstrates how our text mining and data analysis platform are built from both heterogeneous structured and unstructured datasets, and how one can explore the molecular intersection of COVID-19 and CVD based on entity-category associations (i.e., protein-disease associations) as CaseOLAP scores [15].

2.1   Unstructured Text Data Sources

Multiple structured data sources help organize and provide structure to the unstructured text data. The major unstructured data are publication abstracts obtained from PubMed (https://pubmed.ncbi.nlm.nih.gov/), an NLM-maintained database. Structuring that data is enabled by data from the CDC, ICD, MeSH, and the Unified Medical Language System (UMLS) interface. UMLS provides an interface for standardizing the medical terminology (e.g., disease symptoms, comorbidities) using name, synonyms, and abbreviations, which is integrated with diverse and widely-used terminologies (e.g., MeSH, ICD codes, RxNorm, SnomedCT). MeSH (https://

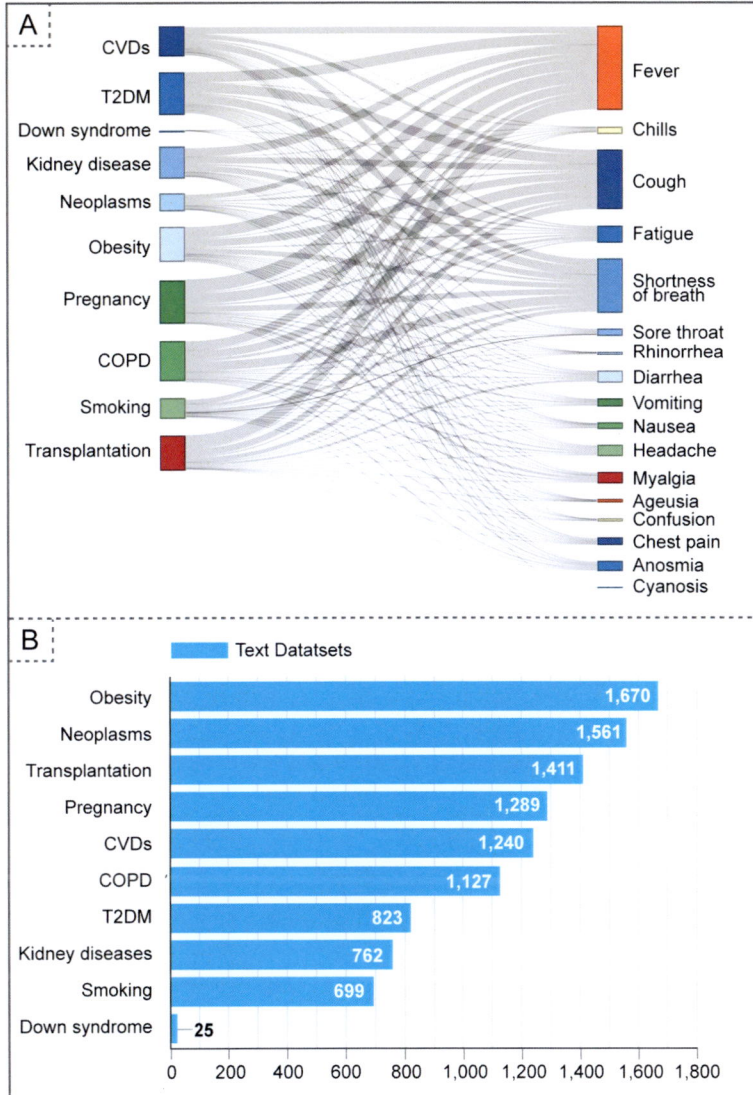

Fig. 4 Relationships of text datasets reporting COVID-19 symptoms and major disease comorbidities. The CDC defined COVID-19 comorbidities were screened in 80,732 publications and 4,588 reports were positively identified; among them, comorbidities and symptoms were jointly reported in 612 publications. Panel **A** displays the relationship between COVID-19 symptoms and major comorbidities. Line thickness corresponds to the frequency a symptom and a comorbidity co-existed. Of the documents reported at least one symptom and one major comorbidity together, the most frequent comorbidities were 'Pregnancy', 'Transplantation', and 'Obesity'; symptoms were 'Fever', 'Cough', and 'Shortness of Breath'. An interactive version of the visualization is available at (https://caseolap.github.io/covid-cvd/plots/major.html). Panel **B** shows a bar chart of publications reporting COVID-19's major comorbidities

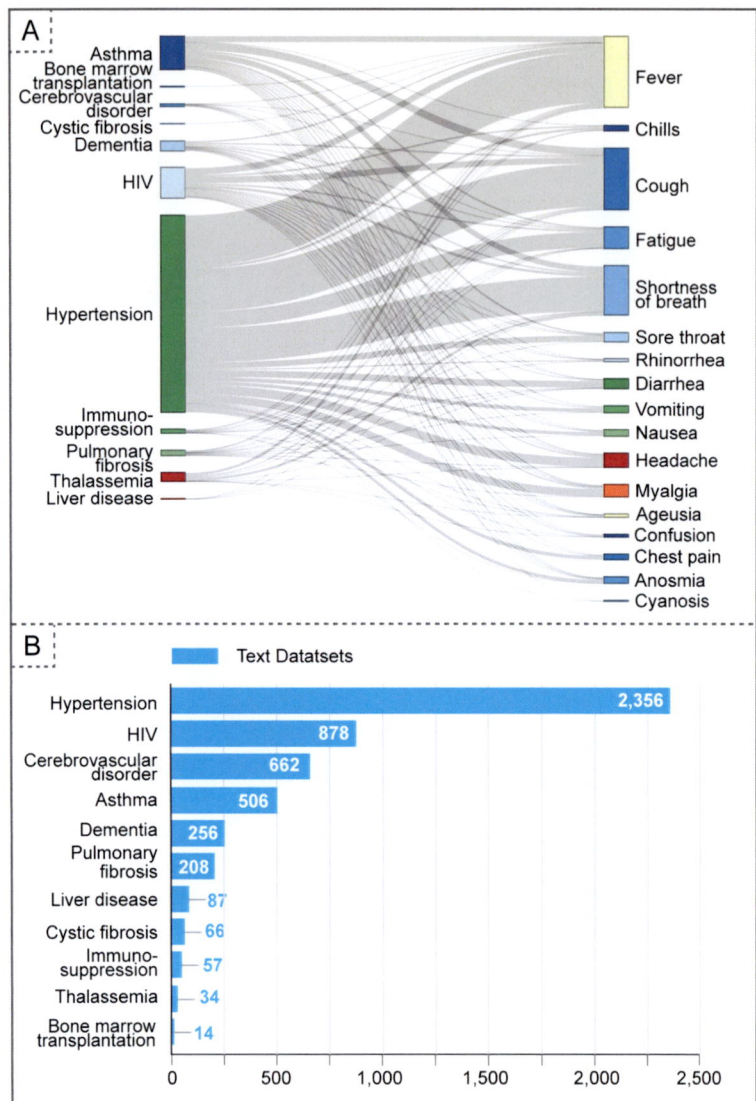

Fig. 5 Relationships of text datasets containing various COVID-19 symptoms and minor comorbidities. In a similar framework as we have done with the major comorbidities, we explored the relationship between COVID-19 symptoms and minor comorbidities. Panel **A** documents the relationship between COVID-19 symptoms and minor comorbidities. Line thickness corresponds to the frequency a symptom and a comorbidity were jointly reported. Of the documents which mention at least one symptom and one minor comorbidity together, the most frequently studied/ reported comorbidities were 'Hypertension', 'HIV', and 'Cerebrovascular'. An interactive visualization is available at (https://caseolap.github.io/covid-cvd/plots/minor.html). Panel **B** reveals the number of publications containing these minor comorbidities for COVID-19

Table 3 Eight categories of CVD

Name	Abbreviation	Major Root Nodes (MeSH)	No. of publications collected
Cardiomyopathies and heart failure	CM	C14.280.238, C14.280.434	212,407
Arrhythmias, cardiac	ARR	C14.280.238	213,331
Heart defects, congenital	CHD	C14.280.400	140,898
Heart valve diseases	VD	C14.280.484	122,036
Myocardial ischemia	IHD	C14.280.647	436,415
Cardiac conduction system disease	CCS	C14.280.123	93,568
Ventricular outflow obstruction	VOO	C14.280.955	36,341
Other heart diseases[a]	OHD	C14.280.195, C14.280.282, C14.280.383, C14.280.470, C14.280.945, C14.280.459, C14.280.720	195,433

The root MeSH terms and root IDs of associated categories are tabulated. Additionally, the number of publications collected for 8 CVDs which were used to construct the Text-cube are also provided
[a]They include Cardiomegaly, Endocarditis, Heart Arrest, Heart Rupture, Ventricular Dysfunction, Heart Neoplasms, Pericarditis

Table 4 Sample of COVID-19-interacting proteins

Uniprot ID	COVID-19 Interacting Protein: Name, Synonyms and Abbreviations
O00203	AP-3 complex subunit beta-1, Adaptor protein complex AP-3 subunit beta-1, Adaptor-related protein complex 3 subunit beta-1, Beta-3A-adaptin, Clathrin assembly protein complex 3 beta-1 large chain
O60885	Bromodomain-containing protein 4, Protein HUNK1
P25440	Bromodomain-containing protein 2, O27.1.1, Really interesting new gene 3 protein
Q6UX04	Spliceosome-associated protein CWC27 homolog, Antigen NY-CO-10, Probable inactive peptidyl-prolyl cis-trans isomerase CWC27 homolog, PPIase CWC27, Serologically defined colon cancer antigen 10
Q86VM9	Zinc finger CCCH domain-containing protein 18, Nuclear protein NHN1
Q8IWA5	Choline transporter-like protein 2, Solute carrier family 44 member 2
O75439	Mitochondrial-processing peptidase subunit beta, EC 3.4.24.64, Beta-MPP, P-52
O95070	Protein YIF1A, 54TMp, YIP1-interacting factor homolog A
P05026	Sodium/potassium-transporting ATPase subunit beta-1, Sodium/potassium-dependent ATPase subunit beta-1
P11310	Medium-chain specific acyl-CoA dehydrogenase, mitochondrial, MCAD, EC 1.3.8.7, Medium chain acyl-CoA dehydrogenase, MCADH

A sample of few COVID-19-interacting proteins with their names, synonyms and abbreviations extracted from UniProt [20] API are provided. For large data files, readers may visit our GitHub repo data file (https://github.com/CaseOLAP/covid-cvd)

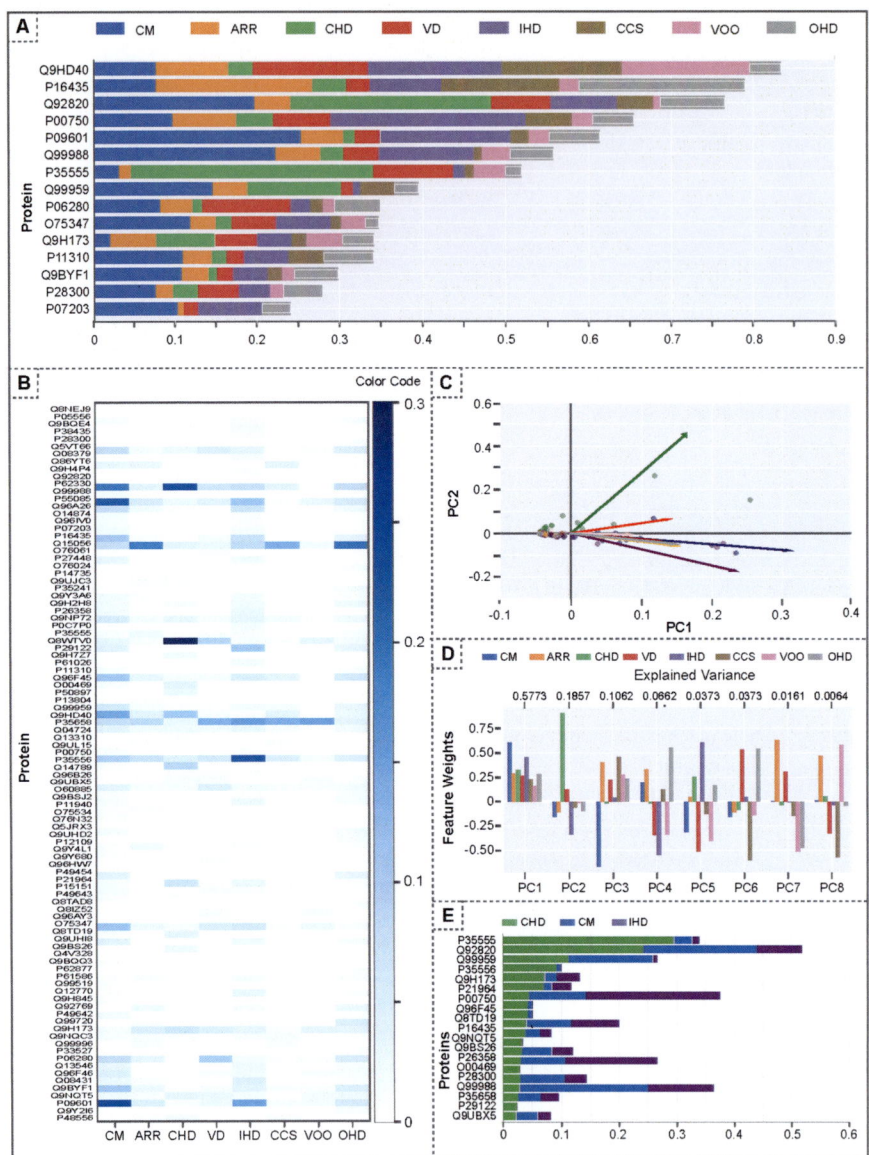

Fig. 6 Elucidating relationships of COVID-19 proteins and their associations across 8 categories of CVDs. Panel **A** shows a stacked bar chart of top 15 CaseOLAP ranked proteins for all 8 CVDs. Each protein is represented in the bar chart (denoted by its UniProt ID on the Y-axis) and it consists of components corresponding to the protein's weight in each category of 8 CVDs. Panel **B** exhibits a heatmap composed of CaseOLAP analysis results in 8 CVDs for all proteins in COVID-19 subproteome. Each protein (each row, UniProt ID) is plotted with each CVD category. The color intensity corresponds to CaseOLAP score; and thus a greater association for that protein in a particular CVD category. Panel **C** details Principal Component Analysis (PCA) performed on a subset of 96 proteins without missing values of their CaseOLAP scores; each of the 8-dimensional

meshb.nlm.nih.gov) and ICD codes (https://icd.codes/) are hierarchically organized codes created by the NLM and WHO respectively, standard medical entity IDs in categories such as "Diseases" in both research and clinical settings.

In the first part of the exploration, (i) the COVID-19 symptoms (Table 1), and (ii) COVID-19 comorbidities (Table 2) informed by the CDC (https://www.cdc.gov/) are prepared with corresponding MeSH terms and synonyms are matched from UMLS. This approach allowed us to explore the publications through indexing documents and searching entities. We extracted 80,732 total publications related to COVID-19 and analyzed them for possible symptom-symptom and symptom-comorbidity relationships. (Figs. 3, 4 and 5).

In the second part of the exploration, we used 8 categories of CVDs (Table 3) and created the document structure called Text-Cube [3] using all available documents in PubMed. We extracted 1,075,200 publications related to CVD and explored the association of COVID-19-interacting proteins with 8 CVDs through CaseOLAP score calculation (Figs. 6 and 7).

MeSH for CVD has 24 categories with over 600 descriptors. In our computations, we prepared specially designed CVD categories with proportionally distributed numbers of documents (Table 3) and implemented the MeSH-based document collection for 8 CVDs to construct the Text-Cube. Each disease's subcategory diseases and conditions (i.e., its descendants in the MeSH tree) are included under the main disease category and contributed to their counts.

2.2 Structured Molecular Data Sources

Multiple structured data sources were used for detailing molecular mechanisms in COVID-19 and CVD, namely the proteins and pathways. The proteins were subsequently scored with the CaseOLAP platform, enabling further exploration of the relevant proteins' associations with one another, their associations with CVDs, and their common pathways.

First, a list of 333 human proteins which have been reported to interact with COVID-19 was assembled (see sample at Table 4). A UniProt [20] API was used to

---

Fig. 6 (continued) vectors correspond to their scores for the 8 CVDs. This 2-dimensional scatterplot uses the 2 PCs of maximum variance (PC1: 57% of variance, PC2: 18% of variance). The dots represent the proteins. The arrowhead vectors represent the 8 CVD vector projections onto the 2-dimensional space. CHD's vector is clearly distinct from all others (CM, ARR, VD, IHD, CCS, VOO and OHD) Panel **D** highlights each PC's linear combination of the 8 CVDs. Each PC is a linear combination of all 8 CVDs. Each bar corresponds to the CVD's contribution to the PC. The first two components, PC1 and PC2, were used to construct the PCA in Fig. 6d. Panel **E** presents a stacked bar chart of the top 20 CaseOLAP ranked proteins for the 3 CVDs, which were the most influential components in the PC1, PC2 (Fig. 6b) and PC3 (Fig. 6d). Each row is designated for each protein (UniProt ID) and is comprised of the CaseOLAP scores for these 3 CVD categories. Thus, the highest ranked proteins in the most influential components of the PCA are displayed. Proteins are ranked by their score in the CHD category

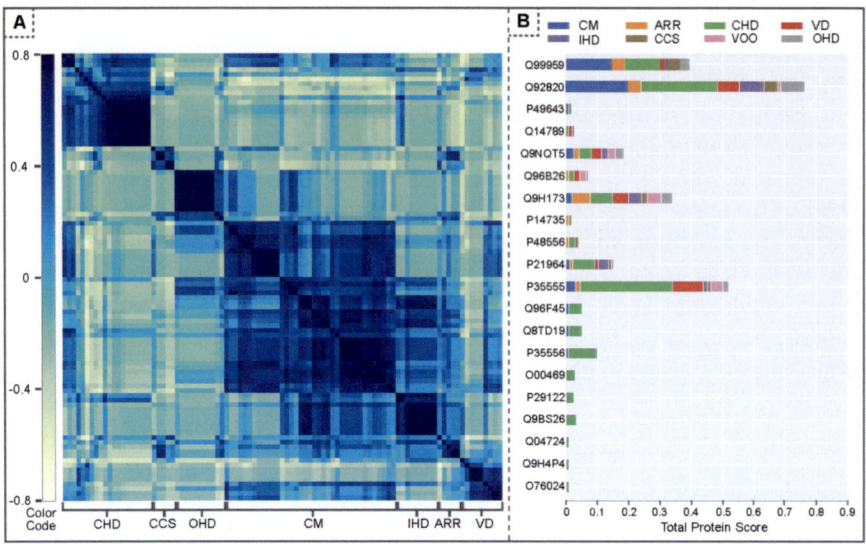

Fig. 7 COVID-19 protein clusters based on CVD associations. In panel **A**, we performed a cluster analysis using the standardized Euclidean distance metric over the 8-dimensional protein scores of each CVD category. Every row or column in the figure represents a protein whereas a tiny square at a specific location (i,j) represents the metric distance of the ith protein from jth protein and vice versa. Clustering results along with a 2-dimensional cluster plot illustrated how the protein associations were clustered across 7 CVD types (namely CHD, CCS, OHD, CM, IHD, ARR and VD) whereas VOO is not noticeably found to have a cluster structure. The distance scores obtained are reflected in a rescaled color pattern ranging from −8.0 to 8.0, as shown in the color legend of the clustering plot. A darker intensity indicates that these proteins were closely clustered together. A hierarchical cluster analysis identified 7 main COVID-19 protein clusters, each roughly pertaining to one CVD group as depicted by the legends on the X-axis. Among 7 identified clusters, CHD is the distinct one whereas CM appears to be the largest among the other clusters. In panel **B**, the stacked horizontal bar charts on the right depict each protein's 8-dimensional score in the CHD cluster. The legend shows the colors for each of the 8 CVDs (i.e., 8 dimensions). The data associated with the cluster visualization is also available at (https://github.com/CaseOLAP/covid-cvd)

obtain the proteins' UniProt IDs, names, synonyms, and abbreviations. The CaseOLAP score for entity-category association was calculated for the proteins in the 8 CVDs. Those scores were 8-dimensional vectors whose dimensions corresponded to the scores in each of the 8 CVD categories.

We searched the high scoring COVID-19-interacting proteins' pathways using Reactome's [21] web application and analyzed them in different ML scenarios: PCA (Fig. 6) and clustering (Fig. 7). We assembled the sets of associated pathways for these proteins into lists. With a unique ID for each protein, we isolated the pathways whose molecular mechanisms are associated with CVD and COVID-19 (Table 5).

Table 5 Highest scoring proteins

Rank (Sum score)	Protein (Uniprot ID)	Pathways	Role
1 (0.834)	O-phosphoseryl-tRNA (Sec) selenium transferase (Q9HD40)	Selenoamino acid metabolism, Metabolism, Metabolism of amino acids and derivatives, Selenocysteine synthesis	Converts O-phosphoseryl-tRNA(Sec) to selenocysteinyl-tRNA(Sec) which is required for synthesizing selenoprotein
2 (0.789)	NADPH-cytochrome P450 reductase (P16435)	Metabolism, Cytochrome P450 – arranged by substrate type, Phase I – Functionalization of compounds, Biological oxidations	Transfers electrons from NADP to heme oxygenase, cytochrome B5, and microsomal cytochrome P450 whose enzymes perform key roles in xenobiotic detoxification and drug interactions
3 (0.765)	Gamma-glutamyl hydrolase (Q92820)	Immune System, Neutrophil degranulation, Innate Immune System	a coenzyme in folic acid and free glutamate production, hydrolyzing the pteroylpolyglutamate's polyglutamate sidechains; it may also affect dietary pteroylpolyglutamate's bioavailability and metabolism and antifolate metabolism
4 (0.654)	Tissue-type plasminogen activator (P00750)	Signaling by PDGF, Signal Transduction, Signaling by Receptor Tyrosine Kinases, Dissolution of Fibrin Clot, Hemostasis	Activates zymogen plasminogen into plasmin. By controlling this plasmin-mediated proteolysis, it affects tissue remodeling and degradation and cell migration (including neuronal migration).
5 (0.613)	Heme oxygenase 1 (P09601)	Signaling by Interleukins, Cytokine Signaling in Immune system, Immune System, Transport of small molecules, Metabolism, Iron uptake and transport, Interleukin-4 and Interleukin-13 signaling, Heme degradation, Insertion of tail-anchored proteins into the endoplasmic reticulum membrane, Metabolism of porphyrins, Protein localization	Converts heme to produce biliverdin, a precursor to bilirubin, which protects cells from excess free heme-induced apoptosis. Heme oxygenase 1 activity is highest in the spleen where senescent erythrocytes are destroyed.
6 (0.557)	Growth/differentiation factor 15 (Q99988)	Unknown	Regulates food intake, energy expenditures, and body weight in response to metabolic- and toxin-induced stress, using neurons to activate part of the brain stem and amygdala. It can also inhibit growth hormone signaling on hepatocytes

(continued)

Table 5 (continued)

Rank (Sum score)	Protein (Uniprot ID)	Pathways	Role
7 (0.518)	Fibrillin-1 (P35555)	Metabolism of proteins, Elastic fibre formation, Regulation of Insulin-like Growth Factor, (IGF) transport and uptake by Insulin-like Growth Factor Binding Proteins (IGFBPs),Degradation of the extracellular matrix, Post-translational protein modification, Integrin cell surface interactions, Post-translational protein phosphorylation, Extracellular matrix organization, Molecules associated with elastic fibers	Plays a role in connective tissue. It helps structural support by providing scaffolding for elastin deposits in blood vessels, regulates tissue homeostasis via growth factor interactions, participates in bone regulation (osteoblast maturation, osteoclastogenesis, osteoclast differentiation and function), and participates in microfibril assembly.
8 (0.394)	Plakophilin-2 (Q99959)	KeratinizationFormation of the cornified envelope,Developmental Biology	May play a role in cell adhesion and junctional plaques, performing binding activities for protein kinase C, ion channels, intermediate filaments, cadherin, and alpha-catenin. Mutations in the Plakophilin-2 gene lead to severe loss of protein function which has been implicated in CM.
9 (0.349)	Alpha-galactosidase A (P06280)	Neutrophil degranulation,Glycosphingolipid metabolism,Immune System,Metabolism of lipids,Sphingolipid metabolism,Metabolism,Innate Immune System	Performs a role in glycosphingolipids degradation in the lysosome
10 (0.347)	Tubulin-specific chaperone A (O75347)	Post-chaperonin tubulin folding pathwayProtein folding,Metabolism of proteins	Part of the folding pathway for tubulin which makes up microtubules
11 (0.342)	Nucleotide exchange factor SIL1 (Q9H173)	Unknown	Necessary for protein translocation and protein folding in the endoplasmic reticulum (ER). It is a nucleotide exchange factor for an ER lumenal chaperone, HSPA5.

Rank (Sum score)	Protein (Uniprot ID)	Pathways	Role
12 (0.341)	Medium-chain specific acyl-CoA dehydrogenase, mitochondrial (P11310)	Beta oxidation of decanoyl-CoA to octanoyl-CoA-CoA.Regulation of lipid metabolism by PPARalpha. Metabolism,Fatty acid metabolism,Beta oxidation of octanoyl-CoA to hexanoyl-CoA, Mitochondrial Fatty Acid Beta-Oxidation,PPARA activates gene expression, mitochondrial fatty acid beta-oxidation of saturated fatty acids, Metabolism of lipids,mitochondrial fatty acid beta-oxidation of unsaturated fatty acids,	Enables energy production from fats by participating in mitochondrial fatty acid beta-oxidation: An aerobic process breaking down fatty acids into acetyl-CoA
13 (0.298)	Angiotensin-converting enzyme 2 (ACE2) (Q9BYF1)	Attachment and Entry,Metabolism of Angiotensinogen to Angiotensins,Attachment and Entry,"SARS-CoV-1 Infection,Potential therapeutics for SARS,Metabolism of proteins,Disease,Infectious disease,SARS-CoV Infections,Peptide hormone metabolism,SARS-CoV-2 Infection,	Regulates blood volume and cardiovascular homeostasis via systemic vascular resistance. ACE2 produces angiotensin 1–9 which has anti-hypertrophic effect in cardiomyocytes. It also produces angiotensin 1–7 which is a beneficial vasodilator and anti-proliferation agent. ACE2 also removes the C-terminal residue from 3 vasoactive peptides: Neurotensin, kinetensin, and des-Arg bradykinin. ACE2 also regulates cell surface expression and catalytic activity of the neutral amino acid transporter SL6A19 on gut epithelial cells' plasma membrane. Notably, ACE2 is a receptor for SARS-CoV-2, COVID-19
14 (0.279)	Protein-lysine 6-oxidase (P28300)	Crosslinking of collagen fibrils,Collagen formation, Extracellular matrix organization,Assembly of collagen fibrils and other multimeric structures,Elastic fibre formation,	Plays a role in aortic wall structure, helps produce fibrous collagen and elastin precursors
15 (0.240)	Glutathione peroxidase 1 (P07203)	Fatty acid metabolism,Detoxification of Reactive Oxygen Species, Arachidonic acid metabolism, Metabolism of nucleotides, Synthesis of 5-eicosatetraenoic acids,Cellular responses to stress,Metabolism, Synthesis of 12-eicosatetraenoic acid derivatives, Metabolism of lipids,Nucleobase catabolism,Synthesis of 15 eicosatetraenoic acid derivatives, Purine catabolism, Cellular responses to external stimuli	Protects erythrocyte-contained hemoglobin from oxidative breakdown

A tabular presentation of highest scoring proteins and their involvement in pathways and associated molecular roles are provided. Note that Pathways were taken from Reactome knowledge base and roles were taken from UniProt [20] database

2.3   CaseOLAP Scoring Platform

The CaseOLAP scoring platform has been described elsewhere in greater detail [3, 15]. In brief, CaseOLAP scores how relevant a particular entity (e.g., a COVID-19-interacting protein) is to a particular category (e.g., a CVD). These user-defined entities and categories are the central metadata needed to implement the CaseOLAP pipeline: building the document structure called Text-Cube and calculating the CaseOLAP score.

The CaseOLAP score is based upon how closely a descriptive phrase matches the target entity phrase (integrity score), an intra-document frequency measure of how prominent an entity is in a document when compared to other phrases within that document (popularity score), and an inter-document frequency measure of how prominent an entity is in a document when compared to other documents with that entity (distinctiveness score) [3]. Here, the entities (i.e., phrases) were 333 COVID-19-interacting proteins scored across the PubMed publications (i.e., the documents) for 8 CVDs.

Each protein in the entity list is assigned a score for each of the 8 CVD categories. Vectorizing each protein in the 8-dimensional CVD space enabled ML (PCA and clustering) opportunities to explore molecular mechanism relationships. These CaseOLAP scores provide information about the molecular interface of COVID-19 and CVD. We made all relevant codes and data publicly available in the GitHub repository (https://github.com/CaseOLAP/covid-cvd).

3   Results and Discussion

In this study, we created a text mining pipeline to better understand associations of CVDs with COVID-19. First, we identified a data corpus of biomedical publications containing COVID-19 symptoms as defined by CDC. Using 333 human COVID-19-interacting proteins as entities and 8 CVDs classified by MeSH as categories, we examined and computed their Context Aware Semantic Analytic Processing (CaseOLAP) scores. Among the COVID-19 subproteome, we identified a subset of 96 proteins with no missing values and implemented machine learning methods to explore the molecular interface of the COVID-19 subproteome and CVDs. More importantly, our results highlighted distinct molecular interfaces of COVID-19 with each category of CVD. For the first time, our cluster analyses determined which COVID-19-interacting proteins are most relevant for each CVD category.

We have validated several findings for COVID-19 symptom rankings, comorbidities, and proteins implicated in CVD via existing published data in individual expert studies; due to the unique text mining approach we have taken to obtain these results, these findings highlight the ability of our platform to effectively extract meaningful information from the collective body of existing research (i.e. 80,000+ PubMed publications mentioning key terms such as COVID-19 symptoms and comorbidities, as well as 1,000,000+ publications mentioning individual proteins in CVDs).

We have provided novel insights into and depictions of the relationships among COVID-19 symptoms, symptoms and comorbidities, and the proteomic molecular interface of COVID-19 and one of its comorbidities, CVD. This relationship will be first explored in part here and further examined in a later work.

3.1 COVID-19 Symptom-Symptom Relationships

We used text mining approaches to identify and count symptoms (Fig. 3) and their pairwise relationships (Fig. 3a and 3d) in all PubMed publications. To overcome the hurdle of different names and abbreviations used for each symptom (Table 1), we used the symptoms' names, synonyms, and abbreviations from UMLS (specifically MeSH, ICD, SnomedCT) as search terms for querying over 80,000 publications on COVID-19. Thus, we counted the number of publications which mentioned each symptom (i.e., symptom frequency) and counted when a symptom was mentioned with another symptom (i.e., symptom-symptom frequency) (Fig. 3b). The symptom frequency rankings (Fig. 3b) roughly correspond to the CDC's symptom rankings. The CDC's list validates our list which we acquired through our massive text mining approach that relied on the aforementioned batches of PubMed publications.

We also added another layer of complexity: the symptom-symptom frequency (Fig. 3a and 3d). This co-mentioning is inferred to correspond to the symptoms co-occurring in COVID-19 patients. Co-occurring symptoms may be due to similar molecular mechanisms. Thus, identifying co-occurring symptoms may provide clues for further exploration of molecular mechanisms in COVID-19, the common genes, proteins, lipids, or pathways. Note: a symptom may be 'double counted' if it appears in a publication with multiple symptoms. For example, if symptoms A, B, and C are in one publication, that publication's symptom A will be in the A-B chord and the A-C chord. Additionally, we depicted the progress of COVID-19 research as measured by the number of monthly PubMed publications on COVID-19 from Feb 2020–Jan 2021, displaying the scientific community's cumulative research efforts over the course of much of the pandemic (Fig. 3c).

3.2 COVID-19 Symptom-Comorbidity Relationships

Our text mining exploration of PubMed publications also identified relationships between COVID-19 symptoms and comorbidities. Our list of comorbidities (Table 2) was taken from the CDC, not created from scratch, but the rankings and symptom-comorbidity relationships were directly informed by all of PubMed's COVID-19 publications (at the time of the study), namely how often the comorbidities were mentioned in their studies. It is possible that researchers may not understand how significant a comorbidity might be in their initial investigation, but over time, the collective work from the scientific community would converge onto the most relevant comorbidities for COVID-19, when findings are becoming established.

COVID-19 symptoms are caused by underlying molecular mechanisms. But as symptoms are not identical across all patients, neither are these molecular mechanisms. Because the molecular mechanisms of COVID-19 and its comorbidities interact differently in the human body, they can affect symptoms and health outcomes differently. Thus, the CDC has categorized COVID-19 comorbidities as major (Fig. 4a, b) or minor (Fig. 5a, b).

Noting that CVD and associated conditions (obesity, hypertension, thalassemia, smoking, diabetes, etc.) are among the reported COVID-19 comorbidities, we further explored how the human proteins proposed to interact with COVID-19 have been studied in 8 different CVD subcategories. This was to better understand the molecular mechanisms at the interface of CVD and COVID-19. For the pairwise comorbidity-symptom counts, box sizes may not perfectly correspond to the absolute number of publications the individual comorbidity or symptom was mentioned in. Our approach also affirmed that CVD and related conditions are COVID-19 comorbidities. This is validated by other reports, such as those by the CDC and WHO, demonstrating the success of text mining approaches on PubMed publications at identifying these comorbidities. This provided the impetus for our subsequent exploration of the proteomic intersection of COVID-19 and CVDs.

3.3   Proteins Implicated in Both COVID-19 and CVDs

Inspired by the success of previous CaseOLAP implementations for biomedical entity-category association [3, 15] the current relevance of COVID-19 and CVDs, and CVDs' comorbidity status for COVID-19 (Fig. 4), we explored associations of 333 COVID-19-interacting proteins (Table 4) in 8 CVDs (Table 3). This is not an analysis of all proteins studied in CVD (i.e. inferred to be implicated in CVD based on the collective body of PubMed publications), but is rather a subset of those proteins which COVID-19 interacts with (Table 4).

Among the 333 COVID-19-interacting proteins, 96 proteins had non-zero CaseOLAP scores across the 8 CVDs (Table 3) (i.e. 237 proteins were not found to be relevant for any of the CVDs). These protein scores are depicted in a heat map (Fig. 6b). The top 15 scoring proteins have their scores displayed in more detail in a stacked horizontal bar chart (Fig. 6a).

All Heart Diseases' Highest Scoring Proteins

Summed across all 8 CVD categories (Fig. 6a), the 15 highest scoring proteins out of the 96 studied proteins were further researched. We found information about these proteins' pathways and molecular roles and presented them in the table (Table 5). High CaseOLAP scores indicate that these proteins are the most studied and mentioned in all heart disease research publications indexed by PubMed. Ideally, these scores should correspond to how relevant a particular protein is to a particular disease.

Principal Component Analysis (PCA)

PCA performs dimensionality reduction of the 8-dimensional scores and helps visualize the scoring pattern variance in an understandable 2-dimensional representation (Fig. 6c, D). The PCA presents the CHD axis as distinct from the other CVDs. The PCs are further visualized based on each CVD dimension's contribution (Fig. 6d). The first PC dimension appears to be the average contribution of all CVDs, but the second dimension is dominated by CHD and IHD, and the third dimension is dominated by CM. Twenty high-scoring proteins were then displayed for these 3 dominating dimensions (CHD, IHD, CM), arranged by highest CHD score (Fig. 6e).

Distinct Nature of CHD and Its Top-Ranking Proteins

We note that in the PCA, the CHD axis appears distinct from all other axes (Fig. 6c) and is a dominant component of PC2 (Fig. 6d). This may be due to CHD's highest scoring proteins: (1) *Fibrillin-1* (UniProt: P35555, score = 0.3), (2) *Gamma-glutamyl hydrolase* (UniProt: Q92820, score = 0.24), and (3) Plakophilin-2 (UniProt: Q99959, score = 0.11). Notably, CHD's top 2 proteins had the 1st highest and 3rd highest score out of any protein in a single disease category (though not the highest sum of all 8 CVD scores). The high scores of these 3 proteins may explain why CHD's PC vector is separated from all other heart disease categories. The reason for this should be explored to further establish if the high scores are indeed from a biologically meaningful reason. We analyzed the top scoring proteins in CHD because of its distinct vector in the PCA and in overall CVD as well as their clustering behaviors. The involved proteins' functions are described in more detail in the next section (see **Result and Discussion Sect. 3**, Fig. 7).

3.4 Exploring Relationships Between Associated Proteins in COVID-19 and CVD

Protein Clustering

Representing the COVID-19-interacting proteins implicated in CVD as points in an 8-dimensional vector space not only enables PCA, but also clustering to identify similarly scoring proteins (i.e. proteins close to one another in the 8-dimensional vector space) through a distance metric [22]. Proteins may score similarly because of similar biomedical functions in CVD. The strength of protein associations is depicted in the cluster analysis (Fig. 7a). Individual protein scores are depicted as stacked horizontal bar charts (Fig. 7b).

The cluster analysis revealed which proteins were associated in each CVD category based on our text mining results. These associated proteins were further explored through the Reactome pathway knowledgebase to narrow down the pathways relevant in CVDs. In other words, we used the graph network structure which connects proteins and pathways to identify the pathways which were connected by proteins from the same cluster in the cluster analysis (i.e. similar CaseOLAP scoring proteins more prominent in a particular CVD). A protein may play a role in multiple pathways, but it could be that only some of those pathways are relevant for a CVD. By identifying if our associated proteins (proteins which were found relevant for a CVD based on our text mining analyses) are also part of the same pathway, the CVD-relevant molecular pathway may be better identified. Here, we present specific results and analysis for CHD, ARR, and OHD respectively. We intend to report further analysis results and the involvement of cardiovascular drug target proteins in a later publication. As a whole, this clustering approach stands as the tool to explore the hidden and less appreciated facts in biomedicine by summarizing millions of publications.

Exploration of CHD Cluster

Using the Reactome pathway knowledgebase, we explored the shared pathways of CHD-related associated proteins as determined by the cluster analysis. The top scoring proteins for CHD are *Fibrillin-1, Gamma-glutamyl hydrolase, Plakophilin-2, Fibrillin-2, Nucleotide exchange factor SIL1, Catechol O-methyltransferase, Tissue-type plasminogen activator, Zinc finger protein 503, Serine/threonine-protein kinase Nek9, NADPH--cytochrome P450 reductase, Exosome complex component RRP40, Endoplasmic reticulum resident protein 44*, and *DNA (cytosine-5)-methyltransferase 1*.

Taking the common pathways from two of the top proteins (which had multiple common pathways), one may infer that for CHD, roles may be played by *Fibrillin-1* and *Fibrillin-2's* potential association with "elastic fiber formation" [23] and "degradation in the extracellular matrix" [24]; "post-translational protein modification" [25] by *26S proteasome non-ATPase regulatory subunit 8* [26] and *Fibrillin-1*; *Exosome complex component RRP40*, and *Exosome complex component RRP43*'s roles in the "unfolded protein response" [27] (with ATF4 and PERK as potentially relevant genes) through "dysfunctional mRNA destabilization" (via "Tristetraprolin, 3' to 5' exoribonuclease", "deadenylation", "Butyrate Response Factor 1", or "KSRP") or "dysfunctional rRNA processing" in the nucleus or cytosol. Or perhaps the "mRNA destabilization" and "rRNA processing" are involved in responses other than the unfolded protein response. These are hypotheses based on the associated proteins' common pathways. Others with domain knowledge can come up with their own hypotheses based off of this data, exploring the literature or performing experiments to address this issue. As mentioned, we intend to publish more results on this analysis in a later publication.

Exploration of ARR Cluster

Two high scoring proteins were found to be associated with "Cell Cycle", "Mitotic", "M Phase", and "Disease" pathways. These pathways would suggest that these 3 proteins play a role in disease resulting from dysfunction in the M phase of the mitotic cell cycle. The 2 proteins were *Golgin subfamily A member 2* (UniProt: Q08379, ARR CaseOLAP = 0.0317) [28] and *A-kinase anchor protein 9* (UniProt: Q99996, ARR CaseOLAP = 0.0183) [29].

Exploration of OHD Cluster

Three high scoring proteins were found to be associated with the "Metabolism" pathway. This pathway shows that the 3 proteins play a role in "Metabolism". The 3 proteins were *NADPH--cytochrome P450 reductase* (UniProt: P16435), *Medium-chain specific acyl-CoA dehydrogenase, mitochondrial* (UniProt: P11310), and *Heme oxygenase 1* (UniProt: P09601).

4 Limitations and Future Work

Understanding disease associated molecular mechanisms is an ultimate goal of personalized medicine. Network medicine [30, 31] represents a leading approach in this area of research. Building Graph Neural Networks (GNN) for disease identifications has proven to be effective [32, 33] to dive deeper into the classifications of various CVD categories, and/or subcategories. When changing the level of abstraction (categories and subcategories), hidden trends may be discovered. CVD categories likely have some common and unique mechanisms. By grouping together more CVDs into one CVD category, we may capture general trends at the expense of insights offered by individual subcategories. Despite this limitation, we succeeded in identifying candidate proteins common and/or unique for each category.

The recent development of smart database systems and accessing data through API queries has opened up new data streaming and analysis pipelines for large data as well as small data scenarios. The future of data analysis pipelines will consist of automated mechanisms to update and access data and data models to achieve optimal interoperability.

When interpreting the PCA, it should be noted that our current analysis is based on the number of COVID-19-interacting proteins and their relevance in CVD, not based on the proteins' relative significance nor their impact in COVID-19. In future studies, ranking proteins could be conducted based on the impact of a COVID-19-interacting protein on COVID-19 outcome if the impacts of the proteins are accurately quantified.

Structuring unstructured data is an ill-defined problem because the desired structure may vary based on a user's needs. Thus, the quantitative tools in these pipelines require human design and domain expertise. CaseOLAP's entity-category association scoring platform is effective to provide some level of quantitative information about biomedical entities over a spectrum of disease categories but falls short on relating that to biological insights. Future work may use a knowledge graph to explore entities' relationships, as well as to define the type of relationships (up-, down-regulated, etc.).

Although the molecular makeup of the cardiac proteome is fairly well known, the subproteomes of each CVD is much less understood. Our work provides new and unique information to the molecular insights of each CVD. Either shared or unique, every relevant protein-CVD relationship reported are aggregated via these CaseOLAP analyses. This represents a new approach to learn about the individual CVD subproteomes. In parallel, each pathway may have multiple relevant proteins to one category of CVD; we attempted to find the disease associated pathways by mapping pathways with multiple relevant proteins. However, this may overlook pathways with only one relevant protein; yet gain the benefit of minimizing false positive pathways.

5   Concluding Remarks

Taken together, our analyses identified candidate proteins from the COVID-19 subproteome and determined the pathways they impinge upon under CVDs, providing potential insight into the molecular processes underlying how CVD, as a comorbidity, may impact COVID-19 outcomes. Moreover, these results demonstrate that our text mining pipeline is a powerful tool for mining and synthesizing the collective knowledge of a biomedical domain for the purpose of understanding molecular processes and disease pathogenesis.

Author Contributions D Sigdel and P Ping designed the project. D Sigdel, M Gupta, A Zhang, D Liem, W Wang, and P Ping did the analyses; D Sigdel, D Wang, and P Ping constructed all figures. D Sigdel, D Steinecke, D Wang, and P Ping completed the writing of the manuscript.

Funding Sources This work was supported by the National Heart, Lung and Blood Institute at the National Institutes of Health [awards R35 HL135772, T32 HL139450, R01 HL146739].

References

1. Wang, L.L., Lo, K.: Text mining approaches for dealing with the rapidly expanding literature on COVID-19. Brief. Bioinform. **22**(2), 781–799 (2021). https://doi.org/10.1093/bib/bbaa296
2. Zhu, R., Tu, X., Huang, J.X.: Chapter 5: Utilizing BERT for biomedical and clinical text mining. In: Lee, K.C., Roy, S.S., Samui, P., Kumar, V. (eds.) Data Analytics in Biomedical Engineering and Healthcare, pp. 73–103. Academic (2021)
3. Liem, D.A., Murali, S., Sigdel, D., Shi, Y., Wang, X., Shen, J., et al.: Phrase mining of textual data to analyze extracellular matrix protein patterns across cardiovascular disease. Am. J. Physiol. Heart Circ. Physiol. **315**(4), H910–H924 (2018). https://doi.org/10.1152/ajpheart.00175.2018
4. Marcos-Zambrano, L.J., Karaduzovic-Hadziabdic, K., Loncar Turukalo, T., Przymus, P., Trajkovik, V., Aasmets, O., et al.: Applications of machine learning in human microbiome studies: a review on feature selection, biomarker identification, disease prediction and treatment. Front. Microbiol. **12**, 634511 (2021). https://doi.org/10.3389/fmicb.2021.634511
5. Deng, F., Huang, J., Yuan, X., Cheng, C., Zhang, L.: Performance and efficiency of machine learning algorithms for analyzing rectangular biomedical data. Lab. Investig. **101**(4), 430–441 (2021). https://doi.org/10.1038/s41374-020-00525-x
6. Cernile, G., Heritage, T., Sebire, N.J., Gordon, B., Schwering, T., Kazemlou, S., et al.: Network graph representation of COVID-19 scientific publications to aid knowledge discovery. BMJ Health Care Inform. **28**(1), e100254 (2021). https://doi.org/10.1136/bmjhci-2020-100254
7. Chen, Q., Allot, A., Lu, Z.: LitCovid: an open database of COVID-19 literature. Nucleic Acids Res. **49**(D1), D1534–D1540 (2021). https://doi.org/10.1093/nar/gkaa952
8. Kumar Das, J., Tradigo, G., Veltri, P., HG, P., Roy, S.: Data science in unveiling COVID-19 pathogenesis and diagnosis: evolutionary origin to drug repurposing. Brief. Bioinform. **22**(2), 855–872 (2021). https://doi.org/10.1093/bib/bbaa420
9. Ge, Y., Tian, T., Huang, S., Wan, F., Li, J., Li, S., et al.: An integrative drug repositioning framework discovered a potential therapeutic agent targeting COVID-19. Signal Transduct. Target. Ther. **6**(1), 165 (2021). https://doi.org/10.1038/s41392-021-00568-6
10. MacLean, F.: Knowledge graphs and their applications in drug discovery. Expert Opin. Drug Discov., 1–13 (2021). https://doi.org/10.1080/17460441.2021.1910673
11. Bansal, M.: Cardiovascular disease and COVID-19. Diabetes Metab. Syndr. **14**(3), 247–250 (2020). https://doi.org/10.1016/j.dsx.2020.03.013
12. Clerkin, K.J., Fried, J.A., Raikhelkar, J., Sayer, G., Griffin, J.M., Masoumi, A., et al.: COVID-19 and cardiovascular disease. Circulation. **141**(20), 1648–1655 (2020). https://doi.org/10.1161/CIRCULATIONAHA.120.046941
13. Mai, F., Del Pinto, R., Ferri, C.: COVID-19 and cardiovascular diseases. J. Cardiol. **76**(5), 453–458 (2020). https://doi.org/10.1016/j.jjcc.2020.07.013

14. Gupta, A.K., Jneid, H., Addison, D., Ardehali, H., Boehme, A.K., Borgaonkar, S., et al.: Current perspectives on coronavirus disease 2019 and cardiovascular disease: a white paper by the JAHA editors. J. Am. Heart Assoc. **9**(12), e017013 (2020). https://doi.org/10.1161/JAHA.120.017013

15. Sigdel, D., Kyi, V., Zhang, A., Setty, S.P., Liem, D.A., Shi, Y., et al.: Cloud-based phrase mining and analysis of user-defined phrase-category association in biomedical publications. J. Vis. Exp. **144**, e59108 (2019). https://doi.org/10.3791/59108

16. Caufield, J.H., Sigdel, D., Fu, J., Choi, H., Guevara-Gonzalez, V., Wang, D., et al.: Cardiovascular Informatics: building a bridge to data harmony. Cardiovasc. Res. **cvab067** (2021). https://doi.org/10.1093/cvr/cvab067

17. Mahmud, M., Kaiser, M.S., McGinnity, T.M., Hussain, A.: Deep learning in mining biological data. Cognit Comput., 1–33 (2021). https://doi.org/10.1007/s12559-020-09773-x

18. Quiroz, J.C., Feng, Y.Z., Cheng, Z.Y., Rezazadegan, D., Chen, P.K., Lin, Q.T., et al.: Development and validation of a machine learning approach for automated severity assessment of COVID-19 based on clinical and imaging data: retrospective study. JMIR Med. Inform. **9**(2), e24572 (2021). https://doi.org/10.2196/24572

19. Peek, N., Holmes, J.H., Sun, J.: Technical challenges for big data in biomedicine and health: data sources, infrastructure, and analytics. Yearb. Med. Inform. **9**, 42–47 (2014). https://doi.org/10.15265/IY-2014-0018

20. The UniProt Consortium: UniProt: the universal protein knowledgebase in 2021. Nucleic Acids Res. **49**(D1), D480–D489 (2021). https://doi.org/10.1093/nar/gkaa1100

21. Jassal, B., Matthews, L., Viteri, G., Gong, C., Lorente, P., Fabregat, A., et al.: The reactome pathway knowledgebase. Nucleic Acids Res. **48**(D1), D498–D503 (2020). https://doi.org/10.1093/nar/gkz1031

22. Murtagh, F., Contreras, P.: Algorithms for hierarchical clustering: an overview. WIREs Data Mining Knowl. Disc. **2**(1), 86–97 (2012). https://doi.org/10.1002/widm.53

23. Mariencheck, M.C., Davis, E.C., Zhang, H., Ramirez, F., Rosenbloom, J., Gibson, M.A., et al.: Fibrillin-1 and fibrillin-2 show temporal and tissue-specific regulation of expression in developing elastic tissues. Connect. Tissue Res. **31**(2), 87–97 (1995). https://doi.org/10.3109/03008209509028396

24. Halper, J., Kjaer, M.: Basic components of connective tissues and extracellular matrix: elastin, fibrillin, fibulins, fibrinogen, fibronectin, laminin, tenascins and thrombospondins. Adv. Exp. Med. Biol. **802**, 31–47 (2014). https://doi.org/10.1007/978-94-007-7893-1_3

25. Reimand, J., Wagih, O., Bader, G.D.: Evolutionary constraint and disease associations of post-translational modification sites in human genomes. PLoS Genet. **11**(1), e1004919 (2015). https://doi.org/10.1371/journal.pgen.1004919

26. Collins, G.A., Goldberg, A.L.: The logic of the 26S proteasome. Cell. **169**(5), 792–806 (2017). https://doi.org/10.1016/j.cell.2017.04.023

27. Grootjans, J., Kaser, A., Kaufman, R.J., Blumberg, R.S.: The unfolded protein response in immunity and inflammation. Nat. Rev. Immunol. **16**(8), 469–484 (2016). https://doi.org/10.1038/nri.2016.62

28. Petrosyan, A.: Unlocking Golgi: why does morphology matter? Biochemistry (Mosc). **84**(12), 1490–1501 (2019). https://doi.org/10.1134/S0006297919120083

29. Liu, Y., Merrill, R.A., Strack, S.: A-kinase anchoring protein 1: emerging roles in regulating mitochondrial form and function in health and disease. Cell. **9**(2), 298 (2020). https://doi.org/10.3390/cells9020298

30. Barabasi, A.L., Gulbahce, N., Loscalzo, J.: Network medicine: a network-based approach to human disease. Nat. Rev. Genet. **12**(1), 56–68 (2011). https://doi.org/10.1038/nrg2918

31. Gysi, D.M., do Valle, I., Zitnik, M., Ameli, A., Gan, X., Varol, O., et al.: Network medicine framework for identifying drug-repurposing opportunities for COVID-19. Proc. Natl. Acad. Sci. U. S. A. **118**(19), e2025581118 (2021). https://doi.org/10.1073/pnas.2025581118

32. Yu, X., Wang, S.H., Zhang, Y.D.: CGNet: a graph-knowledge embedded convolutional neural network for detection of pneumonia. Inf. Process. Manag. **58**(1), 102411 (2021). https://doi.org/10.1016/j.ipm.2020.102411
33. Zhang, Y.-D., Satapathy, S.C., Guttery, D.S., Górriz, J.M., Wang, S.-H.: Improved breast cancer classification through combining graph convolutional network and convolutional neural network. Inf. Process. Manag. **58**(2), 102439 (2021). https://doi.org/10.1016/j.ipm.2020.102439

Clinical Application of iPSC-Derived Cardiomyocytes in Patients with Advanced Heart Failure

Jun Fujita, Shugo Tohyama, Hideaki Kanazawa, Yoshikazu Kishino,
Marina Okada, Sho Tanosaki, Shota Someya, and Keiichi Fukuda

1 Introduction

Heart failure (HF) is the leading cause of death in most developed countries. Patients with chronic HF experience acute exacerbation and their cardiac function gradually deteriorates until they finally progress to advanced heart failure (AHF) [1]. Heart transplantation is a radical intervention and acts as the last resort for patients with AHF. Even though the number of heart transplants has gradually increased, the donor shortage issue has not been resolved [2]. Left ventricular assist devices are currently applied in clinical settings a destination therapy, but they may cause hemorrhage, stroke, and infection [3]. Restoration of cardiac function via the re-muscularization of the damaged heart tissues would be the ultimate treatment for AHF. The replacement of impaired cardiac tissues with regenerated cardiomyocytes (CMs) is a major strategy in cardiac regenerative therapies, so pluripotent stem cells, such as human embryonic stem cells (hESCs) and induced pluripotent stem cells (hiPSCs), which demonstrate a strong potential for CM differentiation *in vitro* [4, 5], may act as a valuable source of functioning CMs. hiPSCs also avoid many of the ethical issues and immunological rejection problems associated with hESCs as they can be reprogrammed from the patient's own somatic cells [6]. Since human iPSCs were reported more than a decade ago, various hiPSC-derivatives have been applied in the clinical setting [7, 8]. Although

J. Fujita (✉)
Department of Cardiology, Keio University School of Medicine, Tokyo, Japan

Endowed Course for Severe Heart Failure Treatment II, Keio University School of Medicine, Tokyo, Japan
e-mail: jfujita@a6.keio.jp

S. Tohyama · H. Kanazawa · Y. Kishino · M. Okada · S. Tanosaki · S. Someya · K. Fukuda
Department of Cardiology, Keio University School of Medicine, Tokyo, Japan

beating CMs in culture dishes easily stimulate the imagination around their potential clinical application in heart failure, it is not always easy to translate *in vitro* results to the clinic. Critical issues in the clinical application of stem cell therapies in cardiac diseases are divided into two categories. First is the preparation of clinical-grade CMs, and the second their transplantation. In the case of the former, the quality of the hiPSCs, their differentiation capacity, the purity of the regenerated CMs, preparation of a large number of CMs, and their maturation for both contraction force and electrical activity, must be resolved prior to their clinical application. In addition, good transplantation strategies are essential for the engraftment and function of CMs *in vivo*. Immunological rejection must be overcome for long-term engraftment and there is a need to resolve the underlying causes of cardiac arrhythmia following CM transplant. In this chapter, we summarize the development of cardiac therapies for the intramyocardial transplantation of cardiac spheroids using human leukocyte antigen (HLA)-matched hiPSCs, and their potential in clinical applications (Fig. 1).

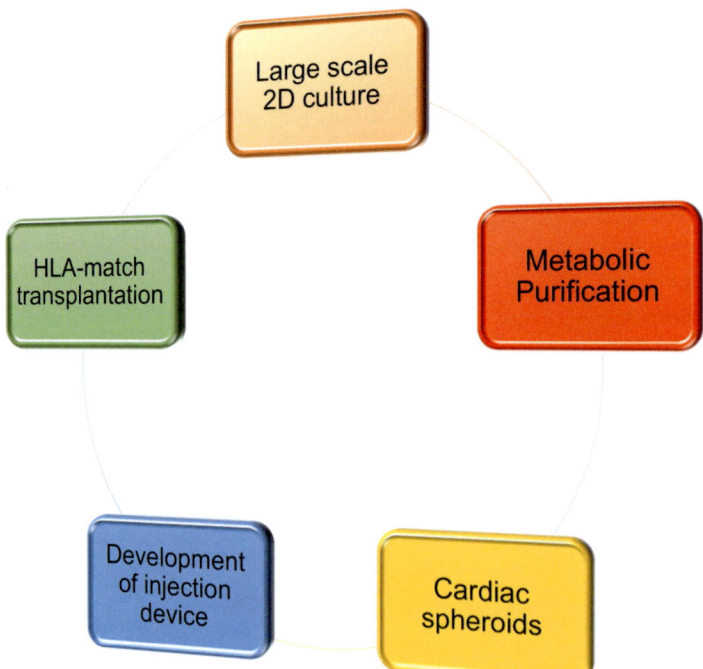

Fig. 1 Core strategies for the clinical application of human induced pluripotent stem cell (hiPSC)-derived cardiomyocytes (CMs). The development of scientific technologies is critical for the safe and effective transplantation of human induced pluripotent stem cell (hiPSC)-derived cardiomyocytes. HLA: human leukocyte antigen

2 Preparation of Clinical Grade hiPSC-Derived Cardiomyocytes

2.1 Preparation of Clinical Grade hiPSCs

In Japan, a master cell bank of clinical-grade hiPSCs was prepared and is maintained at the Center for iPS Cell Research and Application at Kyoto university [9]. Autologous transplantation is an ideal cell transplantation technique. However, it is expensive to establish a patient's hiPSCs using good manufacturing practice (GMP), and it is imperative that all hiPSCs used in clinical studies must have good differentiation capacity without genetic mutation. This means that the identification and evaluation of expanded hiPSCs is critical to successful clinical transplant. Given these restrictions, it makes sense that allogenic transplantation is currently the most widely available cell transplantation technique for hiPSC therapies.

2.2 Large Cell Culture Systems for hiPSC-Derived Cardiomyocyte Expansion

The heart demands significantly more regenerative CMs in order to recover reduced cardiac function than other organs. This is because their proliferation is highly restricted and means that it is necessary to produce a significant quantity of hiPSCs as a cell source.

Large-scale cell culture often relies on floating cell culture systems which use spinner flasks or bioreactors making both technologies common in regenerative medicine laboratories [10]. We developed a massive 2D culture system using multiple layer stack plates and active gas ventilation [11]. This culture technique enables the production of 1.7×10^9 hiPSCs from 1×10^6 hiPSCs in a week. CMs can be sequentially differentiated using glycogen synthase kinase-3β (GSK-3β) and Wnt inhibitors, and approximately 2 billion CMs can be produced in a week. This 2D culture system has several advantages in the preparation of clinical-grade CMs including the fact that it is easy to observe cell morphology and that 2D culture is the conventional method used for hiPSC expansion. Aggregated hiPSCs tend to differentiate into derivatives and the use of 3D culture systems reduces culture medium (energy and oxygen) penetration in the aggregated cells reducing their overall proliferation and health. This is also critical for our purification strategy using metabolic purification. Cell harvesting can be disadvantaged in multi-layer culture plates because cellular detachment can be relatively difficult. We developed a detachment system which relies on resonance vibrations [12], which takes advantage of the specific resonance of the cell culture plates and allows the gentle but efficient detachment of hiPSCs in massive 2D culture systems. The expanded application of hiPSC-based regenerative therapies also relies on an overall reduction in cell culture costs. To this end we went on to develop Stem Fit media®, in conjunction with the

Ajinomoto Corporation, designed to facilitate the large-scale culture of hiPSCs and CMs. The matrix proteins needed for hiPSC culture are also expensive and GMP-grade matrices, such as Laminin-511, are necessary for clinical-grade iPSC expansion. Cell culture surface modification may reduce the costs of the matrix components in large-scale 2D culture systems in the future [13]. Nutritional modifications also help to produce large numbers of hiPSCs. Tryptophan is one of the most commonly used amino acids in hiPSC culture and additional supplementation with tryptophan promotes long-term proliferation and increased cell mass in hPSCs. Metabolome analysis showed that supplementation with tryptophan did not increase kynurenine, which is a major ligand of the aryl hydrocarbon receptor (AhR) and a substrate for the *de novo* synthesis of nicotinamide adenine dinucleotide (NAD), but increased N-formylkynurenine, which is the upstream metabolite of kynurenine. Moreover, supplementation with N-formylkynurenine directly affected hiPSC proliferation [14]. Recently, it was reported that GSK-3β inhibitor (CHIR99021) and low-density passage promote the persistent proliferation of hiPSC-derived CMs [15]. These biotechnologies may resolve the issues surrounding the preparation of CMs for clinical studies.

2.3   Purification of hiPSC-Derived Cardiomyocytes

The most critical step in clinical studies is the purification of hiPSC-derived CMs. This is because hiPSCs demonstrate multipotent capacity, which may result in the production of teratomas. Thus, in an effort to eliminate the remaining stem cells, we focused on the differences in the metabolic characteristics of hiPSCs and CMs. Both transcriptome and metabolome analysis revealed that hiPSCs are highly dependent on glycolysis, similar to the "Warburg effect"described in cancer cells. However, hiPSC-derived CMs have a metabolic signature similar to fetal CMs allowing them to use lactate in place of glucose [16]. This means that the application of glucose-depleted, lactate-supplemented media successfully eliminated the remaining undifferentiated stem cells in hiPSC-derived CMs [17]. Based on the success of this "metabolic purification" we went on to evaluate the vital amino acids in hiPSCs. Four amino acids, glutamine, arginine, serine, and glycine were shown to be preferentially consumed in hiPSCs and we went on to demonstrate that the depletion of these four amino acids and glucose efficiently eliminated undifferentiated hiPSCs from mixed populations within 48 h. Glutamine is the most important amino acid for the survival of hiPSCs and the expression of the metabolic enzymes in the TCA cycle differs between hiPSCs and CMs. The interconversion enzymes ACO2 and IDH2, responsible for the conversion of citrate to α-ketoglutarate in the TCA cycle, are highly expressed in CMs [17]. From the result of metabolome analysis, it was confirmed that hPSCs take advantage of glutaminolysis in addition to glycolysis to drive tricarboxylic acid (TCA) cycle [17] A large-scale targeted

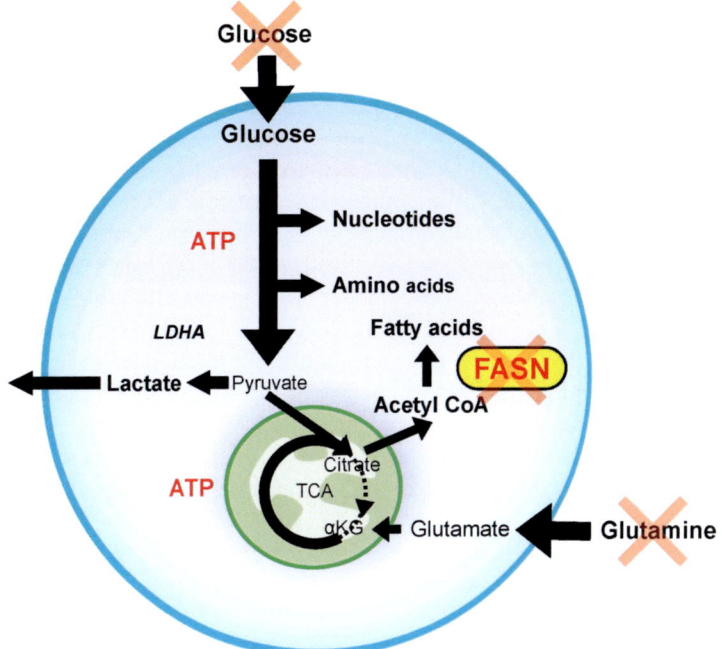

Fig. 2 **Metabolic purification of hiPSC-derived CMs.** Depletion of glucose, glutamine and the FASN inhibitor eliminates any residual hiPSCs in hiPSC-derived CMs

proteomic analysis revealed significant upregulation of *de novo* fatty acid synthesis in undifferentiated hPSCs compared to CMs. Fatty acids produced by *de novo* fatty acid synthesis is essential for survival and proliferation of hPSCs and inhibition of fatty acid synthase (FASN), one of the enzymes involved in *de novo* fatty acid synthesis, selectively induced apoptosis of undifferentiated hPSCs. Orlistat, an antiobesity drug approved by U.S. Food and Drug Administration, has a potent FASN inhibitory effect and hence induced selective elimination of undifferentiated hPSCs [18]. This information was then combined to allow for the "metabolic" purification of large amounts of purified hiPSC-derived CMs, potentially enabling the clinically relevant production of these therapeutic agents (Fig. 2).

3   Transplant Strategies for hiPSC-Derived Cardiomyocytes

The transplantation strategy for hiPSC-derived CMs is a key factor in the success of cell transplantation therapy. Intramyocardial transplantation is ideal for promoting mechanical and electrical coupling between transplanted and recipient CMs, as this

method brings both cell types into direct contact [19]. However, single CM engraftment rates tend to be reasonably low [20]. The use of aggregated CMs significantly
improves their engraftment ratio [21]. Given this we developed a novel type of
microwell plate designed to produce "cardiac spheroids" (CS) from aggregated
hiPSC-derived CMs. Each CS was made up of approximately 1000 CMs and had a
diameter of approximately 150–200 µm [22]. To ensure that the transplanted CMs
were retained more efficiently, we evaluated the use of a gelatin hydrogel during
cardiac transplantation. Gelatin hydrogels not only improved engraftment, but also
enhanced angiogenesis promoting the sustained release of cytokines [23]. Moreover,
a transplant injection device was developed to facilitate the efficient engraftment of
cardiac spheroids. This instrument uses special injection needles with a blind tip
and side holes, which distribute CS more uniformly throughout the myocardium
(Fig. 3). The injection angle was designed so as not to insert the tip of the needle
into the left ventricular cavity and the proportion of retained CMs significantly
improved when using this device to treat swine hearts [22]. Intramyocardial injection of hPSC-derived CMs improved cardiac function in both rodents and swine
[24, 25]. Transplantation of iPSC-derived CMs also improves cardiac function in
monkeys [26]. These preclinical experiments demonstrate the feasibility of direct
injection of hiPSC-derived CMs and suggest that this is ready for human clinical trial.

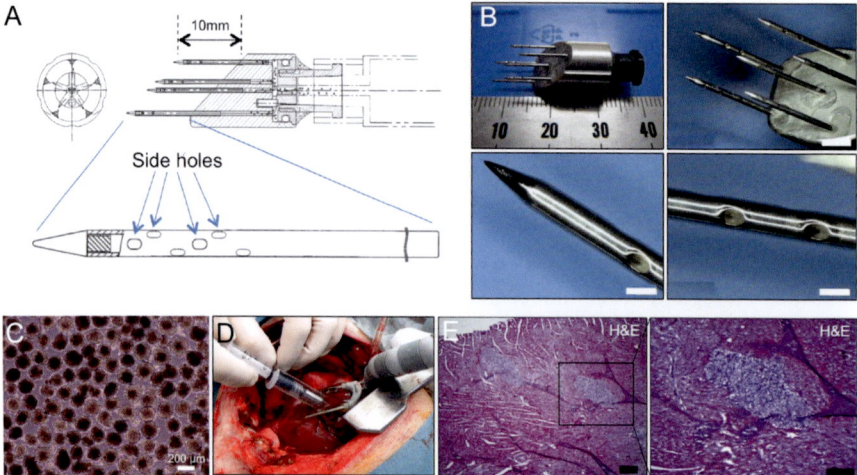

**Fig. 3 Translational approach for the clinical application of hiPSC-derived CMs using a
special injection device.** (**a**) *Design* of a special injection device created to facilitate the transplant
of hiPSC-derived cardiac spheroids. (**b**) This device has a blind edge and side holes. (**c**) Cardiac
spheroids were formed using hiPSC-derived CMs. (**d**) The left ventricular anterior wall was locked
by a tissue stabilizer, and tissue-marking dye was injected using this novel device. (**e**) Human
cardiac spheroids were shown to engraft in a swine heart 1 week after transplantation. Scale
bars = 500 µm for (B), 300 µm for (**e**, left), and 200 µm for (**e**, right). **h** & **e**: Hematoxylin and eosin
staining. Figures adapted from reference [22]

4 Clinical Protocols for Intramyocardial Transplantation of hiPSC-Derived CMs

Many patients with AHF are waiting for heart transplants in Japan. Severe donor shortages have hampered the increase in cardiac transplantation, with a waiting period of more than 4 years (http://www.jsht.jp/registry/japan/). Moreover, people older than 65 years of age are not eligible for cardiac transplant. Over 60% of the cardiac diseases treated with heart transplant in Japan are classified as dilated cardiomyopathies (DCM) [27]. Thus, it is reasonable to use DCM as our model disease. Clinical grade hiPSCs, which are free of transgenes and genomic mutations, are manufactured at the Center for iPS Cell Research and Application at Kyoto University. In order to avoid immunological rejection, the patient's human leukocyte antigen (HLA) is matched to that of the clinical hiPSCs, which incorporate the homozygous haplotypes of the most frequent HLAs in the Japanese population. Our transplantation protocol has been approved by the Health Science Council at the Ministry of Health, Labor, and Welfare of Japan and the recruitment of patients will begin soon.

5 Discussion

Remarkable progress in each step of the creation of novel cardiac regenerative therapies has been achieved in the last 10 years, but there are still many open questions. In particular, maturation, immune response, and arrhythmogenicity should all be resolved prior to intensive clinical application and should dominate the focus of most studies in the near future.

5.1 Maturation of Cardiomyocytes

It is generally agreed that hPSC-derived CMs present with a fetal phenotype. Their myofibrils are not aligned in a sarcomeric structure, and their electrical properties are immature following differentiation. Most of their sodium current properties are determined by the fetal isoform of the Nav1.5 α-subunit of the sodium channel (SCN5A), while the expression of its adult isoform increases with prolonged culture [28]. Rapid delayed-rectifier potassium currents (IKr) are dominant, and the inward-rectifier potassium current (IK1) is barely detected in hiPSC-derived CMs [29]. The transition from glycolysis to oxidative metabolism is also only associated with mature CMs. A lack of T-tubules, which is a special feature of mature muscle cells where they present with L-type calcium channels, delays the contraction of hiPSC-derived CMs. The sarcoplasmic reticulum, which acts in calcium-induced calcium release via the ryanodine receptor, is also immature [30]. Even though long-term culture, *in vivo* environments, mechanical stretch, and/or electrical stimuli may help to mature these CMs, they still fail to reach full maturity and the adult phenotype. Although it is unclear which, younger or adult, CMs are better for cardiac transplant, especially for engraftment, the maturation of CMs seems to be necessary in order to induce strong contraction-forces and the appropriate electrical activities.

5.2 Immunological Rejection

HLA-Matched iPSCs

Long-term engraftment and direct contribution to contraction force are both ideal outcomes for cell-based interventions in cardiac diseases. Immunological reactions between transplanted cells and recipients are key factors in long-term engraftment. HLA matching between patients and hiPSC-derived CMs is a rational strategy for avoiding immunological rejection and reducing the use of immunosuppressive therapies. In fact, matching of HLA-DRs can improve graft survival in heart transplantation [31]. Experiments with monkeys have shown that major histocompatibility complex (MHC)-matched CMs successfully engraft and improve cardiac function in these animal models [26]. A total of 140 unique HLA-A, -B, and -DR homozygous donors would cover 90% of the Japanese population. However, more than 160,000 individuals must be screened to identify many unique haplotypes [32]. Thus, the full coverage of all populations demands another strategy in the future.

Universal Pluripotent Stem Cells

Universal pluripotent stem cells, which can escape attack from the recipient's immune system, are the ultimate strategy for cell transplantation therapy. Knock-out of the key components of the HLA-class I and II (β2-microglobulin and Class II major histocompatibility complex transactivator) complexes enable hiPSC-derivatives to be less immunogenic [33, 34]. However, β2-microglobulin knock-out hESC derivatives enhance the activity of natural killer cells [35]. Various approaches have been developed to overcome the "missing self" using novel gene editing technologies with some showing that the forced expression of HLA-E can be effective to avoid the "missing self" phenomenon [36]. Overexpression of CD47 in hiPSC-derived endothelial cells and CMs can also facilitate escape from the NK cells [33] and the retention of HLA-C in combination with the depletion of HLA-A and B successfully enables innate immune evasion in recipients [37]. The use of universal hiPSCs is expected to enable patients to be free from immunosuppressive agents. Therefore, the clinical application of universal hiPSCs is anticipated.

5.3   Preventive Strategies for Tumor Formation

As previously stated, metabolic purification successfully eliminates any undifferentiated stem cells. However, there is still a small possibility that hiPSC transplant may promote tumor formation, because most experiments describing tumor formation were performed using immunodeficient mice, and there is no experimental data describing the long-term observation of these recipients. In the case of tumor formation, vaccination and immunotherapies may be useful to prevent and eliminate hiPSC-derived tumors. PSC-specific antigens have been researched as targets for cancer immunotherapy for many years. Vaccination of mice with hESCs was reported to suppress the growth of colon carcinomas [38] while murine iPSC vaccines expressing a TLR9 agonist (CpG) and adaptive T cells from murine iPSC-vaccinated mice successfully suppressed the growth of some types of cancer [39]. Based on these findings, an oncofetal antigen, glypican-3, can be selected as a target protein to eliminate undifferentiated hiPSCs ro prevent teratoma formation [40]. Vaccination with hiPSC-specific antigens may provide a firm guarantee for the safe transplantation of hiPSC-derived CMs.

5.4   Electrical Coupling and Arrhythmia

Addition of a fluorescent calcium sensor (G-CaMP) proved that transplanted hESC-CMs can synchronize to contract with the recipient's CMs [19]. Even though there is strong evidence supporting the claim that transplanted hiPSC-CMs can produce electrical couplings, providing support for the application of cell based therapies in AHF, translational studies have reported transient arrhythmogenicity after transplantation with hiPSC-derived CMs in both non-human primates and swine (Table 1) [25, 26, 41, 42]. Ventricular arrhythmia generally emerged within a few days of transplantation and peaked within 1 month. Interestingly, most of these abnormalities disappeared 1–2 months later. The mechanism underlying these arrhythmias is thought to be associated with the re-entrant circuits or automaticity. Electrophysiological studies in swine suggest that the mechanism underlying ventricular arrhythmia was focal automaticity, not macro-reentry [42]. Inflammation due to immunological rejection also triggers arrhythmia [25]. Immaturity of hPSC-derived CMs may also trigger ventricular arrhythmia. Disappearance of inflammation, maturity of hPSC-derived CMs, and/or construction of tight electrical coupling between transplanted CMs and recipient hearts may reduce the emergence of these arrhythmias. Patients with AHF are highly vulnerable to ventricular tachycardia. Thus, the elimination of even acute arrhythmias is critical for the safety of cardiac cell therapies.

Table 1 Arrhythmogenicity after transplantation of PSC-derived cardiomyocytes

	Animal	HF model	Number of cardiomyocytes	VT or AIVR	Approximate duration for ventricular arrhythmia after cell transplantation
Chong, J.J., et al. [41]	Monkey	MI (90 mins) & reperfusion	1×10^9 hESC-CMs	+	4 weeks
Shiba, Y., et al. [26]	Monkey	MI (3 hours) & reperfusion	4×10^8 monkey iPSC-CMs (MHC-match)	+	12 weeks (peak at 2 weeks)
Romagnuolo, R., et al. [42]	Swine	MI (90 mins) & reperfusion	1×10^9 hESC-CMs	+	3 weeks
Kawaguchi, S., et al. [25]	Swine	Cryoinjury	1×10^8 hiPSC-CMs	+	3 weeks

HF heart failure, *VT* ventricular tachycardia, *AIVR* accelerated idioventricular rhythm, *MI* myocardial infarction, *hESC-CM* human embryonic stem cell-derived cardiomyocytes, *hiPSC-CMs* human induced pluripotent stem cell-derived cardiomyocytes, *MHC* major histocompatibility complex

5.5   *Implementation in Cardiac Tissues*

3D cardiac tissue mimics, such as cardiac cell sheets, patches, and spheroids, are used during cell transplantation [25, 43, 44]. The transplantation of mature cardiac tissues with vascular structures may be a better strategy than transplanting pure CMs. Currently, engineered models of the human heart have primarily focused on the elucidation of heart disease and its mechanisms and drug screening as seen in heart-on-a-chip assays [45]. The technologies to form organoids, self-organizing miniature organs, are advanced in several organs and are also expected to help with tissue regeneration [46]. Recently, the successful construction of murine cardiac organoids, which mimic the formation of a four-chamber morphology during cardiac development, was reported [47]. However, the development of spontaneously differentiated human cardiac organoids is not as easy as other organs. Therefore, most experiments have been conducted using a mixture of cardiac tissues, endothelial cells and fibroblasts. In fact, micro-cardiac tissues with endothelial cells and fibroblasts enhance maturation and mimic the disease phenotype *in vitro* [48]. The selection of a specific phenotype in the cardiac tissues may be necessary for cardiac transplantation. In particular, ventricular cardiac tissues appear to be necessary to improve the ejection fraction of the left ventricle. The other issue is that controlling vascular cell growth is difficult using the currently available technologies. Quality control of transplanted tissues will be difficult and will need to include evaluations of the proportion of the cell varieties. Although transplantation strategies must be developed using the information available from engineered cardiac models, mature cardiac tissues may enhance cardiac function. Therefore, transplantation of implemented multicellular cardiac tissues may be a critical strategy for cardiac cell therapies in the future.

6 Conclusion

Cardiac regenerative therapies using hiPSCs are widely expected to improve the prognosis of patients with AHF. The continuing accumulation of scientific evidence will help to develop strategies and protocols to facilitate the successful clinical application of hiPSC-derived CMs.

Acknowledgements This work was supported by a Grant-in-Aid of Scientific Research from the Ministry of Education, Culture, Sports, Science, and Technology (19H03660 [to Dr. Fujita]). Drs. Fujita, Tohyama, Kanazawa, and Fukuda all retain patents related to this work. Drs. Fujita, Tohyama, Kanazawa, and Fukuda own equity in Heartseed, Inc. Dr. Tohyama is an advisor of Heartseed, Inc. and Dr. Fukuda is the co-founder and CEO of Heartseed, Inc., and receives a salary from Heartseed, Inc. All other authors have reported that they have no relationships relevant to the contents of this paper to disclose.

References

1. Goodlin, S.J.: Palliative care in congestive heart failure. J. Am. Coll. Cardiol. **54**(5), 386–396 (2009). https://doi.org/10.1016/j.jacc.2009.02.078
2. Khush, K.K., Cherikh, W.S., Chambers, D.C., Harhay, M.O., Hayes Jr., D., Hsich, E., et al.: The international thoracic organ transplant registry of the International Society for Heart and Lung Transplantation: thirty-sixth adult heart transplantation report - 2019; focus theme: donor and recipient size match. J. Heart Lung Transplant. **38**(10), 1056–1066 (2019). https://doi.org/10.1016/j.healun.2019.08.004
3. Miller, L., Birks, E., Guglin, M., Lamba, H., Frazier, O.H.: Use of ventricular assist devices and heart transplantation for advanced heart failure. Circ. Res. **124**(11), 1658–1678 (2019). https://doi.org/10.1161/CIRCRESAHA.119.313574
4. Kehat, I., Kenyagin-Karsenti, D., Snir, M., Segev, H., Amit, M., Gepstein, A., et al.: Human embryonic stem cells can differentiate into myocytes with structural and functional properties of cardiomyocytes. J. Clin. Invest. **108**(3), 407–414 (2001). https://doi.org/10.1172/JCI12131
5. Zhang, J., Wilson, G.F., Soerens, A.G., Koonce, C.H., Yu, J., Palecek, S.P., et al.: Functional cardiomyocytes derived from human induced pluripotent stem cells. Circ. Res. **104**(4), e30–e41 (2009). https://doi.org/10.1161/circresaha.108.192237
6. Takahashi, K., Tanabe, K., Ohnuki, M., Narita, M., Ichisaka, T., Tomoda, K., et al.: Induction of pluripotent stem cells from adult human fibroblasts by defined factors. Cell. **131**(5), 861–872 (2007). https://doi.org/10.1016/j.cell.2007.11.019
7. Mandai, M., Watanabe, A., Kurimoto, Y., Hirami, Y., Morinaga, C., Daimon, T., et al.: Autologous induced stem-cell-derived retinal cells for macular degeneration. N. Engl. J. Med. **376**(11), 1038–1046 (2017). https://doi.org/10.1056/NEJMoa1608368
8. Takahashi, J.: iPS cell-based therapy for Parkinson's disease: a Kyoto trial. Regen. Ther. **13**, 18–22 (2020). https://doi.org/10.1016/j.reth.2020.06.002
9. Umekage, M., Sato, Y., Takasu, N.: Overview: an iPS cell stock at CiRA. Inflamm. Regen. **39**, 17 (2019). https://doi.org/10.1186/s41232-019-0106-0
10. Kempf, H., Andree, B., Zweigerdt, R.: Large-scale production of human pluripotent stem cell derived cardiomyocytes. Adv. Drug Deliv. Rev. **96**, 18–30 (2016). https://doi.org/10.1016/j.addr.2015.11.016

11. Tohyama, S., Fujita, J., Fujita, C., Yamaguchi, M., Kanaami, S., Ohno, R., et al.: Efficient large-scale 2D culture system for human induced pluripotent stem cells and differentiated cardiomyocytes. Stem Cell Rep. **9**(5), 1406–1414 (2017). https://doi.org/10.1016/j.stemcr.2017.08.025

12. Terao, Y., Kurashina, Y., Tohyama, S., Fukuma, Y., Fukuda, K., Fujita, J., et al.: An effective detachment system for human induced pluripotent stem cells cultured on multilayered cultivation substrates using resonance vibrations. Sci. Rep. **9**(1), 15655 (2019). https://doi.org/10.1038/s41598-019-51944-w

13. Kasai, K., Tohyama, S., Suzuki, H., Tanosaki, S., Fukuda, K., Fujita, J., et al.: Cost-effective culture of human induced pluripotent stem cells using UV/ozone-modified culture plastics with reduction of cell-adhesive matrix coating. Mater. Sci. Eng. C Mater. Biol. Appl. **111**, 110788 (2020). https://doi.org/10.1016/j.msec.2020.110788

14. Someya, S., Tohyama, S., Kameda, K., Tanosaki, S., Morita, Y., Sasaki, K., et al.: Tryptophan metabolism regulates proliferative capacity of human pluripotent stem cells. iScience. **24**(2), 102090 (2021). https://doi.org/10.1016/j.isci.2021.102090

15. Buikema, J.W., Lee, S., Goodyer, W.R., Maas, R.G., Chirikian, O., Li, G., et al.: Wnt activation and reduced cell-cell contact synergistically induce massive expansion of functional human iPSC-derived cardiomyocytes. Cell Stem Cell. **27**(1), 50–63. e5 (2020). https://doi.org/10.1016/j.stem.2020.06.001

16. Tohyama, S., Hattori, F., Sano, M., Hishiki, T., Nagahata, Y., Matsuura, T., et al.: Distinct metabolic flow enables large-scale purification of mouse and human pluripotent stem cell-derived cardiomyocytes. Cell Stem Cell. **12**(1), 127–137 (2013). https://doi.org/10.1016/j.stem.2012.09.013

17. Tohyama, S., Fujita, J., Hishiki, T., Matsuura, T., Hattori, F., Ohno, R., et al.: Glutamine oxidation is indispensable for survival of human pluripotent stem cells. Cell Metab. **23**(4), 663–674 (2016). https://doi.org/10.1016/j.cmet.2016.03.001

18. Tanosaki, S., Tohyama, S., Fujita, J., Someya, S., Hishiki, T., Matsuura, T., et al.: Fatty acid synthesis is indispensable for survival of human pluripotent stem cells. iScience. **23**(9), 101535 (2020). https://doi.org/10.1016/j.isci.2020.101535

19. Shiba, Y., Fernandes, S., Zhu, W.Z., Filice, D., Muskheli, V., Kim, J., et al.: Human ES-cell-derived cardiomyocytes electrically couple and suppress arrhythmias in injured hearts. Nature. **489**(7415), 322–325 (2012). https://doi.org/10.1038/nature11317

20. Hattan, N., Kawaguchi, H., Ando, K., Kuwabara, E., Fujita, J., Murata, M., et al.: Purified cardiomyocytes from bone marrow mesenchymal stem cells produce stable intracardiac grafts in mice. Cardiovasc. Res. **65**(2), 334–344 (2005). https://doi.org/10.1016/j.cardiores.2004.10.004

21. Hattori, F., Chen, H., Yamashita, H., Tohyama, S., Satoh, Y.-S., Yuasa, S., et al.: Nongenetic method for purifying stem cell-derived cardiomyocytes. Nat. Meth. **7**(1), 61–66 (2010) http://www.nature.com/nmeth/journal/v7/n1/suppinfo/nmeth.1403_S1.html

22. Tabei, R., Kawaguchi, S., Kanazawa, H., Tohyama, S., Hirano, A., Handa, N., et al.: Development of a transplant injection device for optimal distribution and retention of human induced pluripotent stem cellderived cardiomyocytes. J. Heart Lung Transplant. **38**(2), 203–214 (2019). https://doi.org/10.1016/j.healun.2018.11.002

23. Nakajima, K., Fujita, J., Matsui, M., Tohyama, S., Tamura, N., Kanazawa, H., et al.: Gelatin hydrogel enhances the engraftment of transplanted cardiomyocytes and angiogenesis to ameliorate cardiac function after myocardial infarction. PLoS One. **10**(7), e0133308 (2015). https://doi.org/10.1371/journal.pone.0133308

24. Caspi, O., Huber, I., Kehat, I., Habib, M., Arbel, G., Gepstein, A., et al.: Transplantation of human embryonic stem cell-derived cardiomyocytes improves myocardial performance in infarcted rat hearts. J. Am. Coll. Cardiol. **50**(19), 1884–1893 (2007). https://doi.org/10.1016/j.jacc.2007.07.054

25. Kawaguchi, S., Soma, Y., Nakajima, K., Kanazawa, H., Tohyama, S., Tabei, R., et al.: Intramyocardial transplantation of human iPS cell-derived cardiac spheroids improves cardiac function in heart failure animals. JACC: Basic Trans. Sci. **6**(3), 239–254 (2021). https://doi.org/10.1016/j.jacbts.2020.11.017

26. Shiba, Y., Gomibuchi, T., Seto, T., Wada, Y., Ichimura, H., Tanaka, Y., et al.: Allogeneic transplantation of iPS cell-derived cardiomyocytes regenerates primate hearts. Nature. **538**(7625), 388–391 (2016). https://doi.org/10.1038/nature19815

27. Fukushima, N., Ono, M., Saiki, Y., Sawa, Y., Nunoda, S., Isobe, M.: Registry report on heart transplantation in Japan (June 2016). Circ. J. **81**(3), 298–303 (2017). https://doi.org/10.1253/circj.CJ-16-0976

28. Veerman, C.C., Mengarelli, I., Lodder, E.M., Kosmidis, G., Bellin, M., Zhang, M., et al.: Switch from fetal to adult SCN5A isoform in human induced pluripotent stem cell-derived cardiomyocytes unmasks the cellular phenotype of a conduction disease-causing mutation. J. Am. Heart Assoc. **6**(7) (2017). https://doi.org/10.1161/JAHA.116.005135

29. Doss, M.X., Di Diego, J.M., Goodrow, R.J., Wu, Y., Cordeiro, J.M., Nesterenko, V.V., et al.: Maximum diastolic potential of human induced pluripotent stem cell-derived cardiomyocytes depends critically on I(Kr). PLoS One. **7**(7), e40288 (2012). https://doi.org/10.1371/journal.pone.0040288

30. Liu, J., Fu, J.D., Siu, C.W., Li, R.A.: Functional sarcoplasmic reticulum for calcium handling of human embryonic stem cell-derived cardiomyocytes: insights for driven maturation. Stem Cells. **25**(12), 3038–3044 (2007). https://doi.org/10.1634/stemcells.2007-0549

31. Ansari, D., Bucin, D., Nilsson, J.: Human leukocyte antigen matching in heart transplantation: systematic review and meta-analysis. Transpl. Int. **27**(8), 793–804 (2014). https://doi.org/10.1111/tri.12335

32. Okita, K., Matsumura, Y., Sato, Y., Okada, A., Morizane, A., Okamoto, S., et al.: A more efficient method to generate integration-free human iPS cells. Nat. Methods. **8**(5), 409–412 (2011). https://doi.org/10.1038/nmeth.1591

33. Deuse, T., Hu, X., Gravina, A., Wang, D., Tediashvili, G., De, C., et al.: Hypoimmunogenic derivatives of induced pluripotent stem cells evade immune rejection in fully immunocompetent allogeneic recipients. Nat. Biotechnol. **37**(3), 252–258 (2019). https://doi.org/10.1038/s41587-019-0016-3

34. Mattapally, S., Pawlik, K.M., Fast, V.G., Zumaquero, E., Lund, F.E., Randall, T.D., et al.: Human leukocyte antigen class I and II knockout human induced pluripotent stem cell-derived cells: universal donor for cell therapy. J. Am. Heart Assoc. **7**(23), e010239 (2018). https://doi.org/10.1161/JAHA.118.010239

35. Wang, D., Quan, Y., Yan, Q., Morales, J.E., Wetsel, R.A.: Targeted disruption of the beta2-microglobulin gene minimizes the immunogenicity of human embryonic stem cells. Stem Cells Transl. Med. **4**(10), 1234–1245 (2015). https://doi.org/10.5966/sctm.2015-0049

36. Gornalusse, G.G., Hirata, R.K., Funk, S.E., Riolobos, L., Lopes, V.S., Manske, G., et al.: HLA-E-expressing pluripotent stem cells escape allogeneic responses and lysis by NK cells. Nat. Biotechnol. **35**(8), 765–772 (2017). https://doi.org/10.1038/nbt.3860

37. Xu, H., Wang, B., Ono, M., Kagita, A., Fujii, K., Sasakawa, N., et al.: Targeted disruption of HLA genes via CRISPR-Cas9 generates iPSCs with enhanced immune compatibility. Cell Stem Cell. **24**(4), 566–578. e7 (2019). https://doi.org/10.1016/j.stem.2019.02.005

38. Li, Y., Zeng, H., Xu, R.H., Liu, B., Li, Z.: Vaccination with human pluripotent stem cells generates a broad spectrum of immunological and clinical responses against colon cancer. Stem Cells. **27**(12), 3103–3111 (2009). https://doi.org/10.1002/stem.234

39. Kooreman, N.G., Kim, Y., de Almeida, P.E., Termglinchan, V., Diecke, S., Shao, N.Y., et al.: Autologous iPSC-based vaccines elicit anti-tumor responses in vivo. Cell Stem Cell. **22**(4), 501–513. e7 (2018). https://doi.org/10.1016/j.stem.2018.01.016

40. Okada, M., Tada, Y., Seki, T., Tohyama, S., Fujita, J., Suzuki, T., et al.: Selective elimination of undifferentiated human pluripotent stem cells using pluripotent state-specific immunogenic antigen Glypican-3. Biochem. Biophys. Res. Commun. **511**(3), 711–717 (2019). https://doi.org/10.1016/j.bbrc.2019.02.094

41. Chong, J.J., Yang, X., Don, C.W., Minami, E., Liu, Y.W., Weyers, J.J., et al.: Human embryonic-stem-cell-derived cardiomyocytes regenerate non-human primate hearts. Nature. **510**(7504), 273–277 (2014). https://doi.org/10.1038/nature13233

42. Romagnuolo, R., Masoudpour, H., Porta-Sanchez, A., Qiang, B., Barry, J., Laskary, A., et al.: Human embryonic stem cell-derived cardiomyocytes regenerate the infarcted pig heart but induce ventricular Tachyarrhythmias. Stem Cell Rep. **12**(5), 967–981 (2019). https://doi.org/10.1016/j.stemcr.2019.04.005

43. Kawamura, M., Miyagawa, S., Miki, K., Saito, A., Fukushima, S., Higuchi, T., et al.: Feasibility, safety, and therapeutic efficacy of human induced pluripotent stem cell-derived cardiomyocyte sheets in a porcine ischemic cardiomyopathy model. Circulation. **126**(11 Suppl 1), S29–S37 (2012). https://doi.org/10.1161/CIRCULATIONAHA.111.084343

44. Ye, L., Chang, Y.H., Xiong, Q., Zhang, P., Zhang, L., Somasundaram, P., et al.: Cardiac repair in a porcine model of acute myocardial infarction with human induced pluripotent stem cell-derived cardiovascular cells. Cell Stem Cell. **15**(6), 750–761 (2014). https://doi.org/10.1016/j.stem.2014.11.009

45. Stein, J.M., Mummery, C.L., Bellin, M.: Engineered models of the human heart: directions and challenges. Stem Cell Rep. (2020). https://doi.org/10.1016/j.stemcr.2020.11.013

46. Hofer, M., Lutolf, M.P.: Engineering organoids. Nat. Rev. Mater., 1–19 (2021). https://doi.org/10.1038/s41578-021-00279-y

47. Lee, J., Sutani, A., Kaneko, R., Takeuchi, J., Sasano, T., Kohda, T., et al.: In vitro generation of functional murine heart organoids via FGF4 and extracellular matrix. Nat. Commun. **11**(1), 4283 (2020). https://doi.org/10.1038/s41467-020-18031-5

48. Giacomelli, E., Meraviglia, V., Campostrini, G., Cochrane, A., Cao, X., van Helden, R.W.J., et al.: Human-iPSC-derived cardiac stromal cells enhance maturation in 3D cardiac microtissues and reveal non-cardiomyocyte contributions to heart disease. Cell Stem Cell. **26**(6), 862–879. e11 (2020). https://doi.org/10.1016/j.stem.2020.05.004

Cell Therapy with Human ESC-Derived Cardiac Cells: Clinical Perspectives

Philippe Menasché

1 Why Pluripotent Stem Cells?

1.1 Rationale for the Use of Cardiac Committed Cells

So far, it is fair to acknowledge that no one can claim that a given cell type has unequivocally demonstrated its superiority over another for inducing heart repair and the attendant improvement in left ventricular function. Nevertheless, the few comparative experimental studies which have been published suggest the interest of using cells which are phenotypically matched to those of the organ to be repaired and thus highlight the potential benefits of cardiac-committed cells. A first comparison of human induced pluripotent stem cell (iPSC)-derived cardiomyocytes versus human mesenchymal stromal cells (MSC) injected 30 minutes after permanent coronary artery ligation in rats reported that the former tended to improve left ventricular function to a greater extent and significantly reduced fibrosis compared with MSC [20]. In a subsequent head-to-head comparison of different human stem cell types, cardiosphere-derived cells were found superior to MSC from bone marrow and adipose tissue and to bone marrow-derived mononuclear cells (BMMNCs) for improving post-infarction cardiac function and cell engraftment in a mouse model also treated just after ligation of the left anterior descending artery [46]. Likewise, human embryonic stem cell (ESC)-derived cardiomyocytes and mesodermal cardiovascular progenitors were found to similarly improve post-infarction systolic function in contrast to BMMNCs in rat hearts treated 4 days after a 60-minute

P. Menasché (✉)
Department of Cardiovascular Surgery, Hôpital Européen Georges Pompidou 20, Paris, France

University of Paris, PARCC, INSERM, F-75015, Paris, France
e-mail: philippe.menasche@aphp.fr

period of coronary artery occlusion [25]. In keeping with these data, the more recent pig study of Ishida et al. [39] shows that in comparison of skeletal myoblasts and MSC, iPSC-derived cardiomyocytes (all cells being of human origin) yielded the best outcomes with regard to regional function, oxidative metabolism, vascular density and limitation of apoptosis. However, an opposite conclusion emerged from another comparative study of iPS derivatives where MSC were found superior to cardiomyocytes for the improvement of cardiac function in heart failure, an effect primarily attributed to their immunomodulatory effects [47]. This discrepancy could be related to very specific features of this protocol: the induction of heart failure by coronary artery occlusion associated with rapid pacing, a setting known for its instability and the intensity of neuro-hormonal stimulation [100]; the use of different PSC sources (ESC for cardiomyocytes and iPSC for MSC); the application of a uniform immunosuppression regimen to MSC and cardiomyocytes whereas the former are known to be more immune evasive, which was indeed reflected in that study by their lower expression of Human Leukocyte Antigen (HLA)-II; the greater susceptibility of ESC-derived cardiomyocytes to an immune response could thus have accounted for their lower rate of survival and their (slightly) inferior functional performance. In brief, these data are difficult to interpret because of all these confounders and do not challenge the concept that an optimal therapeutic benefit likely requires that both transplanted and host cells belong to the same lineage.

In practice, adult tissue sources of cardiac cells are limited to cardiospheres, which are agglomerates of several cell types, predominantly MSC, harvested from the right ventricle by an endomyocardial biopsy [80] with a subsequent intracoronary delivery, and c-*kit*+ cardiac stem cells (CSC), typically grown from a right appendage biopsy taken during a coronary artery bypass operation before being also reinjected into the coronary arteries. However, the first have failed to show benefits in an ischemic cardiomyopathy phase II trial which was prematurely interrupted in April, 2017 for futility; however, the use of these cells in Duchenne muscular dystrophy has yielded an encouraging efficacy signal [84] which needs to be confirmed by the ongoing HOPE-II study, planned to randomize 84 non-ambulatory and ambulatory patients with Duchenne muscular dystrophy to intravenous infusions of either 150 million cardiosphere-derived cells every 3 months for a total of 4 doses or placebo. The use of c-*kit*+ cardiac stem cells has been largely based on preclinical data now considered to be in part fraudulent ('News at a glance', [61]) and there is some consensus that these cells are rather endowed with an angiogenic potential [88]. Four trials are currently registered in the ClinicalTrials.gov website, of which one (CONCERT-HF) which has tested transendocardial injections of autologous MSC, c-*kit*+ CSC, alone or in combination, in patients with ischemic heart failure has yielded mixed results (improved clinical outcomes after cell therapy contrasting with the absence of significant between-group differences in functional indices) [13]. Among the remaining 3 studies, only one, the CHILD trial, which assesses the effects of intramyocardial injection of autologous c-*kit*+ cells in 32 patients with hypoplastic left heart syndrome, is currently recruiting. A second one (TAC-HF-II) planned to test transendocardial injection of autologous MSC alone or in combination with CSC in 55 patients with ischemic left ventricular dysfunction, has not yet started recruiting. The third trial (JOKER) was designed to assess the effects an intracoronary infusion of autologous c-*kit*+ cells expanded from a right appendage biopsy taken during coronary surgery in a small group of 6 patients still presenting a left ventricular ejection fraction <40% after their revascularization; it is reported active but not recruiting. Outcomes of these trials will hopefully shed some light on the place of these cells in the context of cell-based heart repair.

1.2   Interest of PSC as a Source of Cardiac Committed Cells

Given the limitations of these adult sources of cardiac-committed cells, it has looked sound to consider the alternate use of pluripotent stem cells (PSC) to leverage their intrinsic ability to generate lineage-specific cells in response to appropriate cues and thus coax them towards a cardiac differentiation pathway. Since the seminal work of Caspi and coworkers [14] showing that ESC-derived cardiomyocytes improved myocardial performance in chronically infarcted rats, a flurry of studies have confirmed the functional efficacy of PSC cardiac derivatives. This has been a strong incentive for developing multiple and various techniques for PSC scale-up and differentiation, the detailed description of which is beyond the scope of this article (for a recent review, see [44]). Enough is to say that the extensive documentation on the raw products required for a regulatory approval makes highly desirable to use the most straightforward and cost-effective procedures, which has guided our choice of an only two cytokine-based technique described below. Another advantage of PSC is their scalability which is critical particularly if the objective is the "remuscularization" of extensively scarred post-infarction myocardial areas (as discussed below). One could argue that this property is shared by MSC but it is not totally true as there is some evidence that increasing the number of MSC passages can shift their phenotype towards an ageing pattern translating into an impaired functionality of the cells [103].

The ability to control the differentiation pathway of PSC gives the flexibility of "freezing" it at the desired stage and thus provides the option of transplanting early progenitor cells or more mature cardiomyocytes. Each of these cell types has advantages and disadvantages. Early progenitors feature a greater plasticity which could allow them to differentiate in both cardiac and vascular cells and their predominant reliance on anaerobic glycolysis might enhance their survival in a poorly oxygenated environment. This assumption is supported by the findings of Halbach et al. [33] that the highest persistence and grade of electrical coupling of intramyocardially transplanted fetal cardiomyocytes from different developmental stages is achieved by intermediate cells (days 14.5) compared with earlier and later stages (days 9.5 and 18.5, respectively). An additional feature of early progenitor cells is their greater secretory profile [2, 23], which can be an advantage if one relies on a predominant paracrine mechanism of action (see below). At the opposite, an early progenitor cell population can still be "contaminated" by pacemaker cells behaving as foci of automaticity and thus predisposing to arrhythmias. Conversely, this issue is addressed by the use of more mature cardiomyocytes which are thought to be endowed with a greater force-generating potential but, in turn, may be more susceptible to death once transplanted in hypovascularized areas because of their reliance on oxidative phosphorylation. At the end, it is fair to admit that currently the few studies which have compared different stage-specific cardiac differentiated cells have failed to provide unequivocal evidence for the superiority of one type over the other [25, 101]. As a sort of trade-off, an innovative approach has tested a switchable system of *in vivo* differentiation by using ESC-derived cardiovascular progenitor cells containing an Nkx2.5$^{eGFP/w}$ reporter in which a doxycycline (DOX)-inducible MYC (Tet-On-MYC) construct had been inserted. The cells were injected into the border zone of the mouse infarcted myocardium and first expanded *in vivo* using DOX (in the drinking water) and basic Fibroblast Growth Factor (bFGF) given subcutaneously. Subsequently, removal of DOX along with intraperitoneal injections of the WNT pathway inhibitor XAV939 and decreasing bFGF concentrations promoted full differentiation of the progenitors into cardiomyocytes. However, although this protocol allowed to increase graft size and reduce fibrosis, it failed to improve cardiac function, as assessed by magnetic resonance imaging after 90 days [75].

1.3   The ESCORT Trial

Our choice has been to transplant early progenitors differentiated from ESC in the ESCORT trial (NCT02057900) which was a first-in-man safety study of 6 patients with severe ischemic left ventricular dysfunction (median left ventricular ejection fraction: 26%; IQR: 22% to 32%) and in whom the cell therapy treatment was combined with a coronary artery bypass [55]. The trial was based on 10 years of preclinical studies in rodents and nonhuman primates [9, 54]. Pluripotent ESC were first scaled-up to generate a master/working cell bank from which cells were collected, thawed, expanded in a defined medium on clinical-grade feeder cells (irradiated human foreskin fibroblasts) and then committed towards a mesodermal-cardiac lineage by a 4-day exposure to two cytokines (bone morphogenetic protein-2 and a FGF inhibitor) according to Good Manufacturing Practice (GMP) standards. As roughly 44% of the cells only responded to the cardio-instructive cues, a purification step was mandatory. The expression of stage-specific embryonic antigen (SSEA)-1 was then taken as a marker for loss of pluripotency and immunomagnetic sorting using a microbead-coupled anti-SSEA-1 antibody was used for selecting the committed cells. This resulted in the harvest of a highly purified population of SSEA-1$^+$ cells characterized by a knock-down of the pluripotency gene *Nanog* and a parallel upregulation of the cardiac transcription factor *Isl-1*. These two markers, assessed by qPCR, were among the release criteria, with thresholds set as <0.1% and > 5%, respectively, expressed as fold changes relative to the undifferentiated population. Additional lot release criteria included cell viability and purity, with thresholds set at 90% and 95%, respectively, which were fully met for all the patients (median viability and purity rates of 96% [IQR: 96% to 96%] and 97.5% [IQR:

95.5% to 98.7%], respectively). SSEA-1⁺ progenitor cells were then embedded into a fibrin patch generated by first mixing them with fibrinogen followed by addition of thrombin to induce rapid polymerization of the gel (fibrinogen and thrombin were components of a clinically used surgical glue). Finally, during the surgical procedure, a 20 cm² piece of autologous pericardium was harvested and sutured to the epicardium along one-half the borders of the infarct area, thereby creating a pocket into which the cell-loaded fibrin patch was simply slipped. The free edge of the pericardial flap was then stitched to the remaining half of the infarct circumference, thereby enclosing the fibrin patch and ensuring its stability. The use of the pericardium was based on the assumption that it could act as a natural bioreactor providing growth factors to the underlying cellular graft and thus contributing to enhance early cell survival (Figs. 1 and 2). A median dose of SSEA-1⁺ *Isl*-1 cardiovascular progenitors of 8.2 million (IQR: 5–10 million) was delivered without any adverse intraoperative events. The primary end point was safety at 1 year, primarily assessed on (1) cardiac teratoma or remote tumor tracked by whole body computed tomography and fluorine-18 deoxyglucose positron emission tomography scans, (2)

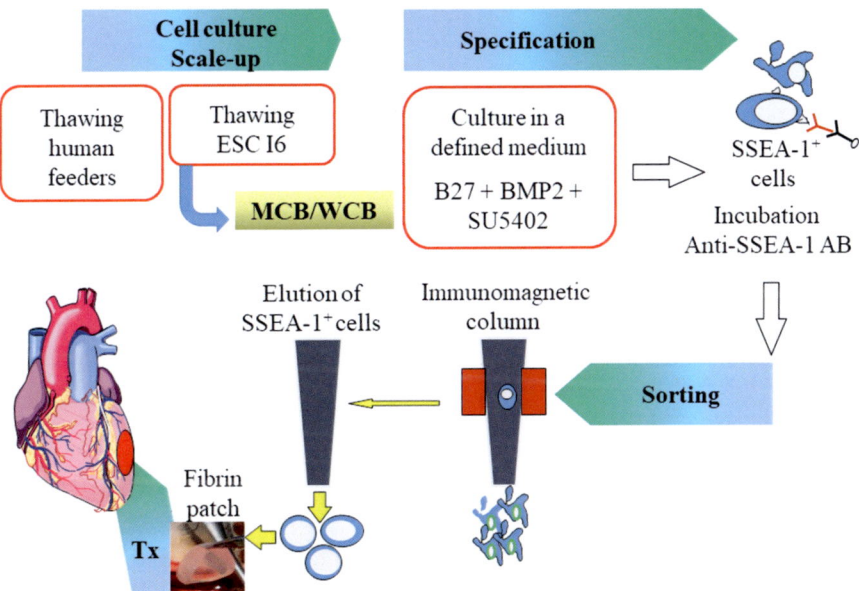

Fig. 1 Summary of the protocol in the ESCORT trial. Human Embryonic Stem Cells (ESC) from the I6 cell line were expanded on human feeders to generate a Master/Working Cell Bank (MCB/WCB). Expanded pluripotent stem cells (scale-up) were then cardiac-committed (specification) by a 4-day exposure to Bone Morphogenetic Protein (BMP)-2 and a Fibroblast Growth Factor inhibitor (SU5402) in B27 medium. Committed cells express the Stage-Specific Embryonic Antigen (SSEA)-1 indicating their loss of pluripotency and could thus be immune-magnetically sorted using an anti SSEA-1 antibody. The SSEA-1 enriched cardiovascular progenitor cell population was then embedded in a fibrin patch which was transplanted onto the epicardium of the infarct area. Tx: transplantation

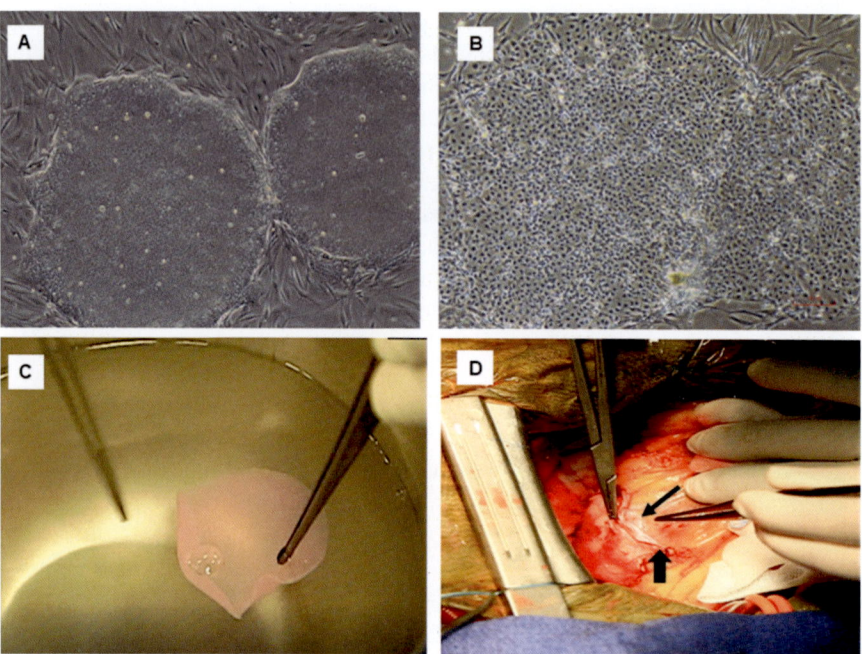

Fig. 2 Main steps of the procedure in the ESCORT trial. (**a**) Pluripotent ESC of the I6 cell line. (**b**) Cardiovascular progenitors at the completion of the 4-day specification step. (**c**) Fibrin patch loaded with the cardiovascular progenitors (intra-operative picture showing the rinsing of the patch before its implantation in the patient). (**d**) Final step: the cell-loaded patch has been delivered onto the epicardium of the infarct area and is partly covered by a pericardial flap already sutured along one-half the infarct circumference, thereby creating a pocket (between the flap and the epicardium) inside which the patch has been slid; the long and thin arrow indicates the border of the patch. The short and wider arrow indicates the suture line of the pericardial flap to the epicardium. Once the cell-loaded fibrin patch seats within the pocket, this suture line will be completed along the remaining one-half of the infarct circumference to enclose it completely, thereby ensuring its stability while providing some trophic support

arrhythmias, detected by serial interrogations of the cardioverter-defibrillator implanted in all patients, and (3) alloimmunization, assessed by the presence of donor-specific antibodies. All patients had an uneventful post-operative course, except for one who died shortly after the operation from multiple comorbidities. With a follow-up currently ranging from 4 and one-half to 7 years, no patient presented an adverse event that could be related to the cells and/or the patch. While it would be meaningless to draw conclusions regarding efficacy, enough is to say that an encouraging signal was provided by a significant improvement of the wall motion of the cell/patch-treated segments during follow-up with a score that decreased from 4.2 ± 0.8 at baseline to 2.5 ± 0.4 at 1 year ($p = 0.004$ by the mixed model ANOVA on ranks). Of note, in 3 of the 4 patients who contributed these 1-year data, the treated segments had not been revascularized. Since we were not expecting a long-term cellular engraftment and primarily relied on a paracrine mechanism of action (see below), patients were only immunosuppressed transiently and while the initial planning was to give the drugs for 2 months, the duration was shortened to 1 month from the second patient onward. Drugs were given at a relatively low dosing (target trough levels of cyclosporine: 100 to150 ng/ml; mycophenolate mofetil, 2 g/day) since our pre-operative mixed lymphocyte reaction assays had shown that SSEA-1$^+$ cells are weakly immunogenic.

1.4 Other PSC Clinical Trials

Other investigators have made the different choice of transplanting PSC-derived cardiomyocytes at a later stage of differentiation (although their persistent fetal-like phenotype precludes their assimilation to *bona fide* myocardium-resident cardiomyocytes) and have switched to iPSC as the source cells for practicality and/or ethical reasons. Once differentiated, iPSC-derived cardiomyocytes share with ESC the ability to improve the function of infarcted hearts [45] but also the lack of long term engraftment [63]. The use of iPSC has been aggressively promoted by those who oppose ESC for religious reasons with the premise that they could be differentiated from the patient's own somatic cells, thereby obviating the use of immunosuppression. This argument is no longer tenable since there is a consensus that iPSC for clinical purposes should rather be harvested from healthy donors, i.e., in an allogeneic mode, to improve safety and potency and decrease costs. This is actually the case for two trials in patients with ischemic heart failure: one in Japan (jRCT2053190081) aims at grafting *allogeneic* iPSC-derived cardiomyocytes in 10 patients with an ejection fraction ≤35% under the form of cell sheets prepared according to a well-documented technology [32] and as a stand-alone procedure. Its 3-patient roll-in phase is now listed as terminated (UMIN000032989). The second trial (NCT04982081), in China, should include 20 patients with congestive heart failure and ejection fractions less than 40% in whom 100 or 400 million *allogeneic* iPSC-derived cardiomyocytes will be delivered by an endocardial catheter. This study is currently recruiting. A third trial planned to include 3 patients with chronic heart failure and in whom *autologous* iPSC-derived cardiomyocytes would be delivered intravenously has been registered in ClinicalTrials.gov in November 2018 (NCT03759405) but has surprisingly not yet started recruiting. Another study has started early this year in Germany, in which the plan is to graft, as a stand-alone procedure, engineered heart constructs [86] made of iPSC-derived cardiomyocytes and collagen in patients with end-stage heart failure.

1.5 The Issue of Dosing

A critically relevant issue raised by transplantation of PSC-derived cardiomyocytes, regardless of their ESC or iPSC source, is their optimal dosing. Several factors have to be taken into account, including the mass of lost myocardium, particularly if the objective is its remuscularization (as discussed below), the rate of cell attrition and the possible proliferation of the surviving cells although the latter factor is of unlikely clinical relevance because of the consistent finding of the absence of sustained cell engraftment. In a meta-analysis of stem cells at large in large animal studies (primarily performed in pigs), cell doses were extremely variable, ranging from 5×10^5 to 25×10^6 for c-kit CSC, from 1.3×10^6 to 1×10^7 for cardiospheres and also averaged 1×10^7 for Sca-1+ CSC. Interestingly, in the only dose-escalating study of cardiospheres [102], the highest dose (150 million) was found superior to the lower ones when the assessment was made 4 weeks after treatment but the benefit on global and regional left ventricular function (versus a control group) was lost in a subsequent cohort receiving this high dose but assessed at a later time point (8 weeks). In a Phase II dose-escalation study of allogeneic mesenchymal precursor cells in patients with ischemic or nonischemic heart failure [66], the best outcomes were yielded by the highest dose (150 million versus 25 and 75 million) and a similar dose-dependent response (better with 100 versus 20 million cells) has been reported in the TRIDENT trial in which allogeneic bone marrow-derived-MSC were delivered in patients with ischemic cardiomyopathy [26]. These data are consistent with those of a meta-analysis of 914 MSC clinical trials (for all types of indications) in which the minimal effective dose ranged between 100 and 150 million cells [40]. Admittedly, however, none of these studies have used PSC derivatives. Looking exclusively at those transplanted in nonhuman primate models generates conflicting results. Chong et al. [18] reported an extensive remuscularization of the infarcted myocardium injected with 1×10^9 ESC-derived cardiomyocytes but the small sample size precluded any meaningful functional assessment. Conversely, despite injecting a similar cell number, Romagnuolo et al. [71] failed to show a benefit on infarct size or global left ventricular function (with no correlation between cardiomyocyte purity and graft size) after 4 weeks. In contrast, Liu et al. [50] reported a functional benefit following transplantation of a slightly lower dose (7.5×10^8) of ESC-derived cardiomyocytes and in the study of Shiba et al. [78], a post-transplantation improvement of contractile function at 4 and 12 weeks could even be yielded by an almost two-fold lower dose dose (4×10^8 iPSC-derived cardiomyocytes). Finally, in another study using ESC-derived cardiovascular progenitors, a much lower dose (1×10^7) also improved heart function provided the immunosuppression regimen was appropriate [108]. Put together, these data suggest that the highest doses are not necessarily the most effective, particularly in view of the risk of arrhythmias outlined in the next section, and although the issue of the optimal dosing is still open, they tend to support the earlier experimental findings that even though increasing the dose of ESC translates into an increased graft size, this may not be reflected by an improvement in heart function [90].

2 How Do PSC-Derived Cardiac Cells Work?

2.1 The "Remuscularization" Hypothesis

A logical objective of using PSC-derived cardiac cells is to structurally replace those which have been irreversibly damaged and thus to generate a neo-myocardium which can contribute to improve pump function. This is indeed the basis of the "remuscularization" concept which, despite being intellectually sound and attractive, raises two major clinically relevant challenges: maintenance of long term cell survival and arrhythmias.

Cell Survival

If one targets the generation of a new myocardial tissue, it is expected that the grafted cells will remain engrafted and alive over time. Unfortunately, cells tend to die rapidly following transplantation, which has led to the investigation of multiple empowering strategies, primarily based on chemical or physical (heat shock) pre-conditioning and genetic engineering [57], but these approaches have been challenging to translate clinically, most likely because of their complexity, cost and difficulties to meet regulatory standards. In practice, three main factors of cell death have been identified. The first is the loss of cell anchorage to an extracellular matrix which occurs at the time of their usual dissociation before injection; this can be addressed by incorporation of the cells in a biomaterial which has a dual interest: (1) it provides a three-dimensional template conducive to extracellular matrix secretion enhancing cell cohesiveness, and (2) it acts as a shielding structure that increases cell retention. It is beyond the scope of this article to discuss the choice of biomaterials. Suffice is to say that if they are considered for catheter-based delivery, they should feature shear-thinning properties allowing their injection followed by an *in situ* gelation to improve retention of the matrix-encapsulated cells [5] whereas they should have mechanical properties allowing to generate an easily manipulable patch if they are poised to an intraoperative epicardial application, as in the ESCORT trial and others. A second factor of cell death is the hostile nature of the environment they are implanted in, with a mix of inflammatory, hypovascularized and scarred areas which makes challenging for the transplanted cells to survive. One way of addressing this issue is to co-transplant them with supportive vascular and/or stromal cells to provide trophic and structural support. The critical role of non cardiac cells (at a roughly 30% ratio) for increasing the therapeutic potential of iPSC-derived cardiomyocytes has been established [38] and further confirmed recently by the ability of ESC-derived epicardial cells co-transplanted with ESC-derived cardiomyocytes to enhance cardiac graft size and heart function in athymic rats treated 4 days after a 60-minute coronary artery ligation [7]. The study by Gao et al. [27] is of even greater clinical relevance in that it used a large animal (porcine) model to establish the functional benefits of a composite patch construct made of iPSC-derived cardiomyocytes, endothelial and smooth muscle cells. In the clinics, these co-transplantation are technically doable as exemplified by the CONCERT-HF and the Japanese iPS sheet trials but they may complicate the cell manufacturing process. A third, and possible, the most challenging cause of cell death, is rejection of these allogeneic PSC derivatives. Currently, only drug-based immunosuppression is used to prevent rejection but despite the persisting uncertainties regarding the optimal drug regimen, it is admitted that these drugs are fraught with side-effects which may lead to their discontinuation with an attendant loss of the grafted cells [31]. A variant of this approach could be a "biological" immunomodulation by co-transplantation of MSC which have been shown, at least in a subcutaneous implantation model, to control allogeneic iPSC-CM rejection via regulatory T cells and cell-cell contact with activated lymphocytes [104]; however, because MSC are immune evasive and not immune privileged, they can only extend the survival of the co-transplanted iPSC-derived cardiomyocytes without preventing their ultimate disappearance [104].

Arrhythmias

A major concern raised by nonhuman primate studies in which ESC-derived cardiomyocytes have been transplanted has been the occurrence of ventricular arrhythmias [18, 19, 50, 71, 78], some of which were life-threatening. Different mechanisms have been hypothesized but rather then re-entry induced by tracks of slow conduction, it seems that focal activation at the graft/host interface is a key trigger of these events [50, 71]. Of note, arrhythmias have usually been observed early, i.e., during the first post-transplantation weeks with a progressive diminution onwards which could reflect a progressive *in situ* maturation of the graft towards a ventricular-like pattern. This supports the transplantation of an homogeneous population of cells featuring a mature electrophysiological phenotype but also emphasizes the importance of ensuring that the graft is appropriately purged from all cells which could still feature a pace-maker phenotype.

2.2 The Paracrine Hypothesis

Aside from "remuscularization", a second mechanism whereby PSC-derived cardiac cells could act is paracrine signalling, i.e., the release of factors harnessing endogenous repair pathways [28].

Evidence for a Paracrine Mechanism of Action

This hypothesis is strongly supported by three lines of reasoning. First, there is a consistent temporal discrepancy between the physical presence of cells in the transplanted tissue and the functional outcomes, i.e. an improvement in heart function is commonly demonstrated at a time where all grafted cells have disappeared. This applies to PSC cardiac derivatives as well. Thus, Riegler et al. [69] found no difference in cardiac function between chronically infarcted rats transplanted with human ESC-derived cardiomyocytes mixed with collagen and those in which cells had previously been made nonviable by irradiation, thereby suggesting that cell engraftment was not directly responsible for functional improvements. Likewise, differences in ESC graft size have been shown not to translate into differences in the preservation of post-infarction heart function which further supports the idea that remuscularization may not be the key driver of the cell-associated therapeutic benefit [52, 90]. More recently, ESC-derived cardiovascular progenitor cells were reported to improve function of infarcted and adequately immunosuppressed non-human primates at a time where cells are no longer detectable [108]. Admittedly, another macaque study has reported the persistence of substantial grafts until 3 months after transplantation, possibly because of the use of more differentiated cardiomyocytes, but the scarcity of their connexin-43 expression still makes uncertain a functionally effective coupling with host cardiomyocytes [49] and this concern is strengthened by the finding of scar tissue isolating ESC- and iPSC-cardiomyocyte grafts from the host myocardium of infarcted rats, which makes unlikely their direct contribution to cardiac contractility [45, 90]. Second, cells are known to release a myriad of cytokines, growth factors and other biologics, many of them being packaged in extracellular vesicles (exosomes and microparticles) which can modulate the function of recipient cells through the delivery of their cargo [65] and this mechanism also pertains to vesicles released by PSC [105]. Extracellular vesicles can then shift recipient cell signalling pathways towards cardiac repair through multiple mechanisms, primarily mitigation of apoptosis, inflammation and fibrosis and stimulation of angiogenesis [29, 107], while the re-induction of a mitotic cycle of native cardiomyocytes leading to an increased contractile cell pool remains more controversial. Third, the extracellular vesicle-enriched secretome of stem cell-derived cardiac cells (or of MSC) recapitulates (and even sometimes outweighs) the beneficial effects of their parental cells, an observation which has now been made across a wide variety of preclinical cardiac [1, 23, 41, 70] and noncardiac disease models (reviewed in [22]) and is consistent with the finding of an overlap of the microRNA profiles between iPSC-cardiomyocytes [73] or MSC [76] and their respective exosomal content.

Relying on a paracrine mechanism of action implies that cells behave as platforms releasing bioactive molecules and will remain only transiently in the grafted tissue; however, as nobody still knows exactly how long is enough for the cells to release the factors underpinning their therapeutic benefits, the use of adjunctive biomaterials to extend their residency time can still be justified [15, 34, 94]. Importantly, rejection does not need to be prevented any longer; it should only be delayed, which implies a short-duration of immunosuppression and consequently reduces the risk of drug-induced side effects. We previously mentioned that in the ESCORT trial, immunosuppressive drugs were only given for 1 month with good tolerance and the Japanese trial entailing the use of iPSC-derived cardiomyocyte cell sheets has likewise planned a limited period (3 months) of immunosuppression.

Use of the PSC Secretome

Advantages

The assumption that most, if not all, of the cardioprotective effects of stem cells can be duplicated by the exclusive use of their secretome has logically led to consider the latter as the only therapeutics that could be delivered. This approach has distinct clinically relevant advantages, including (1) a standardized pharma-like manufacturing process, (2) a likely lack of immunogenicity, provided the cells of origin themselves express little, if any, of the Human Leukocyte Antigens (HLA) I and II, which is the case for iPSC-derived cardiovascular progenitors [48], and (3) a functional stability under cryo-storage compatible with an off-the-shelf availability.

Challenges

However, the translational use of the secretome requires to address at least four main challenges, the first of which is the choice of the parental cells. Although the RNA and protein cargo of undifferentiated ESC [42] and iPSC [12] has shown cardio-protective effects both *in vitro* and in animal models of myocardial infarction, the clinical use of such a secretome would likely raise safety issues because of its possible enrichment with pluripotency markers [68], which rather leads to privilege more differentiated cells as the source material. In this context, and in parallel to what has been shown for cells, the bioactivity of the secretome of cardiac-committed cells outweighs that of noncardiac cells [8], possibly because the released factors are better tailored to the specific biology of the target cardiac tissue; even within the cardiac cell population, the secretome released by early progenitor cells seems to feature superior cardioprotective properties [2, 43] and these data are in line with our observations [23] and those of others [53] that untreated adult cardiomyocytes secrete minimal amounts of exosomes. These findings strongly rationalize the choice of PSC as the parental cells because of the dual possibility of differentiating them into cardiac-committed cells and stopping the differentiation process at an early developmental stage. For this reason, our next clinical trial will entail the administration of an extracellular vesicle-enriched secretome of iPSC-derived cardiovascular progenitor cells [23]. A second challenge is the mode of delivery of the secretome. If direct intramyocardial injections are considered, a one-shot delivery without any additional carrier exposes to a rapid wash-out and an attendant loss of efficacy. Thus, the incorporation of the cell-produced biologics into biomaterials enabling a sustained release [15, 94], much like for traditional stem cell therapy, can be an effective means of extending their retention and improving outcomes and this concept applies to the secretome of PSC-derived cardiac cells [49]. Here, a distinct advantage of cell-derived biologics is that they can be successfully embedded into scaffolds suitable for long-term cryopreservation and immediate availability, thereby overcoming survival and storage issues associated with the use of cells [36]. However, a more appealing means of leveraging the acellular nature of the secretome is to deliver it intravenously. Although a limited amount of the injectate may then reach the target organ, the intravenous route is yet associated with a functional benefit which has been demonstrated across a wide variety of preclinical models of acute myocardial infarction [21, 87], Duchenne [70] and chemotherapy-induced [56, 79, 81, 91] cardiomyopathies, as well as in non-cardiac disease models. One postulated mechanism of action of these intravenously injected EV-enriched secretomes is that following their predominant uptake by macrophages [59] and liver sequestration [96], they could act like cells through a systemic modulation of inflammation triggering tissue-protective signals which are then conveyed to the target organ by the circulating cells [93]. It is clear that the intravenous route has clinically relevant advantages: it is simple as it does not require costly equipment and facilities; more importantly, it is noninvasive and as such allows repeated administrations, which could be critical for optimizing the therapeutic benefit, as suggested by the finding that repeated administrations of transplanted cells are superior to the single administration of an equivalent cumulative dose [83].

The last two translational challenges raised by the use of the cellular secretome are its degree and method of purification as well as the characterization of its content. However, these issues are not specific for PSC used as the parental cells and their discussion is beyond the scope of this chapter (for reviews, see [3, 85]).

2.3 A Common Mechanism-Independent Issue: Safety

Regardless of the putative mechanism of action, safety is obviously a key issue because the presence in the cell product of still undifferentiated and/or transformed cells could lead to their uncontrolled proliferation and give rise to a teratoma or a terato-carcinoma. The opponents of ESC have obviously rushed on this risk to dishonestly demonize ESC in contempt of the actual data which show that although the first five patients to receive ESC derivatives, in this case oligodendrocytes for spinal cord injury, are now at 8 years of their treatment without any warning signal (Press release from Asterias Biotherapeutics Inc., January 24, 2019). Likewise, the first patient of the ESCORT trial has now 7 years of follow-up and is doing clinically well without any abnormality on imaging studies (computerized tomography and ^{18}F-FDG positron emission tomography scans) documented so far. Nevertheless, it is obviously critical to purify the differentiated cells to ensure that they are no longer contaminated with unwanted and still pluripotent cells.

Multiple purification methods have been developed (reviewed in [6, 58]), of which only a sorting based on surface markers is currently suitable for large-scale GMP-compliant applications. In the ESCORT trial, we used SSEA-1 as a marker for cells which were no longer undifferentiated and magnetic bead-bound anti-SSEA-1 antibodies for selecting them. Even though this mode of immunomagnetic positive selection has been used for years for enriching in hematopoietic cell populations, a negative mode of selection targeting markers for pluripotency [82] is more appealing in that it would allow the final cell therapy product for patient use to remain magnetic bead-free. Other strategies based on microfluidics or field-flow fractionation are currently investigated and might be of interest if they can be scaled-up under GMP conditions. Of note, selection of the cardiac-committed cells is primarily relevant to the use of cardiovascular progenitors where stage-specific markers have to be used (*Isl-1* in the ESCORT trial) since an extended period of culture is expected to yield an almost pure population of cardiomyocytes whose identity can be confirmed by a battery of markers (Flk-1, PDGFR- α, signal-regulatory protein alpha ([SIRPA] and vascular cell adhesion molecule 1 [VCAM1]) and which, in principle, does not harbour undifferentiated cells any longer.

However, regardless of the cell culture strategy, it is mandatory to ensure that the final product which will be delivered to the patient has no tumorigenic potential. This first implies *in vitro* checking for its genetic stability to eliminate the occurrence of cancer-driving mutations that can be induced by culture conditions and time in culture [30, 64]. Several quality control metrics have been proposed to ensure that oncogenic networks have not been activated (reviewed in [4]. As an example, the cardiovascular SSEA-1+ cells used in the ESCORT trial were assessed by karyotyping, fluorescence in situ hybridization (FISH) and array comparative genomic hybridization (aCGH). Of note, iPSC may be more prone to genetic/epigenetic alterations than ESC because of the potential of mutations already present in the parental cells from which they are reprogrammed [10, 64], with the caveat that it can be difficult to establish a correlation between genetic modifications of the cells and the actual risk that will form a tumour *in vivo*. Maybe, in the future, the organs-on-chip technology will provide a human-like micro-environment more suitable for an accurate risk prediction. The identity of the cardiac-committed product also needs to be tested by lineage-specific markers, usually by flow cytometry and/or gene sequencing, in particular to ensure that pluripotency genes have been appropriately knocked down. It may also be advisable to enrich the risk analysis with assays evaluating the potential of the cells to cause genotoxicity and chromosomal damage.

Despite the utility of these *in vitro* tests, tumorigenicity studies in animals still represent a major pillar of the preclinical safety assessment. Several factors have then to be taken into consideration, which include the number of cells to be inoculated, the duration of follow-up, the site of transplantation and the limit of detection of the relevant assays [74]. The choice of the animal model is particularly critical, keeping in mind that most studies entail transplanting human cells into rodents and, as such, should be interpreted cautiously since xenotransplantation is less likely to induce teratomas than allografts [24]. This has been the main argument for us to test

Rhesus ESC-derived cardiovascular progenitors in infarcted Rhesus monkeys with the premise that such an intraspecies transplantation would sensitize the potential occurrence of a tumour [11]. Importantly, these teratoma experiments should include spiking with different ratios of undifferentiated/lineage-committed cells to identify the threshold above which a tumour occurs and thus helps defining lot release criteria [67]. This information is a critical component of the risk assessment. Finally, off-target adverse effects have to be ruled out by biodistribution studies.

3 Perspectives

3.1 *Remuscularization*

For those who remain committed to use PSC-derived cardiac-committed cells as physical and expectedly permanent substitutes for the lost cardiomyocytes, there are at least three areas of research which need to be prioritized. The first is to define the optimal stage of differentiation at which cells should be transplanted. As mentioned above, there are no robust arguments favouring early progenitors versus more mature cardiomyocytes but, assuming that the latter could more efficiently increase pump function, it is mandatory to address the risk of arrhythmias by optimizing the maturation state of the transplanted cardiomyocytes and thus mitigate an electrical mismatch at the graft/host interface [71]; this can be achieved by a variety of strategies including co-culture with endothelial cells, prolongation of the culture time, control of the composition, topography and stiffness of the extracellular matrix, mechanical and electrical stimulation [44]. The second area of research pertains to the development of strategies allowing an immune tolerance of these allogeneic cells in order to avoid the use of high doses of immunosuppressive drugs and their cohort of adverse effects. In this setting, the use of HLA-haplotyped cell lines [60] could be helpful by allowing a better match between the grafted cells and the immune profile of the recipient and is clinically doable as HLA-matched cells are currently transplanted in a Parkinson trial (https://upload.umin.ac.jp/cgi-open-bin/ctr_e/ctr_view.cgi?recptno=R000038278). However, the most conceptually attractive option could be the use of gene editing to generate cell lines no longer expressing HLA-I and -II antigens but engineered to express molecules preventing cell destruction by Natural Killer cells [98], provided that the efficacy and safety of these universal cell lines can be validated. In the context of these efforts to "lighten" immunosuppressive regimens, another approach, possibly easier to implement in the clinics, could be the use of regulatory T cells or even their derived exosomes which have been shown to successfully extend liver graft survival [16]. Finally, even though the success of pro-survival approaches could enable a prolonged engraftment of the cells, it is likely that their number will progressively decrease over time. This assumption is supported by the finding that cardiac function was significantly improved in mice injected with ESC-derived cardiomyocytes at 4 weeks after transplantation but that this benefit was not retained at 3 months [89], thereby suggesting that maintenance of a therapeutic effect likely requires repeated administrations. In the clinics, this can only be achieved, easily and safely, by intravenous injections. Admittedly, however, intravenously injected ESC-derived endothelial cells or iPSC-derived neural stem cells have been therapeutically effective in ischemic hindlimb [35] and amyotrophic lateral sclerosis [62] models, respectively, without causing short-term adverse events, but one could be concerned about the safety of a systemic delivery of PSC derivatives and it is uncertain that such a strategy would be approved by the regulators (it might be the reason why the Chinese trial mentioned above and which had planned intravenous infusions of iPSC-derived cardiomyocytes in 3 heart failure patients has not yet started 18 months after its official registration). In this context, an innovative approach has been proposed which consists of implanting an epicardial device connected to a subcutaneous port which can be periodically replenished with cells. The extent to which this system, which has shown promising results in a rodent model of myocardial infarction [95], can be translated clinically remains to be established.

3.2 Paracrine Signalling

Reliance on a paracrine mechanism of action logically leads to consider to exclusively deliver the cellular secretome even though its effects on *in vitro* potency assays may not be representative of what happens when it is directly released by the parental cells *in vivo*. This limitation, however, may not be so critical as demonstrated almost 10 years ago by Timmers et al. [87] who showed, in a pig model of myocardial infarction, the protective effects of the intravenously infused conditioned medium (only clarified and concentrated) of ESC-derived MSC. Further confirmation has been brought more recently in a rat model of myocardial infarction in which a single intramyocardial injection of the total conditioned media released by neonatal human CSC was significantly more effective functionally than either the parental cells of their purified exosomal fraction [77]. Nevertheless, one area of research is to identify the key drivers of the biological effects of the cellular secretome and many different miRNAs and proteins have been reported as the most effective candidates. While it is tempting to identify a short list of compounds in the perspective of synthetizing them as a multi-targeted drug, it could indeed be counterproductive to deconstruct the cargo of the secretome because of the multiplicity of its components and their possible interactions leading to synergistic effects, notwithstanding the challenge of artificially recapitulating a complex repertoire of bioactive molecules. Another reason for keeping the whole extracellular vesicle payload intact is that it can be shuttled in the recipient cells without the risk of degradation because of the protection afforded by the lipid bilayer of the vesicles whereas it is unlikely it would still be the case if its individual components had to be delivered individually. A second area of research pertains to the means of "priming" the parental cells during the culture period to enrich the extracellular vesicle package with proteins and other biologics that can bolster their biological effects [97, 99, 106]. A third clinically relevant perspective pertains to the optimized delivery of the cellular secretome. Although using source cells which share the same repertoire of surface receptors as those of the target tissue can yet be therapeutically beneficial, as discussed above, a greater benefit could likely be yielded by a more selective targeting of the injectate towards the heart, which can be achieved by two different approaches. The first consists of engineering the parental cells to make them over-expressing receptors expected to traffic to their extracellular vesicles and whose display on the external part of the vesicular membrane can enhance recognition by target cells, docking and vesicle uptake. This strategy has been shown successful for promoting homing of intravenously injected exosomes derived from cardiac pro-genitor cells overexpressing CXCR4 to enhance interaction of this receptor with its SDF-1α ligand which is upregulated in ischemic tissues [21]. This approach, how-ever, is challenged by the risk of peptide degradation [37], hence the interest of an alternate approach based on the direct engineering of the secreted vesicles by anchorage of cardiac-specific peptides to their membrane [92] or modifications of their glysosylation pattern to promote binding to endothelial selectins [51, 72]. A last clinically relevant issue is dosing. There is consistent evidence that the effects of extracellular vesicles are dose-dependent [41] while the apparently low number of miRNA carried by exosomes [17] likely requires large amounts of them to be transferred to convey a significant biological effect. Different metrics can be consid-ered here such as an absolute number of particles, a protein concentration or a "cell equivalent", i.e., the amount of particles assumed to be released by a given number of the parental cells but no consensus has emerged yet.

In conclusion, while iPSC look more attractive for disease modelling and drug screening, both ESC and iPSC share the same potential for inducing heart repair and possibly regeneration. For reasons related to practicality, iPSC-derived cardiac-committed cells, regardless of their stage of differentiation, may provide a better cost-effective model, particularly if they are primarily used as *ex vivo* biofactories producing a biologically active secretome which would then represent the therapeu-tics delivered to the patient. This approach would allow to leverage the cardiopro-tective potential of PSC derivatives while overcoming the safety issues that might be associated with their direct delivery. Future clinical studies will tell whether such an expectation can be really met. In the meantime, the observation that by August, 2021, the registry for human PSC-based cell therapies (https://hpscreg.eu) had listed 83 studies provides compelling evidence that regardless of the cell source (ESC or iPSC), the clinical indication and the delivery strategy, PSC-derived lineage-specific cells have now integrated the armamentarium of therapies against a wide variety of diseases.

References

1. Adamiak, M., et al.: Induced pluripotent stem cell (iPSC)-derived extracellular vesicles are safer and more effective for cardiac repair than iPSCs. Circ. Res. **122**(2), 296–309 (2018). https://doi.org/10.1161/CIRCRESAHA.117.311769
2. Agarwal, U., et al.: Experimental, systems, and computational approaches to understanding the MicroRNA-mediated reparative potential of cardiac progenitor cell–derived exosomes from pediatric PatientsNovelty and significance. Circ. Res. **120**(4), 701–712 (2017). https://doi.org/10.1161/CIRCRESAHA.116.309935
3. Andriolo, G., et al.: Exosomes from human cardiac progenitor cells for therapeutic applications: development of a GMP-grade manufacturing method. Front. Physiol. **9**, 1169 (2018). https://doi.org/10.3389/fphys.2018.01169
4. Assou, S., Bouckenheimer, J., De Vos, J.: Concise review: assessing the genome integrity of human induced pluripotent stem cells: what quality control metrics? Stem Cells (Dayton, Ohio). **36**(6), 814–821 (2018). https://doi.org/10.1002/stem.2797
5. Ban, K., et al.: Cell therapy with embryonic stem cell-derived cardiomyocytes encapsulated in injectable Nanomatrix gel enhances cell engraftment and promotes cardiac repair. ACS Nano. **8**(10), 10815–10825 (2014). https://doi.org/10.1021/nn504617g
6. Ban, K., Bae, S., Yoon, Y.-S.: Current strategies and challenges for purification of cardiomyocytes derived from human pluripotent stem cells. Theranostics. **7**(7), 2067–2077 (2017). https://doi.org/10.7150/thno.19427
7. Bargehr, J., et al.: Epicardial cells derived from human embryonic stem cells augment cardiomyocyte-driven heart regeneration. Nat. Biotechnol. **37**(8), 895–906 (2019). https://doi.org/10.1038/s41587-019-0197-9
8. Barile, L., et al.: Cardioprotection by cardiac progenitor cell-secreted exosomes: role of pregnancy-associated plasma protein-a. Cardiovasc. Res. **114**(7), 992–1005 (2018). https://doi.org/10.1093/cvr/cvy055
9. Bellamy, V., et al.: Long-term functional benefits of human embryonic stem cell-derived cardiac progenitors embedded into a fibrin scaffold. J. Heart Lung Transplant. **34**(9), 1198–1207 (2015). https://doi.org/10.1016/j.healun.2014.10.008
10. Ben-David, U., Benvenisty, N.: 'The tumorigenicity of human embryonic and induced pluripotent stem cells', *nature reviews*. Cancer. **11**(4), 268–277 (2011). https://doi.org/10.1038/nrc3034
11. Blin, G., et al.: A purified population of multipotent cardiovascular progenitors derived from primate pluripotent stem cells engrafts in postmyocardial infarcted nonhuman primates. J. Clin. Invest. **120**(4), 1125–1139 (2010). https://doi.org/10.1172/JCI40120
12. Bobis-Wozowicz, S., et al.: Human induced pluripotent stem cell-derived microvesicles transmit RNAs and proteins to recipient mature heart cells modulating cell fate and behavior: hiPSC-MVs transmit RNAs and proteins to heart cells. Stem Cells. **33**(9), 2748–2761 (2015). https://doi.org/10.1002/stem.2078
13. Bolli, R., et al.: A phase II study of mesenchymal stromal cells and c-kit positive cardiac stem cells, alone or in combination, in patients with ischaemic heart failure: the CCTRN CONCERT-HF trial. Eur J Heart Fail. **23**(4), 661–674 (2021). https://doi.org/10.1002/ejhf.2178
14. Caspi, O., et al.: Transplantation of human embryonic stem cell-derived cardiomyocytes improves myocardial performance in infarcted rat hearts. J. Am. Coll. Cardiol. **50**(19), 1884–1893 (2007). https://doi.org/10.1016/j.jacc.2007.07.054
15. Chen, C.W., et al.: Sustained release of endothelial progenitor cell-derived extracellular vesicles from shear-thinning hydrogels improves angiogenesis and promotes function after myocardial infarction. Cardiovasc. Res. **114**(7), 1029–1040 (2018). https://doi.org/10.1093/cvr/cvy067

16. Chen, L., et al.: Exosomes derived from T regulatory cells suppress CD8+ cytotoxic T lymphocyte proliferation and prolong liver allograft survival. Med. Sci. Monit. Int. Med. J. Exp. Clin. Res. **25**, 4877–4884 (2019). https://doi.org/10.12659/MSM.917058

17. Chevillet, J.R., et al.: Quantitative and stoichiometric analysis of the microRNA content of exosomes. Proc. Natl. Acad. Sci. **111**(41), 14888–14893 (2014). https://doi.org/10.1073/pnas.1408301111

18. Chong, J.J.H., et al.: Human embryonic-stem-cell-derived cardiomyocytes regenerate non-human primate hearts. Nature. **510**(7504), 273–277 (2014a). https://doi.org/10.1038/nature13233

19. Chong, J.J.H., et al.: Human embryonic-stem-cell-derived cardiomyocytes regenerate non-human primate hearts. Nature. **510**(7504), 273–277 (2014b). https://doi.org/10.1038/nature13233

20. Citro, L., et al.: Comparison of human induced pluripotent stem-cell derived cardiomyocytes with human mesenchymal stem cells following acute myocardial infarction. PloS One. **9**(12), e116281 (2014). https://doi.org/10.1371/journal.pone.0116281

21. Ciullo, A., et al.: Exosomal expression of CXCR4 targets Cardioprotective vesicles to myocardial infarction and improves outcome after systemic administration. Int. J. Mol. Sci. **20**(3), 468 (2019). https://doi.org/10.3390/ijms20030468

22. Desgres, M., Menasché, P.: Clinical translation of pluripotent stem cell therapies: challenges and considerations. Cell Stem Cell. **25**(5), 594–606 (2019). https://doi.org/10.1016/j.stem.2019.10.001

23. El Harane, N., et al.: Acellular therapeutic approach for heart failure: in vitro production of extracellular vesicles from human cardiovascular progenitors. Eur. Heart J. **39**(20), 1835–1847 (2018). https://doi.org/10.1093/eurheartj/ehy012

24. Erdö, F., et al.: Host-dependent tumorigenesis of embryonic stem cell transplantation in experimental stroke. J. Cereb. Blood Flow Metab. **23**(7), 780–785 (2003). https://doi.org/10.1097/01.WCB.0000071886.63724.FB

25. Fernandes, S., et al.: Comparison of human embryonic stem cell-derived cardiomyocytes, cardiovascular progenitors, and bone marrow mononuclear cells for cardiac repair. Stem Cell Rep. **5**(5), 753–762 (2015). https://doi.org/10.1016/j.stemcr.2015.09.011

26. Florea, V., et al.: Dose comparison study of allogeneic mesenchymal stem cells in patients with ischemic cardiomyopathy (the TRIDENT study). Circ. Res. **121**(11), 1279–1290 (2017). https://doi.org/10.1161/CIRCRESAHA.117.311827

27. Gao, L., et al.: Large cardiac muscle patches engineered from human induced-pluripotent stem cell–derived cardiac cells improve recovery from myocardial infarction in swine. Circulation. **137**(16), 1712–1730 (2018). https://doi.org/10.1161/CIRCULATIONAHA.117.030785

28. Garbern, J.C., Lee, R.T.: Cardiac stem cell therapy and the promise of heart regeneration. Cell Stem Cell. **12**(6), 689–698 (2013). https://doi.org/10.1016/j.stem.2013.05.008

29. Garikipati, V.N.S., et al.: Extracellular vesicles and the application of system biology and computational modeling in cardiac repair. Circ. Res. **123**(2), 188–204 (2018). https://doi.org/10.1161/CIRCRESAHA.117.311215

30. Goldring, C.E.P., et al.: Assessing the safety of stem cell therapeutics. Cell Stem Cell. **8**(6), 618–628 (2011). https://doi.org/10.1016/j.stem.2011.05.012

31. Guan, X., et al.: 'Transplantation of human induced pluripotent stem cell-derived cardiomyocytes improves myocardial function and reverses ventricular remodeling in infarcted rat hearts', *Stem Cell Research & Therapy*, 11(1), p. 73. (2020). https://doi.org/10.1186/s13287-020-01602-0

32. Guo, R., et al.: Stem cell-derived cell sheet transplantation for heart tissue repair in myocardial infarction. Stem Cell Res Ther. **11**(1) (2020). https://doi.org/10.1186/s13287-019-1536-y

33. Halbach, M., et al.: Cell persistence and electrical integration of transplanted fetal cardiomyocytes from different developmental stages. Int. J. Cardiol. **171**(3), e122–e124 (2014). https://doi.org/10.1016/j.ijcard.2013.12.115

34. Han, C., et al.: Human umbilical cord mesenchymal stem cell derived exosomes encapsulated in functional peptide hydrogels promote cardiac repair. Biomater. Sci. **7**(7), 2920–2933 (2019). https://doi.org/10.1039/C9BM00101H

35. Huang, N.F., et al.: Embryonic stem cell-derived endothelial cells engraft into the ischemic hindlimb and restore perfusion. Arterioscler. Thromb. Vasc. Biol. **30**(5), 984–991 (2010). https://doi.org/10.1161/ATVBAHA.110.202796
36. Huang, K., et al.: An off-the-shelf artificial cardiac patch improves cardiac repair after myocardial infarction in rats and pigs. Sci. Trans. Med. **12**(538), eaat9683 (2020). https://doi.org/10.1126/scitranslmed.aat9683
37. Hung, M.E., Leonard, J.N.: Stabilization of exosome-targeting peptides via engineered glycosylation. J. Biol. Chem. **290**(13), 8166–8172 (2015). https://doi.org/10.1074/jbc.M114.621383
38. Iseoka, H., et al.: Pivotal role of non-cardiomyocytes in electromechanical and therapeutic potential of induced pluripotent stem cell-derived engineered cardiac tissue. Tissue Eng. Part A. **24**(3–4), 287–300 (2018). https://doi.org/10.1089/ten.TEA.2016.0535
39. Ishida, M., et al.: Transplantation of human-induced pluripotent stem cell-derived cardiomyocytes is superior to somatic stem cell therapy for restoring cardiac function and oxygen consumption in a porcine model of myocardial infarction. Transplantation. **103**(2), 291–298 (2019). https://doi.org/10.1097/TP.0000000000002384
40. Kabat, M., et al.: Trends in mesenchymal stem cell clinical trials 2004-2018: is efficacy optimal in a narrow dose range? Stem Cells Transl. Med. **9**(1), 17–27 (2020). https://doi.org/10.1002/sctm.19-0202
41. Kervadec, A., et al.: Cardiovascular progenitor–derived extracellular vesicles recapitulate the beneficial effects of their parent cells in the treatment of chronic heart failure. J. Heart Lung Transplant. **35**(6), 795–807 (2016). https://doi.org/10.1016/j.healun.2016.01.013
42. Khan, M., et al.: Embryonic stem cell-derived exosomes promote endogenous repair mechanisms and enhance cardiac function following myocardial infarction. Circ. Res. **117**(1), 52–64 (2015). https://doi.org/10.1161/CIRCRESAHA.117.305990
43. Könemann, S., et al.: Cardioprotective effect of the secretome of Sca-1+ and Sca-1− cells in heart failure: not equal, but equally important? Cardiovasc. Res. (2019). https://doi.org/10.1093/cvr/cvz140
44. Le, M.N.T., Hasegawa, K.: 'Expansion culture of human pluripotent stem cells and production of cardiomyocytes', *bioengineering (Basel, Switzerland)*, 6(2). (2019). https://doi.org/10.3390/bioengineering6020048
45. Lee, W.H., et al.: Comparison of non-coding RNAs in exosomes and functional efficacy of human embryonic stem cell- versus induced pluripotent stem cell-derived cardiomyocytes. Stem Cells (Dayton, Ohio). **35**(10), 2138–2149 (2017). https://doi.org/10.1002/stem.2669
46. Li, T.-S., et al.: Direct comparison of different stem cell types and subpopulations reveals superior paracrine potency and myocardial repair efficacy with cardiosphere-derived cells. J. Am. Coll. Cardiol. **59**(10), 942–953 (2012). https://doi.org/10.1016/j.jacc.2011.11.029
47. Liao, S., et al.: Potent immunomodulation and angiogenic effects of mesenchymal stem cells versus cardiomyocytes derived from pluripotent stem cells for treatment of heart failure. Stem Cell Res. Ther. **10**(1) (2019). https://doi.org/10.1186/s13287-019-1183-3
48. Lima Correa, B., et al.: Extracellular vesicles from human cardiovascular progenitors trigger a reparative immune response in infarcted hearts. Cardiovasc. Res. (2020). https://doi.org/10.1093/cvr/cvaa028
49. Liu, B., et al.: Cardiac recovery via extended cell-free delivery of extracellular vesicles secreted by cardiomyocytes derived from induced pluripotent stem cells. Nat. Biomed. Eng. **2**(5), 293–303 (2018a). https://doi.org/10.1038/s41551-018-0229-7
50. Liu, Y.-W., et al.: Human embryonic stem cell-derived cardiomyocytes restore function in infarcted hearts of non-human primates. Nat. Biotechnol. **36**(7), 597–605 (2018b). https://doi.org/10.1038/nbt.4162
51. Lo, C.Y., et al.: Cell surface glycoengineering improves selectin-mediated adhesion of mesenchymal stem cells (MSCs) and cardiosphere-derived cells (CDCs): pilot validation in porcine ischemia-reperfusion model. Biomaterials. **74**, 19–30 (2016). https://doi.org/10.1016/j.biomaterials.2015.09.026

52. Luo, J., et al.: Targeting survival pathways to create infarct-spanning bridges of human embryonic stem cell-derived cardiomyocytes'. J. Thorac. Cardiovasc. Surg. **148**(6), 3180–3188.e1 (2014). https://doi.org/10.1016/j.jtcvs.2014.06.087
53. Malik, Z.A., et al.: 'Cardiac myocyte exosomes: stability, HSP60, and proteomics', *American journal of physiology*. Heart Circ. Physiol. **304**(7), H954–H965 (2013). https://doi.org/10.1152/ajpheart.00835.2012
54. Menasché, P., et al.: Towards a clinical use of human embryonic stem cell-derived cardiac progenitors: a translational experience. Eur. Heart J. **36**(12), 743–750 (2015). https://doi.org/10.1093/eurheartj/ehu192
55. Menasché, P., et al.: Transplantation of human embryonic stem cell-derived cardiovascular progenitors for severe ischemic left ventricular dysfunction. J. Am. Coll. Cardiol. **71**(4), 429–438 (2018). https://doi.org/10.1016/j.jacc.2017.11.047
56. Milano, G., et al.: Intravenous administration of cardiac progenitor cell-derived exosomes protects against doxorubicin/trastuzumab-induced cardiac toxicity. Cardiovasc. Res. (2019). https://doi.org/10.1093/cvr/cvz108
57. Mohsin, S., et al.: Empowering adult stem cells for myocardial regeneration. Circ. Res. **109**(12), 1415–1428 (2011). https://doi.org/10.1161/CIRCRESAHA.111.243071
58. Moon, S.-H., et al.: From bench to market: preparing human pluripotent stem cells derived cardiomyocytes for various applications. Int. J. Stem Cells. **10**(1), 1–11 (2017). https://doi.org/10.15283/ijsc17024
59. Morishita, M., et al.: Pharmacokinetics of exosomes-an important factor for elucidating the biological roles of exosomes and for the development of exosome-based therapeutics. J. Pharm. Sci. **106**(9), 2265–2269 (2017). https://doi.org/10.1016/j.xphs.2017.02.030
60. Neofytou, E., et al.: Hurdles to clinical translation of human induced pluripotent stem cells. J. Clin. Invest. **125**(7), 2551–2557 (2015). https://doi.org/10.1172/JCI80575
61. News at a glance: Science (New York, N.Y.). **362**(6412), 268–270 (2018). https://doi.org/10.1126/science.362.6412.268
62. Nizzardo, M., et al.: Minimally invasive transplantation of iPSC-derived ALDHhiSSCloVLA4+ neural stem cells effectively improves the phenotype of an amyotrophic lateral sclerosis model. Hum. Mol. Genet. **23**(2), 342–354 (2014). https://doi.org/10.1093/hmg/ddt425
63. Okano, S., Shiba, Y.: Therapeutic potential of pluripotent stem cells for cardiac repair after myocardial infarction. Biol. Pharm. Bull. **42**(4), 524–530 (2019). https://doi.org/10.1248/bpb.b18-00257
64. Oliveira, P.H., da Silva, C.L., Cabral, J.M.S.: Concise review: genomic instability in human stem cells: current status and future challenges. Stem Cells (Dayton, Ohio). **32**(11), 2824–2832 (2014). https://doi.org/10.1002/stem.1796
65. Patil, M., et al.: The art of intercellular wireless communications: exosomes in heart disease and therapy. Front. Cell Dev. Biol. **7**, 315 (2019). https://doi.org/10.3389/fcell.2019.00315
66. Perin, E.C., et al.: A phase II dose-escalation study of allogeneic mesenchymal precursor cells in patients with ischemic or nonischemic heart FailureNovelty and significance. Circ. Res. **117**(6), 576–584 (2015). https://doi.org/10.1161/CIRCRESAHA.115.306332
67. Priest, C.A., et al.: Preclinical safety of human embryonic stem cell-derived oligodendrocyte progenitors supporting clinical trials in spinal cord injury. Regen. Med. **10**(8), 939–958 (2015). https://doi.org/10.2217/rme.15.57
68. Ratajczak, J., et al.: Embryonic stem cell-derived microvesicles reprogram hematopoietic progenitors: evidence for horizontal transfer of mRNA and protein delivery. Leukemia. **20**(5), 847–856 (2006). https://doi.org/10.1038/sj.leu.2404132
69. Riegler, J., et al.: Human engineered heart muscles engraft and survive long term in a rodent myocardial infarction model. Circ. Res. **117**(8), 720–730 (2015). https://doi.org/10.1161/CIRCRESAHA.115.306985
70. Rogers, R.G., et al.: Disease-modifying bioactivity of intravenous cardiosphere-derived cells and exosomes in mdx mice. JCI Insight. **4**(7) (2019). https://doi.org/10.1172/jci.insight.125754

71. Romagnuolo, R., et al.: Human embryonic stem cell-derived cardiomyocytes regenerate the infarcted pig heart but induce ventricular Tachyarrhythmias. Stem Cell Rep. 12(5), 967–981 (2019). https://doi.org/10.1016/j.stemcr.2019.04.005

72. Sackstein, R.: Glycosyltransferase-programmed stereosubstitution (GPS) to create HCELL: engineering a roadmap for cell migration. Immunol. Rev. 230(1), 51–74 (2009). https://doi.org/10.1111/j.1600-065X.2009.00792.x

73. Santoso, M.R., et al.: Exosomes from induced pluripotent stem cell–derived cardiomyocytes promote autophagy for myocardial repair. J. Am. Heart Assoc. 9(6) (2020). https://doi.org/10.1161/JAHA.119.014345

74. Sato, Y., et al.: Tumorigenicity assessment of cell therapy products: the need for global consensus and points to consider. Cytotherapy. 21(11), 1095–1111 (2019). https://doi.org/10.1016/j.jcyt.2019.10.001

75. Schwach, V., et al.: Expandable human cardiovascular progenitors from stem cells for regenerating mouse heart after myocardial infarction. Cardiovasc. Res. (2019). https://doi.org/10.1093/cvr/cvz181

76. Shao, L., et al.: MiRNA-sequence indicates that mesenchymal stem cells and exosomes have similar mechanism to enhance cardiac repair. Biomed. Res. Int. 2017, 1–9 (2017). https://doi.org/10.1155/2017/4150705

77. Sharma, S., et al.: A deep proteome analysis identifies the complete Secretome as the functional unit of human cardiac progenitor cells. Circ. Res. 120(5), 816–834 (2017). https://doi.org/10.1161/CIRCRESAHA.116.309782

78. Shiba, Y., et al.: Allogeneic transplantation of iPS cell-derived cardiomyocytes regenerates primate hearts. Nature. 538(7625), 388–391 (2016). https://doi.org/10.1038/nature19815

79. Singla, D.K., et al.: Embryonic stem cells improve cardiac function in doxorubicin-induced cardiomyopathy mediated through multiple mechanisms. Cell Transplant. 21(9), 1919–1930 (2012). https://doi.org/10.3727/096368911X627552

80. Smith, R.R., et al.: Regenerative potential of cardiosphere-derived cells expanded from percutaneous endomyocardial biopsy specimens. Circulation. 115(7), 896–908 (2007). https://doi.org/10.1161/CIRCULATIONAHA.106.655209

81. Sun, X., et al.: Intravenous mesenchymal stem cell-derived exosomes ameliorate myocardial inflammation in the dilated cardiomyopathy. Biochem. Biophys. Res. Commun. 503(4), 2611–2618 (2018). https://doi.org/10.1016/j.bbrc.2018.08.012

82. Tang, C., et al.: An antibody against SSEA-5 glycan on human pluripotent stem cells enables removal of teratoma-forming cells. Nat. Biotechnol. 29(9), 829–834 (2011). https://doi.org/10.1038/nbt.1947

83. Tang, X.-L., et al.: Repeated administrations of cardiac progenitor cells are superior to a single Administration of an Equivalent Cumulative Dose. J. Am. Heart Assoc. 7(4) (2018). https://doi.org/10.1161/JAHA.117.007400

84. Taylor, M., et al.: Cardiac and skeletal muscle effects in the randomized HOPE-Duchenne trial. Neurology. 92(8), e866–e878 (2019). https://doi.org/10.1212/WNL.0000000000006950

85. Théry, C., et al.: Minimal information for studies of extracellular vesicles 2018 (MISEV2018): a position statement of the International Society for Extracellular Vesicles and update of the MISEV2014 guidelines. J. Extracell. Vesic. 7(1), 1535750 (2018). https://doi.org/10.1080/20013078.2018.1535750

86. Tiburcy, M., et al.: Defined engineered human myocardium with advanced maturation for applications in heart failure modeling and repair. Circulation. 135(19), 1832–1847 (2017). https://doi.org/10.1161/CIRCULATIONAHA.116.024145

87. Timmers, L., et al.: Human mesenchymal stem cell-conditioned medium improves cardiac function following myocardial infarction. Stem Cell Res. 6(3), 206–214 (2011). https://doi.org/10.1016/j.scr.2011.01.001

88. van Berlo, J.H., et al.: C-kit+ cells minimally contribute cardiomyocytes to the heart. Nature. 509(7500), 337–341 (2014). https://doi.org/10.1038/nature13309

89. van Laake, L.W., et al.: Human embryonic stem cell-derived cardiomyocytes survive and mature in the mouse heart and transiently improve function after myocardial infarction. Stem Cell Res. 1(1), 9–24 (2007). https://doi.org/10.1016/j.scr.2007.06.001

90. van Laake, L.W., et al.: Improvement of mouse cardiac function by hESC-derived cardiomyocytes correlates with vascularity but not graft size. Stem Cell Res. **3**(2–3), 106–112 (2009). https://doi.org/10.1016/j.scr.2009.05.004
91. Vandergriff, A.C., et al.: Intravenous cardiac stem cell-derived exosomes ameliorate cardiac dysfunction in doxorubicin induced dilated cardiomyopathy. Stem Cells Int. **2015**, 1–8 (2015). https://doi.org/10.1155/2015/960926
92. Vandergriff, A., et al.: Targeting regenerative exosomes to myocardial infarction using cardiac homing peptide. Theranostics. **8**(7), 1869–1878 (2018). https://doi.org/10.7150/thno.20524
93. Walker, P.A., et al.: Bone marrow–derived stromal cell therapy for traumatic brain injury is neuroprotective via stimulation of non-neurologic organ systems. Surgery. **152**(5), 790–793 (2012). https://doi.org/10.1016/j.surg.2012.06.006
94. Waters, R., et al.: Stem cell-inspired secretome-rich injectable hydrogel to repair injured cardiac tissue. Acta Biomater. **69**, 95–106 (2018). https://doi.org/10.1016/j.actbio.2017.12.025
95. Whyte, W., et al.: Sustained release of targeted cardiac therapy with a replenishable implanted epicardial reservoir. Nat. Biomed. Eng. **2**(6), 416–428 (2018). https://doi.org/10.1038/s41551-018-0247-5
96. Wiklander, O.P.B., et al.: Extracellular vesicle in vivo biodistribution is determined by cell source, route of administration and targeting. J. Extracell. Vesic. **4**, 26316 (2015)
97. Wiklander, O.P.B., et al.: Advances in therapeutic applications of extracellular vesicles. Sci. Trans. Med. **11**(492), eaav8521 (2019). https://doi.org/10.1126/scitranslmed.aav8521
98. Xu, H., et al.: Targeted disruption of HLA genes via CRISPR-Cas9 generates iPSCs with enhanced immune compatibility. Cell Stem Cell. (2019a). https://doi.org/10.1016/j.stem.2019.02.005
99. Xu, R., et al.: Exosomes derived from pro-inflammatory bone marrow-derived mesenchymal stem cells reduce inflammation and myocardial injury via mediating macrophage polarization. J. Cell. Mol. Med. **23**(11), 7617–7631 (2019b). https://doi.org/10.1111/jcmm.14635
100. Yarbrough, W.: Large animal models of congestive heart failure: a critical step in translating basic observations into clinical applications. J. Nucl. Cardiol. **10**(1), 77–86 (2003). https://doi.org/10.1067/mnc.2003.16
101. Ye, J., et al.: Treatment with hESC-derived myocardial precursors improves cardiac function after a myocardial infarction. PloS One. **10**(7), e0131123 (2015). https://doi.org/10.1371/journal.pone.0131123
102. Yee, K., et al.: Allogeneic cardiospheres delivered via percutaneous transendocardial injection increase viable myocardium, decrease scar size, and attenuate cardiac dilatation in porcine ischemic cardiomyopathy. PloS One. **9**(12), e113805 (2014). https://doi.org/10.1371/journal.pone.0113805
103. Yin, J.Q., Zhu, J., Ankrum, J.A.: Manufacturing of primed mesenchymal stromal cells for therapy. Nat. Biomed. Eng. **3**(2), 90–104 (2019). https://doi.org/10.1038/s41551-018-0325-8
104. Yoshida, S., et al.: Syngeneic mesenchymal stem cells reduce immune rejection after induced pluripotent stem cell-derived allogeneic cardiomyocyte transplantation. Sci. Rep. **10**(1), 4593 (2020). https://doi.org/10.1038/s41598-020-58126-z
105. Yuan, A., et al.: Transfer of microRNAs by embryonic stem cell microvesicles. PloS One. **4**(3), e4722 (2009). https://doi.org/10.1371/journal.pone.0004722
106. Yuan, O., et al.: Exosomes derived from human primed mesenchymal stem cells induce mitosis and potentiate growth factor secretion. Stem Cells Dev. **28**(6), 398–409 (2019). https://doi.org/10.1089/scd.2018.0200
107. Zhao, J., et al.: Mesenchymal stromal cell-derived exosomes attenuate myocardial ischaemia-reperfusion injury through miR-182-regulated macrophage polarization. Cardiovasc. Res. **115**(7), 1205–1216 (2019). https://doi.org/10.1093/cvr/cvz040
108. Zhu, K., et al.: Lack of remuscularization following transplantation of human embryonic stem cell-derived cardiovascular progenitor cells in infarcted nonhuman primates. Circ. Res. **122**(7), 958–969 (2018). https://doi.org/10.1161/CIRCRESAHA.117.311578

Index

Printed by Books on Demand, Germany